T0334632

This book deals with flows over propellers operating behind ships, and the hydrodynamic forces and moments which the propeller generates on the shaft and on the ship hull.

The first part of the text is devoted to fundamentals of the flow about hydrofoil sections (with and without cavitation) and about wings. It then treats propellers in uniform flow, first via advanced actuator disc modelling, and then using lifting-line theory. Pragmatic guidance is given for design and evaluation of performance, including the use of computer modelling. The second part covers the development of unsteady forces arising from operation in non-uniform hull wakes. First, by a number of simplifications, various aspects of the problem are dealt with separately until the full problem of a non-cavitating, wide-bladed propeller in a wake is treated by a new and completely developed theory. Next, the complicated problem of an intermittently cavitating propeller in a wake and the pressures and forces it exerts on the shaft and on the ship hull is examined. A final chapter discusses the optimization of efficiency of compound propulsors. The authors have taken care to clearly describe physical concepts and mathematical steps. Appendices provide concise expositions of the mathematical techniques used.

The book will be of interest to students, research workers and professional engineers (naval architects) in propeller dynamics.

CAMBRIDGE OCEAN TECHNOLOGY SERIES 3

HYDRODYNAMICS OF SHIP PROPELLERS

Cambridge Ocean Technology Series
1. Faltinsen: Sea Loads on Ships and Offshore Structures
2. Burcher & Rydill: Concepts in Submarine Design
3. Breslin & Andersen: Hydrodynamics of Ship Propellers

HYDRODYNAMICS OF SHIP PROPELLERS

John P. Breslin
Professor Emeritus, Department of Ocean Engineering,
Stevens Institute of Technology

and

Poul Andersen
Department of Ocean Engineering,
The Technical University of Denmark

CAMBRIDGE
UNIVERSITY PRESS

PUBLISHED BY THE PRESS SYNDICATE OF THE UNIVERSITY OF CAMBRIDGE
The Pitt Building, Trumpington Street, Cambridge CB2 1RP, United Kingdom

CAMBRIDGE UNIVERSITY PRESS
The Edinburgh Building, Cambridge CB2 2RU, United Kingdom
40 West 20th Street, New York, NY 10011–4211, USA
10 Stamford Road, Oakleigh, Melbourne 3166, Australia

First published 1994
Reprinted 1996
First paperback edition 1996

A catalogue record for this book is available from the British Library

Library of Congress Cataloguing in Publication data

Breslin, John P.
Hydrodynamics of ship propellers / John P. Breslin, Poul Andersen.
p. cm. – (Cambridge ocean technology series; 3)
Includes bibliographical references and index.
1. Propellers. 2. Ships–Hydrodynamics. I. Andersen, Poul,
1951– . II. Title. III. Series.
VM753.B68 1993
623.8′73–dc20 93–26511 CIP

ISBN 0 521 41360 5 hardback
ISBN 0 521 57470 6 paperback

Transferred to digital printing 2003

Contents

Preface xi

Notation xiv

Abbreviations xxiv

1 **Brief review of basic hydrodynamic theory** 1

Continuity 1

Equations of motion 2

Velocity fields induced by basic singularities 7

Vorticity 17

2 **Properties of distributions of singularities** 26

Planar distributions in two dimensions 26

Non-planar and planar distributions in three dimensions 33

3 **Kinematic boundary conditions** 42

4 **Steady flows about thin, symmetrical sections in two dimensions** 46

The ogival section 51

The elliptical section 54

Generalization to approximate formulae for families of two-dimensional hydrofoils 57

A brief look at three-dimensional effects 62

5 **Pressure distributions and lift on flat and cambered sections at small angles of attack** 66

The flat plate 66

Cambered sections 74

6 **Design of hydrofoil sections** 86

Application of linearized theory 87

Application of non-linear theory 103

7 **Real fluid effects and comparisons of theoretically and experimentally determined characteristics** 111

Phenomenological aspects of viscous flows 111

Experimental characteristics of wing sections and comparisons with theory 117

8 **Cavitation** 128
Historical overview 128
Prediction of cavitation inception 130
Cavitating sections 140
Partially cavitating hydrofoils 142
Modification of linear theory 151
Supercavitating sections 156
Unsteady cavitation 159

9 **Actuator disc theory** 162
Heavily loaded disc 166
Lightly loaded disc 187

10 **Wing theory** 196

11 **Lifting-line representation of propellers** 207
Induced velocities from vortex elements 209
Generalization to a continuous radial variation of circulation 219
Induction factors 222
Forces acting on the blades and the equation for the circulation density 224

12 **Propeller design via computer and practical considerations** 227
Criteria for optimum distributions of circulation 227
Optimum diameter and blade-area-ratio determinations 235
Calculation procedures 239
Pragmatic considerations 252

13 **Hull-wake characteristics** 262
Analysis of the spatial variation of hull wakes 264
Temporal wake variations 270

14 **Pressure fields generated by blade loading and thickness in uniform flows; comparisons with measurements** 272
Pressure relative to fixed axes 272
Comparisons with measurements 281

15 **Pressure fields generated by blade loadings in hull wakes** 290

16 **Vibratory forces on simple surfaces** 301

17 **Unsteady forces on two-dimensional sections and hydrofoils of finite span in gusts** 315
Two-dimensional sections 315
Unsteady lift on hydrofoils of finite span 327
Implications for propellers 332

18 **Lifting-surface theory** **334**
Overview of extant unsteady theory 334
Blade geometry and normals 337
Linear theory 340
A potential-based boundary-element procedure 368

19 **Correlations of theories with measurements** **374**

20 **Outline of theory of intermittently cavitating propellers** **387**
A basic aspect of the pressure field generated by unsteady cavitation 388
Pressure field due to cavitating propeller 393
Numerical solution of the intermittently-cavitating propeller problem 403
Comparison of calculated and observed transient blade cavitation and pressures 404

21 **Forces on simple bodies generated by intermittent cavitation** **411**
Hull forces without solving the diffraction problem 418

22 **Pressures on hulls of arbitrary shape generated by blade loading, thickness and intermittent cavitation** **425**
Representation of hulls of arbitrary shape in the presence of a propeller and water surface 425
Correlation of theory and measurements 435
Correlations of theory and measurements for non-cavitating conditions 451
Summary and conclusion 451

23 **Propulsor configurations for increased efficiency** **454**
A procedure for optimum design of propulsor configurations 456
Optimized loadings on compound propulsor configurations 462
Flow-conditioning devices 477
Summary and conclusion 482

Appendices **484**
A Inversion of the airfoil integral equations 484
B The Kutta-Joukowsky theorem 490
C The mean value of the radial velocity component induced by a helical vortex at downstream infinity 494
D Conservation of circulation 496
E Method of characteristics 498
F Boundary conditions imposed by water surface at high and low frequencies 500

Mathematical compendium **503**
1 Taylor expansion 503
2 Dirac's δ-function 504
3 Green's identities and Green function 505
4 Evaluation of integrals with Cauchy- and Hadamard-type singular-kernel functions 507
5 Fourier expansions of $1/R$ 513
6 Properties of the Legendre function $Q_{n-\frac{1}{2}}$ 518
7 Outline of calculus of variations 522
8 Table of airfoil integrals 523

References 527

Authors cited 547

Sources of figures 550

Index 551

Preface

This book reflects the work of a great number of researchers as well as our own experience from research and teaching of hydrodynamics and ship-propeller theory over a combined span of more than 60 years. Its development began in 1983-84 during the senior author's tenure as visiting professor in the Department of Ocean Engineering, The Technical University of Denmark, by invitation from Professor Sv. Aa. Harvald. During this sabbatical year he taught a course based on his knowledge of propeller theory garnered over many years as a researcher at Davidson Laboratory and professor at Stevens Institute of Technology. Written lecture notes were required, so we were soon heavily engaged in collecting material and writing a serial story of propeller hydrodynamics with weekly publications. As that large audience consisted of relatively few masters and doctoral students but many experienced naval architects, it was necessary to show mathematical developments in greater detail and, in addition, to display correlations between theory and practical results.

Encouraged by Professor P. Terndrup Pedersen, Department of Ocean Engineering, The Technical University of Denmark, we afterwards started expanding, modifying and improving the notes into what has now become this book. In the spirit of the original lecture notes it has been written primarily for two groups of readers, viz. students of naval architecture and ship and propeller hydrodynamics, at late undergraduate and graduate levels, and practicing naval architects dealing with advanced propulsion problems. It is our goal that such readers, upon completion of the book, will be able to understand the physical problems of ship-propeller hydrodynamics, comprehend the mathematics used, read past and current literature, interpret calculation and experimental findings and correlate theory with their own practical experiences.

To make reading as easy as possible the mathematical concepts and derivations which might have caused trouble for those readers of a more practical background have been explained and executed in far greater detail than found in the literature. Physical interpretations are given throughout together with explanations of the procedures and results in engineering terms and with simple solutions of practical utility wherever possible. We hope that the book in this form will be equally suitable as a text in university courses, a guide for self-tuition and a reference book in ship-design offices.

The subject matter is broadly divided into two parts. In the first, basic hydrodynamics is outlined with comprehensive applications to the construction of practical representations of the steady performance of hydrofoils, with and without cavitation, wings and propellers. Here lifting-line theory is described, including propeller design and analysis via computer and pragmatic considerations from actual performance. The last part addresses the unsteady forces on propellers in wakes via lifting-surface theory as well as propeller-induced vibratory forces on simple, nearby boundaries and upon ship hulls. Both non-cavitating and cavitating propellers are treated. In the final chapter a rational procedure for the optimization of compound propulsors for increased efficiency is described. Throughout the book, in addition to the theoretical developments, the results of calculations are correlated with experimental findings. Remarks and developments that the reader may wish to skip in his first reading are set in small print. No exercises are provided; to achieve proficiency, the reader, after initial study of the text, should derive the results independently.

An immense pleasure, when writing this book, has been to experience the interest and help from colleagues, institutions and companies all over the world. They generously spent their time answering our questions and supplied us with material, including photographs and figures, with permission to reproduce them in the text. These sources are acknowledged in the figure captions. We are very grateful for this assistance without which this book would have been much more incomplete and less useful. We are particular indebted to Dr. W. van Gent, Maritime Research Institute Netherlands; Professor M. D. Greenberg, University of Delaware; Mr. C.-A. Johnsson, SSPA Maritime Consulting AB; Professor J. E. Kerwin and Dr. S. A. Kinnas, Massachusetts Institute of Technology. Our sincere thanks are also due to Mr. J. H. McCarthy, David Taylor Research Center; Dr. K. Meyne, Ostermann Metallwerke; Dr. W. B. Morgan, David Taylor Research Center; Mr. P. Bak Olesen, A.P. Møller; and Mr. H. Yagi, Mitsui Engineering and Shipbuilding Co., Ltd. for help and support and to Professor R. Eatock Taylor, Oxford University, for his effective proposal of our manuscript to Cambridge University Press. We also wish to express our gratitude to present and former colleagues at the Department of Ocean Engineering, The Technical University of Denmark. They include Professor Emeritus Sv. Aa. Harvald and Professor P. Terndrup Pedersen who initiated vital parts of the entire process and later together with Professor J. Juncher Jensen, Head of Department, gave us encouragement and support. Invaluable help was provided by the Staff; Ms. L. Flicker typed the lecture-notes version of the manuscript and later versions were typed by Ms. V. Jensen.

We acknowledge the financial support of F. L. Smidth & Co. A/S who, on the occasion of their 100th-year anniversary, sponsored the first author's stay as visiting professor. Later support was provided by The Danish Technical Research Council under their Marine Design Programme.

Lyngby, Denmark
October 1992

John P. Breslin
Poul Andersen

Notation

The following list of symbols is provided partly as an aid to the reader who wants to use this text as a reference book and read selected chapters. The list contains mainly globally used symbols while many other symbols, including those distinguished by subscript, are defined locally. The notation is not entirely consistent, symbols being used with different definitions, however, rarely in the same sections. Practical usage has been given priority. For this reason ITTC notation has only been partly used.

The coordinate systems are as follows: For two-dimensional flows the x-axis is horizontal, generally displayed in figures as pointing to the right, with the y-axis vertical and positive upwards. Incoming flow is along the x-axis but opposite in direction. For three-dimensional flows the x-axis is horizontal, with a few exceptions coinciding with the propeller axis and generally displayed in figures as pointing to the right. The y-axis is also horizontal, pointing to port and the z-axis is vertical, pointing upwards. As in the two-dimensional case the incoming flow is along the x-axis but opposite in direction. Moreover, a cylindrical system is used. Its x-axis coincides with that of the cartesian system while the angle is measured from the vertical (z-axis), positive in the direction of rotation of a right-handed propeller.

For the two-dimensional case this orientation of axes is in contrast to that used by aerodynamicists (who take the incoming flow along the positive x-axis). However, it is consistent with the three-dimensional definition as well as with the long tradition in naval architecture that the ship is viewed from starboard and the bow consequently is to the right hand.

A_c	cavity sectional area	
A_E/A_0	expanded-area ratio	
$A_{\vert m\vert}$	propagation amplitude function	
A_R	aspect ratio	$= \dfrac{\text{span}^2}{\text{area}}$
a	semi-chord	$= \dfrac{c}{2}$
a	NACA mean-line designation	

b	semi–span (of wing)	
b_T	first basic coefficient	$= \dfrac{TN^2}{\rho U_A{}^4}$
b_Q	second basic coefficient	$= \dfrac{QN^3}{\rho U_A{}^5}$
C_D	(sectional) drag coefficient	$= \dfrac{D}{\frac{1}{2}\rho U^2 c}$
C_D^w	drag coefficient, wing	$= \dfrac{D^w}{\frac{1}{2}\rho U^2 \cdot \text{area}}$
C_L	(sectional) lift coefficient	$= \dfrac{L}{\frac{1}{2}\rho U^2 c}$
$\dfrac{\partial C_L}{\partial \alpha}$	lift–curve slope, lift rate, lift effectiveness	
C_{L_i}	ideal lift coefficient	
C_L^w	lift coefficient, wing	$= \dfrac{L^w}{\frac{1}{2}\rho U^2 \cdot \text{area}}$
$C_{M_{c/4}}$	(sectional) moment coefficient	$= \dfrac{M_{c/4}}{\frac{1}{2}\rho U^2 c^2}$
C_p	(Euler) pressure coefficient	$= \dfrac{p - p_\infty}{\frac{1}{2}\rho U^2}$
ΔC_p	dimensionless pressure difference over section	
$C_{p\min}$	minimum pressure coefficient	
$\Delta C_{p\min}$	dimensionless minimum pressure difference over section	
C_{Th}	thrust–loading coefficient	$= \dfrac{T}{\frac{1}{2}\rho U_A^2 \pi \frac{D^2}{4}}$
c	chord	$= 2a$
D	propeller diameter	
D	drag	
D^w	drag of wing	
D_i^w	induced drag	
d	depth	

F	force
F_a, F_r, F_t	axial, radial and tangential force components
F_r	Froude number $= \dfrac{\text{velocity}}{\sqrt{g \cdot \text{length}}}$
F_x, F_y, F_z	x-, y- and z-directed force components
f	camber (maximum) of section
f_a, f_r, f_t	axial, radial and tangential partial-force operators
G	Green function
g	acceleration due to gravity
$\mathbf{g}$	acceleration due to gravity vector
$H(x)$	Heaviside step function
$H_n^{(1)}(k)$, $H_n^{(2)}(k)$	Hankel functions of the first and second kind of order n
h	helicoidal coordinate of helicoidal-normal coordinate system (h,n)
h_l, h_t	lengths to leading and trailing edges in helicoidal coordinates
h_l^c, h_t^c	helicoidal coordinates of cavity leading and trailing edges
$I_n(k)$	modified Bessel function of the first kind of order n
Im	imaginary part of
I^*	integral
$\mathbf{i}$, $\mathbf{j}$, $\mathbf{k}$	direction vectors in x-, y- and z-directions
$\mathbf{i}_a$, $\mathbf{i}_r$, $\mathbf{i}_t$	direction vectors in axial, radial and tangential directions
i_a, i_r, i_t	axial, radial and tangential induction factors
J	advance ratio, free stream $= \dfrac{U}{ND}$
J_A	advance ratio, behind ship $= \dfrac{U_A}{ND}$
$J_n(k)$	Bessel function of the first kind of order n
J^*	integral
$\mathbf{j}$	– see **i**, **j**, **k**

K	constant in inverted airfoil equation	
K_F	force coefficient	$= \frac{F}{\rho N^2 D^4}$
K_M	moment coefficient	$= \frac{M}{\rho N^2 D^5}$
K_N	normal-force coefficient	$= \frac{F_N}{\rho N^2 D^4}$
K_T	thrust coefficient	$= \frac{T}{\rho N^2 D^4}$
K_Q	torque coefficient	$= \frac{Q}{\rho N^2 D^5}$
$K_n(k)$	modified Bessel function of the second kind of order n	
K_p	pressure coefficient	$= \frac{p}{\rho N^2 D^2}$
k	reduced frequency	$= \frac{\omega c}{2U}$
$\mathbf{k}$	– see **i**, **j**, **k**	
L	lift	
L_c	lift due to cavitation	
L^w	lift of wing	
ℓ	Lagrange multiplier, or inverse of	
ℓ_c	cavity length	
M	moment	
M	source strength	
$M_{c/4}$	moment of section about quarter-chord point	
M_x, M_y, M_z	x-, y- and z-directed moment components	
m	source strength, distribution	
m_c	cavity source strength, distribution	
N	normal force	
N	rate of revolutions	$= \frac{\omega}{2\pi}$

n	normal coordinate of helicoidal-normal coordinate system (h,n)	
n_a, n_r, n_t	axial, radial and tangential components of normal vector	
n_x, n_y, n_z	x-, y- and z-directed components of normal vector	
n	normal vector	
O(k)	order of magnitude of k	
P	blade pitch	
P_A	fluid pitch, average inflow	$= \frac{2\pi U_A}{\omega}$
P_f	fluid pitch	$= \frac{2\pi U}{\omega}$
P_f^*	fluid pitch parameter	$= \frac{P_f}{2\pi}$
p	pressure	
p_{at}	atmospheric pressure	
p_c	pressure in fluid due to cavitation; pressure within cavity	
p_f^*	fluid pitch parameter	$= \frac{\omega}{U}$
p_A^*	fluid pitch parameter	$= \frac{\omega}{U_A}$
p_{min}	minimum pressure	
p_v	vapor pressure	
Δp	pressure difference over section or blade	
$\Delta p_n(h',r')$	Fourier components of $\Delta p(h',r';\gamma_0)$	
Q	torque	
$Q_{n-1/2}(\bar{Z})$	associated Legendre function of the second kind of degree n − 1/2 and zero order	
q	order of harmonic relative to blade frequency (blade-frequency order)	
q	velocity (=|**q**|)	
q	torque ratio (between propulsor components)	
q	velocity vector	

R	radius of propeller
R	distance between dummy point and field point
$\boldsymbol{R}$	vector from dummy point to field point
R_c	radius of cavity center
R_e	Reynolds number $= \dfrac{\text{velocity} \cdot \text{length}}{\nu}$
Re	real part of
R_h	radius of hub
r	– see x, r, γ
r_c	radius of lowest instantaneous radial extent of cavity on blade
r_l	leading-edge or nose radius
S	surface
S_c	surface of cavity over blade
T	thrust
t	maximum thickness of section
t	thrust-deduction fraction
t	time
U	parallel inflow velocity (along negative x-axis)
U	ship speed
U_A	mean inflow to propeller disc, behind ship
$U_a(r)$	circumferentially mean inflow to propeller disc, radius r
u, v, w	x-, y- and z-directed components of velocity, generally superimposed by body
u_a, u_r, u_t	axial, radial and tangential components of velocity, generally superimposed by body
u_{an}, u_{rn}, u_{tn}	harmonic components of u_a, u_r, u_t of order n
u_a^w, u_r^w, u_t^w	axial, radial and tangential components of superimposed wake velocities
u_{an}^c, u_{an}^s, u_{tn}^c, u_{tn}^s	wake velocity harmonics of u_a^w and u_t^w of order n, cosine and sine components
u_{an}^w, u_{rn}^w, u_{tn}^w	wake velocity harmonics of u_a^w, u_r^w, u_t^w of order n

V	volume
$V(r)$	resulting inflow to blade section due to translation and rotation
V_c	cavity volume on propeller blade
V_i	critical or (cavitation-) inception speed
v	– see u, v, w
w	wake fraction
w	– see also u, v, w
$w(r)$	circumferentially mean wake fraction at radius r
x, r, γ	cylindrical coordinates. x-axis positive forward coinciding generally with propeller axis. γ measured from vertical, positive in direction of rotation of right-handed propeller.
x, y	cartesian coordinates for two-dimensional flow. x-axis positive to the right in negative inflow direction. y-axis vertical and upwards.
x, y, z	cartesian coordinates. x-axis positive forward coinciding generally with propeller axis and shown in figures to the right. y-axis positive to port and z-axis vertical and upwards.
x_{ac}	position of aerodynamic center
x_l, x_t	leading- and trailing-edge coordinates, corresponding to helicoidal coordinates h_l and h_t
x_r	rake
x_s	rake induced by skew
$Y_n(k)$	Bessel function of the second kind of order n
y	– see x, y and x, y, z
y_f	camber ordinate
$\mathcal{Z}$	argument of associated Legendre function
Z	number of propeller blades
z	– see also x, y and x, y, z
z	complex variable $= x + iy$

α	angle of incidence or attack	
α_i	ideal angle of attack	
α_{in}	induced angle	
α_0	angle of zero lift	
β	fluid pitch angle	$= \tan^{-1} \dfrac{U}{\omega r}$
β_A	fluid pitch angle, mean inflow	$= \tan^{-1} \dfrac{U_A}{\omega r}$
β_a	fluid pitch angle, local inflow	$= \tan^{-1} \dfrac{U_a}{\omega r}$
β_i	induced pitch angle	
β_p	blade pitch angle	$= \tan^{-1} \dfrac{P}{2\pi r}$
Γ	circulation	
Γ	Gamma function	
γ	angle, – see also x, r, γ	
γ	strength of vortex distribution (vortex density)	
γ_0	blade-position angle, key blade	$= \omega t$
γ_ν	blade-position angle, ν-th blade	$= \dfrac{2\pi\nu}{z} + \gamma_0$
∇	gradient operator	
∇^2	Laplace operator	
$\nabla\cdot$	divergence operator	
$\nabla\times$	curl operator	
$\delta(x)$	Dirac delta-function	
δ_{mom}	boundary-layer momentum thickness	
δT, δQ	variation of T, Q, etc.	
ε_q	parameter, = 1 for q = 0, = 2 for q ≠ 0	
ζ	complex variable	$= \xi + i\eta$
ζ	– see also $\boldsymbol{\zeta}$	
$\boldsymbol{\zeta}$	vorticity vector, $= \xi\mathbf{i} + \eta\mathbf{j} + \zeta\mathbf{k}$	

η	cavity ordinate	
η	propeller efficiency	$= \frac{J}{2\pi}\frac{K_T}{K_Q}$
η	– see also ζ, $\boldsymbol{\zeta}$	
η_{ideal}	ideal efficiency	
μ	dynamic viscosity	
ν	blade index number, = 0, ..., Z – 1. $\nu = 0$ ~ key blade	
ν	kinematic viscosity	
ξ	– see ζ, $\boldsymbol{\zeta}$	
ρ	mass density of water	
Σ	small surface region	
Σ_x, Σ_y, Σ_z	strengths of x–, y– and z–directed point dipoles	
σ	cavitation number	$= \frac{p - p_c}{\frac{1}{2}\rho U^2}$
σ, σ_n	strength of normally directed dipoles	
σ_N	cavitation number	$= \frac{p - p_v}{\frac{1}{2}\rho N^2 D^2}$
σ_{U_A}	cavitation number	$= \frac{p - p_v}{\frac{1}{2}\rho U_A^2}$
σ_x, σ_y, σ_z	strengths of x–, y– and z–directed distributed dipoles	
$\sigma_{0.7R}$	cavitation number	$= \frac{p - p_v}{\frac{1}{2}\rho V(0.7R)^2}$
τ	semi–thickness, body, section or blade section	
Φ, ϕ	velocity potentials	
ϕ_c	cavity velocity potential	
ϕ_s	source velocity potential	
ϕ_v	vortex velocity potential	
ψ	local angle for point on blade relative to blade–position angle γ_ν	
ψ	(Stokes') stream function	

ψ_j	velocity potential of j-th motion	
ψ_S	skew angle	
ω	angular velocity of propeller	$= 2\pi N$
$\breve{a}$	dimensionless form of a	
$\bar{a}$	mean value of a	
$\tilde{a}$	amplitude of a	
x', x''	dummy variable; dummy-point coordinate in distinction to field-point coordinate x	
f'	derivative of function f with respect to argument	
$\frac{D}{Dt}$	substantive derivative	
$⨍$	Cauchy principal-value integral	
$⨎$	finite-part integral	
$(p)_{qZ}$	qZ-th amplitude function of p (qZ-th blade frequency)	

Vector symbols and operators are printed **boldfaced**; the NACA mean-line designation, **a**, is the only other symbol so printed.

Abbreviations

ATTC	American Towing Tank Conference
CETENA	Centro per gli Studi di Tecnica Navale
DTMB	David Taylor Model Basin – later DTRC
DTRC	David Taylor Research Center
HSVA	Hamburgische Schiffbau-Versuchsanstalt
INA	Institution of Naval Architects – later RINA
ITTC	International Towing Tank Conference
L.E.	leading edge
MARIN	Maritime Research Institute Netherlands
MIT	Massachusetts Institute of Technology
NACA	National Advisory Committee for Aeronautics – later NASA
NASA	National Air and Space Administration
NPL	National Physical Laboratory
PUF	Propeller Unsteady Force (MIT computer program)
RINA	The Royal Institution of Naval Architects
SNAME	The Society of Naval Architects and Marine Engineers
SSPA	SSPA Maritime Consulting AB – (SSPA: Statens Skeppsprovningsanstalt)
T.E.	trailing edge
TMB	Taylor Model Basin – later DTRC
VWS	Versuchsanstalt für Wasserbau und Schiffbau

1 Brief Review of Basic Hydrodynamic Theory

An extensive, highly mathematical literature exists dealing with fluid-mechanical aspects of ship propellers.

Invariably, the mathematical developments are only outlined, impeding easy comprehension even by knowledgeable readers. Our aim is to elucidate the mathematical theory in much greater detail than is generally available in extant papers. In this context, the first three chapters are provided as aids for those who have not had extensive practice in the application of classical hydrodynamical theory to flows induced in fluids by the motions of bodies. The fluid of interest is water which is taken to be incompressible and inviscid. Modifications arising from viscosity are described in a later chapter (Chapter 7) through reference to experimental observations.

This review begins with the derivation of the concept of continuity or conservation of mass at all points in sourceless flow and proceeds to the development of the Euler equations of motion. In the restricted but important class of irrotational motions (zero vorticity) Laplace's equation for the velocity potential is obtained. The remainder of this chapter is devoted to derivations of fundamental solutions of Laplace's equation in two and three dimensions.

It is emphasized that these first two chapters are necessarily limited in scope, being directed to our needs in subsequent chapters. There are many excellent books which should be consulted for those seeking greater depth and broader description of hydrodynamic theory. Among these we suggest Batchelor (1967), Lamb (1963), Lighthill (1986), Milne-Thomson (1955), and Yih (1988), and Newman (1977) for modern applications.

CONTINUITY

Consider a general, three-dimensional flow field whose vector velocity is defined by

$$\mathbf{q} = \mathbf{i}u(x,y,z,t) + \mathbf{j}v(x,y,z,t) + \mathbf{k}w(x,y,z,t) \tag{1.1}$$

for an incompressible fluid.

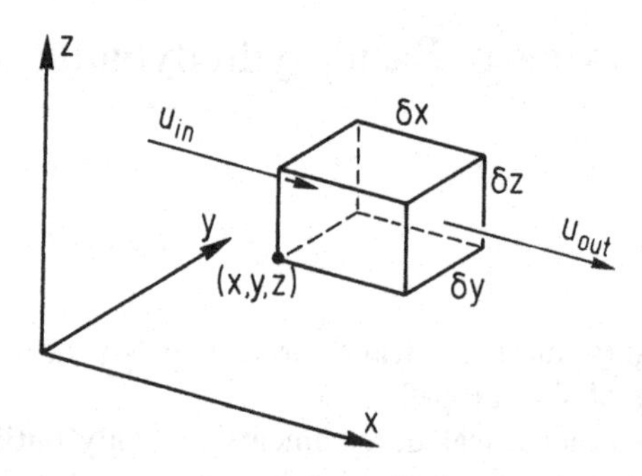

Figure 1.1 Flow through a fluid volume element.

Net flux through a differential element in the x-direction is

$$u_{in}\delta y\,\delta z - u_{out}\delta y\,\delta z$$

Now as u = u(x,y,z,t), we can expand in a Taylor expansion about x to x+δx, cf. Mathematical Compendium, Section 1, Equation (M1.5), p. 504

$$u = u(x,y,z,t) + \frac{\partial u}{\partial x}((x + \delta x) - x) + \ldots \tag{1.2}$$

$$u_{out} = u_{in} + \frac{\partial u}{\partial x}\,\delta x + \text{higher order terms.}$$

Then the net flux in the x-direction is

$$u\,\delta y\,\delta z - \left[u + \frac{\partial u}{\partial x}\,\delta x\right]\delta y\,\delta z = -\frac{\partial u}{\partial x}\,\delta x\,\delta y\,\delta z$$

Similar contributions in the y- and z-directions yield

$$-\left[\frac{\partial u}{\partial x} + \frac{\partial v}{\partial y} + \frac{\partial w}{\partial z}\right]\delta x\delta y\delta z = \text{total net flux/unit time} \tag{1.3}$$

If the fluid is sourceless at all such elements, then the total net flux ≡ 0 and as the volume $\delta x\delta y\delta z$ is arbitrary we have for an incompressible fluid

$$\frac{\partial u}{\partial x} + \frac{\partial v}{\partial y} + \frac{\partial w}{\partial z} = 0 \tag{1.4}$$

EQUATIONS OF MOTION

For an incompressible, inviscid fluid, applying forces due to pressure p to the faces of a rectangular fluid parallelepiped as above, and including the components of the extraneous forces per unit volume, we have from Newton's second law applied to an elemental volume

$$(\text{Mass}) \cdot (\text{Acceleration}) = \text{Net external force}$$

Then in the x-direction

$$(\rho \delta x \delta y \delta z) \frac{D}{Dt} u = p \delta y \delta z - \left[p + \frac{\partial p}{\partial x} \delta x \right] \delta y \delta z + F_x \delta x \delta y \delta z \tag{1.5}$$

$$\frac{D}{Dt} = \frac{\partial}{\partial t} + \frac{\partial x}{\partial t} \frac{\partial}{\partial x} + \frac{\partial y}{\partial t} \frac{\partial}{\partial y} + \frac{\partial z}{\partial t} \frac{\partial}{\partial z} \tag{1.6}$$

and F_x, F_y, F_z are the components of the extraneous forces (such as gravity) per unit volume, and those imposed locally by lifting bodies.

Here the derivatives of the fluid-particle coordinates are the particle velocities

$$\frac{\partial x}{\partial t} = u \quad ; \quad \frac{\partial y}{\partial t} = v \quad ; \quad \frac{\partial z}{\partial t} = w \tag{1.7}$$

Hence (1.5) reduces to (considering all components)

$$\begin{aligned}
\frac{\partial u}{\partial t} + u \frac{\partial u}{\partial x} + v \frac{\partial u}{\partial y} + w \frac{\partial u}{\partial z} &= -\frac{1}{\rho} \frac{\partial p}{\partial x} + \frac{F_x}{\rho} \\
\frac{\partial v}{\partial t} + u \frac{\partial v}{\partial x} + v \frac{\partial v}{\partial y} + w \frac{\partial v}{\partial z} &= -\frac{1}{\rho} \frac{\partial p}{\partial y} + \frac{F_y}{\rho} \\
\frac{\partial w}{\partial t} + u \frac{\partial w}{\partial x} + v \frac{\partial w}{\partial y} + w \frac{\partial w}{\partial z} &= -\frac{1}{\rho} \frac{\partial p}{\partial z} + \frac{F_z}{\rho}
\end{aligned} \tag{1.8}$$

Now as $q^2 = u^2 + v^2 + w^2$ and as $u \, \partial u/\partial x = 1/2 \, \partial u^2/\partial x$, this suggests adding and subtracting terms in each equation to introduce

$$\frac{\partial}{\partial x} \left[\frac{1}{2} q^2 \right], \quad \frac{\partial}{\partial y} \left[\frac{1}{2} q^2 \right], \quad \text{and} \quad \frac{\partial}{\partial z} \left[\frac{1}{2} q^2 \right].$$

Thus the first of (1.8) becomes

$$\frac{\partial u}{\partial t} + u \frac{\partial u}{\partial x} + v \frac{\partial v}{\partial x} + w \frac{\partial w}{\partial x} + v \left[\frac{\partial u}{\partial y} - \frac{\partial v}{\partial x} \right] + w \left[\frac{\partial u}{\partial z} - \frac{\partial w}{\partial x} \right] = -\frac{1}{\rho} \frac{\partial p}{\partial x} + \frac{F_x}{\rho} \tag{1.9}$$

Now recalling the definition of the vorticity vector

$$\zeta = \begin{vmatrix} \mathbf{i} & \mathbf{j} & \mathbf{k} \\ \dfrac{\partial}{\partial x} & \dfrac{\partial}{\partial y} & \dfrac{\partial}{\partial z} \\ u & v & w \end{vmatrix} = \xi \mathbf{i} + \eta \mathbf{j} + \zeta \mathbf{k} \tag{1.10}$$

Then

$$\xi = \frac{\partial w}{\partial y} - \frac{\partial v}{\partial z} \; ; \; \eta = -\frac{\partial w}{\partial x} + \frac{\partial u}{\partial z} \; ; \; \zeta = \frac{\partial v}{\partial x} - \frac{\partial u}{\partial y} \tag{1.11}$$

We see that the coefficients of v and w are $-\zeta$ and η, respectively. Hence, (1.9) becomes (and the other components by analogy)

$$\begin{aligned} \frac{\partial u}{\partial t} + \frac{\partial}{\partial x}\left[\frac{1}{2}q^2\right] + w\eta - v\zeta &= -\frac{1}{\rho}\frac{\partial p}{\partial x} + \frac{F_x}{\rho} \\ \frac{\partial v}{\partial t} + \frac{\partial}{\partial y}\left[\frac{1}{2}q^2\right] + u\zeta - w\xi &= -\frac{1}{\rho}\frac{\partial p}{\partial y} + \frac{F_y}{\rho} \\ \frac{\partial w}{\partial t} + \frac{\partial}{\partial z}\left[\frac{1}{2}q^2\right] + v\xi - u\eta &= -\frac{1}{\rho}\frac{\partial p}{\partial z} + \frac{F_z}{\rho} \end{aligned} \tag{1.12}$$

These equations both in the form of (1.8) and (1.12) are referred to as ***Euler Equations.***

The products of velocity and vorticity components can be identified by considering

$$\mathbf{q} \times \zeta = \begin{vmatrix} \mathbf{i} & \mathbf{j} & \mathbf{k} \\ u & v & w \\ \xi & \eta & \zeta \end{vmatrix} \tag{1.13}$$

Then the x-component, for example, is

$$(\mathbf{q} \times \zeta)_x = v\zeta - w\eta$$

and hence our equations of motion can be written in vector form

$$\frac{\partial \mathbf{q}}{\partial t} + \nabla\left[\frac{p}{\rho} + \frac{1}{2}q^2\right] - \mathbf{q} \times \zeta = \frac{\mathbf{F}}{\rho} \tag{1.14}$$

where

$$\nabla = \mathbf{i}\frac{\partial}{\partial x} + \mathbf{j}\frac{\partial}{\partial y} + \mathbf{k}\frac{\partial}{\partial z} \quad \text{; the gradient operator}$$

$$\mathbf{q} = \mathbf{i}u + \mathbf{j}v + \mathbf{k}w$$

Steady Irrotational Motion

Here we limit attention to irrotational flows for which $\zeta \equiv \mathbf{0}$, and initially, to steady flows, i.e., $\partial \mathbf{q}/\partial t = \mathbf{0}$.

The condition or restriction to irrotational flows yields

$$\xi = \frac{\partial w}{\partial y} - \frac{\partial v}{\partial z} = 0 \ ; \ \eta = \frac{\partial u}{\partial z} - \frac{\partial w}{\partial x} = 0 \ ; \ \zeta = \frac{\partial v}{\partial x} - \frac{\partial u}{\partial y} = 0 \tag{1.15}$$

These conditions are necessary and sufficient[1] for the motion to be represented by a velocity potential function whose spatial derivatives yield the velocity components in the specified direction.

Thus, let $\phi = \phi(x,y,z)$ be defined as a velocity potential function such that

$$\frac{\partial \phi}{\partial x} = u \ ; \ \frac{\partial \phi}{\partial y} = v \ ; \ \frac{\partial \phi}{\partial z} = w \tag{1.16}$$

Then we see that condition $\zeta = \mathbf{0}$ is satisfied, i.e., $\partial(\partial\phi/\partial z)/\partial y - \partial(\partial\phi/\partial y)/\partial z = 0$ at all points except singular points, and so on for η and ζ.

Then the ***continuity equation*** requires that at all field points

$$\frac{\partial^2 \phi}{\partial x^2} + \frac{\partial^2 \phi}{\partial y^2} + \frac{\partial^2 \phi}{\partial z^2} = 0 \qquad \text{or} \quad \nabla^2 \phi = 0 \tag{1.17}$$

i.e., ϕ must be a solution of ***Laplace's Equation*** and, in addition, must meet kinematic conditions on the body boundaries and "radiation" conditions at infinity.

Using that the flow is irrotational, i.e. $\zeta = \mathbf{0}$, and assuming that the force density $\mathbf{F}$ can be expressed as the negative gradient of a function Ω, i.e.,

$$\mathbf{F} = -\nabla \Omega \tag{1.18}$$

then (1.14) becomes, for steady flow,

$$\nabla\left[\frac{p}{\rho} + \frac{1}{2}q^2 + \frac{\Omega}{\rho}\right] = \mathbf{0} \tag{1.19}$$

which implies that

$$\frac{(p + \Omega)}{\rho} + \frac{1}{2}q^2 = C \qquad \text{;(a constant on all stream surfaces)} \tag{1.20}$$

[1] assuming the components to be continuously differentiable; Kellogg (1967), p. 69 and sequel.

which is referred to as the ***Bernoulli Equation*** (after Daniel Bernoulli (1700 – 1783)).

The most commonly encountered external force is that due to gravity for which $\Omega = \rho gz$ with z vertical and positive upward. (In this text gravity will be generally ignored but we shall be concerned with the effect of forces acting on the fluid imposed by lifting surfaces which are opposite and equal to those applied by the fluid to those surfaces).

The constant C can be evaluated in terms of the ambient pressure at a distance, say upstream where the velocity **q** is also known. So that

$$\frac{p}{\rho} + \frac{1}{2} q^2 = \frac{p_\infty}{\rho} + \frac{1}{2} q_\infty^2 + \frac{(\Omega_\infty - \Omega)}{\rho} \tag{1.21}$$

where $q^2 = (\partial\phi/\partial x)^2 + (\partial\phi/\partial y)^2 + (\partial\phi/\partial z)^2$, in rectangular coordinates.

For a body fixed in a stream (moving parallel to the negative x-axis) the total potential is of the form

$$\phi = -Ux + \phi_b \tag{1.22}$$

and for a body which produces weak disturbances, such that

$$u_b = \frac{\partial\phi_b}{\partial x} << U \quad ; \quad v_b = \frac{\partial\phi_b}{\partial y} << U \quad ; \quad w_b = \frac{\partial\phi_b}{\partial z} << U$$

and $\quad u_b, v_b, w_b \longrightarrow 0 \quad$ for $\quad \sqrt{x^2 + y^2 + z^2} \longrightarrow \infty$

then Bernoulli's Equation (1.21) reduces to (upon dropping Ω)

$$\frac{p}{\rho} + \frac{1}{2}\left\{\left[-U + u_b\right]^2 + v_b^2 + w_b^2\right\} = \frac{p_\infty}{\rho} + \frac{1}{2} U^2 \tag{1.23}$$

Neglecting squares of body-generated perturbations, we secure the linearized pressure-velocity relation

$$\frac{p - p_\infty}{\rho} = u_b U \tag{1.24}$$

dividing through by $U^2/2$

$$\frac{p - p_\infty}{\frac{1}{2}\rho U^2} = 2\left[\frac{u_b}{U}\right] \tag{1.25}$$

The left side is usually designated by C_p, the Euler pressure coefficient

$$C_p(x,y,z) = 2\,\frac{u_b(x,y,z)}{U} \tag{1.26}$$

Unsteady Irrotational Flows

When the velocity field varies with both time and spatial position, then (1.14) becomes (dropping **F** for convenience and using $\mathbf{q} = \nabla\phi$)

$$\nabla\left[\frac{\partial\phi}{\partial t} + \frac{p}{\rho} + \frac{1}{2}q^2\right] = \mathbf{0}$$

whence

$$\frac{\partial\phi}{\partial t} + \frac{1}{2}q^2 + \frac{p}{\rho} = f(t), \text{ a function of time at most;} \tag{1.27}$$

where the velocity potential and hence q and p are functions of x,y,z and time t. The right side appears to be an inconvenient unknown. As Sir James Lighthill (1986) points out a uniform pressure $\rho f(t)$ throughout the fluid is of limited dynamic significance, and moreover f(t) can be removed by taking a redefined velocity potential

$$\Phi = \phi - \rho\int_0^t f(t')dt' \tag{1.28}$$

Then clearly $\nabla\Phi = \nabla\phi = \mathbf{q}$ so the function of time f(t) is a reflection of the definition of the velocity potential function which is specified, save for the addition of any function of time.

We see from this that the pressure p is composed of the *transient* pressure $-\rho\,\partial\phi/\partial t$ and the *dynamic* pressure $1/2\ \rho q^2$. Linearization with respect to a main stream –U as in the foregoing gives

$$p - p_\infty = \rho u_b U - \rho\frac{\partial\phi_b}{\partial t} \tag{1.29}$$

$$p - p_\infty = \rho\left[U\frac{\partial}{\partial x} - \frac{\partial}{\partial t}\right]\phi_b(x,y,z,t) \tag{1.30}$$

We shall make considerable use of the linearized Equations (1.24) and (1.30).

VELOCITY FIELDS INDUCED BY BASIC SINGULARITIES

The Source (In two dimensions)

Consider fluid to be emanating from each element of the z-axis in a uniform fashion. Then the flow is independent of z and we can look at a unit section about the origin, i.e., z = 0.

Let the radial velocity at any r be u_r. Then if

$$M \frac{\text{cubic units}}{\text{linear unit}}/\text{sec} = M \frac{\text{square units}}{\text{sec}}$$

are appearing at x = y = 0, it follows that through any circle with center at the origin, the same flow rate is obtained. Hence,

$$2\pi r u_r = M$$

or

$$u_r = \frac{M}{2\pi r} \tag{1.31}$$

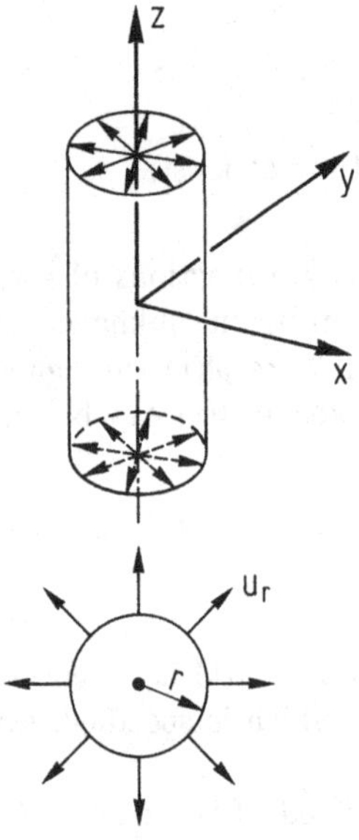

Figure 1.2 Flow due to a source.

Since this flow is free of vorticity, we can define a velocity potential ϕ_s such that

$$\frac{\partial \phi_s}{\partial r} = u_r = \frac{M}{2\pi r} = \frac{d\phi}{dr} \tag{1.32}$$

Integrating we get $\phi_s = M/2\pi \ln r +$ constant.

If this function is to be useful, it must be shown to satisfy Laplace's Equation

$$\frac{\partial \phi_s}{\partial x} = \frac{M}{2\pi}\frac{\partial}{\partial x} \ln \sqrt{x^2+y^2} = \frac{M}{2\pi}\frac{\partial r}{\partial x}\frac{\partial}{\partial r} \ln r = \frac{M}{2\pi}\frac{x}{r^2} = \frac{M}{2\pi}\frac{x}{x^2+y^2}$$

$$\frac{\partial^2 \phi_s}{\partial x^2} = \frac{M}{2\pi}\left[\frac{(x^2+y^2)(1)-x(2x)}{(x^2+y^2)^2}\right] = \frac{M}{2\pi}\left[\frac{y^2-x^2}{r^4}\right] \;;$$

then as x and y enter symmetrically

$$\frac{\partial^2 \phi_s}{\partial y^2} = \frac{M}{2\pi}\frac{(x^2-y^2)}{r^4}$$

Hence for all x,y except x = y = 0,

$$\frac{\partial^2 \phi_s}{\partial x^2} + \frac{\partial^2 \phi_s}{\partial y^2} = 0 \tag{1.33}$$

As Laplace's Equation is linear, then any sum of solutions is also a solution.

The source can be located anywhere, say, at x = x', y = y'. Then

$$\phi_S = \frac{M}{2\pi} \ln \sqrt{(x - x')^2 + (y - y')^2} \qquad (1.34)$$

A distribution of sources over a portion of the x-axis is secured by replacing M by m(x')dx, (m has the dimension length/second) and integrating

$$\phi_S = \frac{1}{2\pi} \int_{-a}^{a} m(x') \ln \sqrt{(x - x')^2 + y^2}\, dx' \qquad (1.35)$$

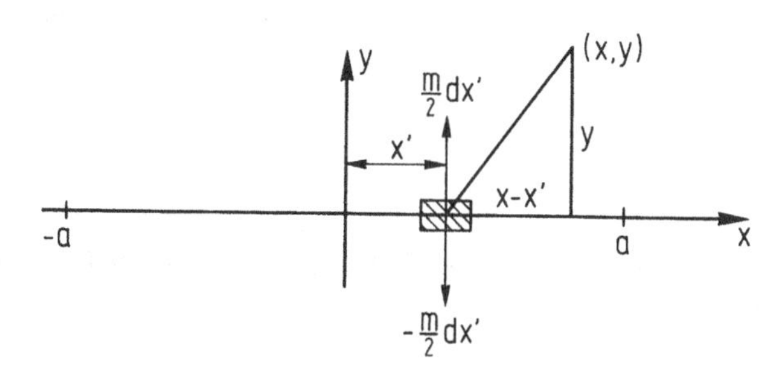

Figure 1.3 Distribution of sources over a portion of the x-axis.

The Vortex (In two dimensions)

Consider a circular flow about the origin such that $\int_0^{2\pi} u_t(r)\, r d\theta = \Gamma$, the circulation, a constant. Then

$$u_t(r) = \frac{\Gamma}{2\pi r} \qquad (1.36)$$

Define $\phi_{vortex} = \phi_v$ by

$$\frac{1}{r}\frac{\partial \phi_v}{\partial \theta} = u_t = \frac{\Gamma}{2\pi r}$$

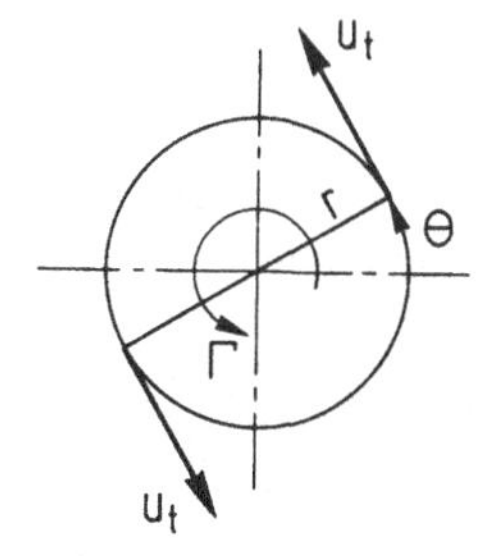

Figure 1.4 Flow due to a vortex.

Integrating

$$\phi_v = \frac{\Gamma}{2\pi}\,\theta + \text{a constant} \qquad (1.37)$$

Again, the constant can be ignored as we are only interested in derivatives of ϕ_v.

In rectangular coordinates,

$$\phi_V = \frac{\Gamma}{2\pi}\tan^{-1}\frac{y}{x} \quad ; \quad -\pi < \tan^{-1}\frac{y}{x} < \pi$$

To make ϕ_V single valued we define it in the region excluding the cut, cf. Figure 1.5. Then on the upper bank of the cut

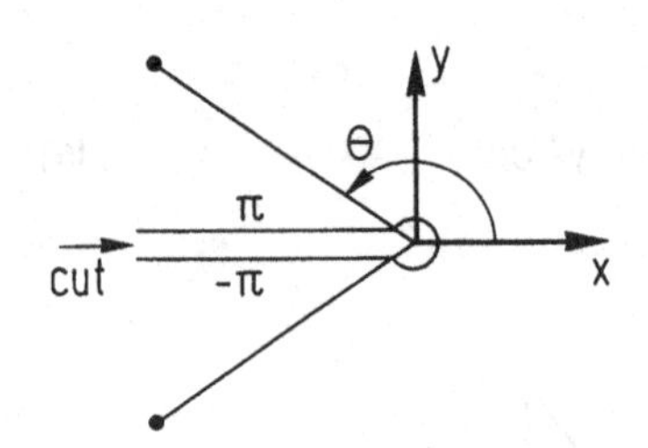

Figure 1.5 Cut along negative x–axis.

$$\phi_{V+} = \frac{\Gamma}{2} \quad (\theta = \pi)$$

and on the lower bank

$$\phi_{V-} = -\frac{\Gamma}{2} \quad (\theta = -\pi)$$

∴ the jump in ϕ_V across the cut is

$$\Delta\phi_V = \phi_{V+} - \phi_{V-} = \Gamma \tag{1.38}$$

To show that ϕ_V satisfies the Laplace Equation, consider the form in polar coordinates, namely

$$\left[\frac{\partial^2}{\partial r^2} + \frac{1}{r}\frac{\partial}{\partial r} + \frac{1}{r^2}\frac{\partial^2}{\partial\theta^2}\right]\phi = 0 \tag{1.39}$$

As ϕ_V is independent of r, we have

$$\frac{1}{r^2}\frac{\partial^2}{\partial\theta^2}\left[\frac{\Gamma}{2r}\theta\right]$$

which is clearly zero for all $r \neq 0$.

The Dipole (In two dimensions)

The dipole[2] is defined to be the limit of the sum of a source and a sink as they are brought together in such a way that the product of their common strength and the distance between them is a constant defined to be the dipole moment strength. Thus, dipoles have magnitude and directivity.

Consider a source located at $x = 0$, $y = \varepsilon$ and a sink of equal strength at the origin, Figure 1.6.

Then the sum of these potentials is

$$\phi = \frac{M}{2\pi}\ln\sqrt{x^2 + (y - \varepsilon)^2} - \frac{M}{2\pi}\ln\sqrt{x^2 + y^2} \tag{1.40}$$

[2] dipoles are often referred to as doublets.

Write

$$\sqrt{x^2 + (y - \varepsilon)^2}$$

$$= \sqrt{x^2 + y^2 - 2\varepsilon y + \varepsilon^2}$$

$$\approx \sqrt{x^2+y^2}\sqrt{1 - \frac{2\varepsilon y}{x^2 + y^2} + \ldots}$$

as $\varepsilon \longrightarrow 0$;

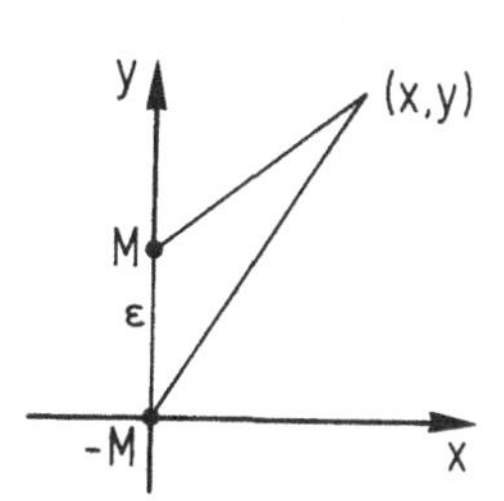

Figure 1.6 Source and sink of equal strength.

then

$$\phi = \frac{M}{2\pi} \ln \frac{\sqrt{x^2+(y-\varepsilon)^2}}{\sqrt{x^2+y^2}} = \frac{M}{2\pi} \ln \left[1 - \frac{1}{2}\left[\frac{2\varepsilon y}{x^2 + y^2}\right] + ..\right] \tag{1.41}$$

The expansion of the logarithm gives, in general,

$$\ln(1 - z) = -\left[z + \frac{z^2}{2} + \ldots\right] \qquad |z| < 1 \tag{1.42}$$

Then to order ε

$$\phi \longrightarrow -\frac{M\varepsilon}{2\pi} \frac{y}{x^2 + y^2} \tag{1.43}$$

Defining $M\varepsilon = \Sigma_y$, the y-directed dipole strength, we have

$$\phi_{y\text{-dipole}} = \phi_{yd} = -\frac{\Sigma_y}{2\pi} \frac{y}{x^2 + y^2} \tag{1.44}$$

This result may be obtained more adroitly by differentiating the unit source displaced along the y-axis.

Thus, for a source of unit strength,

$$\left[\phi_{yd}\right]_1 = \frac{1}{2\pi} \frac{\partial}{\partial y'} \ln \sqrt{x^2 + (y - y')^2}\, \Big|_{y'=0}$$

$$= -\frac{1}{2\pi} \frac{y - y'}{x^2 + (y - y')^2}\Big|_{y'=0} = -\frac{1}{2\pi} \frac{y}{x^2 + y^2} \tag{1.45}$$

For a dipole of strength Σ_y, then

$$\phi_{yd} = -\frac{\Sigma_y}{2\pi} \frac{y}{x^2 + y^2} \tag{1.46}$$

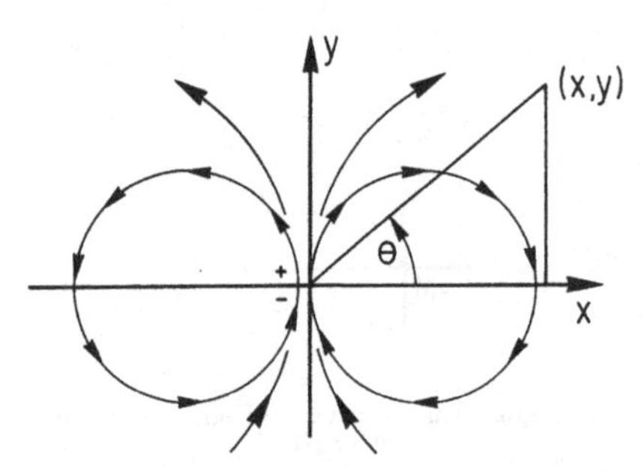

Figure 1.7 Flow due to a vertical upward-directed dipole.

This gives a vertically upward-directed dipole at the origin.

In polar coordinates

$$\phi_{yd} = -\frac{\Sigma_y \sin\theta}{2\pi r} \tag{1.47}$$

then $\partial\phi_{yd}/\partial r = \Sigma_y \sin\theta/2\pi r^2$, showing that the radial velocity is an odd function of θ directed outward (away from the "source" in the region $0 < \theta < \pi$ and inward – toward the "sink" in the region $-\pi < \theta < 0$), cf. Figure 1.7. This agrees with our expectations.

For a vertically downward directed dipole, change the sign.

Consider a line distribution of downward dipoles along the negative x-axis having uniform strength σ_0. The potential of this array is

$$\phi = \frac{y}{2\pi}\sigma_0 \int_{-\infty}^{0} \frac{dx'}{(x - x')^2 + y^2} \tag{1.48}$$

This integrates to

$$= -\frac{y}{2\pi}\frac{\sigma_0}{y}\tan^{-1}\frac{(x - x')}{y}\Bigg|_{-\infty}^{0}$$

$$= -\frac{\sigma_0}{2\pi}\left\{\tan^{-1}\frac{x}{y} - \frac{\pi}{2}\right\}$$

$$= -\frac{\sigma_0}{2\pi}\{-\theta\} = \frac{\sigma_0}{2\pi}\theta \tag{1.49}$$

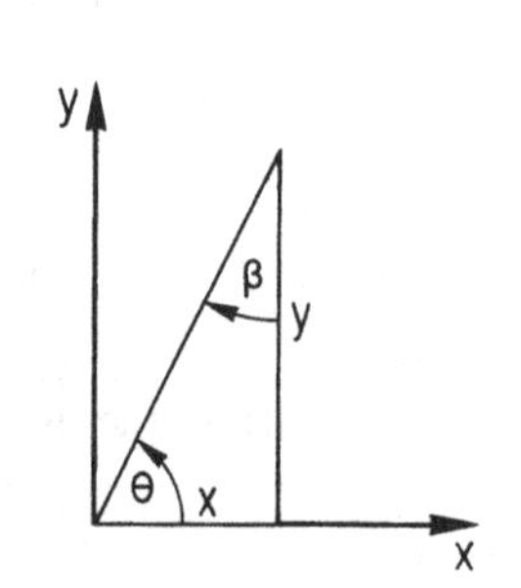

$\beta + \theta = \frac{\pi}{2} \quad \tan^{-1}\frac{x}{y} = \frac{\pi}{2} - \theta$

Figure 1.8

This is identical to the potential of a vortex at the origin by taking $\sigma_0 = \Gamma$.

Thus the potential of a vortex is equivalent to that of a line distribution of downward dipoles along the cut whose strengths are equal to Γ. Since the dipole is derivable from a source (by directed differentiation) we see that the source is the basic singularity or the *fountainhead of classical hydrodynamics.*

The Point Source (In three dimensions)

The basic solution or ***Green function*** (cf. Mathematical Compendium, Section 3, p. 505) for the Laplace Equation in a boundless fluid is

$$\phi(x,y,z,t) = -\frac{M(t)}{4\pi\sqrt{x^2 + y^2 + z^2}} \tag{1.50}$$

where M is the source strength depending at most on time t and independent of the coordinates of any field point (x,y,z). The physical representation of this is as follows:

Imagine fluid to be appearing in a spherically symmetrical manner at the origin of coordinates at a rate of M cubic units per unit time. Then by continuity the flow passing across any spherical boundary of radius R with center at the origin must satisfy

$$4\pi R^2 u_r = M \tag{1.51}$$

where u_r is the radial velocity component normal to the spherical surface (of radius R). Thus

$$u_r(R) = \frac{M}{4\pi R^2} \tag{1.52}$$

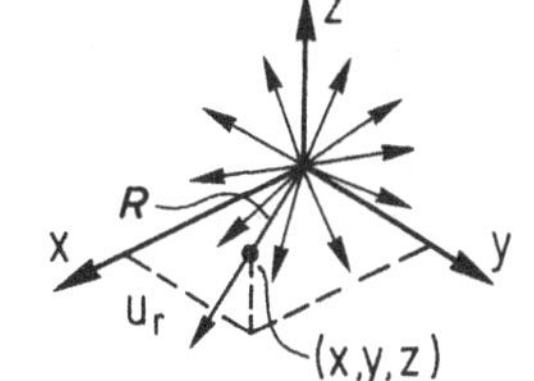

Figure 1.9 Flow due to a source.

As this flow is free of vorticity we can replace u_r by the radial derivative of a velocity potential function $\phi = \phi(R)$ so that

$$u_r = \frac{\partial \phi}{\partial R} = \frac{M}{4\pi R^2}$$

Integrating

$$\phi(R) = -\frac{M}{4\pi R} + \text{a constant} \tag{1.53}$$

As $\phi(R)$ must vanish as $R \longrightarrow \infty$, the constant is zero.

Finally, in rectangular coordinates $R = \sqrt{x^2 + y^2 + z^2}$ and (1.50) is obtained.

If the point at which the fluid is appearing is placed at x = x', y = y', z = z', then the velocity potential is

$$\phi(x,y,z;x',y',z') = -\frac{M}{4\pi\sqrt{(x-x')^2 + (y-y')^2 + (z-z')^2}} \tag{1.54}$$

This function is seen to be singular if the field point P(x,y,z) is moved to the source point Q(x',y',z'), i.e. $\phi \longrightarrow \infty$ which is undefined.

To show that this potential satisfies Laplace's Equation

$$\frac{\partial^2 \phi}{\partial x^2} + \frac{\partial^2 \phi}{\partial y^2} + \frac{\partial^2 \phi}{\partial z^2} = 0 \tag{1.55}$$

we may proceed to calculate each term (omitting $M/4\pi$)

$$\frac{\partial \phi}{\partial x} = \frac{(x - x')}{R^3}$$

where R is now the denominator of (1.54). Then

$$\frac{\partial^2 \phi}{\partial x^2} = \frac{1}{R^3} - \frac{3(x - x')^2}{R^5}$$

As y and z enter in the same way we get by inspection

$$\left[\frac{\partial^2}{\partial x^2} + \frac{\partial^2}{\partial y^2} + \frac{\partial^2}{\partial z^2}\right]\frac{1}{R} = \frac{3}{R^3} - \frac{3}{R^5}\left[(x-x')^2 + (y-y')^2 + (z-z')^2\right] \tag{1.56}$$

and this is seen to vanish everywhere except for $x = x'$, $y = y'$, $z = z'$. This is not surprising since the Laplace Equation is a re-statement of the continuity requirement which is violated at the source.

In general the equation satisfied by the point source is

$$\nabla^2\phi = M\ \delta(x - x')\ \delta(y - y')\ \delta(z - z') \tag{1.57}$$

where δ is the Dirac delta function, see the Mathematical Compendium, Section 2, p. 504.

The point source singularity can be generalized to line, surface and volume distributions. To secure a line distribution one replaces the source strength M by m(s)ds where m is cubic units / linear unit / unit of time and s is arc length. For example, a line distribution of sources along $x = y = 0$, the z-axis between $z = -a$ and $z = a$ has the velocity potential

$$\phi(x,y,z) = -\frac{1}{4\pi}\int_{-a}^{a} \frac{m(z')}{\sqrt{x^2 + y^2 + (z - z')^2}}\, dz' \tag{1.58}$$

For a uniform distribution $m(z') = m_0$, a constant, the integral can be carried out to give

$$\phi(r,z) = \frac{m_0}{4\pi} \ln\left\{\frac{z - a + \sqrt{(z - a)^2 + r^2}}{z + a + \sqrt{(z + a)^2 + r^2}}\right\} \tag{1.59}$$

where $r^2 = x^2 + y^2$.

If we extend the line from $-\infty$ to ∞ we should expect to recapture the potential of the two-dimensional source. However, the integral diverges logarithmically yielding formally $\ln(0/\infty) \longrightarrow -\infty$. To achieve the finite part we may proceed as follows:

$$-\frac{4\pi\phi}{m_0} = \int_{-a}^{z} \frac{1}{\sqrt{r^2 + (z-z')^2}}\, dz' + \int_{z}^{a} \frac{1}{\sqrt{r^2 + (z'-z)^2}}\, dz' \tag{1.60}$$

$$= -\ln r + \ln\left[z + a + \sqrt{r^2 + (z + a)^2}\right]$$

$$+ \ln\left[a - z + \sqrt{r^2 + (a - z)^2}\right] - \ln r \tag{1.61}$$

Then for large a

$$= -2\ln r + 2\ln 2a \tag{1.62}$$

Now as $a \longrightarrow \infty$ the last term is infinite but as it is independent of x and y or r, it is a constant which can be dropped. Hence

$$\lim_{a \to \infty} \phi(r,z) = \frac{m_0}{2\pi} \ln r \tag{1.63}$$

which is the potential of the two-dimensional source of strength density m_0.

The Point Dipole (In three dimensions)

As in two dimensions, the three-dimensional dipole is generated by drawing together a point source and a point sink along a directed line and passing to a limit as the distance between them is made to approach zero in a defined manner. For example, consider a point source at $x = 0$ and a point sink, a negative source, at $x = -\varepsilon$. Then their sum is

$$\phi = \phi_{so} + \phi_{si} = -\frac{M}{4\pi}\left[\frac{1}{\sqrt{x^2 + r^2}} - \frac{1}{\sqrt{(x + \varepsilon)^2 + r^2}}\right] \tag{1.64}$$

and keeping terms of order ε

$$\frac{1}{\sqrt{x^2 + r^2 + 2\varepsilon x}} = \frac{1}{\sqrt{x^2 + r^2}\sqrt{1 + \dfrac{2\varepsilon x}{x^2 + r^2}}}$$

Expanding by the binomial expansion theorem

$$= \frac{1}{\sqrt{x^2 + r^2}} \left[1 - \frac{1}{2}\left[\frac{2\varepsilon x}{x^2 + r^2}\right] + O(\varepsilon^2) \ldots\right] \tag{1.65}$$

Put this in (1.64), then

$$\lim_{\varepsilon \to 0} \phi = \phi_d =$$

$$-\lim_{\varepsilon \to 0} \frac{M}{4\pi} \left[\frac{1}{\sqrt{x^2 + r^2}} - \frac{1}{\sqrt{x^2 + r^2}} + \frac{\varepsilon x}{[x^2 + r^2]^{3/2}} + O(\varepsilon^2)\right]$$

Defining the dipole moment strength by

$$\Sigma_x = M\varepsilon \tag{1.66}$$

we obtain

$$\phi_d = -\frac{\Sigma_x x}{4\pi\ [x^2 + r^2]^{3/2}} \tag{1.67}$$

as the potential of an x-directed dipole at the origin with axis in the positive x-direction.

This can be achieved directly by differentiation of the source. To secure the positively directed x-dipole at the origin place a source at $x = x'$, $y = z = 0$. Then

$$\phi_d = \frac{\partial}{\partial x'} \left\{-\frac{M}{4\pi} \frac{1}{\sqrt{(x - x')^2 + r^2}}\right\} \Bigg|_{x' = 0}$$

where the derivative is evaluated at $x' = 0$ *after* differentiation. Then

$$\phi_d = -\frac{M}{4\pi}\left[-\frac{1}{2}\right] \frac{2(x - x')(-1)}{[(x - x')^2 + r^2]^{3/2}} \Bigg|_{x' = 0}$$

Then formally replacing M by Σ_x we have (1.67).

In general to achieve a dipole whose axis has the direction cosines n_x, n_y, n_z at any point (x',y',z') apply the following operation

$$\phi_d = -\frac{\Sigma}{4\pi}\left[n_x\frac{\partial}{\partial x'} + n_y\frac{\partial}{\partial y'} + n_z\frac{\partial}{\partial z'}\right] \frac{1}{\sqrt{(x-x')^2 + (y-y')^2 + (z-z')^2}} \tag{1.68}$$

where $\Sigma_x = n_x\Sigma$, $\Sigma_y = n_y\Sigma$, $\Sigma_z = n_z\Sigma$

Dipoles may be distributed on lines and surfaces in the same manner as sources.

VORTICITY

We have seen previously that by manipulating the equations of motion (Equations (1.8) to (1.14)), terms arise which, together, are identified as components of a so-called vorticity vector

$$\boldsymbol{\zeta} = \xi\mathbf{i} + \eta\mathbf{j} + \zeta\mathbf{k}$$

the components of which are given by velocity gradients, i.e.,

$$(x) \rightarrow (z) \rightarrow (y)$$

$$\xi = \frac{\partial w}{\partial y} - \frac{\partial v}{\partial z} \; ; \; \eta = \frac{\partial u}{\partial z} - \frac{\partial w}{\partial x} \; ; \; \zeta = \frac{\partial v}{\partial x} - \frac{\partial u}{\partial y} \tag{1.69}$$

$$(x) \rightarrow (y) \rightarrow (z)$$

Here we see that the operations are in cyclic order and the numerators are in acyclic or reverse alphabetic order.

To give a more physically based exposition of the rôle of vorticity we can examine the excursions in the velocity components between two neighboring points (x,y,z) and $(x + \delta x, y + \delta y, z + \delta z)$. Thus the changes in the velocities over these infinitely small distances are formally

$$\begin{aligned}
\delta u &= \frac{\partial u}{\partial x}\delta x + \frac{\partial u}{\partial y}\delta y + \frac{\partial u}{\partial z}\delta z \\
\delta v &= \frac{\partial v}{\partial x}\delta x + \frac{\partial v}{\partial y}\delta y + \frac{\partial v}{\partial z}\delta z \\
\delta w &= \frac{\partial w}{\partial x}\delta x + \frac{\partial w}{\partial y}\delta y + \frac{\partial w}{\partial z}\delta z
\end{aligned} \tag{1.70}$$

Our previous experience may perhaps suggest the formation of an array or matrix and that some insight might result if this array is made symmetric, in part, about the diagonal as indicated in (1.70).

Taking the first equation in (1.70) and observing the *coefficients* of δx in the second and third equations we add and subtract terms in the first equation as follows

$$\delta u = \frac{\partial u}{\partial x}\delta x + \left[\frac{\partial u}{\partial y} + \frac{1}{2}\frac{\partial v}{\partial x}\right]\delta y + \left[\frac{\partial u}{\partial z} + \frac{1}{2}\frac{\partial w}{\partial x}\right]\delta z - \frac{1}{2}\frac{\partial v}{\partial x}\delta y - \frac{1}{2}\frac{\partial w}{\partial x}\delta z$$

The vorticity components can be obtained by the following alteration and analogous manipulations of the remaining two equations of (1.70) to give

$$\begin{aligned}
\delta u &= \frac{\partial u}{\partial x}\delta x + \frac{1}{2}\left[\frac{\partial u}{\partial y} + \frac{\partial v}{\partial x}\right]\delta y + \frac{1}{2}\left[\frac{\partial u}{\partial z} + \frac{\partial w}{\partial x}\right]\delta z + \frac{1}{2}\overset{\eta}{\left[\frac{\partial u}{\partial z} - \frac{\partial w}{\partial x}\right]}\delta z - \frac{1}{2}\overset{\zeta}{\left[\frac{\partial v}{\partial x} - \frac{\partial u}{\partial y}\right]}\delta z \\
\delta v &= \frac{1}{2}\left[\frac{\partial u}{\partial y} + \frac{\partial v}{\partial x}\right]\delta x + \frac{\partial v}{\partial y}\delta y + \frac{1}{2}\left[\frac{\partial w}{\partial y} + \frac{\partial v}{\partial z}\right]\delta z + \frac{1}{2}\zeta\,\delta x - \frac{1}{2}\xi\,\delta z \\
\delta w &= \frac{1}{2}\left[\frac{\partial u}{\partial z} + \frac{\partial w}{\partial x}\right]\delta x + \frac{1}{2}\left[\frac{\partial w}{\partial y} + \frac{\partial v}{\partial z}\right]\delta y + \frac{\partial w}{\partial z}\delta z + \frac{1}{2}\xi\,\delta y - \frac{1}{2}\eta\,\delta x
\end{aligned} \tag{1.71}$$

Thus the motion of a small element is composed of three parts:

i. Translation as a whole with components u,v,w.

ii. Straining as given by the first three terms in each row of the array.

iii. Rotations of the element as given by the last two terms in each equation; the components of angular velocity being $\xi/2$, $\eta/2$, $\zeta/2$ (termed 'mean rotations' by Cauchy in 1841, (cf. Lamb (1963)).

Although we limit our attention to flows in which the vorticity is generally zero we shall find it useful to represent lifting surfaces by isolated vorticity on the body and into the wake.

It is to be noted that the vorticity in analogy with the velocity field is divergence-free, i.e.,

$$\nabla \cdot \boldsymbol{\zeta} = \frac{\partial \xi}{\partial x} + \frac{\partial \eta}{\partial y} + \frac{\partial \zeta}{\partial z} = 0 \tag{1.72}$$

as may be seen by carrying out this operation on (1.69).

To calculate the velocity field $\mathbf{q}_V$ induced by any given distribution of vorticity in three (or two) dimensions we may proceed as follows:

We have seen that the velocity field $\mathbf{q}_V$ and vorticity vector $\boldsymbol{\zeta}$ are related by

$$\nabla \times \mathbf{q}_V = \boldsymbol{\zeta} \tag{1.73}$$

Assume that there exists a vector potential $\mathbf{B}_V$ such that

$$\nabla \times \mathbf{B}_V = \mathbf{q}_V \tag{1.74}$$

Thus $\mathbf{q}_V$ is mutually perpendicular to the vector operator ∇ and the vector potential $\mathbf{B}_V$.

Then the field equation for $\mathbf{B}_V$ is got by replacing $\mathbf{q}_V$ in (1.73) i.e.,

$$\nabla \times \nabla \times \mathbf{B}_V = \zeta \tag{1.75}$$

The vector triple product is equivalent to

$$\nabla (\nabla \cdot \mathbf{B}_V) - \nabla^2 \mathbf{B}_V = \zeta$$

and if we can assume for the present that the divergence of $\mathbf{B}_V$ vanishes everywhere i.e.,

$$\nabla \cdot \mathbf{B}_V \equiv 0 \tag{1.76}$$

then the field equation for $\mathbf{B}_V$ is

$$\nabla^2 \mathbf{B}_V = - \zeta(x,y,z) \tag{1.77}$$

which is an inhomogeneous Laplace Equation named after ***Poisson***. The solution of this equation is known to be (cf. Milne-Thomson (1955))

$$\mathbf{B}_V = \frac{1}{4\pi} \int_V \frac{\zeta(x',y',z')}{R} \, dV' \tag{1.78}$$

where

$$R = \sqrt{(x - x')^2 + (y - y')^2 + (z - z')^2}$$

and

$$\int_V (\ ) dV'$$

is a volume integral ($dV' = dx'dy'dz'$) throughout the entire volume in which the vorticity is distributed. This is essentially a volumetric distribution of sources of vector strength density ζ over all dummy points (x',y',z').

With hindsight we can see that (1.78) is a solution of (1.77) because of the previously observed behavior of the Laplacian of the source potential $1/R$ (see Equation (1.57)), i.e.

$$\nabla^2 \frac{1}{R} = - 4\pi\, \delta(x - x')\, \delta(y - y')\, \delta(z - z')$$

so that the δ-functions upon integration pick out ζ at $x' = x$, $y' = y$, $z' = z$.

To demonstrate the requirement which must be met to ensure compliance with our assumption (1.76) we can operate on (1.78) with

$$\nabla = \mathbf{i}\frac{\partial}{\partial x} + \mathbf{j}\frac{\partial}{\partial y} + \mathbf{k}\frac{\partial}{\partial z}$$

to get

$$\nabla \cdot \mathbf{B}_V = \frac{1}{4\pi}\int_V \zeta(x',y',z') \cdot \nabla\left[\frac{1}{R}\right] dV' \tag{1.79}$$

Now as R involves (x,y,z) and (x',y',z') as differences respectively,

$$\nabla \frac{1}{R} = -\nabla' \frac{1}{R} \ , \quad \nabla' = \mathbf{i}\frac{\partial}{\partial x'} + \mathbf{j}\frac{\partial}{\partial y'} + \mathbf{k}\frac{\partial}{\partial z'}$$

Then

$$\zeta(x') \cdot \nabla' \frac{1}{R} = \nabla' \cdot \left[\frac{\zeta}{R}\right] - \frac{1}{R}\nabla' \cdot \zeta \tag{1.80}$$

But the last term is zero (see (1.72)) and hence

$$\nabla \cdot \mathbf{B}_V = -\frac{1}{4\pi}\int_V \nabla' \cdot \left[\frac{\zeta}{R}\right] dV' \tag{1.81}$$

which by Gauss' theorem (see Mathematical Compendium, Section 3, Equation (M3.1), p. 505) yields

$$\nabla \cdot \mathbf{B}_V = -\frac{1}{4\pi}\int_S \frac{\zeta \cdot \mathbf{n}'}{R} dS' \quad ; \ \mathbf{n}' \text{ is normal to surface S} \tag{1.82}$$

Thus $\nabla \cdot \mathbf{B}_V$ will be zero if the vorticity vector at the bounding surface is normal to the surface i.e., $\zeta \cdot \mathbf{n} \equiv 0$. This will always be the case in all applications considered herein.

We may now utilize (1.78) and (1.74) to arrive at a formula for the velocity induced by an arbitrary distribution of vorticity. Thus

$$\mathbf{q}_V = \frac{1}{4\pi}\int_V \nabla \times \frac{\zeta(x',y',z')}{R} dV' \tag{1.83}$$

which is equivalent to

$$\mathbf{q}_v = \frac{1}{4\pi}\int_V \frac{\zeta(x',y',z')\times \boldsymbol{R}}{R^3}\,dV' \tag{1.84}$$

The *practical* application of this is to vorticity distributions along lines and surfaces. In line distributions ζ is zero everywhere except along the line and hence the volume integral reduces to a line integral which *must be closed* or end normal to boundaries. The vorticity vector is everywhere tangent to the curve and is positive in the direction which produces circulation about the line element in the sense of the right-hand rule for rotation or moment vectors.

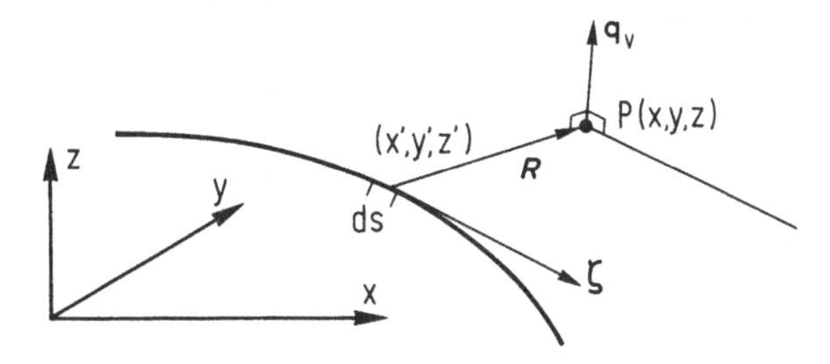

Figure 1.10 Velocity induced in point P by differential line element ds in Q.

The local induced velocity at P due to ζds is mutually perpendicular to $\boldsymbol{R}$ and ζ in the direction obtained by the right-hand rule obtained by rotating ζ into $\boldsymbol{R}$. The velocity at P must be obtained by integrating completely around the vortex filament. If the vortex extends to infinity then it may be considered to be closed by adding a circuit at infinite radius which closes the curve but makes no contribution.

Thus for vorticity distributed over a closed curve, cf. Figure 1.10

$$\mathbf{q}_v = \frac{1}{4\pi}\int_s \frac{\zeta(x',y',z'(x',y'))\times \boldsymbol{R}}{R^3}\,ds \tag{1.85}$$

This formula is called the ***Biot-Savart law*** and was first deduced experimentally by Biot and Savart in 1820 (cf. Yih (1988)) as the magnetic vector induced by a steady current flowing in a closed conducting wire.

To apply this and recapture the two-dimensional formula deduced earlier consider vorticity of strength Γ distributed along the z-axis from $-\infty$ to ∞. Then $ds = dz'$, $x' = y' = 0$, $\zeta = 0\,\mathbf{i} + 0\,\mathbf{j} + \Gamma\,\mathbf{k}$ and $\boldsymbol{R} = x\mathbf{i} + y\mathbf{j} + (z-z')\mathbf{k}$. Then

$$\mathbf{q}_v = \frac{1}{4\pi}\int_{-\infty}^{\infty} \frac{\begin{vmatrix} \mathbf{i} & \mathbf{j} & \mathbf{k} \\ 0 & 0 & \Gamma \\ x & y & z - z' \end{vmatrix} dz'}{[x^2 + y^2 + (z - z')^2]^{3/2}} \tag{1.86}$$

The x-component is (expanding by the minor of **i**)

$$u = -\frac{y\Gamma}{4\pi}\int_{-\infty}^{\infty} \frac{dz'}{[r^2 + (z - z')^2]^{3/2}}$$

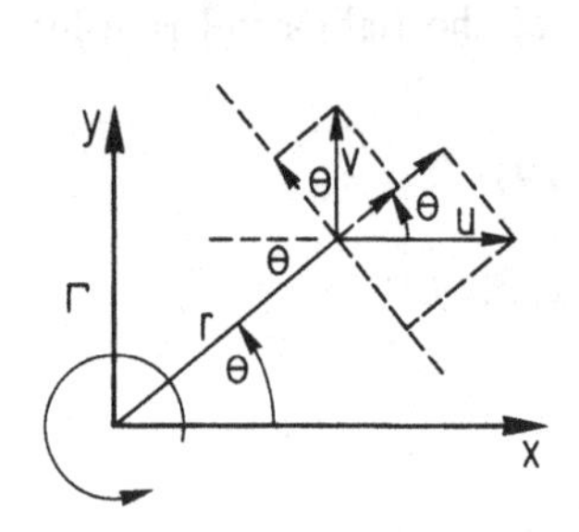

Figure 1.11 Radial and tangential velocities.

$$= \frac{y\Gamma}{4\pi}\frac{1}{r^2}\frac{(z - z')}{\sqrt{r^2 + (z - z')^2}}\Bigg|_{-\infty}^{\infty}$$

$$= -\frac{\Gamma}{2\pi}\frac{y}{r^2} = -\frac{\Gamma}{2\pi}\frac{\sin\theta}{r} \tag{1.87}$$

The y-component involves

$$-\begin{vmatrix} 0 & \Gamma \\ x & z - z' \end{vmatrix} = \Gamma x$$

and the integral is the same yielding

$$v = \frac{\Gamma}{2\pi}\frac{x}{r^2} = \frac{\Gamma}{2\pi}\frac{\cos\theta}{r}$$

The tangential velocity is given formally by

$$u_t = v\cos\theta - u\sin\theta$$

$$= \frac{\Gamma}{2\pi}\left[\frac{\cos^2\theta + \sin^2\theta}{r}\right] = \frac{\Gamma}{2\pi r} \tag{1.88}$$

the two-dimensional result.

The radial component is

$$u_r = v\sin\theta + u\cos\theta = \frac{\Gamma}{2\pi}(\cos\theta\sin\theta - \sin\theta\cos\theta)$$

$$= 0 \text{ which checks.} \tag{1.89}$$

Finally, we note that the z-component is zero as it must be since it involves the determinant which is the minor of **k**, namely

$$\begin{vmatrix} 0 & 0 \\ x & y \end{vmatrix} \equiv 0$$

It is necessary to add that along any vortex filament the vorticity is constant.

An important theorem due to ***Stokes***:

- ***The velocity potential induced by any line vortex of strength κ is given by the potential of a distribution of dipoles of strength κ normal to any surface bounded or "ringed" by the closed vortex filament.***

Thus the velocity potential of a circular ring vortex of strength κ about the origin whose plane is perpendicular to the x-axis and whose radius is R is given by a uniform distribution of x-directed dipoles over the disc x = 0, $0 \le r \le R$, $0 \le \gamma \le 2\pi$ cf. Figure 1.12. Explicitly, using Equation (1.68) with direction cosines $n_x = 1$, $n_y = 0$, $n_z = 0$, the potential of positively directed x-dipoles over the disc is, using cylindrical coordinates,

$$\phi(x,r,\gamma) = -\frac{\kappa}{4\pi}\frac{\partial}{\partial x'}\int_0^{2\pi} d\gamma' \int_0^R \frac{r'dr'}{\sqrt{(x-x')^2+r^2+r'^2-2rr'\cos(\gamma-\gamma')}}\Bigg|_{x'=0} \tag{1.90}$$

Evaluation is not possible in terms of elementary functions for all field points, P. However, along the x-axis a simple result can be secured.

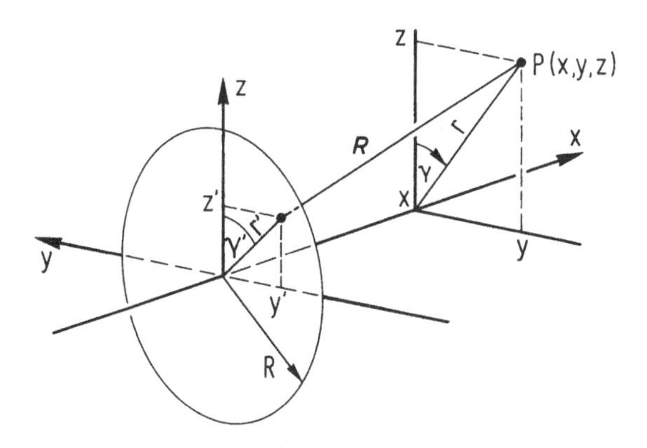

Figure 1.12 Circular disc with x-directed dipoles.

Carrying out the x'-differentiation and setting r = 0,

$$\phi(x,0) = -\frac{\kappa}{4\pi} x \int_0^{2\pi} d\gamma' \int_0^R \frac{r'\ dr'}{[x^2 + r'^2]^{3/2}} \; ; \tag{1.91}$$

integrating over r' and γ'

$$\phi(x,0) = \frac{\kappa}{2} \left[\frac{x}{\sqrt{x^2 + r'^2}} \right]_0^R = \frac{\kappa}{2} \left\{ \frac{x}{\sqrt{x^2 + R^2}} - \frac{x}{|x|} \right\} \tag{1.92}$$

where $|x| = \sqrt{x^2}$.

We note in passing that ϕ is discontinuous across the disc i.e.

$$\lim_{x \to 0_+} \phi(x,0) = -\frac{\kappa}{2} \quad \text{as} \quad \lim_{x \to 0_+} \frac{x}{|x|} = +1$$

and

$$\lim_{x \to 0_-} \phi(x,0) = +\frac{\kappa}{2} \quad \text{as} \quad \lim_{x \to 0_-} \frac{x}{|x|} = -1$$

This jump in the potential, namely

$$\phi(0_+,0) - \phi(0_-,0) = -\kappa \tag{1.93}$$

is true all over the disc and is a basic property of surface distributions of dipoles. More about this later.

To calculate $u = \partial\phi/\partial x$ say for $x > 0$

$$u = \frac{\kappa}{2} \frac{\partial}{\partial x} \left\{ \frac{x}{\sqrt{x^2 + R^2}} - 1 \right\} \tag{1.94}$$

$$u = \frac{\kappa R^2}{2[x^2 + R^2]^{3/2}} \tag{1.95}$$

We now compare this with what is obtained from Biot-Savart, cf. (1.85).

$$\begin{aligned} \zeta ds &= 0\mathbf{i} + \kappa dy'\mathbf{j} + \kappa dz'\mathbf{k} \\ &= 0\mathbf{i} - \kappa R \cos\gamma' d\gamma' \mathbf{j} \\ &\quad - \kappa R \sin\gamma' d\gamma' \mathbf{k} \\ \boldsymbol{R} &= x\mathbf{i} + (y-y')\mathbf{j} + (z-z')\mathbf{k} \\ &= x\mathbf{i} - (r \sin\gamma - R \sin\gamma')\mathbf{j} \\ &\quad + (r \cos\gamma - R \cos\gamma')\mathbf{k} \end{aligned}$$

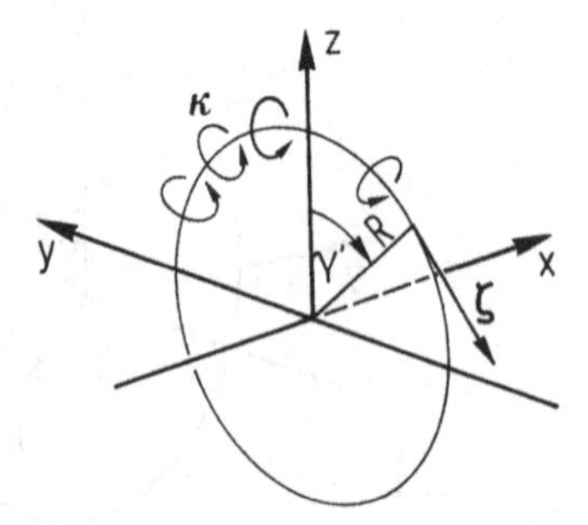

Figure 1.13 Circular ring vortex.

$$R(x,r,\gamma;0,R,\gamma') = |\boldsymbol{R}|$$

$$\begin{aligned} ds &= +\, R d\gamma' \\ dy' &= -\, R \cos\gamma' d\gamma' \\ dz' &= -\, R \sin\gamma' d\gamma' \end{aligned}$$

Then

$$\mathbf{q} = \frac{\kappa}{4\pi}\int_0^{2\pi} \frac{\begin{vmatrix} \mathbf{i} & \mathbf{j} & \mathbf{k} \\ 0 & -R\cos\gamma' d\gamma' & -R\sin\gamma' d\gamma' \\ x & -(r\sin\gamma - R\sin\gamma') & r\cos\gamma - R\cos\gamma' \end{vmatrix}}{R(x,r,\gamma;0,R,\gamma')^3} \tag{1.96}$$

The x-component is

$$u = \frac{\kappa R}{4\pi}\int_0^{2\pi} \frac{(-r\cos\gamma\cos\gamma' + R\cos^2\gamma' - r\sin\gamma\sin\gamma' + R\cdot\sin^2\gamma')}{R(x,r,\gamma;0,R,\gamma')^3}\, d\gamma' \tag{1.97}$$

Recalling that $\cos(\gamma - \gamma') = \cos\gamma \cos\gamma' + \cos\gamma \sin\gamma'$ we have in general

$$u(x,r,\gamma) = \frac{\kappa R}{4\pi}\int_0^{2\pi} \frac{(R - r\cos(\gamma-\gamma'))d\gamma'}{[x^2+r^2+R^2-2Rr\cos(\gamma-\gamma')]^{3/2}} \tag{1.98}$$

We may reduce the complication of this integral by noting that it is the negative of the partial derivative of $1/R$ with respect to R

$$u(x,r,\gamma) = -\frac{\kappa R}{4\pi}\frac{\partial}{\partial R}\int_0^{2\pi} \frac{d\gamma'}{\sqrt{x^2 + r^2 + R^2 - 2Rr\cos(\gamma - \gamma')}}$$

Evaluating along $r = 0$

$$u(x,0) = -\frac{\kappa R}{2}\frac{\partial}{\partial R}\left\{\frac{1}{\sqrt{x^2 + R^2}}\right\} = \frac{\kappa R^2}{2[x^2 + R^2]^{3/2}} \tag{1.99}$$

This is seen to agree with (1.95) and while not proving the theorem of the equivalence of vortex filaments and distributions of normal dipoles over surfaces spanning the vortices, the foregoing exercise does provide support.

We may now turn to an examination of the properties of surface distributions of singularities in two and three dimensions.

2 Properties of Distributions of Singularities

In this chapter we determine the basic behavior of the velocity fields of the various singular solutions of Laplace's equation when they are distributed or "smeared" along lines and surfaces of finite extent. Their properties are particularly important on the lines and surfaces as they will be repeatedly used to generate approximate flow envelopes about thin hydrofoil sections in two dimensions and about slender bodies and lifting surfaces in three space dimensions.

PLANAR DISTRIBUTIONS IN TWO DIMENSIONS

Source Distributions

Source distributions are useful in generating section shapes symmetrical about the long axis. It is therefore important to understand the connection between the source density and the velocity components induced by the entire distribution of sources.

From the foregoing, the potential of a line of sources in two dimensions is from (1.35)

$$\phi(x,y) = \frac{1}{2\pi}\int_{-a}^{a} m(x') \ln \sqrt{(x - x')^2 + y^2}\, dx' \tag{2.1}$$

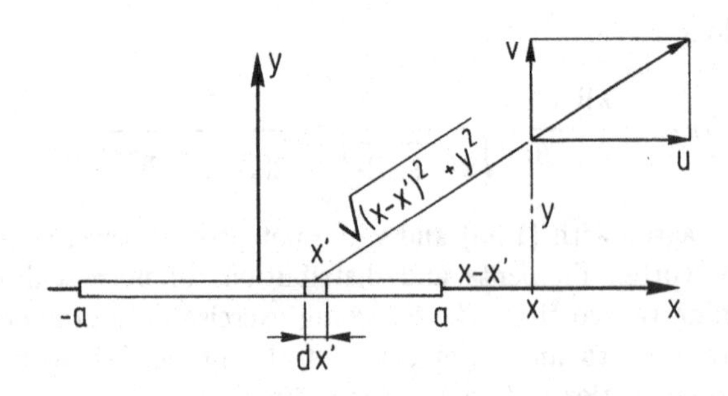

Figure 2.1 Line distribution of sources.

The *axial* component of velocity is

$$u = \frac{\partial \phi}{\partial x} = \frac{1}{2\pi} \int_{-a}^{a} m(x') \frac{\partial}{\partial x} \ln \sqrt{(x-x')^2 + y^2}\, dx' \tag{2.2}$$

$$= \frac{1}{2\pi} \int_{-a}^{a} \frac{m(x')(x - x')}{(x - x')^2 + y^2}\, dx' \tag{2.3}$$

We note that $u(y) = u(-y)$, an even function as expected physically.

It will be seen later that we shall make use of $u(x,0)$. Setting $y = 0$ in (2.3) provides formally

$$u(x,0) = \frac{1}{2\pi} ⨍_{-a}^{a} \frac{m(x')}{x - x'}\, dx' \tag{2.4}$$

This integral does not exist in the ordinary sense because $1/(x - x')$ becomes singular and hence requires a special definition as devised by Cauchy and termed by him, "principe valeur" or principal value[3]. The bar indicates the ***Cauchy principal value***, often designated in the literature by P.V. $\int(\ )$ or $P\int(\ )$.

From the foregoing we learn

- ***The axial component of a source distribution along the x-axis is symmetric in y or an even function of y.***

- ***On the axis itself the axial component depends on the entire distribution, i.e., all source elements contribute.***

The "*athwartship*" or y-component (from 2.1)

$$v(x,y) = \frac{y}{2\pi} \int_{-a}^{a} \frac{m(x')}{(x - x')^2 + y^2}\, dx' \tag{2.5}$$

Here we observe that $v(x,y) = -v(x,y)$, i.e. it is an odd function of y.

When we ask what is the value of v on the distribution, i.e., on $-a \leq x \leq a$, $y = 0$ at first we may say zero because $y \to 0$ in the numerator. However at $y = 0$ the kernel function becomes $(x - x')^{-2}$ and the integrand tends to "blow up". Thus we suspect a "battle" going on between the vanishing of the numerator and the infinity of the integrand as x' passes through x and so we have to "creep up" on this limit by the use of a transformation.

[3] See Mathematical Compendium, Section 4, p. 507 and sequel.

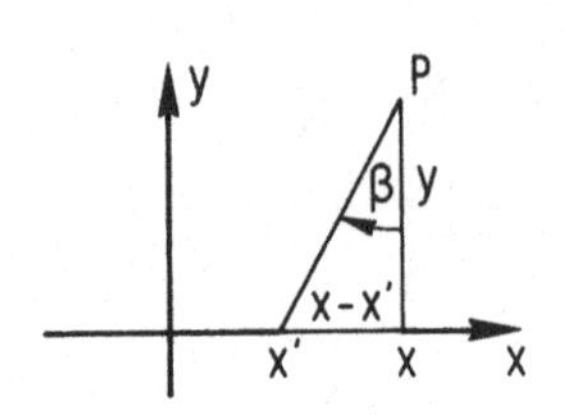

Figure 2.2

Let

$$x - x' = y \tan\beta$$

Then

$$dx' = -y \sec^2\beta d\beta$$

$$\beta = \tan^{-1} \frac{x - x'}{y}$$

Then (2.5) becomes

$$v(x,y) = -\frac{y}{2\pi} \int_{\tan^{-1}\frac{x+a}{y}}^{\tan^{-1}\frac{x-a}{y}} \frac{m(x - y\tan\beta)\, y \sec^2\beta}{y^2 \sec^2\beta} d\beta \tag{2.6}$$

Now consider the field point P to descend from above, so $y \to 0_+$. As $x < a$ we see that $\tan^{-1}(x-a)/y \to -\pi/2$ and $\tan^{-1}(x+a)/y \to \pi/2$. $m(x-y\tan\beta) \to m(x)$ independent of β and hence

$$\lim_{y \to 0_+} v(x,y) = -\frac{m(x)}{2\pi} \int_{\pi/2}^{-\pi/2} d\beta = \frac{m(x)}{2} \tag{2.7}$$

Analogously as $y \to 0_-$, $\tan^{-1}(x-a)/y \to \pi/2$ and $\tan^{-1}(x+a)/y \to -\pi/2$ and hence

$$\lim_{y \to 0_-} v(x,y) = -\frac{m(x)}{2} \tag{2.8}$$

Thus

- ***The velocity normal to the source sheet jumps and the value depends only on the source density at that point and not upon an integral over the distribution.***

The source density is given by

$$m(x) = v(x,0_+) - v(x,0_-) = \Delta v(x) \tag{2.9}$$

Thus if there is some way of obtaining Δv then $m(x)$ is known. It should also be noted that for $|x| > a$

$$v(x,0) \equiv 0 \tag{2.10}$$

These results can be secured by purely physical arguments. We can see that for small y, source-elements at a distance from the adjacent element can induce only very weak normal velocity and as $y \to 0_{\pm}$ we see that the flow through an elementary control surface is simply as depicted in Figure 2.3.

By continuity we have

$$v_+(x)\delta x = \frac{m(x)}{2}\delta x \quad \text{or}$$

$$v_-(x)\delta x = -\frac{m(x)}{2}\delta x$$

$$\therefore\ v_+(x) = \frac{m(x)}{2} \quad ;$$

$$v_-(x) = -\frac{m(x)}{2} \qquad (2.11)$$

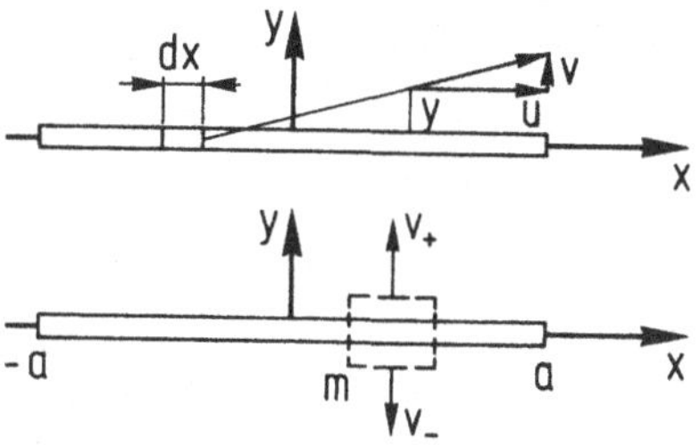

Figure 2.3

From the vertical asymmetry it is obvious that for all points $y = 0$, $|x| > a$ the vertical component is zero.

It is also to be noted that as P is moved far with respect to 2a then

$$u \longrightarrow \frac{1}{2\pi}\frac{x\int_{-a}^{a} m(x')dx'}{x^2 + y^2} \longrightarrow 0 \qquad (2.12)$$

$$v \longrightarrow u \longrightarrow 0 \qquad (2.13)$$

Vertical Dipole Distributions

The potential induced by a line distribution of vertical dipoles along the x-axis is

$$\phi(x,y) = \frac{1}{2\pi}\frac{\partial}{\partial y'}\int_{-a}^{a} \sigma_y(x') \ln\sqrt{(x-x')^2 + (y-y')^2}\,\Big|_{y'=0} dx' \qquad (2.14)$$

$$\phi(x,y) = -\frac{y}{2\pi}\int_{-a}^{a}\frac{\sigma_y(x')}{(x - x')^2 + y^2}dx' \qquad (2.15)$$

As (2.15) has the same structure as the transverse velocity developed by sources (2.5) then the behavior of the *potential* is the same, save for sign, as we seek $\phi(x,0_+)$ and $\phi(x,0_-)$. Thus

$$\phi(x,0_+) = -\frac{\sigma_y}{2} \ ; \ \phi(x,0_-) = +\frac{\sigma_y}{2} \ ; \ |x| \le a \qquad (2.16)$$

$$\phi(x,0_+) - \phi(x,0_-) = -\sigma_y = \Delta\phi, \text{ the jump in } \phi. \tag{2.16a}$$

$$\text{For } |x| > a \quad \phi(x,0_\pm) \equiv 0 \tag{2.17}$$

As $\phi(x,y)$ is an odd function of y, the y-derivative is an even function of y and hence the vertical or transverse velocity is continuous through the dipole sheet. Thus

$$v(x,y) = \frac{\partial\phi}{\partial y} = -\frac{1}{2\pi}\int_{-a}^{a}\frac{[(x-x')^2 - y^2]}{[(x-x')^2 + y^2]^2}\sigma_y(x')dx'$$

and

$$v(x,0_\pm) = -\frac{1}{2\pi}\int_{-a}^{a}\frac{\sigma_y(x')}{(x-x')^2}dx' = \frac{1}{2\pi}\frac{\partial}{\partial x}\int_{-a}^{a}\frac{\sigma_y(x')}{(x-x')}dx' \tag{2.18}$$

where the last integral is to be taken as given by its Cauchy principal value. The axial component is

$$u(x,y) = \frac{y}{\pi}\int_{-a}^{a}\frac{(x-x')\,\sigma_y(x')}{[(x-x')^2 + y^2]^2}dx' \tag{2.19}$$

It is clear that $u(x,0_\pm) \equiv 0$ for $|x| > a$, i.e. outside of the sheet along $y = 0$ since the integrand is non-singular there. However for $|x| < a$ the integrand is seen to be highly singular along $y = 0$. To investigate the behavior as $y \to 0$ we may write (2.19) as

$$u = -\frac{1}{2\pi}\frac{\partial}{\partial x}\left\{y\int_{-a}^{a}\frac{\sigma_y(x')}{(x-x')^2 + y^2}dx'\right\} \tag{2.20}$$

and now the quantity in brackets again exhibits the same jump properties as the potential, yielding

$$u(x,0_+) = -\frac{1}{2}\frac{\partial\sigma_y(x)}{\partial x} \quad ; \quad u(x,0_-) = \frac{1}{2}\frac{\partial\sigma_y(x)}{\partial x} \tag{2.21}$$

Thus

- ***The tangential velocity of the vertical dipole distribution exhibits a jump across the sheet***

or

$$\frac{\partial\phi(x,0_+)}{\partial x} - \frac{\partial\phi(x,0_-)}{\partial x} = -\frac{\partial\sigma_y}{\partial x} \tag{2.22}$$

A most useful application of distributions of normal dipoles is to represent the pressure induced at any field point by an arbitrary pressure jump, Δp, along the distribution i.e.

$$p(x,y) = -\frac{y}{2\pi}\int_{-a}^{a} \frac{\Delta p(x')}{(x - x')^2 + y^2}\,dx' \tag{2.23}$$

Hence

$$p(x,0_+) = -\frac{\Delta p(x)}{2} \quad ; \quad p(x,0_-) = +\frac{\Delta p(x)}{2} \tag{2.24}$$

Axial or x-Directed Dipole Distributions

The potential is

$$\phi(x,y) = -\frac{1}{2\pi}\int_{-a}^{a} \sigma_x(x')\frac{\partial}{\partial x'}\ln\sqrt{(x - x')^2 + y^2}\,dx' \tag{2.25}$$

$$\phi(x,y) = \frac{1}{2\pi}\int_{-a}^{a} \frac{(x - x')\,\sigma_x(x')}{(x - x')^2 + y^2}\,dx' \tag{2.25a}$$

This is seen to be an even function of y which along $y = 0$ is

$$\phi(x,0) = \frac{1}{2\pi}\,⨍_{-a}^{a} \frac{\sigma_x(x')}{x - x'}\,dx' \tag{2.26}$$

which indicates that all the dipoles contribute to the potential at any x and for $|x| < a$, the integral has a Cauchy kernel.

To achieve the x-component of induced velocity we foresee that differentiating (2.25a) with respect to x will yield a highly singular kernel for $|x| < a$ and $y = 0$ so we may go back to (2.25) which can be written as a derivative with respect to x by simply changing the sign. Then

$$u(x,y) = \frac{1}{2\pi}\frac{\partial^2}{\partial x^2}\int_{-a}^{a} \sigma_x(x')\ln\sqrt{(x - x')^2 + y^2}\,dx'$$

$$= \frac{1}{2\pi}\frac{\partial}{\partial x}\int_{-a}^{a} \frac{(x - x')\,\sigma_x(x')}{(x - x')^2 + y^2}\,dx' \tag{2.27}$$

$$\text{Then } u(x,0_\pm) = -\frac{1}{2\pi}\int_{-a}^{a}\left\{\frac{\partial}{\partial x'}\frac{1}{(x-x')}\right\}\sigma_x(x')dx' \;;\; \frac{\partial}{\partial x} = -\frac{\partial}{\partial x'} \tag{2.28}$$

This may be integrated by parts to yield

$$-\frac{1}{2\pi}\left\{\frac{\sigma_x(x')}{x-x'}\bigg|_{-a}^{a} - \int_{-a}^{a}\frac{1}{x-x'}\frac{d\sigma_x(x')}{dx'}dx'\right\}$$

If, as is often the case, $\sigma_x(\pm a) = 0$ then the integrated terms vanish and

$$u(x,0_\pm) = \frac{1}{2\pi}\int_{-a}^{a}\frac{\sigma_x'(x')}{x-x'}dx' \tag{2.29}$$

with $\sigma_x' = d\sigma_x/dx'$. Thus all the dipoles contribute.

The vertical or y-component is seen also to yield a highly singular kernel along $y = 0$ for $|x| < a$ by differentiation of (2.25a) with respect to y. We may do the following:

$$\frac{\partial\phi}{\partial y} = v(x,y) = \frac{1}{2\pi}\frac{\partial^2}{\partial y\partial x}\int_{-a}^{a}\sigma_x \ln\sqrt{(x-x')^2+y^2}\,dx' \tag{2.30}$$

and exchanging the orders of differentiation

$$v(x,y) = \frac{1}{2\pi}\frac{\partial}{\partial x}\left\{y\int_{-a}^{a}\frac{\sigma_x dx'}{(x-x')^2+y^2}\right\} \tag{2.31}$$

and as $y \to 0$ the limit of the bracketed term which is now familiar yields, (cf. (2.20))

$$v(x,0_\pm) = \pm\frac{1}{2}\frac{\partial\sigma_x(x)}{\partial x} \;;\; |x| < a \tag{2.32}$$

Thus on the distribution v is discontinuous across the sheet of axial dipoles and is given by the local x-derivative of the dipole density as indicated by (2.32). Thus if $v(x,0_+)$ is specified (as it is when the offsets of a given symmetrical section are given which will be shown later) then σ_x can be found by integration.

Vortex Distributions

We saw earlier that the potential of an infinitely long line vortex (the vorticity vector being perpendicular to the x,y-plane) is (when located at the origin) given by

$$\phi = \frac{\Gamma}{2\pi} \tan^{-1} \frac{y}{x}$$

A distribution along the x-axis is secured by replacing x by x − x' and Γ by $\gamma(x')dx'$. Thus

$$\phi(x,y) = \frac{1}{2\pi} \int_{-a}^{a} \gamma(x') \tan^{-1} \frac{y}{x - x'} dx' \tag{2.33}$$

Then

$$u(x,y) = -\frac{y}{2\pi} \int_{-a}^{a} \frac{\gamma(x')}{(x - x')^2 + y^2} dx' \tag{2.34}$$

and

$$v(x,y) = \frac{1}{2\pi} \int_{-a}^{a} \frac{(x - x')\,\gamma(x')}{(x - x')^2 + y^2} dx' \tag{2.35}$$

These components are seen to be directly comparable to the y- and x-components of source distributions (see Equations (2.5) and (2.3) except for the sign of (2.34)). The reason is that they arise from conjugate functions. Thus we can immediately infer their behavior on the vortex sheet. Thus

$$u(x,0_{\pm}) = \begin{cases} \mp \dfrac{\gamma(x)}{2} & |x| < a \qquad (2.36) \\ 0 & |x| > a \qquad (2.37) \end{cases}$$

$$v(x,0) = \frac{1}{2\pi} \int_{-a}^{a} \frac{\gamma(x')dx'}{x - x'} \quad \text{all } x \tag{2.38}$$

Thus u depends only on the local value of γ whereas v depends on the entire distribution. For $|x| < a$ the integral is a Cauchy principal value.

NON-PLANAR AND PLANAR DISTRIBUTIONS IN THREE DIMENSIONS

Source Distributions

Consider a curved surface over which sources are distributed having a strength density m cubic units per second per square unit of area. Thus m has the dimensions of velocity (linear units/sec.).

The potential of such a distribution is

$$\phi = -\frac{1}{4\pi}\int_S \frac{m(x',y',z'(x',y'))}{\sqrt{(x-x')^2 + (y-y')^2 + (z-z')^2}}\, dS' \tag{2.39}$$

The primed coordinates are not independent being related through the equation of the surface S which may be written

$$Z(x',y',z') = z' - \zeta(x',y') = 0 \tag{2.40}$$

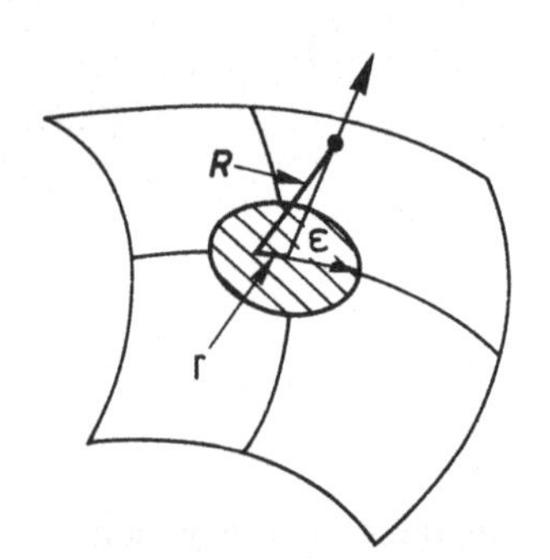

Figure 2.4 Integration over small circular region Σ.

We may seek the velocity induced normal to the surface at any point on the surface by the concerted action of all the sources. To accomplish this we break up the integration into two parts, one over a small circular region, Σ, indicated by a cylinder of radius ε with axis along the local normal. In that region as $\varepsilon \to 0$, m is uniform and is given by the value at the point $n = r = 0$, cf. Figure 2.4.

Then

$$\frac{\partial\phi}{\partial n} = -\frac{1}{4\pi}\int_{S_0} m\frac{\partial}{\partial n}\frac{1}{R}\, dS' - \frac{1}{4\pi}\int_{\Sigma} m\frac{\partial}{\partial n}\frac{1}{R}\, dS' \tag{2.41}$$

where S_0 is the remainder of the surface excluding the region Σ. Concentrating on the last integral we may write dS' in polar coordinates and take R as the distance from a point on the local normal to any point at radius r in the circle of radius ε on the surface.

Then

$$I = -\frac{1}{4\pi}\int_{\Sigma} m\frac{\partial}{\partial n}\frac{1}{R}\, dS' \Rightarrow -\frac{1}{4\pi}\int_0^{2\pi} d\theta\int_0^{\varepsilon} m\frac{\partial}{\partial n}\frac{1}{\sqrt{n^2 + r^2}}\, r\,dr$$

$$= \frac{2\pi}{4\pi}\int_0^{\varepsilon} \frac{nr}{[n^2 + r^2]^{3/2}}\, m\,dr = -\frac{m(x,y)}{2}\left[\frac{n}{\sqrt{n^2 + r^2}}\right]_{r=0}^{\varepsilon}$$

m passes through the integral because it is assumed to be regular and as $\varepsilon \to 0$ becomes the value at the point. Inserting the limits we get

$$I = -\frac{m(x,y)}{2}\left[\frac{n}{\sqrt{n^2+\varepsilon^2}} - \frac{n}{|n|}\right] \quad ; \quad |n| = \sqrt{n^2} \tag{2.42}$$

Now holding ε fixed we descend to the surface making $n \to 0_+$

$$\therefore \quad I = \frac{m}{2} \tag{2.42a}$$

and as $n \to 0_-$, $n/|n| \to -1$ yielding

$$I = -\frac{m}{2} \tag{2.42b}$$

Thus

$$v_{n\pm} = \pm\frac{m}{2} - \frac{1}{4\pi}\int_S m\,\frac{\partial}{\partial n}\frac{1}{R}\,dS' \;;\ \text{on the surface} \tag{2.43}$$

Now

$$\frac{\partial}{\partial n}\frac{1}{R} = -\frac{\cos(\mathbf{n},\boldsymbol{R})}{R^2} \tag{2.44}$$

where the angle $(\mathbf{n},\boldsymbol{R})$ is that between the normal $\mathbf{n}$ and the vector $\boldsymbol{R}$ drawn from the foot of the normal to any variable point on the surface. When the surface is *planar* the angle $(\mathbf{n},\boldsymbol{R}) \equiv \pi/2$ and the integral term vanishes. However regardless of the curvature of the surface we have that

$$\left[\frac{\partial\phi}{\partial n}\right]_+ - \left[\frac{\partial\phi}{\partial n}\right]_- = m \quad \text{the source density} \tag{2.45}$$

so when the jump in the normal velocity is specified then the source density is known.

Thus for the distribution of sources

- ***The potential itself is continuous through the surface.***
- ***The tangential velocity is also continuous.***
- ***The normal velocity is discontinuous.***

The presence of the integral term in (2.43) can be understood physically because on the curved surface the sources *elsewhere* can contribute to the normal velocity at any point in addition to the contribution of the local element as depicted in Figure 2.5.

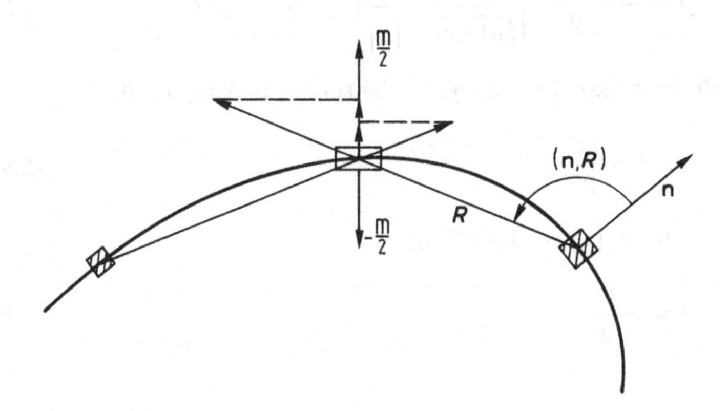

Figure 2.5 Normal velocity induced by source element.

Distributions of Normal Dipoles

We have seen that dipoles can be "constructed" by differentiating the source potential with respect to the dummy variables and in the direction desired. Hence for normal dipoles on a curved surface we have

$$\phi(x,y,z) = -\frac{1}{4\pi}\int_S \sigma_n \frac{\partial}{\partial n'}\left[\frac{1}{R}\right] dS' \qquad (2.46)$$

Here σ_n has the dimension length²/second.

But since field point coordinates (x,y,z) and dummy point coordinates (x',y',z') always enter as differences within R we can write

$$\frac{\partial}{\partial n'} = -\frac{\partial}{\partial n}$$

and then the potential of the surface distribution of normal dipoles has the same structure as that of the *normal velocity* induced by a surface distribution of *sources* (except for sign)

$$\therefore \quad \phi = \frac{1}{4\pi}\int_S \sigma_n \frac{\partial}{\partial n}\left[\frac{1}{R}\right] dS' \qquad (2.47)$$

We may then write that

$$\phi_{\pm} = \mp\frac{\sigma_n}{2} + \frac{1}{4\pi}\int_S \sigma_n \frac{\partial}{\partial n}\frac{1}{R}\, dS' \qquad (2.48)$$

or

$$\phi_{\pm} = \mp\frac{\sigma_n}{2} - \frac{1}{4\pi}\int_S \sigma_n \frac{\partial}{\partial n'}\frac{1}{R}\,dS' \tag{2.49}$$

Again if the surface is planar the integral term vanishes.
Thus for the distribution of ***normal dipoles***

- ***The normal velocity is continuous through the surface.***
- ***The tangential velocity is discontinuous through the surface.***

Planar Distributions of Vorticity

By the theorem of Stokes, p. 23, the potential of a vortex array is that due to a distribution of dipoles whose axes are normal to any surface bounded by the array. Consider the "horseshoe" vortex filament as shown in Figure 2.6. The closing leg is at x = − ∞ and is generally not shown – hence the name horseshoe.

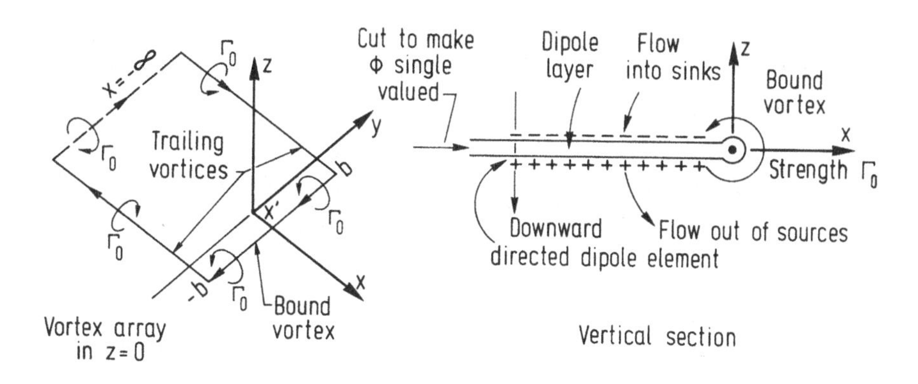

Figure 2.6 Horseshoe vortex.

The velocity potential of this array is ($\kappa = \Gamma_0$ in (1.90))

$$\phi(x,y,z) = -\frac{1}{4\pi}\left[-\frac{\partial}{\partial z'}\right]\int_{-\infty}^{x'} dx''\int_{-b}^{b}\frac{\Gamma_0}{\sqrt{(x-x'')^2 + (y-y')^2 + (z-z')^2}}\bigg|_{z'=0} dy' \tag{2.50}$$

where $-\partial/\partial z'$ is due to the direction cosine, $n_x = 0$, $n_y = 0$, $n_z = -1$, as the angle between the positive z-axis and the dipole axis is π. This gives the flow pattern consistent with the direction of the vorticity vector shown in the figure.

To achieve a spanwise variation of vorticity we may pile on an ensemble of horseshoe vortices each with a different circulation and thus Γ_0 is replaced by $\Gamma(y')$ in the limit of a continuous distribution, cf. Figure 2.7.

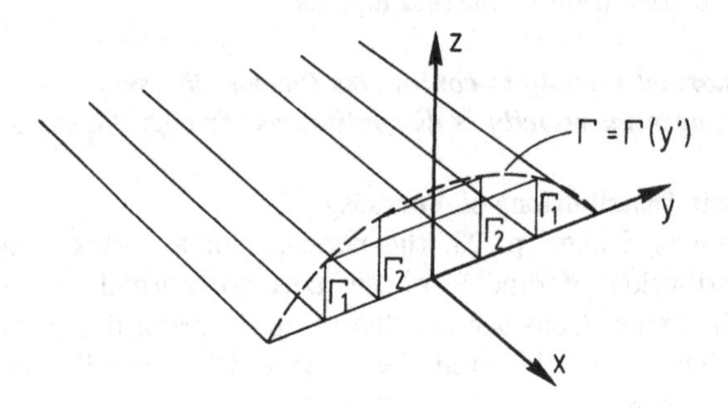

Figure 2.7 Spanwise variation of vorticity.

As the vorticity along each trailing filament is independent of x'' we can integrate (2.50) over x'' after differentiation with respect to z' and replacing Γ_0 by $\Gamma(y')$ to achieve

$$\phi(x,y,z) = -\frac{z}{4\pi}\int_{-b}^{b}\frac{\Gamma(y')}{(y-y')^2+z^2}\left[\frac{x-x''}{\sqrt{(x-x'')^2+(y-y')^2+z^2}}\right]_{x''=-\infty}^{x'} dy' \tag{2.51}$$

$$= -\frac{z}{4\pi}\int_{-b}^{b}\frac{\Gamma(y')}{(y-y')^2+z^2}\left[\frac{x-x'}{\sqrt{(x-x')^2+(y-y')^2+z^2}}-1\right] dy' \tag{2.52}$$

We now can observe certain properties

i. Far upstream, i.e., $x \to +\infty$ we see that the first term in the bracket passes to unity and

$$\lim_{x\to+\infty}\phi(x,y,z) = 0 \tag{2.53}$$

ii. Far downstream, $x \to -\infty$, the same term passes to minus unity. Hence

$$\lim_{x\to-\infty}\phi(x,y,z) = +\frac{z}{2\pi}\int_{-b}^{b}\frac{\Gamma(y')}{(y-y')^2+z^2}\,dy' \tag{2.54}$$

This is seen to be a distribution of vertical dipoles in $z = 0$ in the two dimensions y and z. To show this as the potential of a vorticity distribution we integrate by parts. Let

$$u = \Gamma(y') \quad ; \quad du = \Gamma'(y')dy'$$

$$dv = \frac{dy'}{(y - y')^2 + z^2} \quad ; \quad v = \frac{1}{z}\tan^{-1}\frac{z}{y - y'}$$

Then

$$\lim_{x \to -\infty} \phi(x,y,z) = \frac{z}{2\pi}\left\{\frac{\Gamma(y')}{z}\tan^{-1}\frac{z}{y-y'}\Bigg|_{-b}^{b} - \frac{1}{z}\int_{-b}^{b}\Gamma'(y')\tan^{-1}\left[\frac{z}{y-y'}\right]dy'\right\}$$

Physically $\Gamma(\pm b) = 0$ so the integrated terms vanish and

$$\phi(-\infty,y,z) = -\frac{1}{2\pi}\int_{-b}^{b}\Gamma'(y')\tan^{-1}\left[\frac{z}{y - y'}\right]dy' \tag{2.55}$$

which is a y'-distribution of vorticity $\gamma(y') = \Gamma'(y')$ in $z' = 0$ between $-b$ and b, cf. p. 10. Thus the vorticity in the wake of a non-uniformly loaded bound vortex is given by the gradient of the circulation, i.e. $d\Gamma/dy'$.

A Surface of Bound Vorticity

To generalize the foregoing to produce a planar surface over which vorticity is bound we integrate over the location of the bound vortex x' and replace $\Gamma(y')dy'$ by $\gamma(x',y')dx'dy'$. Thus generalizing (2.52) for a rectangular planform $-a < x < a$, $-b < y < b$

$$\phi = -\frac{z}{4\pi}\int_{-b}^{b}\frac{dy'}{(y-y')^2 + z^2}\int_{-a}^{a}\gamma(x',y')\left[\frac{x - x'}{\sqrt{(x-x')^2 + (y-y')^2 + z^2}} - 1\right]dx' \tag{2.56}$$

To show the regions where ϕ jumps through $z = 0$ we can employ the same substitution as in two dimensions (p. 28). Here let $y - y' = z\tan\beta$. Then we see, as before, that the y'-integral is transformed and that in the radical $(y - y')^2 + z^2 = z^2\sec^2\beta$. Passing to the limit the y'-integral yields $\mp\pi$ for $-b < y < b$, otherwise zero, as $z \to 0_{\pm}$ and

$$\lim_{z \to 0_{\pm}} \phi = \mp\frac{1}{4}\int_{-a}^{a}\gamma(x',y)\left[\frac{x - x'}{|x - x'|} - 1\right]dx' \tag{2.57}$$

Thus, if $x > a$ then $x - x' > 0$ and $(x-x')/(|x-x'|) = 1$.

Hence

$$\lim_{\substack{z \to 0_\pm \\ x > a}} \phi = 0 \qquad (2.58)$$

For $-a < x < a$, break up the integral, i.e., $\int_{-a}^{a} = \int_{-a}^{x}(\ ) + \int_{x}^{a}(\ )$. In the first of these $(x-x')/(|x-x'|) = 1$ as $x' < x$ and it does not contribute. In the second $(x-x')/(|x-x'|) = -1$ so we have

$$\lim_{\substack{z \to 0_\pm \\ -a< x <a}} \phi = \pm\frac{1}{2}\int_{x}^{a} \gamma(x',y)dx' \qquad (2.59)$$

Finally for $-\infty < x < -a$, $(x-x')/(|x-x'|) - 1 \equiv -2$ and we have for all such x

$$\lim_{\substack{z \to 0_\pm \\ -\infty< x <a}} \phi = \pm\frac{1}{2}\int_{-a}^{a} \gamma(x',y)dx' \qquad (2.60)$$

From the foregoing we can deduce that the x-component $u(x,y,0_\pm)$ has no jump forward of $x = a$ and that for $-a < x <a$

$$u_\pm = \pm\frac{1}{2}\frac{\partial}{\partial x}\int_{x}^{a} \gamma(x',y)\ dx' = \mp\frac{\gamma(x,y)}{2} \quad ; \quad |x| < a \qquad (2.61)$$

Then for $-\infty < x < -a$

$$u_\pm = \frac{\partial}{\partial x}\phi_\pm = \pm\frac{1}{2}\frac{\partial}{\partial x}\int_{-a}^{a} \gamma(x',y)dx' \equiv 0 \quad |x| > a \qquad (2.62)$$

Thus

- ***u(x,y,z) only has a jump in way of the bound vorticity.***

Similarly, we can exhibit the jump in the transverse velocity v. Thus

$$v_\pm(y) = \frac{\partial}{\partial y}\phi(x,y,0_\pm) = \begin{cases} 0 & a < x < \infty \quad (2.63) \\ \pm\dfrac{1}{2}\displaystyle\int_{x}^{a}\frac{\partial}{\partial y}\gamma(x',y)dx' & -a < x < a \quad (2.64) \end{cases}$$

$$v_{\pm}(y) = \pm\frac{1}{2}\int_{-a}^{a} \frac{\partial}{\partial y}\,\gamma(x',y)\,dx' \qquad -\infty < x < -a \tag{2.65}$$

Thus the transverse velocity exhibits a full jump abaft $x = -a$. As $\gamma(x',y)$ is a decreasing function of y from $y = 0$ to $y = b$ then $\partial\gamma/\partial y < 0$ and we see that *the transverse flow is inward on the upper and outward on the lower side* as indicated in Figure 2.8.

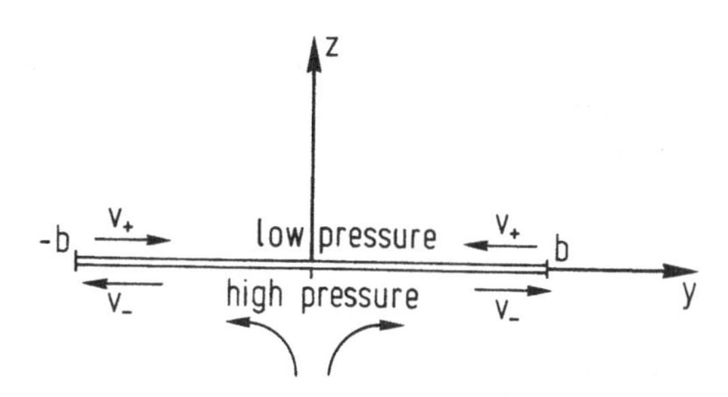

Figure 2.8 Transverse velocities above and below vortex distribution.

3 Kinematic Boundary Conditions

The kinematic condition on a non-porous moving boundary is that the fluid particles move with both the normal and the tangential motion of the boundary. In a non-viscous fluid the particles slip tangentially and acquire the normal velocity at each point of a moving, non-porous boundary, otherwise there would be a flow of fluid across the boundary. This can be stated as a mathematical condition on any surface described by the equation F(x,y,z,t) = 0 by

$$\frac{D}{Dt} F(x,y,z,t) = 0 \tag{3.1}$$

or

$$\frac{\partial F}{\partial t} + \frac{\partial x}{\partial t}\frac{\partial F}{\partial x} + \frac{\partial y}{\partial t}\frac{\partial F}{\partial y} + \frac{\partial z}{\partial t}\frac{\partial F}{\partial z} = 0 \tag{3.2}$$

As the instantaneous coordinates of any fluid particle are (x(t), y(t), z(t)) then $u = \partial x/\partial t$; $v = \partial y/\partial t$; $w = \partial z/\partial t$ and hence the condition (3.2) may be expressed by

$$\frac{\partial F}{\partial t} + u\frac{\partial F}{\partial x} + v\frac{\partial F}{\partial y} + w\frac{\partial F}{\partial z} = 0 \tag{3.3}$$

If we now divide through by

$$N = \sqrt{\left[\frac{\partial F}{\partial x}\right]^2 + \left[\frac{\partial F}{\partial y}\right]^2 + \left[\frac{\partial F}{\partial z}\right]^2} \tag{3.4}$$

we have

$$\frac{1}{N}\frac{\partial F}{\partial t} + u\,\frac{1}{N}\frac{\partial F}{\partial x} + v\,\frac{1}{N}\frac{\partial F}{\partial y} + w\,\frac{1}{N}\frac{\partial F}{\partial z} = 0 \tag{3.5}$$

and recalling from the differential geometry of surfaces that the direction cosines of the normal to any surface at any point (x,y,z) are given by

$$(n_x,n_y,n_z) = \left[\frac{1}{N}\frac{\partial F}{\partial x}, \frac{1}{N}\frac{\partial F}{\partial y}, \frac{1}{N}\frac{\partial F}{\partial z}\right] \tag{3.6}$$

we see that (3.5) takes the form

$$\frac{1}{N}\frac{\partial F}{\partial t} + n_x u + n_y v + n_z w = 0 \tag{3.7}$$

or in vector notation

$$\frac{1}{N}\frac{\partial F}{\partial t} + \mathbf{q}\cdot\mathbf{n} = 0 \tag{3.8}$$

where

$$\mathbf{q} = (u,v,w) \quad ; \quad \mathbf{n} = (n_x,n_y,n_z)$$

When the surface does not change shape nor position with time, $\partial F/\partial t \equiv 0$ and we have the familiar condition

$$\mathbf{q}\cdot\mathbf{n} = 0 \text{ on S, for } \partial F/\partial t = 0 \tag{3.9}$$

To fix our ideas and to interpret the boundary condition in physical terms, let us consider a body restrained from translation in a uniform current $-U$ and having otherwise imposed arbitrary motion which varies with time. If the disturbance velocity due to the body and its motions is (u_b, v_b, w_b) then substituting for u, v and w in (3.3)

$$u = u_b - U, \quad v = v_b \quad \text{and} \quad w = w_b \tag{3.10}$$

$$\frac{\partial F}{\partial t} + (u_b - U)\frac{\partial F}{\partial x} + v_b\frac{\partial F}{\partial y} + w_b\frac{\partial F}{\partial z} = 0 \tag{3.11}$$

and expressing the vertical coordinate to the surface as $z = f(x,y,t)$ we see F has the form

$$F(x,y,z,t) = z - f(x,y,t) = 0 \tag{3.12}$$

We can evaluate the partial derivatives of F as

$$\frac{\partial F}{\partial t} = -\frac{\partial f}{\partial t}; \quad \frac{\partial F}{\partial x} = -\frac{\partial f}{\partial x}; \quad \frac{\partial F}{\partial y} = -\frac{\partial f}{\partial y}; \quad \frac{\partial F}{\partial z} = 1 \tag{3.13}$$

then (3.11) becomes

$$-\frac{\partial f}{\partial t} - (u_b - U)\frac{\partial f}{\partial x} - v_b\frac{\partial f}{\partial y} + w_b = 0 \tag{3.14}$$

We see that this condition is linear in u_b, v_b and w_b. Solving for the requirement on the vertical component w_b

$$w_b = \frac{\partial f}{\partial t} + (u_b - U)\frac{\partial f}{\partial x} + v_b\frac{\partial f}{\partial y} \tag{3.15}$$

Thus the vertical velocity is made up of the vertical rate of change of the surface f(x,y,t) with time plus the convective terms coupling with the slope of the surface in the x- and y-directions.

Figure 3.1 indicates the physical manifestation of the terms on the right side of (3.15).

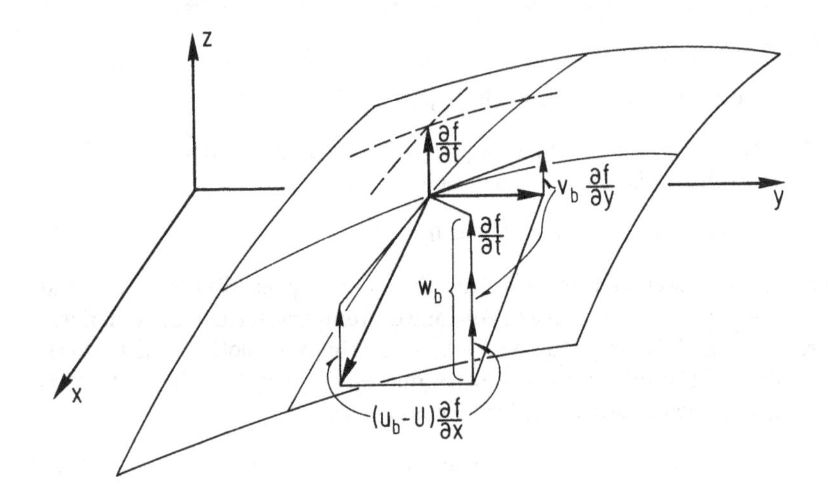

Figure 3.1 Components of vertical velocity.

In our applications the transverse velocity component v_b is always small compared to U and the slope $\partial f/\partial y$ is small so that the product $v_b \partial f/\partial y$ is of second order and can be neglected in a first order theory.

As we have seen in the linearized form of the pressure equation (1.24), u_b is taken to be small with respect to U although this is not uniformly true in the leading and trailing edge regions and as a result this approximation will lead to unrealistic results. Proceeding as is usual in first order theories the term $u_b \partial f/\partial x$ is also neglected so the approximate condition on w_b is

$$w_b = \frac{\partial f}{\partial t} - U\frac{\partial f}{\partial x} \tag{3.16}$$

Thus for irrotational flow, this boundary condition on the disturbance potential is

$$\frac{\partial \phi}{\partial z} = \frac{\partial f}{\partial t} - U\frac{\partial f}{\partial x} \quad \text{on} \quad z = f(x,y,t). \tag{3.17}$$

In a completely fixed surface $\partial f/\partial t = 0$.

In two dimensions where the surface may be described by $y = \eta(x,t)$, then the condition (with the analogous restrictions) is

$$\frac{\partial \phi}{\partial y} = \left[\frac{\partial}{\partial t} - U\frac{\partial}{\partial x}\right]\eta \quad ; \quad y = \eta(x,t) \tag{3.18}$$

We shall see later that the requirement that the derivative of the potential function be evaluated on the surface is replaced when the surface is everywhere close to the axis of the section.

If a body translates in a quiescent fluid with components (U,V,W) then the kinematic condition on the body boundary is

$$\frac{\partial \phi}{\partial n} = n_x U + n_y V + n_z W = \mathbf{n}\cdot\mathbf{Q} \tag{3.19}$$

There are other boundary conditions involving pressure referred to as dynamic conditions such as occur on the boundaries of cavities and the water surface where the pressure is spatially constant. These will be dealt with as they arise.

4 Steady Flows About Thin, Symmetrical Sections in Two Dimensions

To understand the application of singularities to the generation of flows about sections of interest consider at first, in a heuristic fashion, the flow generated by placing a source in a uniform flow.

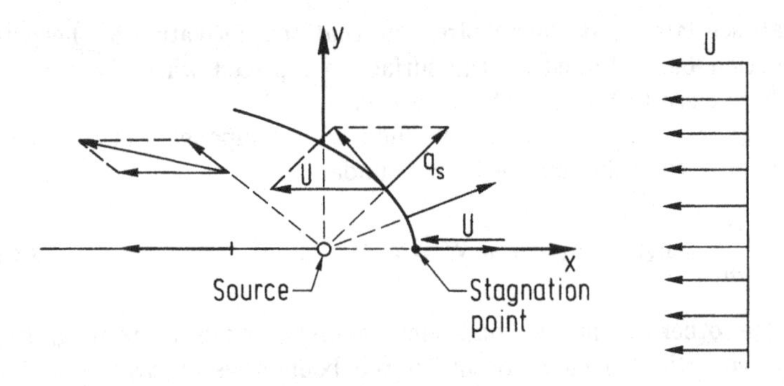

Figure 4.1 Source in uniform flow.

At some distance, x_S, along $y = 0$ the x-flow component from the source flow will just balance the free stream velocity $-U$ yielding a stationary point or stagnation point. The particles will follow the direction of the resultant velocity at all points producing a streamline pattern. There will be one dividing streamline beginning at the stagnation point which will, above and below the axis of symmetry, separate the source flow from the outer flow.

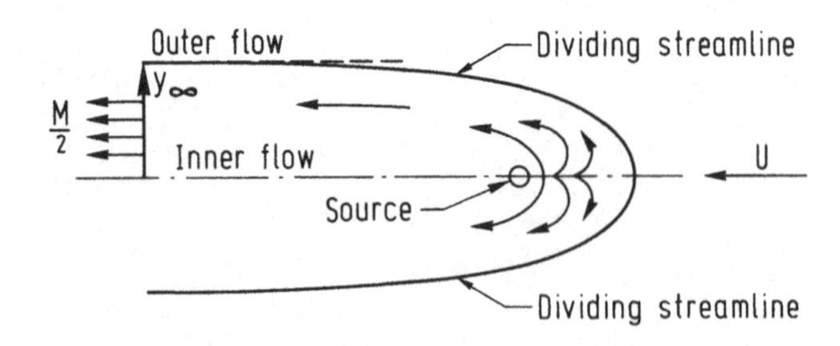

Figure 4.2 Inner and outer flow separated by dividing streamline.

As our interest is in the velocity and pressure along the surface of sections we can see, heuristically at least, that such singularities will lie within the form and hence all components will always be bounded in the region of interest unless we allow the singularities to extend to the leading edge as we shall see later. To calculate this dividing stream surface it is much easier to employ a function called the stream function ψ which is orthogonal to the velocity potential function.

Streamlines produced by a source alone are simply radial lines.

Figure 4.3 Streamlines produced by a source (left) and by parallel flow (right).

The quantity of flow per unit time between the x-axis and any radial streamline at angle θ is, for the source

$$\psi_s = \frac{M\theta}{2\pi} = \frac{M}{2\pi}\tan^{-1}\frac{y}{x} \tag{4.1}$$

where the arctangent is defined as lying between 0 and π in $y > 0$ and 0 and $-\pi$ for $y < 0$.

$$u_r = \frac{1}{r}\frac{\partial}{\partial\theta}\psi_s = \frac{M}{2\pi r} \tag{4.2}$$

and the tangential velocity is

$$u_t = \frac{\partial}{\partial r}\psi_s = 0 \qquad (\psi_s \text{ is independent of r}) \tag{4.3}$$

For the free stream the quantity of flow per unit time between the x-axis and any streamline y is

$$\psi_{fs} = -Uy = -Ur\sin\theta \tag{4.4}$$

Thus the combined stream function

$$\psi = \psi_s + \psi_{fs} = 0 \text{ on the dividing streamline.} \tag{4.5}$$

Hence

$$\frac{M}{2\pi}\tan^{-1}\frac{y}{x} - Uy = 0 \tag{4.6}$$

Unfortunately (4.6) is a transcendental equation which can only be solved by numerical methods for selected values of M/U. We shall be content with finding three points on this streamline which will allow us to sketch it. To calculate the location of the leading edge we proceed as follows:

The x-component is $\partial\psi/\partial y$ and the y-component, $-\partial\psi/\partial x$. To calculate the stagnation point,

$$\frac{\partial\psi}{\partial y} = u = \frac{M}{2\pi}\frac{x}{(x^2 + y^2)} - U = 0$$

From symmetry of the flow about y = 0, this point lies on y = 0. Then

$$\therefore\ x = x_s = \frac{M}{2\pi U} \tag{4.7}$$

A second point is deduced as follows:

The offset abreast of the source is got where for x = 0, $\tan^{-1} y/x = \pi/2$

$$\therefore\ y_0 = \frac{M}{4U} = \frac{\pi}{2}x_s \quad \text{or} \quad \frac{y_0}{x_s} = \frac{\pi}{2} \tag{4.8}$$

A third "point" is established by the asymptotic half breadth far downstream for which

$$\tan^{-1}\frac{y}{x} \longrightarrow \pi \quad \text{or} \quad y_\infty = \frac{M}{2U} = \pi x_s \quad \text{or} \quad \frac{y_\infty}{x_s} = \pi \tag{4.9}$$

Then the variation of the half breadths can be sketched as in Figure 4.4.

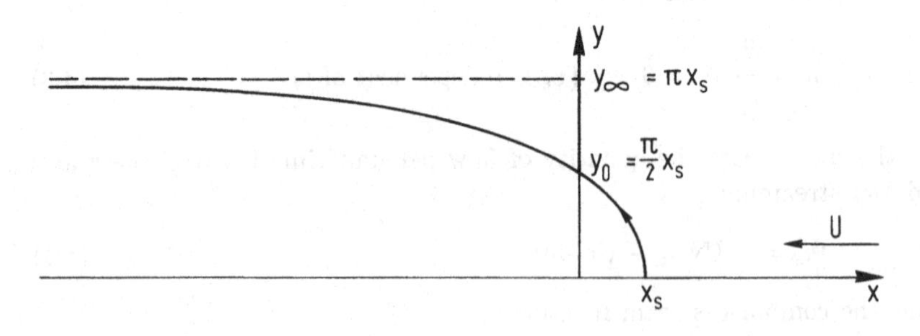

Figure 4.4 Half—breadth of Rankine half—body (lower half is the mirror image in y = 0).

A closed Rankine oval is generated by the combination of a source and an equal sink in a uniform stream cf. Figure 4.5.

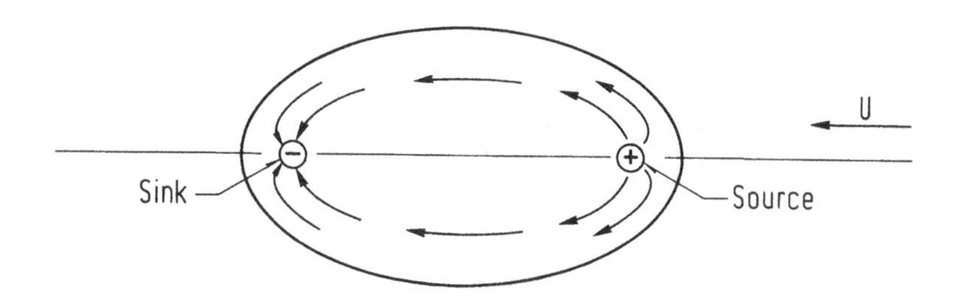

Figure 4.5 Rankine oval.

The important point is that the flows from the source and sink in the uniform stream generate an envelope which separates the source and sink from the region of interest, namely the envelope and the exterior.

It is necessary to point out that generation of forms by internal singularities does not guarantee that any specific arbitrarily selected shape can be so generated. This is guaranteed by distributing singularities on the given contour but at the expense of considerable computational labor.

Let us suppose that a thin symmetrical profile is given and we wish to calculate the flow about it in a uniform stream parallel to the long axis. We assume that the streamline form can be generated by a source-sink distribution along the axis of symmetry. Thus if we take

$$\phi = \frac{1}{2\pi}\int_{-a}^{a} m(x') \ln \sqrt{(x - x')^2 + y^2}\, dx' \tag{4.10}$$

Then the boundary condition on the given shape

$$y = \pm\, \tau(x) \tag{4.11}$$

on the upper side (as we have seen earlier, Equation (3.18)), requires that

$$\frac{\partial\phi}{\partial y} = \left[\frac{\partial\phi}{\partial x} - U\right]\frac{\partial\tau}{\partial x} \tag{4.12}$$

As the section is thin $\partial\phi/\partial x$ is much less than U except in the vicinity of the ends where $\partial\phi/\partial x$ must approach U. However if we keep $\partial\phi/\partial x$ we shall obtain an integral equation for the unknown source strength m(x'). To avoid this complication we drop $\partial\phi/\partial x$ and keep our eyes open for trouble!

Using (4.10), (4.12) becomes

$$\frac{y}{2\pi}\int_{-a}^{a}\frac{m(x')}{(x-x')^2+y^2}\,dx'\Bigg|_{y=\tau(x)} = -U\tau' \qquad |x| < a \tag{4.13}$$

where $\tau' = \partial\tau/\partial x$.

We now argue that if the section is thin the lateral velocity induced by the sources on $y = \tau(x)$ is very nearly that given on $y = 0$. Hence we evaluate the left side of (4.13) on $y = 0$ and as we have seen in Equations (2.5) and (2.7) we achieve

$$\frac{m(x)}{2} = -U\tau'(x) \tag{4.14}$$

Using this in (4.10) yields the approximate potential for an arbitrary, thin, symmetrical section to be

$$\phi = -\frac{U}{\pi}\int_{-a}^{a}\tau'(x')\ln\sqrt{(x-x')^2+y^2}\,dx' \tag{4.15}$$

It is important to note that the variable in τ and the derivative are changed to the dummy variable when inserted to (4.10).

To calculate the linearized pressure distribution recall from Equation (1.24) that

$$\frac{p-p_\infty}{\rho} = U\frac{\partial\phi}{\partial x} \tag{4.16}$$

Hence

$$\frac{p-p_\infty}{\rho} = -\frac{U^2}{\pi}\int_{-a}^{a}\frac{(x-x')\,\tau'(x')}{(x-x')^2+y^2}\,dx' \tag{4.17}$$

for all (x,y). For points on the boundary, $y = \tau(x)$, and here again we seek the simplification of evaluating on the source sheet by taking $y = 0$.

Now at last the reader may see why the values upon the singularity distributions were sought in the foregoing!

Thus the change in pressure from the ambient to the local value on the section is approximately

$$C_p(x) = \frac{p-p_\infty}{\frac{1}{2}\rho U^2} = -\frac{2}{\pi}\int_{-a}^{a}\frac{\tau'(x')}{x-x'}\,dx' \tag{4.18}$$

where the integral is now a Cauchy principal value, cf. Mathematical Compendium, Section 4, p. 508–509.

Let us now investigate the results for some simple forms for which the exact theoretical solutions are known, namely the ogival and elliptical sections.

THE OGIVAL SECTION

A symmetrical ogival section (or symmetrical lens) is composed of two circular arcs as depicted in Figure 4.6. The exact equation for the circular arc in terms of the coordinates x and y is cumbersome. For these sections the ogive can be very closely approximated by a parabola whose equation is

$$y = \tau(x) = \tau_0\left[1 - \left[\frac{x}{a}\right]^2\right] \quad ; \quad \tau' = -\frac{2x\tau_0}{a^2} \tag{4.19}$$

where τ_0 is the maximum semi-thickness and a is the halfchord.

The Approximate Pressure Distribution

To obtain the approximate pressure distribution we insert (4.19) into (4.18) to obtain

$$C_P(x) = \frac{4}{\pi}\frac{\tau_0}{a^2}\int_{-a}^{a}\frac{x'}{x - x'}\,dx'$$

To simplify, let $\check{t} = \tau_0/a$, the thickness ratio and replace x' by $a\check{x}'$ and x by $a\check{x}$ where $\check{x}'$ and $\check{x}$ are fractions of the semi-chord a. Thus we have

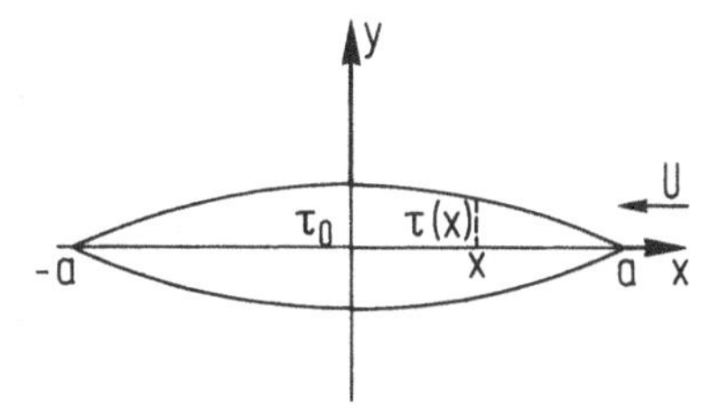

Figure 4.6 Ogival section.

$$C_P(\check{x}) = \frac{4}{\pi}\check{t}\int_{-1}^{1}\frac{\check{x}'}{\check{x} - \check{x}'}\,d\check{x}' \tag{4.20}$$

where we keep in mind that all linear dimensions are fractions of the semichord. This integral can be evaluated from (M8.2) of the Mathematical Compendium, Section 8, Table of Airfoil Integrals. Thus

$$C_P(\check{x}) = \frac{4}{\pi}\check{t}\left[\check{x}\ln\frac{1 + \check{x}}{1 - \check{x}} - 2\right] \tag{4.21}$$

We observe immediately that this function becomes infinite at ± 1, the leading and trailing edges. Ignoring this total lack of realism for the present we evaluate at the midpoint, $\tilde{x} = 0$, where the pressure coefficient should be a minimum and find that

$$C_{p_{min}}(0) = -\frac{8}{\pi}\frac{\tau_0}{a} \tag{4.22}$$

It turns out that in spite of our several assumptions, (namely, $\partial\phi/\partial x << U$ everywhere in the boundary condition and in Bernoulli's equation and that evaluations be made on $y = 0$ rather than on $y = \tau$ and finally that the circular arc can be replaced by a parabola), this result is very close to that from exact theory. We shall also see that (4.21) closely fits the entire exact distribution except at $x = \pm 1$ where it is unbounded. We may now turn to the exact solution for comparison.

The Exact Solution for the Ogival Section

As this is not a treatise on theoretical hydrodynamics (although up to now it may seem so!) we shall not derive the precise inviscid flow about this section but lift the results from Milne-Thomson (1955). There, on p. 170, the square of the velocity along a symmetrical ogive in a uniform stream U is given, which we can cast into the exact equation for C_p, viz.,

$$C_p = 1 - \frac{q^2}{U^2} = 1 - \frac{16}{n^4}\left[\frac{\cosh\eta - \cosh\xi}{\cosh\frac{2\eta}{n} - \cos\frac{2\xi}{n}}\right]^2 \tag{4.23}$$

where q is the resultant speed for any general point in the field. Here n is a parameter which when less than 2, the motion applies to the ogival section in a uniform stream. For our interest

$$n = 2\left[1 - \frac{2}{\pi}\tan^{-1}\breve{t}\right] \tag{4.24}$$

and for all points along the ogive specified by $\breve{t}$

$$\xi = \xi_0 = \frac{n\pi}{2} \tag{4.25}$$

ξ and η are coaxial coordinates defined in Figure 4.7.

From geometry

$$\xi + \theta_2 + \pi - \theta_1 = \pi \quad \therefore \quad \xi = \theta_1 - \theta_2 \tag{4.26}$$

and η is defined by

$$\eta = \ln \frac{r_2}{r_1}$$

If P moves along the arc of any circle passing through AB then the parameter ξ is a constant. The relations between dimensionless x,y and ξ,η are

$$x = \frac{\sinh \eta}{\cosh \eta - \cos \xi}\ ;$$

$$y = \frac{\sin \xi}{\cosh \eta - \cos \xi} \qquad (4.27)$$

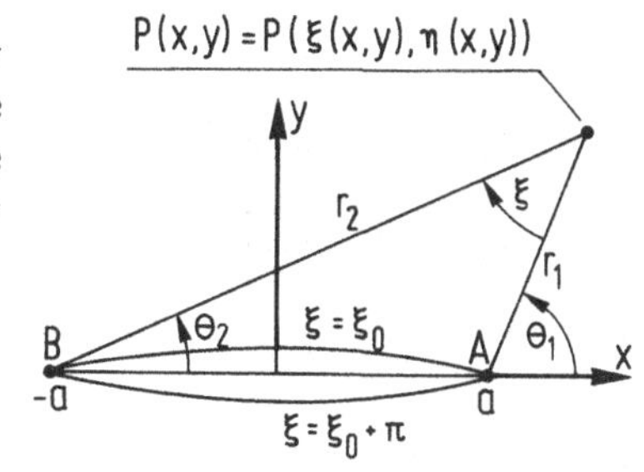

Figure 4.7 Coaxial coordinates.

To evaluate (4.23) on an ogive specified by the thickness ratio $\check{t}$ we set

$$\xi = \xi_0 = \frac{\pi}{2} n = (\pi - 2 \tan^{-1} \check{t})$$

using n from (4.24) and select arbitrary values of η from zero to "large" values ("large" being 10). Then from (4.27) we can find the x-wise location for each η. For the mid-length location $r_2 = r_1$, hence $\eta = \ln r_1/r_2 = 0$. Then

$$C_{pmin}(0,\check{t}) = 1 - \frac{1}{\left[1 - \frac{2}{\pi} \tan^{-1}\check{t}\right]^4} \left[\frac{1 - \cos(\pi - 2 \tan^{-1}\check{t})}{1 + 1}\right] \qquad (4.28)$$

Expanding this for $\tan^{-1} \check{t} << \pi/2$ we find

$$C_{pmin}(0,\check{t}) = -\frac{8}{\pi} \tan^{-1} \check{t} - \left[1 + \frac{40}{\pi^2}\right]\left[\tan^{-1} \check{t}\right]^2 + \dots \qquad (4.29)$$

Comparison with Linearized $C_{pmin}(0,0)$

As in the range of interest $\tan^{-1} \check{t}$ is well approximated by $\check{t}$ we see that our linearized theory gives the correct leading term (cf. (4.22)). This argument applies only to sections for $\check{t} < 0.25$.

What is more surprising is that the linearized theory, Equation (4.21), fits the exact distribution over the entire length of the chord as may be seen in Figure 4.8 except at $x = \pm 1$ where the linearized theory becomes logarithmically infinite. (Note that only the forward half is shown because the shape is symmetrical about $x = 0$). In any event our interest is in the minimum pressure (where C_p is most negative) and at this juncture we can ignore the singular behavior at $x = \pm 1$.

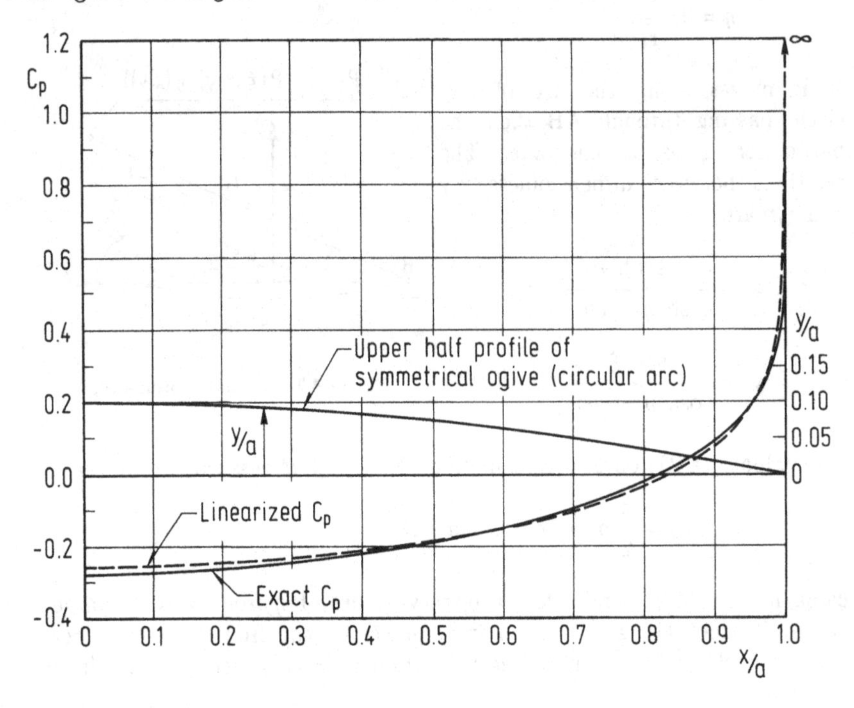

Figure 4.8 Comparison of linearized pressure coefficient distribution with exact solution for a symmetrical ogival section at zero angle of attack; thickness/chord = 0.10.

THE ELLIPTICAL SECTION

Writing (4.18) in dimensionless form and noting that the equation for an ellipse of thickness t is, in variables made non-dimensional by a, the semichord

$$\check{\tau} = \check{t}\sqrt{1 - \check{x}^2}$$

then the approximate pressure coefficient variation with $\check{x}$ is given by

$$C_p(\check{x}) = \frac{2}{\pi}\check{t}\int_{-1}^{1}\frac{\check{x}'d\check{x}'}{\sqrt{1 - \check{x}'^2}\,(\check{x} - x')} \tag{4.30}$$

This integral is evaluated using (M8.7) in the Mathematical Compendium, Section 8, Table of Airfoil Integrals, thus

$$C_P(\check{x}) \equiv -2\check{t} \quad \text{for all } -1 < \check{x} < 1 \tag{4.31}$$

It is perhaps surprising that although this integral appears to be a function of $\check{x}$ it is independent of $\check{x}$, the value being $-\pi$. Thus the linearized theory evaluated along $y = 0$ gives a perfectly flat pressure distribution which is a linear function of $\check{t}$ with a *coefficient which is less negative than that of all other sections having a radius of curvature equal to and less than that of the ellipse.*

We now turn to the results from the exact theoretical solution for the flow about elliptical sections of any thickness ratio in a uniform stream at zero angle of attack.

The Exact Solution for the Elliptical Section

Milne-Thomson (1955), p. 163, Equation 2 provides the pressure distribution on any elongated ellipse at an angle of attack $\alpha = 0$ and a trivial manipulation yields,

$$C_P(\eta(\check{x})) = 1 - \left[\frac{1 + \check{t}}{1 - \check{t}}\right]\left[\frac{1 - \cos 2\eta}{\cosh 2\xi_0 - \cos 2\eta}\right] \tag{4.32}$$

Here $\xi = \xi_0$ places us on the contour to which the hyperbolae η = a constant are orthogonal. Fixing the thickness ratio determines ξ_0 by the relation

$$\xi_0 = \sinh^{-1}\left[\frac{\check{t}}{\sqrt{1 - \check{t}^2}}\right] = \frac{1}{2}\ln\left[\frac{1 + \check{t}}{1 - \check{t}}\right] \tag{4.33}$$

The connection between η and $\check{x}$ for points on an ellipse $\xi = \xi_0$ = a constant is

$$\sin\eta = \frac{\check{y}}{\sqrt{1 - \check{x}^2}\,\sinh\,\xi_0} = \frac{\check{y}}{\check{t}} \tag{4.34}$$

and as $\check{y} = \check{t}\sqrt{1 - \check{x}^2}$

$$\eta = \sin^{-1}\left[\sqrt{1-\check{x}^2}\,\right] \tag{4.35}$$

Thus η is the eccentric angle of any point P on an ellipse as displayed in Figure 4.9.

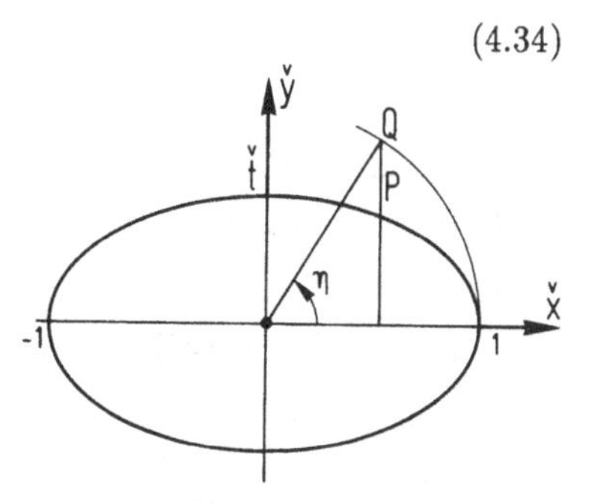

Figure 4.9

We see that η varies from 0 to π as P moves from 1 to -1 along the ellipse fixed by $\check{t}$.

Thus at $(0,\check{t})$, $\eta = \pi/2$ and (4.32) using $\cosh \xi_0 = 1/\left[\sqrt{1 - \check{t}^2}\right]$ and $\cosh 2\xi_0 = 2\cosh^2 \xi_0 - 1$ reduces to the amazingly simple formula

$$C_{p_{min}} = -2\check{t} - \check{t}^2 \tag{4.36}$$

Comparison with Linearized $C_p(0,0)$

Thus again we see that the linearized theory gives the same leading term (cf. (4.31)). It is to be noted that the contribution of the quadratic term is exceedingly small for thin sections being $\check{t}/2$ relative to the linear term. Hence for $\check{t} = 0.10$, the error is 5 per cent. However, comparison of the linear theory with experimental pressure-distribution measurements shows closer correlation for $C_{p_{min}}$ than with exact theory. This is because the viscous flow in the boundary layer has the effect of reducing the curvature in such regions and hence the experimental $C_{p_{min}}$ is less negative than exact theory.

We see from Figure 4.10 that the linearized $C_p(x,0)$ very well approximates the exact distribution from between $\check{x} = \pm 0.8$ or to within 10 per cent of the chord from the leading and trailing edges. It fails in this neighborhood of the leading edge because the ellipse has a blunt leading edge.

Let us now turn to a generalization of the foregoing and compare $C_{p_{min}}$ from results of calculations available for a variety of airfoil sections, which we should more aptly refer to as hydrofoil sections.

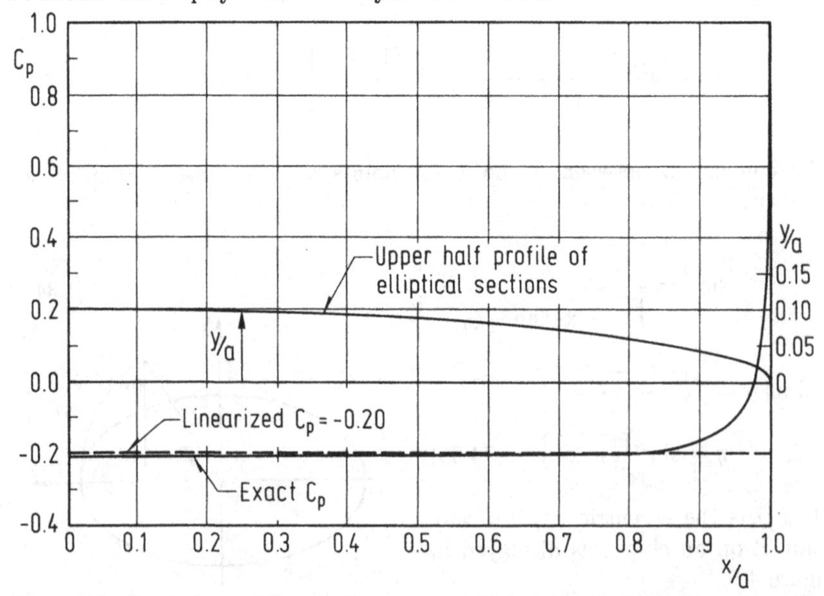

Figure 4.10 Comparison of linearized pressure coefficient distribution with exact solution for an elliptical section at zero angle of attack; thickness/chord = 0.10.

GENERALIZATION TO APPROXIMATE FORMULAE FOR FAMILIES OF TWO-DIMENSIONAL HYDROFOILS

Some further insight into the mechanism which produces the minimum pressure on symmetrical sections at zero angle of attack may be obtained by expanding the section slope function $\partial\tau/\partial x$ about the location of maximum thickness in a Taylor series.

This takes the form in the variable x (cf. Mathematical Compendium, Section 1, p. 504)

$$\frac{\partial\tau(x)}{\partial x} = \tau'(x) = \tau'(x_m) + \frac{\tau''(x_m)}{1!}(x-x_m) + \frac{\tau'''}{2!}(x-x_m)^2 + \dots \tag{4.37}$$

where $'$ indicates a derivative with respect to x and the arguments are the values of x at which the derivatives are evaluated (after differentiation!).

The first term is zero as x_m is the point where the thickness is a maximum. Substituting (4.37) in (4.18) (replacing x by $\check{x}$) and adding and subtracting $\check{x}$ in the numerator terms yield simple integrals giving

$$C_P(\check{x}) \approx -\frac{2}{\pi}\left\{-2\check{\tau}'' - (\check{x} - \check{x}_m)\left[\check{\tau}' + \frac{\check{\tau}'''}{2}(\check{x} - \check{x}_m)\right] \ln\frac{1-\check{x}}{1+\check{x}} + \check{\tau}'''(2\check{x}_m - \check{x}) + ..\right\} \tag{4.38}$$

where all the derivatives are evaluated at $\check{x} = \check{x}_m$.

Evaluating (4.38) at the maximum point $\check{x} = \check{x}_m$

$$C_{P\min}(\check{x}_m) = -\frac{2}{\pi}\left\{-\check{\tau}''(\check{x}_m) + \check{x}_m\check{\tau}'''(\check{x}_m) + \dots\right\}$$

This shows that the minimum pressure depends primarily on the radius of curvature at $\check{x}_m$ plus higher order terms since from calculus the dimensionless curvature $\check{\kappa}$ is

$$\check{\kappa} = \frac{\check{\tau}''}{[1 + \check{\tau}'^2]^{3/2}}$$

and at points where $\check{\tau}'(\check{x}) = 0$ as at $\check{x} = \check{x}_m$

$$\check{\tau}''(\check{x}_m) = \check{\kappa} = \frac{1}{\text{radius of curvature } / \text{ a}}$$

Thus the curvature imposed by the thickness distribution is a dominant mechanism in producing the minimum pressure. Indeed, when the curvature is constant as in the case of the ogive, then $\check{r}''(0) = -2\check{t}$, and as all higher derivatives are zero, we recapture our previous result, $C_{pmin} = -(8/\pi)\check{t}$. However, this analysis is far from complete because it fails to give a correct leading term in the case of an ellipse. We can only conclude that for sections having blunt leading edges the first order theory is inadequate and moreover we cannot expect to capture the effect of the distribution of section slopes by a Taylor expansion about x_m. Indeed the linearized theory for an ellipse as given by the integral in (4.30) suggests strong contributions from the leading and trailing edges by the presence of the weighting factor $1/\sqrt{1 - \check{x}'^2}$ which becomes square-root singular at $\check{x}' = \pm 1$.

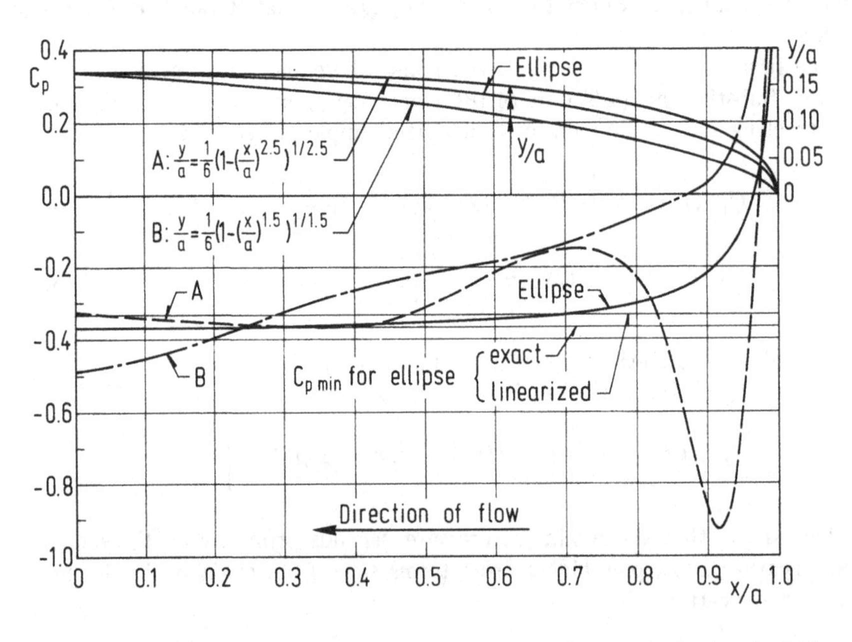

Figure 4.11 Comparison of pressure distributions about three ovals for length–thickness ratio of 6.0.

From: Freeman, H.B. (1942). *Calculated and Observed Speeds of Cavitation About Two- and Three-Dimensional Bodies in Water.* Report 495. Washington, D.C.: The David W. Taylor Model Basin — United States Navy.
By courtesy of David Taylor Research Center, USA.

Indeed the shape of the leading edge of sections has a pronounced effect on the pressure distributions as shown in Breslin & Landweber (1961). For a family of ovals

$$\check{r} = (1 - \check{x}^n)^{1/n}$$

Freeman (1942)[4] has shown that for n < 2 (less blunt than an ellipse) the minimum pressure occurs at midlength but is always more negative than for the ellipse. For n > 2 (more blunt than an ellipse) the minimum pressure occurs very close to the leading edge and is again more negative than that for an ellipse. These results are displayed in Figure 4.11. Data from measurements made in the NACA (now NASA) windtunnels show the effect of leading-edge curvature and suggests that a curvature slightly greater than that at the nose of an ellipse yields a small improvement over the ellipse. See Figures 4.12 and 4.13. For our purposes, the ellipse will be taken as the best possible (i.e., having the least negative $C_{p_{min}}$).

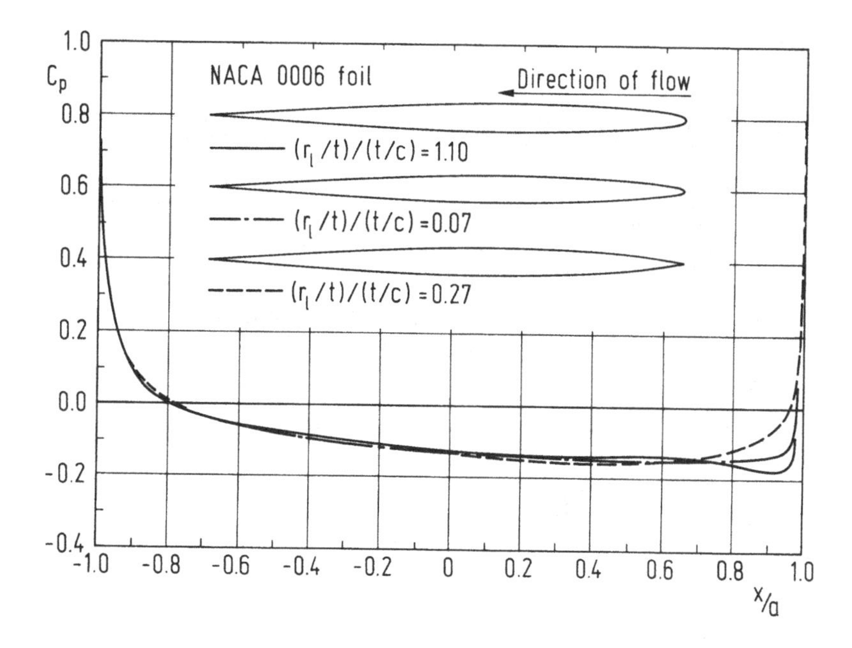

Figure 4.12 Variation of pressure distribution near leading edge with change of nose radius r_l for NACA 0006 foil (from NACA TN 3172) at zero angle of attack.

From: Breslin, J.P. & Landweber, L. (1961). A manual for calculation of inception of cavitation on two- and three-dimensional forms. *SNAME T&R Bulletin*, no. 1–21. New York, N.Y.: SNAME.

[4] Based on calculations by L. Landweber at DTRC.

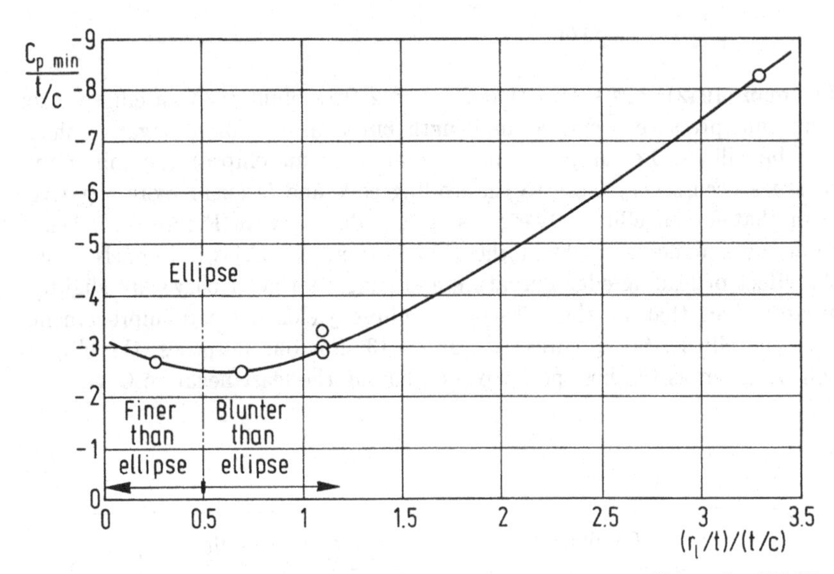

Figure 4.13 Influence of nose radius r_l on minimum pressure coefficient for NACA 4- digit foils (From NACA TN 3172) at zero angle of attack.

From: Breslin, J.P. & Landweber, L. (1961). A manual for calculation of inception of cavitation on two- and three-dimensional forms. *SNAME T&R Bulletin*, no. 1–21. New York, N.Y.: SNAME.

By courtesy of SNAME, USA.

As a generalization of (4.18) we see that we may postulate that

$$C_{p\min}(\check{x}_m) = -A\check{t} - B\check{t}^2 \tag{4.39}$$

where A and B are functionals of the shape distribution. Specifically

$$A(\check{x}_m) = -\frac{2}{\pi}\int_{-1}^{1}\frac{\check{t}'(\check{x}')}{\check{x}_m - \check{x}'}\,d\check{x}' \tag{4.40}$$

and B is similarly some definite integral of the distribution of section shape. To test this hypothesis we have plotted results from exact numerical calculations for families of sections as generally given in Abbott & von Doenhoff (1959). A tabulation of the coefficients (from Breslin & Landweber (1961)) is given Table 4.1 and a graphical comparison is provided in Figure 4.14. It is clear from this figure that the assumed relation (4.39) fits the results from exact evaluations over the range of thickness ratios $0 < \check{t} \leq 0.3$. The elliptical section is the best but of course is not suitable

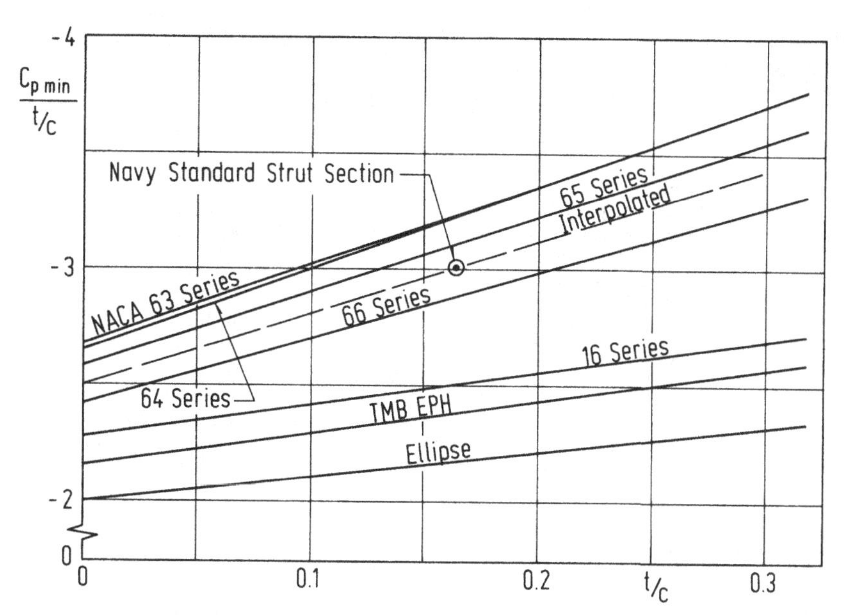

Figure 4.14 Variation of minimum pressure coefficient with thickness ratio for several foil sections at zero angle of attack.

From: Breslin, J.P. & Landweber, L. (1961). A manual for calculation of inception of cavitation on two– and three–dimensional forms. *SNAME T&R Bulletin*, no. 1–21. New York, N.Y.: SNAME.

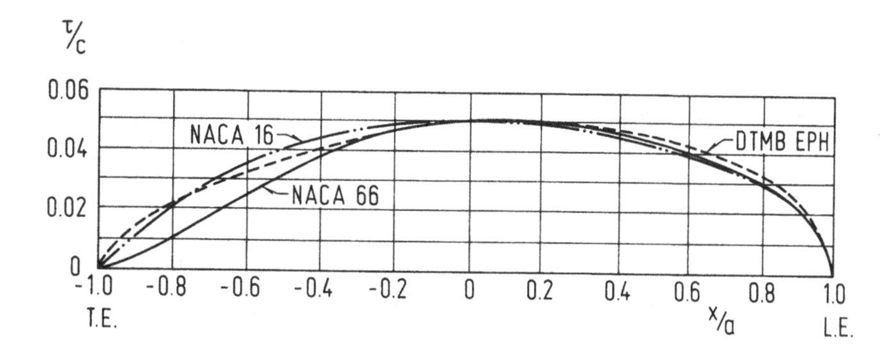

Figure 4.15 Examples of several thickness distributions.

From: Breslin, J.P. & Landweber, L. (1961). A manual for calculation of inception of cavitation on two– and three–dimensional forms. *SNAME T&R Bulletin*, no. 1–21. New York, N.Y.: SNAME.

for use as a hydrofoil because the trailing edge is too blunt giving rise to viscous pressure drag in excess of sections having sharp trailing edges[5].

The geometry of some of these sections is shown in Figure 4.15.

Section	A	B
Ellipse	2.00	1.00
DTMB-EPH	2.15	1.31
NACA-16 Series	2.28	1.30
NACA-66 Series	2.42	2.73
Navy Std. Strut Section	2.50*	3.0*
Double Circular Arc	2.55 (= $8/\pi$)	1.43
NACA-65 Series	2.58	3.14
NACA-64 Series	2.65	3.39
NACA-63 Series	2.67	3.33
NACA 4 and 5 Digit	3.50**	0

* Interpolated by use of value of $C_{p_{min}}$ at one thickness ratio, hence approximate.
** These sections have pressure minima relatively close to the leading edge and have large nose radii; as a result, they do not conform to this formulation too well, the listed value of A being valid in the interval $0.08 < \check{t} < 0.2$.

Table 4.1 Coefficients A and B in Equation (4.39), $C_{p_{min}} = -A\check{t} - B\check{t}^2$ for various foil sections.

From: Breslin, J.P. & Landweber, L. (1961). A manual for calculation of inception of cavitation on two- and three-dimensional forms. *SNAME T&R Bulletin*, no. 1–21, New York, N.Y.: SNAME.

By courtesy of SNAME, USA.

A BRIEF LOOK AT THREE-DIMENSIONAL EFFECTS

To apply the foregoing to propeller blades one is assuming that the blade sections at different radii are hydrodynamically uncoupled, i.e., the radial flow component and the radial pressure gradient are ignored. This is of course not the case particularly for ship propellers where the blade aspect ratio is of the order of unity. To grasp some sense of the effects of three-dimensionality on the pressures induced by symmetrical sections of finite span we may consider a non-cambered, rectangular wing with thickness.

We treat at first a wing having ogival or lenticular sections whose thickness is independent of the spanwise coordinate z. To generate this form in a uniform flow $-U$ we distribute three-dimensional sources over the plane $z \equiv 0$ in $|x| \leq a$; $|y| \leq b$ (2b being the span). This distribution induces the potential

$$\phi(x,y,z) = -\frac{1}{4\pi}\int_{-a}^{a} dx' \int_{-b}^{b} \frac{m(x',y')}{\sqrt{(x-x')^2 + (y-y')^2 + z^2}}\, dy' \quad (4.41)$$

[5] C.-A. Johnsson at SSPA finds no discernible penalty in the use of elliptical trailing edges for propeller blade sections.

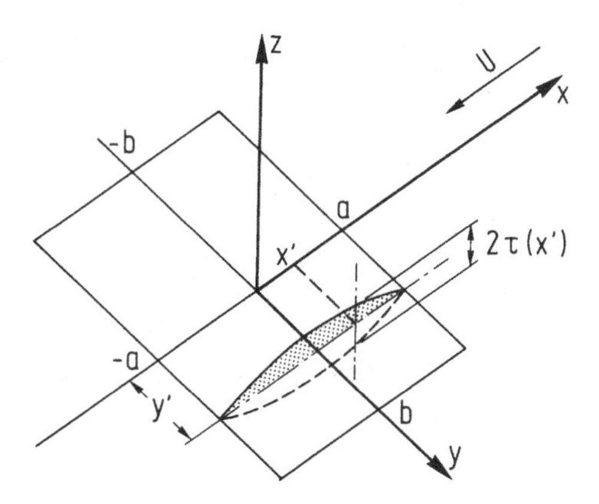

Figure 4.16 Non-cambered, rectangular wing with thickness.

As we have seen previously (cf. Equation (2.43)) a planar distribution of sources yields a normal velocity on the source sheet given by $\pm$ m(x,y)/2. The boundary condition on ϕ is then evaluated as in two dimensions by

$$\frac{\partial\phi(x,y,0)}{\partial z} = \frac{m(x,y)}{2} = -U\tau'(x,y) \tag{4.42}$$

Here we take the semi-thickness $\tau(x)$ i.e. independent of y for the sake of simplicity to demonstrate the effect. Replacing m in (4.41) by use of (4.42) gives

$$\phi(x,y,z) = \frac{U}{2\pi}\int_{-a}^{a} dx' \int_{-b}^{b} \frac{\tau' dy'}{\sqrt{(x-x')^2 + (y-y')^2 + z^2}} \tag{4.43}$$

For the ogival sections $\tau(x) = \tau_0(1-(x/a)^2)/a$ and $\tau' = -2\tau_0 x/a^2$, then ϕ becomes

$$\phi = -\frac{U}{\pi}\frac{\tau_0}{a^2}\int_{-a}^{a} dx'x' \int_{-b}^{b} \frac{dy'}{\sqrt{(x-x')^2 + (y-y')^2 + z^2}} \tag{4.44}$$

The pressure coefficient is

$$C_p = 2\frac{1}{U}\frac{\partial\phi}{\partial x} = \frac{2}{\pi}\frac{\tau_0}{a^2}\int_{-a}^{a} dx'x' \int_{-b}^{b} \frac{(x-x')}{((x-x')^2 + (y-y')^2 + z^2)^{3/2}}\, dy' \tag{4.45}$$

For purposes of demonstration we may evaluate along x $\equiv$ 0 and as before evaluate on the sheet z = 0. The y'-integral can be effected to yield

$$C_p(0,0,0) = -\frac{4}{\pi}\frac{\tau_0 b}{a^2}\int_{-a}^{a}\frac{dx'}{\sqrt{x'^2 + b^2}} = -\frac{8}{\pi} t \frac{b}{a} \sinh^{-1}\frac{a}{b} \qquad (4.46)$$

We may note immediately that as the semi-span b is made large then $\sinh^{-1} a/b \approx a/b - (1/6)(a/b)^3 + \ldots$ and

$$\lim_{b/a\to\infty} C_p(0,0,0) = -\frac{8}{\pi}\frac{\tau_0}{a}\lim_{b/a\to\infty}\left[1 - \frac{1}{6}\left[\frac{a}{b}\right]^2 \ldots\right] = -\frac{8}{\pi}\frac{\tau_0}{a} \qquad (4.47)$$

the two-dimensional result (cf. (4.22)) as the aspect ratio b/a becomes infinite. For any finite aspect ratio, the ratio of the 3-D to the 2-D coefficients is

$$R_{3:2} = \frac{(C_{p\min})_3}{(C_{p\min})_2} = \frac{b}{a}\sinh^{-1}\frac{a}{b} = \frac{b}{a}\ln\left[\frac{a}{b} + \sqrt{1 + \left[\frac{a}{b}\right]^2}\,\right] \qquad (4.48)$$

Values for several values of the aspect ratio are presented in Table 4.2.

$A_R = b/a$	$R_{3:2}$
2	0.962
1.5	0.938
1	0.881
0.5	0.722
0.25	0.524

Table 4.2 Ratios of 3-D to 2-D minimum pressure coefficients for various aspect ratios.

Thus in the vicinity of the aspect ratio of wide-bladed propellers the pressure coefficient *in the center* is reduced by the order of 10 per cent as compared to the two-dimensional value. The three-dimensional "relief" will be more pronounced as we approach the wing tips. To examine this ratio along $x = 0$ and $0 \leq y \leq b$, we find

$$R_{3:2}(y/a) = \frac{1}{2}\left\{\left[\frac{b}{a} - \frac{y}{a}\right]\sinh^{-1}\frac{1}{\frac{b}{a} - \frac{y}{a}} + \left[\frac{b}{a} + \frac{y}{a}\right]\sinh^{-1}\frac{1}{\frac{b}{a} - \frac{y}{a}}\right\} \qquad (4.49)$$

and values are presented for an aspect ratio of unity ($b/a = 1.0$) in Table 4.3.

Hence the reduction in pressure, relative to two dimensions is seen to vary slowly until one gets to 0.6 of the semi-span. Thereafter the decay is rapid. It is interesting to note that if the wing were semi-infinite the ratio at the tip is 0.50. Thus in this regard the wing of aspect ratio unity appears to be nearly semi-infinite at the tip where we see a ratio of 0.481.

y/a	$R_{3:2}(y/a)$
0.0	0.881
0.2	0.874
0.4	0.850
0.6	0.802
0.8	0.708
1.0	0.481

Table 4.3 Ratios of 3-D to 2-D pressure coefficients along the span (for aspect ratio of unity).

Of course the thickness ratios on propeller blades decrease from about 0.25 at the root to about 0.02 or so at the tip. The effect of taper in thickness of the wing could be examined by computer. This is left for the interested reader! It is conjectured that for such cases the effect of three-dimensional "relief" will be more pronounced than found in the foregoing.

For a wing having uniform elliptical sections the integration leads to the Elliptic function of the first kind. We find at the center that

$$R_{3:2} = \frac{2}{\pi} \frac{1}{\sqrt{1 + (a/b)^2}} F\left[\frac{2}{\pi}, \frac{a/b}{\sqrt{1 + (a/b)^2}}\right] \tag{4.50}$$

where F is the Elliptic function of the first kind.

As a check we can find[6] that as $a/b \to 0$, $R_{3:2} \to 1.0$.

For aspect ratio of unity

$$R_{3:2} = \frac{2}{\pi} \frac{1}{\sqrt{2}} F\left[\frac{\pi}{2}, 0.707\right] = 0.834 \text{ at } x = 0 \tag{4.51}$$

Thus the three-dimensional effect is considerably greater for elliptical than for ogival sections where this ratio at $b/a = 1.0$ was found to be 0.881.

We can conclude that the use of two-dimensional-section results in estimating minimum pressures on propeller blade sections is conservative and because of three-dimensional reductions one could use somewhat larger thickness ratios (with of course consequent increase in propeller weight) to achieve sections which are more tolerant to angles of attack.

We now turn to the calculation of pressure distribution on two-dimensional sections having camber and at small angles of attack.

[6] See for example Abramowitz & Stegun (1972).

5 Pressure Distributions and Lift on Flat and Cambered Sections at Small Angles of Attack

The non-symmetrical flow generated by flat and cambered laminae at angles of attack is at first modelled by vorticity distributions via classical linearized theory. Here, in contrast to the analysis of symmetrical sections, we encounter integral equations in the determination of the vorticity density because the local transverse component of flow at any one point depends upon the integrated or accumulated contributions of all other elements of the distribution. Pressure distributions at non-ideal incidence yield a square-root-type infinity at the leading edge because of the approximations of first order theory. Lighthill's (1951) leading edge correction is applied to give realistic pressure minima at non-ideal angles of incidence.

Our interest in pressure minima of sections is due to our concern for cavitation which can occur when the total or absolute pressure is reduced to the vapor pressure of the liquid at the ambient temperature. Since cavitation may cause erosion and noise it should be avoided or at least mitigated which may possibly be done by keeping the minimum pressure above the vapor pressure. This corresponds to maintaining the (negative) minimum-pressure coefficient $C_{p_{min}}$ higher than the negative of the cavitation index.

At this point we shall not go deeper into the details of cavitation which is postponed until Chapter 8. Instead we shall continue our theoretical development with flat and cambered sections.

THE FLAT PLATE

We now seek the pressure distributions and the lift on sections having zero thickness but being cambered and, in general, set at any arbitrary (but small) angle of attack to the free stream, $-U$. Consider a flat plate at small angle α.

Then we might envisage a flow pattern as sketched in Figure 5.1 as a plausible one with stagnation points S and S', S being on the lower side and S' on the upper side. The flow around the leading edge is strongly curved (very small radius of curvature) resulting in high local velocity and hence (by Bernoulli) develops very low pressure on the upper side, particularly near the leading edge. (In a real fluid the flow about such a sharp leading edge will separate from the upper side, the extent of separation being dependent on the angle of attack. Here we do not account for this.) Conversely on the lower side the fluid is slowed and the pressure is higher than the ambient pressure. Thus the pressure on the upper side may be written

$$\frac{p_+ - p_\infty}{\frac{1}{2}\rho U^2} = 1 - \left[\frac{(u_+ - U)^2 + v_+^2}{U^2}\right] = C_{p_+} \tag{5.1}$$

and on the lower side

$$\frac{p_- - p_\infty}{\frac{1}{2}\rho U^2} = 1 - \left[\frac{(u_- - U)^2 + v_-^2}{U^2}\right] = C_{p_-} \tag{5.2}$$

where u and v are the pertubation components in the x- and y-directions.

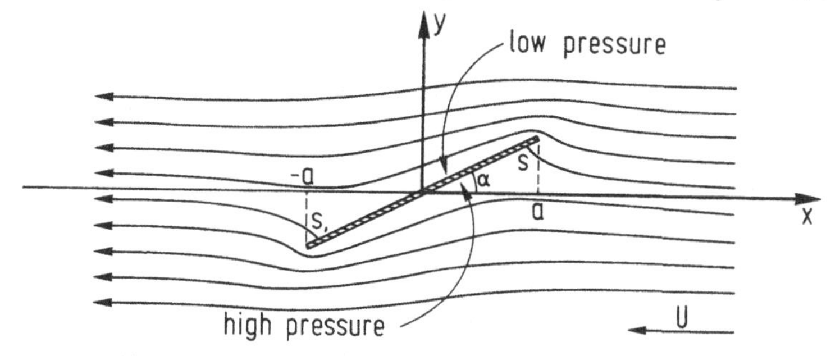

Figure 5.1 Conjectural sketch of flow around flat plate.

In way of the plate the flow must be tangential to the plate, or the slope of the flow must be equal to the slope of the plate, i.e.

$$\frac{v_+}{u_+ - U} = \frac{dy_+}{dx} = \frac{d}{dx}(x \tan\alpha) \approx \alpha \tag{5.3}$$

and

$$\frac{v_-}{u_- - U} = \frac{dy_-}{dx} \approx \alpha \tag{5.4}$$

As the slope of the plate is the same on both sides then

$$v_+u_- - Uv_+ = v_-u_+ - Uv_- \tag{5.5}$$

Now as both $u_\pm$ and $v_\pm$ are of order α, then v_+u_- and v_-u_+ are of order α^2. This statement is not true everywhere (as in symmetrical flow) and fails in the neighborhood of stagnation points. Thus this approximation is not uniformly valid. Adopting this as being sufficiently true gives from (5.5)

$$v_+ = v_- \tag{5.6}$$

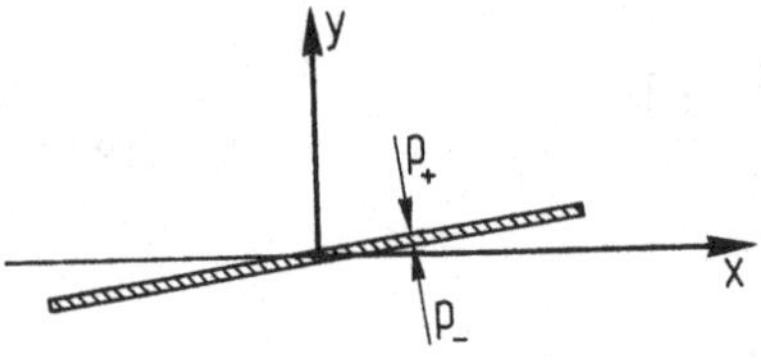

Hence, to first order the vertical perturbation flow is symmetric.

We can now use (5.6) in (5.1) and (5.2) to calculate the difference in pressure between the upper and lower sides i.e. from Figure 5.2, the net *upward* pressure difference at any fixed x is $p_- - p_+$ or

Figure 5.2

$$C_{p_-} - C_{p_+} = \Delta C_p = \frac{(u_+ - U)^2 - (u_- - U)^2}{U^2}$$

$$= 2\,\frac{(u_- - u_+)}{U} + \frac{u_+^2 - u_-^2}{U^2} \tag{5.7}$$

To be consistent with neglect of terms of order α^2 we must drop the quadratic terms and then

$$\Delta C_p \approx 2\,\frac{(u_- - u_+)}{U} \tag{5.8}$$

To generate this flow by a distribution of singularities we must look for that type for which the vertically induced component is an even function of y. We have seen that source distributions yield asymmetrical vertical velocities (recall that $v(x,0_\pm) = \pm\, m/2$). Thus source distributions will not provide the needed symmetry in y, i.e. $v(x,y) = v(x,-y)$.

From our previous considerations we see that normal dipoles or vortices yield vertical or normal velocities along the line of their distribution which are even functions of the normal. Thus for our flat plate we choose a distribution of two-dimensional vortices whose potential is (cf. (2.33))

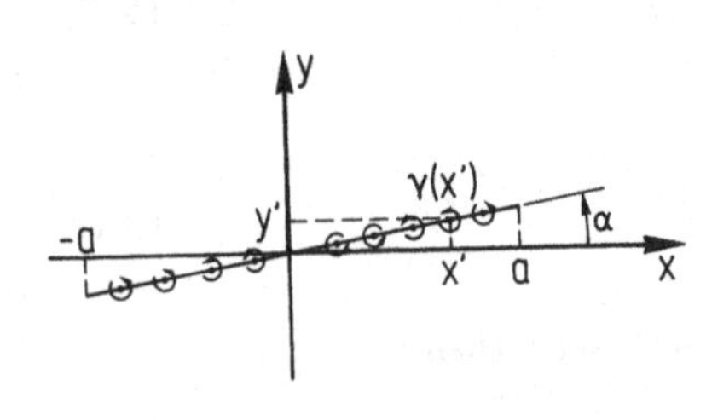

Figure 5.3 Vortex distribution on inclined flat plate.

$$\phi = \frac{1}{2\pi}\int_{-a}^{a} \gamma(x')\,\tan^{-1}\left[\frac{y - y'}{x - x'}\right] dx' \tag{5.9}$$

where, in principle, the vortices are distributed along the inclined plate, so $y' = \alpha x'$. Then the vertical velocity at any (x,y), induced by this distribution of vorticity, is

$$\frac{\partial\phi}{\partial y} = v(x,y) = \frac{1}{2\pi}\int_{-a}^{a}\frac{(x-x')\ \gamma(x')}{(x-x')^2 + (y-\alpha x')^2}\,dx' \tag{5.10}$$

If we apply the boundary condition from (5.3) (neglecting u_+) on the plate $y = \alpha x$, we get the requirement on the unknown vortex density γ that, in general

$$v(x,\alpha x) = \frac{1}{2\pi}\int_{-a}^{a}\frac{(x-x')\ \gamma(x')}{(x-x')^2\ (1+\alpha^2)}\,dx' = -U\frac{dy}{dx};\ |x| \le a$$
$$= -U\alpha \quad \text{, on flat plate}$$
$$(5.11)$$

We observe that by applying the boundary condition on the plate we have introduced terms of order α^2 and also higher order terms since $(1 + \alpha^2)^{-1} = 1 - \alpha^2 + \alpha^4 - ....$ As this is inconsistent with our prior neglect of terms of $O(\alpha^2)$ we see that to first order the vortices should be placed on $y' \equiv 0$ and the evaluations made not on $y = \alpha x$ but on $y = 0$, the fluid reference line. Thus taking $\alpha = 0$ in the left side of (5.11) implies the requirement on γ to be

$$2\pi v(x,0) = \int_{-a}^{a}\frac{\gamma(x')}{x - x'}\,dx' = -2\pi U y' \approx -2\pi U\alpha \tag{5.12}$$

Thus all the vorticity distribution makes a contribution and we are confronted with a singular integral equation of the first kind having a Cauchy kernel $(x - x')^{-1}$. Fortunately this equation can be inverted analytically by a number of techniques. A development of the following solution for this equation is relegated to Appendix A (p. 484) in order not to impede our progress here with undue mathematical details. The general solution of

$$\int_{a_1}^{a_2}\frac{\gamma(x')}{x - x'}\,dx' = h(x) \qquad a_1 < x < a_2 \tag{5.13}$$

where h(x) is a known function is, from (A.20) p. 489

$$\gamma(x) = -\frac{1}{\pi^2\sqrt{(x-a_1)(a_2-x)}}\left[\int_{a_1}^{a_2}\frac{\sqrt{(x'-a_1)(a_2-x')}}{x-x'}\,h(x')dx' + \pi K\right] \tag{5.14}$$

Here K is a constant. Thus there are infinitely many solutions, one for each K, and hence the solution is not unique. The reason for this is that the homogeneous equation, i.e., for $h = 0$

$$\int_{a_1}^{a_2} \frac{\gamma_0(x')}{x - x'} dx' = 0 \tag{5.15}$$

has the infinity of solutions

$$\gamma_0 = \frac{K}{\pi\sqrt{(x - a_1)(a_2 - x)}} \tag{5.16}$$

and as we can always add the homogeneous solution, we have the general solution in the form of (5.14).

To expedite the further discussion we specialize (5.14) by taking $a_1 = -a$, $a_2 = a$. We may obtain a physical interpretation of K as follows. Let us integrate both sides of (5.14) over the section to secure the circulation Γ. Thus

$$\Gamma = \int_{-a}^{a} \gamma(x)dx$$

$$= -\frac{1}{\pi^2}\int_{-a}^{a} \frac{dx}{\sqrt{a^2 - x^2}} \int_{-a}^{a} \frac{\sqrt{a^2 - x'^2}}{x - x'} h(x')\, dx' - \frac{K}{\pi}\int_{-a}^{a} \frac{dx}{\sqrt{a^2 - x^2}}$$

In the first term, interchanging the orders of integration, we find

$$\int_{-a}^{a} \frac{dx}{\sqrt{a^2 - x^2}\,(x - x')} = 0$$

by (M8.6) in the Table of Airfoil Integrals, Section 8 of the Mathematical Compendium. In the last integral let $x = a \sin\alpha$ and the integral is seen to be π. Hence,

$$\Gamma = -K \tag{5.17}$$

This gives a hint as to how to select K or Γ in order to pick out of the infinity of solutions, one which may be of practical use. Use of (5.17) gives

$$\gamma(x) = -\frac{1}{\pi^2}\frac{1}{\sqrt{a^2 - x^2}}\left[\int_{-a}^{a} \frac{\sqrt{a^2 - x'^2}}{x - x'} h(x')dx' - \pi\Gamma\right] \tag{5.18}$$

Now we know from the properties of vorticity sheets that

$$u_- = \gamma/2 \quad \text{and} \quad u_+ = -\gamma/2 \quad \text{(cf. Equation (2.36))} \tag{5.19}$$

or

$$u_- - u_+ = \gamma \tag{5.20}$$

The Russian scientist Kutta (1910) noticed from experiments that the flow at small angles about airfoils with sharp trailing edges always left the trailing edge smoothly, i.e. the flow does not go around the trailing edge as in the heuristic sketch of Figure 5.1. Kutta then reasoned that the circulation must be such as to move S' to the trailing edge, or alternatively $u_-(-a) = u_+(-a)$ giving from (5.20)

$$\gamma(-a) = 0 \tag{5.21}$$

If the flow would not leave the sharp trailing edge smoothly theory would predict infinite velocities around this edge. The condition that the velocity must be finite is generally known as the ***Kutta condition***.

To apply (5.21) to (5.18) where $\gamma(x)$ is seen to be square-root singular at $x = \pm a$ we require that $\gamma(x)$ so approaches zero in order that

$$\lim_{x \to -a} \gamma(x)\sqrt{a+x} = -\frac{1}{\pi^2} \lim_{x \to -a} \frac{1}{\sqrt{a-x}} \left[\int_{-a}^{a} \frac{\sqrt{a^2-x'^2}}{x-x'} h(x')dx' - \pi\Gamma \right] = 0 \tag{5.22}$$

Hence the circulation necessary to achieve this condition is

$$\Gamma = -\frac{1}{\pi} \int_{-a}^{a} \sqrt{\frac{a - x'}{a + x'}}\, h(x')dx' \tag{5.23}$$

Insert (5.23) into (5.18) and we have for any x

$$\gamma(x) = -\frac{1}{\pi^2\sqrt{a^2 - x^2}} \int_{-a}^{a} h(x') \left[\frac{\sqrt{a^2 - x'^2}}{x - x'} + \sqrt{\frac{a - x'}{a + x'}} \right] dx'$$

$$= -\frac{1}{\pi^2\sqrt{a^2 - x^2}} \int_{-a}^{a} \sqrt{a - x'}\, h(x') \left[\frac{a + x' + x - x'}{(x - x')\sqrt{a + x'}} \right] dx'$$

$$= -\frac{1}{\pi^2} \sqrt{\frac{a + x}{a - x}} \int_{-a}^{a} \sqrt{\frac{a - x'}{a + x'}} \frac{h(x')}{x - x'} dx' \tag{5.24}$$

We see that this result goes to zero at $x = -a$ as required.

The integrand becomes $-\sqrt{a - x'}/(a + x')^{3/2}$ and hence is singular at $x' = -a$ which raises a question about the convergence to zero; however, the factor $\sqrt{a + x}$ as $x \to -a$ prevails.

We also observe that in general, $\gamma(a) \to \infty$ as a square-root singularity. In order for $\gamma(x)$ to be regular at $x = a$ as well as at $x = -a$ we can require that

$$\lim_{x \to a} \sqrt{a-x}\,\gamma(x) = -\frac{1}{\pi^2} \lim_{x \to a} \sqrt{a+x} \int_{-a}^{a} \sqrt{\frac{a - x'}{a + x'}} \frac{h(x')}{x - x'} dx' = 0 \tag{5.25}$$

This places the requirement on $h(x')$ that

$$\int_{-a}^{a} \frac{h(x')}{\sqrt{a^2 - x'^2}} dx' = 0 \tag{5.26}$$

This is called an integral condition on h which is obviously satisfied by all h which are odd functions of x'. Thus to have bounded results at both ends restricts $h(x)$ to satisfy (5.26). We may now add (5.26) to (5.24) (since it is zero) to provide

$$\gamma(x) = \frac{1}{\pi^2} \sqrt{\frac{a + x}{a - x}} \int_{-a}^{a} h(x') \left\{ -\sqrt{\frac{a - x'}{a + x'}} \frac{1}{x - x'} + \frac{1}{\sqrt{a^2 - x'^2}} \right\} dx'$$

which reduces to

$$\gamma(x) = -\frac{1}{\pi^2} \sqrt{a^2 - x^2} \int_{-a}^{a} \frac{h(x')\, dx'}{\sqrt{a^2 - x'^2}\,(x - x')} \tag{5.27}$$

Now we see that γ is regular everywhere with the proviso that $h(x')$ satisfies (5.26).

We may now complete the development of the lift on the flat plate for which $h(x)$ has the special value $-2\pi U\alpha$, (cf. Equations (5.12) and (5.13)). Then (5.24) becomes

$$\gamma(x) = \frac{2}{\pi} U\alpha \sqrt{\frac{a + x}{a - x}} \int_{-a}^{a} \sqrt{\frac{a - x'}{a + x'}} \frac{dx'}{x - x'} \tag{5.28}$$

The integral can be evaluated by multiplying numerator and denominator by $\sqrt{a - x}$ yielding two integrals, one being zero by (M8.6) and the other π by (M8.6) and (M8.7) of the Table of Airfoil Integrals. Thus (5.28) simplifies to

$$\gamma(x) = 2U\alpha \sqrt{\frac{a + x}{a - x}} \tag{5.29}$$

As $\gamma(x)$ is square-root singular at the leading edge, one may wonder if this solution is at all useful. To investigate this in regard to lift we integrate over x to secure the circulation,

$$\Gamma = \int_{-a}^{a} \gamma(x)dx = 2U\alpha \int_{-a}^{a} \sqrt{\frac{a + x}{a - x}}\, dx \tag{5.30}$$

and upon multiplying above and below by $\sqrt{a + x}$ we see that one integral is an odd function of x and is hence zero by inspection and the other yields πa or

$$\Gamma = 2\pi a U\alpha$$

By ***Kutta-Joukowsky's theorem*** (cf. Appendix B, p. 490) the lift is at right angles to the velocity $-U$ acting on each vorticity element and is given by

$$L = \rho U\Gamma = 2\pi\rho a U^2\alpha \tag{5.31}$$

Defining the lift coefficient in the usual way as based on the chord $c = 2a$

$$C_L = \frac{L}{\frac{1}{2}\rho c U^2} = 2\pi\alpha \tag{5.32}$$

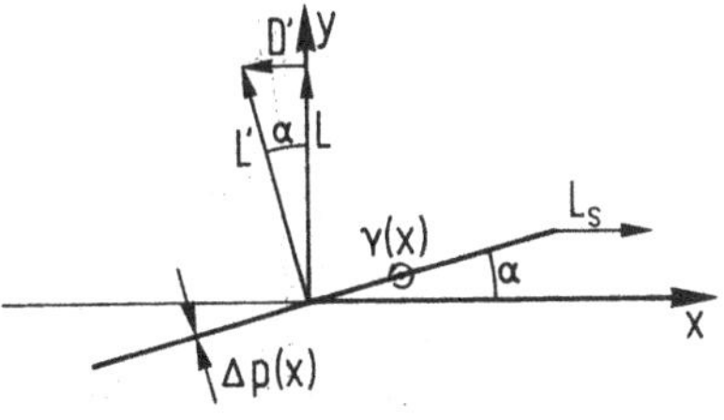

Figure 5.4 Leading–edge suction force.

One may well feel it is a great paradox that there is no drag, in particular when considering the flow around the inclined flat plate. From (5.8) and (5.20) we have the pressure difference $\Delta p(x) = \rho U\gamma(x)$. When integrating this from the trailing edge to the leading edge we have the same integral as in (5.30) and we obtain the same result as in (5.31) for the lift, but we should expect this lift to be perpendicular to the plate. Since the angle of incidence is small there will be no numerical difference to first order between this force and its vertical projection but it will have a component in the negative x-direction, a drag $D' = L'\alpha = 2\pi\rho a U^2\alpha^2$. The leading-edge suction force L_s will be opposite and equal to this drag to balance it, cf. Figure 5.4. The sharp leading edge (where in contrast to the trailing edge no Kutta condition applies) causes a strongly curved flow around this edge. As a consequence of this the flow has a low pressure over this region which then produces the leading-edge suction force. A more detailed derivation of this force is given by Sparenberg (1984).

The result given by (5.32), which was obtained at the turn of this century (actually $C_L = 2\pi \sin\alpha$ from non-linear theory), had a pronounced impact on the prevalent attitudes of scientists and engineers who had regarded inviscid theory to be only a "playground" for mathematicians since it produced the uniform result that the drag on all non-lifting (in both two and three dimensions) and lifting bodies (in two dimensions) was (and still is, of course) zero in steady flow. The comparison of (5.32) and the corresponding result for cambered sections with experimental data convinced most of these "doubting Thomases" of a new practical utility for inviscid flow theory. We shall review this comparison after dealing with the more general and much more useful case of cambered sections at angles of attack.

CAMBERED SECTIONS

Let the camber of a section as measured from the nose-tail line be $y_f(x)$. The maximum of which (to conform to ITTC notation) will always be designated by f, the physical camber. We may see this geometrically in Figure 5.5.

The equation of the cambered line at angle α is clearly

$$y = \alpha x + y_f \tag{5.33}$$

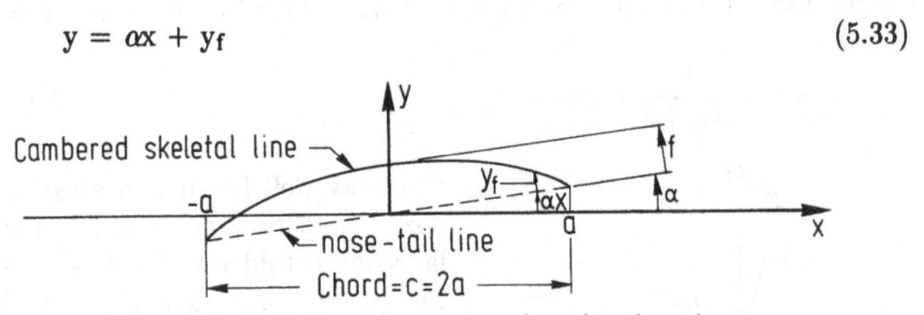

Figure 5.5 Definition of geometry of cambered section.

Then the function h(x) in the general form for the solution of the singular integral equation is, from (5.12) and (5.13),

$$h(x) = -\,2\pi(\alpha + y_f')U \tag{5.34}$$

and hence (5.24) becomes

$$\gamma(x) = \frac{2}{\pi} U \sqrt{\frac{a + x}{a - x}} \int_{-a}^{a} \sqrt{\frac{a - x'}{a + x'}} \frac{(\alpha + y_f')}{x - x'} dx' \tag{5.35}$$

Now for a given camber slope function we can always find an attitude α called the ***ideal angle*** α_i such that $\gamma(a)$ is finite. Thus using (5.26) together with (5.34) we deduce this angle to be

$$\alpha_i = -\frac{1}{\pi} \int_{-a}^{a} \frac{y_f'}{\sqrt{a^2 - x'^2}} dx' = -\frac{2}{\pi}\frac{f}{c} \int_{-1}^{1} \frac{\check{y}_f'}{\sqrt{1 - \check{x}'^2}} d\check{x}' \tag{5.36}$$

$$\check{x}' = x'/a \quad ; \quad \check{y}_f = y_f/f \quad ; \quad a = c/2$$

We see immediately that all camber lines which are even functions of x' (about $x' = 0$) will have y_f' as odd function and hence for all such camber lines, $\alpha_i = 0$. Other camber curves may also provide $\alpha_i = 0$ but, in general, α_i is a small positive angle given by the operation indicated in (5.36).

Employing (5.27), the vorticity distribution is, at the ideal angle α_i,

$$\gamma_i(x) = \frac{2}{\pi} U \sqrt{a^2 - x^2} \int_{-a}^{a} \frac{(\alpha_i + y_f')\ dx'}{\sqrt{a^2 - x'^2}\ (x - x')}$$

$$= \frac{2}{\pi} U \sqrt{a^2 - x^2} \int_{-a}^{a} \frac{y_f'\ dx'}{\sqrt{a^2 - x'^2}\ (x - x')} \tag{5.37}$$

as the first integral is zero.

This is the distribution for the so-called "*shockless entry*" of the flow at the leading edge. Thus, at this special attitude, the pressure distribution is bounded everywhere and both the lift and pressure distribution from this approximate theory turn out to be of practical use.

For a camber curve given by the circular arc as approximated by the "flat" parabola,

$$y_f = f\left[1 - \left[\frac{x}{a}\right]^2\right] \tag{5.38}$$

$$y_f' = -2fx/a^2 \tag{5.39}$$

As y_f is an even function, $\alpha_i = 0$ (cf. (5.36)) and (5.37) evaluates to

$$\frac{\gamma_i}{U} = 8\frac{f}{c}\sqrt{1-\left[\frac{x}{a}\right]^2} \tag{5.40}$$

and hence γ_i is distributed elliptically and is proportional to the camber ratio, f/c, c being the total chord.

The *ideal lift coefficient*, C_{L_i}, at this design or ideal condition is found to be

$$C_{L_i} = 4\pi\frac{f}{c} \tag{5.41}$$

Thus for a typical design lift coefficient of 0.20, the camber ratio f/c = 0.0159. Hence a small camber produces a sizeable C_{L_i}. (A camber ratio of $1/4\pi = 0.0796$ produces an ideal lift coefficient of unity).

In an arbitrarily cambered foil operated at any angle α, we may subtract and add α_i in the integrand of (5.35) and obtain the lift coefficient as the following sum

$$C_L = 2\pi(\alpha - \alpha_i) + C_{L_i} \tag{5.42}$$

where C_{L_i} is the ideal, or design, C_L at $\alpha = \alpha_i$. Here the general expression for C_{L_i} is, by integrating (5.37),

$$C_{L_i} = -4\frac{f}{c}\int_{-1}^{1}\frac{\check{x}'\check{y}_f\, d\check{x}'}{\sqrt{1-\check{x}'^2}} \tag{5.43}$$

An angle of interest is the *angle of zero lift* $\alpha = \alpha_0$ which from (5.42) by setting $C_L = 0$, gives

$$\alpha_0 = \alpha_i - \frac{C_{L_i}}{2\pi} \tag{5.44}$$

or solving for α_i

$$\alpha_i = \alpha_0 + \frac{C_{L_i}}{2\pi} \tag{5.45}$$

Hence one can express (5.42) by eliminating α_i in favor of α_0 to obtain

$$C_L = 2\pi(\alpha - \alpha_0) \tag{5.46}$$

By using (5.44), (5.36) and (5.43) the angle of zero lift can be calculated from

$$\alpha_0 = -\frac{2}{\pi}\frac{f}{c}\int_{-1}^{1}\sqrt{\frac{1-\check{x}'}{1+\check{x}'}}\,\check{y}_f'\,d\check{x}' \tag{5.47}$$

It is seen that the angle of zero lift is largely dependent on the slope of the camber in the region of the trailing edge where the weighting factor $\sqrt{(1-\check{x}')/(1+\check{x}')}$ becomes square-root infinite at $\check{x}' = -1$.

As α_0 is negative the graph of the lift coefficient against angle of attack α is of the form of Figure 5.6 for various camber ratios.

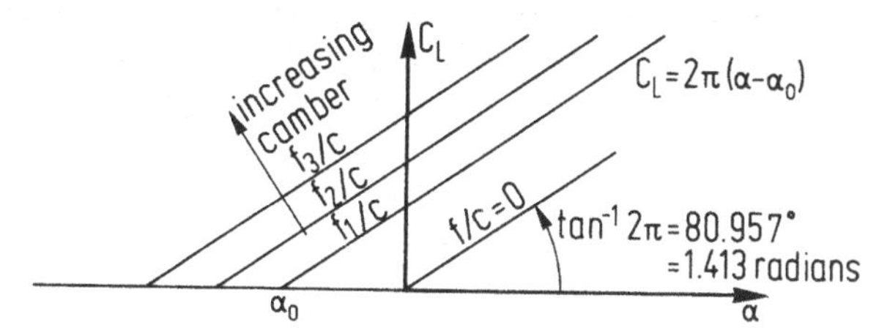

Figure 5.6 Lift coefficient C_L against angle of attack α.

It is important to note that the rate of change of lift coefficient with angle of attack α for all sections, cambered or non-cambered, is

$$\frac{dC_L}{d\alpha} = 2\pi \tag{5.48}$$

The variation of C_L with α becomes non-linear and decreases sharply at angles of the order 8–16 degrees because of real fluid effects. Comparison of (5.48) with experimental results and some discussion of the effect of real fluid properties will be given in Chapter 7.

The pressure distribution on cambered sections can be obtained from (5.35) by recalling that

$$C_{p+} = \frac{2u_+}{U} = -\frac{2}{U}\frac{\gamma}{2} = -\frac{\gamma}{U} \tag{5.49}$$

on the upper or suction side and the negative of this on the lower or pressure side. By subtracting and adding α_i within (5.35) we can secure (after some manipulation),

$$C_{p+} = -2(\alpha - \alpha_i)\sqrt{\frac{a+x}{a-x}} - \frac{2}{\pi}\sqrt{a^2 - x^2}\int_{-a}^{a}\frac{y_f'\,dx'}{\sqrt{a^2 - x'^2}\,(x - x')} \tag{5.50}$$

We see that at and near the leading edge, $x = a = c/2$, this expression gives totally unrealistic results arising from the square-root singularity there.

This flaw can be corrected as shown by Lighthill (1951), who carried out an analysis to second order in powers of a small parameter ε, i.e., the camber, thickness and angle of attack are taken to be of order ε. We shall be content to take only the first order correction from Lighthill, who found the *surface* velocity, q, (as shown in Figure 5.7) in our notation and coordinate system to be

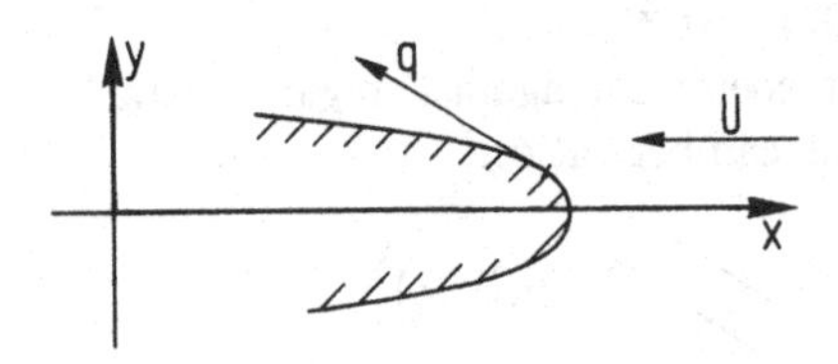

Figure 5.7 Surface velocity q.

$$\frac{q}{U} = \sqrt{\frac{a - x}{a - x + \left[\frac{r_l}{2}\right]}} \left\{ -1 \mp \sqrt{\frac{a + x}{a - x}} (\alpha - \alpha_i) \right.$$

$$\left. \mp \frac{\sqrt{a^2 - x^2}}{\pi} \int_{-a}^{a} \frac{y_f' \, dx'}{\sqrt{a^2 - x'^2}\,(x - x')} - \frac{1}{\pi} \int_{-a}^{a} \frac{T'}{x - x'} dx' \right\} \tag{5.51}$$

where r_l is the leading-edge radius of curvature and the upper signs apply to the upper (suction) side and the lower signs to the lower (pressure) side. The last term is of course the contribution due to thickness.

In terms of the resultant surface velocity, q, the pressure coefficient is exactly

$$C_P = 1 - \left[\frac{q}{U}\right]^2 \tag{5.52}$$

The minimum pressure for even small angular excursions $\alpha - \alpha_i$ will be dominated by the angle-of-attack term in (5.51) and the location of the minimum will be very nearly at the leading edge. To calculate the minimum C_P we can differentiate (5.52) with respect to x and set the derivative to zero.

$$\frac{\partial C_P}{\partial x} = -\frac{2}{U^2} q \frac{\partial q}{\partial x} = 0$$

thus
$$\frac{\partial q}{\partial x} = 0 \tag{5.53}$$

For this purpose we may take only the first two terms in (5.51), i.e., for $\alpha - \alpha_i \neq 0$ and taking only the upper side

$$\frac{q}{U} \approx -\left\{\frac{\sqrt{a - x} + \sqrt{a + x}\,(\alpha - \alpha_i)}{\sqrt{a - x + \frac{r_l}{2}}}\right\} \tag{5.54}$$

Then

$$\frac{1}{U}\frac{\partial q}{\partial x} = 0$$

yields (after some algebraic manipulations) the following expression for the value of x say x_m at which the extreme value occurs

$$\frac{x_m}{a} = \frac{1 - \dfrac{r_l^2}{(4a + r_l)^2(\alpha - \alpha_i)^2}}{1 + \dfrac{r_l^2}{(4a + r_l)^2(\alpha - \alpha_i)^2}} \tag{5.55}$$

This can be simplified by expanding the denominator via the binomial theorem provided that

$$\frac{r_l^2}{(4a + r_l)^2(\alpha - \alpha_i)^2} \approx 0.10 << 1.0 \tag{5.56}$$

Then

$$\frac{x_m}{a} = 1 - \frac{2r_l^2}{(4a + r_l)^2(\alpha - \alpha_i)^2} + O\left[\frac{r_l}{4a(\alpha - \alpha_i)}\right]^3 \tag{5.57}$$

and following Lighthill we can replace $4a + r_l$ by $4a$ (since r_l is of the order of $0.01(2a)$), giving

$$\frac{x_m}{a} = 1 - \frac{1}{2(\alpha - \alpha_i)^2}\left[\frac{r_l}{2a}\right]^2$$

$$a - x_m = \frac{a}{2(\alpha - \alpha_i)^2}\left[\frac{r_l}{2a}\right]^2 \tag{5.58}$$

If in (5.54) we set $x = x_m$ and write $a + x_m \approx 2a$ then

$$\frac{q_{max}}{U} = -\frac{1 + \dfrac{\sqrt{2a}\,(\alpha - \alpha_i)}{\sqrt{a - x_m}}}{\sqrt{1 + \dfrac{r_l}{2(a - x_m)}}} \tag{5.59}$$

and using (5.58)

$$\frac{q_{max}}{U} = -\frac{1 + 2\dfrac{c}{r_l}(\alpha - \alpha_i)^2}{\sqrt{1 + 2\dfrac{c}{r_l}(\alpha - \alpha_i)^2}} = -\sqrt{1 + 2\frac{c}{r_l}(\alpha - \alpha_i)^2} \tag{5.60}$$

r_l being the radius of curvature of the nose and c the chord. (This result agrees with Equation (47) of Lighthill (1951) when note is taken there that $1/4F_0^2 = r_l/2$). The minimum pressure is then approximately, using (5.60) in (5.52)

$$C_{pmin} \approx 1 - \left[\frac{q_{max}}{U}\right]^2 = -2\frac{c}{r_l}(\alpha - \alpha_i)^2 \tag{5.61}$$

As many thickness distributions have dimensionless radii of curvature proportional to $(t/c)^2$, i.e.

$$\frac{r_l}{c} = k\left[\frac{t}{c}\right]^2 \tag{5.62}$$

then (5.61) can be written

$$C_{pmin} = -\frac{2}{k}\frac{(\alpha - \alpha_i)^2}{(t/c)^2} \approx -4\frac{(\alpha - \alpha_i)^2}{(t/c)^2} \tag{5.63}$$

as k is nearly 1/2. (For an ellipse k can be found to be 1/2).

These simple formulae show that the minimum pressure varies with the angle excursion, $(\alpha - \alpha_i)^2$, and directly as the dimensionless nose curvature or inversely as the square of the thickness-chord ratio. Clearly the smaller the curvature (or the larger the nose radius) the more tolerant will be the section to excursions in the angle from the ideal angle.

This result requires from (5.56) that the angular excursion be such that

$$\alpha - \alpha_i >> \frac{r_l}{2c} = \frac{k}{2}\left[\frac{t}{c}\right]^2 \tag{5.64}$$

and as k is of the order of 1/2 and the thickness ratio $t/c = O(1/10)$

$$\alpha - \alpha_i >> \frac{1}{400} \text{ radians (0.143 degrees)}$$

Hence, the foregoing applies for $\alpha - \alpha_i \gtrsim 1.5°$. (For $t/c = 0.05$, $(\alpha - \alpha_i)$ should be ≥ 0.38 degrees).

In general, the modified pressure coefficient is (cf. (5.51))

$$C_p = 1 - \frac{(a - x)}{a - x + \frac{r_l}{2}} \left\{ -1 \mp \sqrt{\frac{a + x}{a - x}} (\alpha - \alpha_i) \mp \frac{\sqrt{(a^2 - x^2)}}{\pi} \int_{-a}^{a} \frac{y_f' \, dx'}{\sqrt{(a^2 - x'^2)} \, (x - x')} - \frac{1}{\pi} \int_{-a}^{a} \frac{\tau'}{x - x'} dx' \right\}^2 \tag{5.65}$$

where terms are to be retained of consistent order. Here we can see that $C_p(a)$ (at the leading edge) is 1 as in exact calculations and that as the distance from the leading edge, $a - x$, is increased (decreasing x) so that as $a - x$ becomes large with respect to $r_l/2$, i.e., for distances of the order of only $10(0.0025\ c) = 0.025\ c$, the result returns to that from first order theory by retaining only the first two terms in $\{\ \}^2$ in (5.65) or

$$C_p \approx \mp 2 \sqrt{\frac{a + x}{a - x}} (\alpha - \alpha_i) \mp \frac{\sqrt{a^2 - x^2}}{\pi} \int_{-a}^{a} \frac{y_f' \, dx'}{\sqrt{a^2 - x'^2} \, (x - x')} - \frac{2}{\pi} \int_{-a}^{a} \frac{\tau' \, dx'}{(x - x')} \tag{5.66}$$

and in the central portions ($x \to 0$) this simplifies to

$$C_p = \mp 2 \sqrt{\frac{a + x}{a - x}} (\alpha - \alpha_i) \mp \left[\frac{\partial C_p}{\partial C_{L_i}}\right] C_{L_i} - A \frac{t}{c} \tag{5.67}$$

where the coefficient $\partial C_p/\partial C_{L_i}$ is a functional of the camber distribution (≈ 0.56 as will be seen[7]) and A is a functional of the thickness distribution (as has been shown, p. 60).

To ascertain the usefulness of Lighthill's minimum pressure coefficient as given by (5.61) we compare it with exact (computer evaluated) minima provided as, for example, by Brockett (1966). There we have envelope curves for various sections from which we choose some of those pertaining to modified NACA-66 sections and a BuShips section. In the NACA-66 thickness family $r_l/c = 0.448\,(t/c)^2$, then Lighthill's minimum becomes

$$C_{pmin} = -\,0.001358\,\frac{(\alpha - \alpha_i)^2}{(t/c)^2} \tag{5.68}$$

where $\alpha - \alpha_i$ is *now in degrees.* Values obtained from (5.68) are shown in Figure 5.8 for the symmetrical section for which $\alpha_i \equiv 0$. Here we see that Lighthill's values from (5.68) are less negative than those from the exact calculations and that the discrepancies increase with increasing thickness ratio. This is owing to the neglect of the contribution of the distributed thickness. We may expect that a lower bound (most negative) could be achieved by adding the linearized C_{pmin} applicable at (or near) to the maximum thickness as deduced in Chapter 4, i.e., $-At/c$ where values of A are given in Table 4.1 (p. 62) for various thickness families. With this *arbitrary* addition, noting that A = 2.42 for the NACA-66 series, (5.68) may be replaced by

$$C_{pmin} \approx -\,0.001358\,\frac{(\alpha - \alpha_i)^2}{(t/c)^2} - 2.42\left[\frac{t}{c}\right]\ ;\ \text{(For NACA-66)} \tag{5.69}$$

Comparison with the exact minimum C_p envelopes in Figure 5.8 shows amazingly close fits (like a glove!) for $0.02 \lesssim t/c \lesssim 0.10$, particularly for

$$\alpha - \alpha_i \geq 10\,\frac{r_l}{2c} = 5k\left[\frac{t}{c}\right]^2 = 5(0.448)\left[\frac{t}{c}\right]^2 = 2.24\left[\frac{t}{c}\right]^2 \text{radians}$$
$$= 128\,(t/c)^2 \text{ degrees} \tag{5.70}$$

for which the Lighthill part is valid. These values are shown on the curves. The largest fractional discrepancies between values from the approximation (5.69) and the exact values are seen to be in the vicinity of the "bilges" of the "buckets" or the "knees" of the curves. In that region (5.69) gives conservative values, i.e., more negative than the exact curves.

---

[7] The contribution due to ideal lift coefficient, C_{L_i}, is explained in Chapter 6.

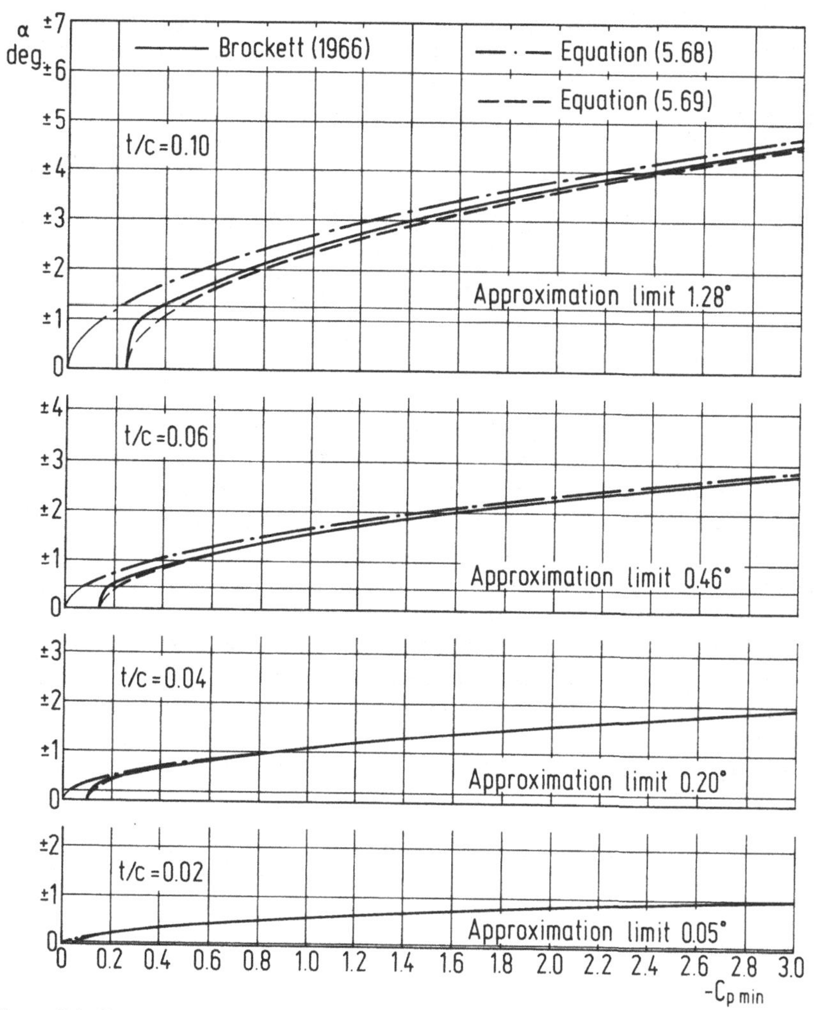

Figure 5.8 Comparison of minimum pressure envelopes for NACA-66 sections (TMB Modified Nose and Tail) with zero camber, Brockett (1966) with Lighthill's C_{pmin}, Equation (5.68) and a C_{pmin} estimated by Equation (5.69).

Note: Values for $\alpha > 0$ include the Lighthill term which requires, in principle, that α is greater than the indicated levels. Values for α less than these levels are nonetheless seen to be generally good.

(Symmetrical profiles: $\alpha_i = 0$)

Present results added to those of Brockett, from Brockett, T. (1966). *Minimum Pressure Envelopes for Modified NACA-66 Sections with NACA a = 0.8 Camber and BuShips Type I and Type II Sections*, Hydromechanics Laboratory, Research and Development Report 1790. Washington, D.C.: David Taylor Model Basin — Department of the Navy. By courtesy of David Taylor Research Center, USA.

A further check of our recipe (the concoction of (5.69) is a bit of cooking!) is made to another section for which computer-generated envelopes of C_{pmin} are provided in Brockett (1966). There the dimensionless nose-radius of curvature of the BuShips Type I-Section is given as

$$r_l/c = 0.489\,(t/c)^2 \tag{5.71}$$

Then using the value of A = 2.50 from Table 4.1 (for the U.S. Navy Standard strut section which has the same C_{pmin} as BuShips Type I) we obtain the following approximation for the BuShips Type I-Section

$$C_{pmin} = -\,0.001246\,\frac{(\alpha - \alpha_i)^2}{(t/c)^2} - 2.50\left[\frac{t}{c}\right] \tag{5.72}$$

The fit of this relation with the envelopes for this section is displayed in Figure 5.9. Here we see excellent agreement for t/c = 0.02 and 0.04. For larger thickness ratios the approximation is seen to be slightly non-conservative i.e. values which are less negative. However, this is not significant, especially when we recall from Chapter 4 that three-dimensional effects are such as to give significant reductions in C_{pmin}.

Although additional correlations with exact two-dimensional envelopes for symmetrical sections are necessary, such will not be attempted at this juncture, as they provide suitable exercises for interested readers!

Comparisons with envelopes for cambered sections are displayed in the next chapter. We may now turn to the design of sections to provide prescribed pressure distributions and lift coefficients.

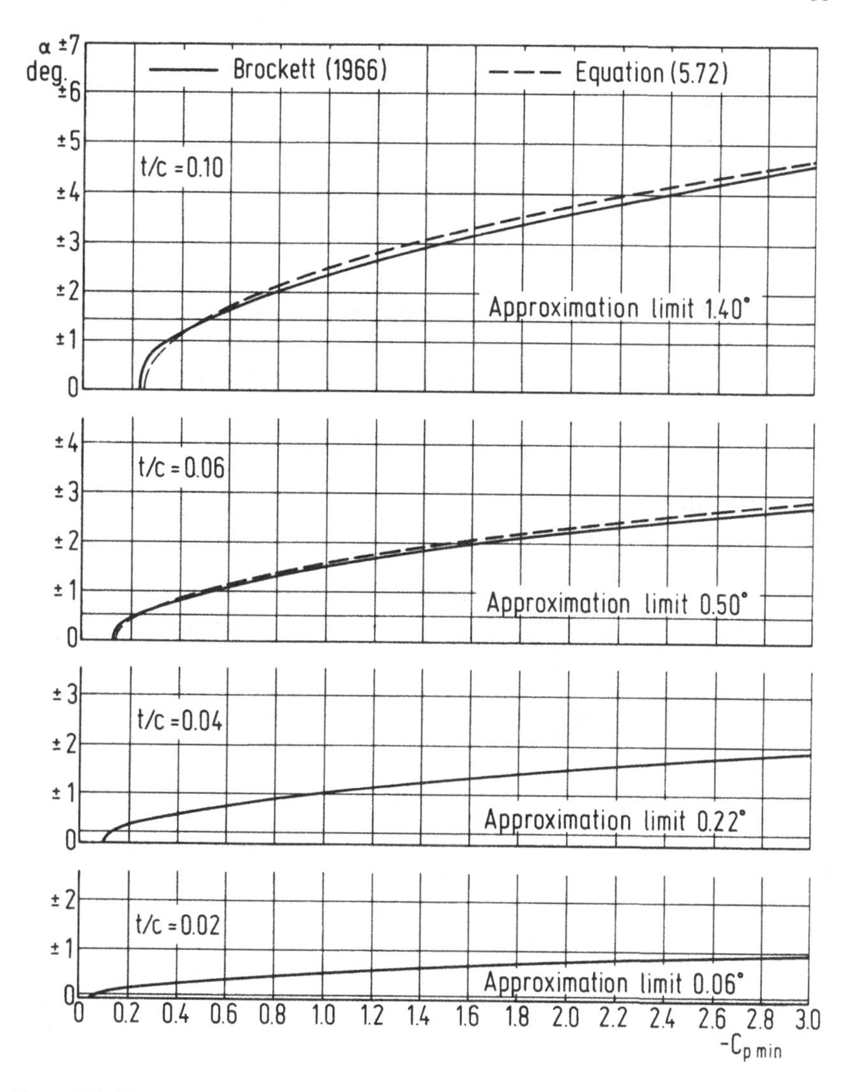

Figure 5.9 Comparison of minimum pressure envelopes for the BuShips Type I and Type II Sections with zero camber, Brockett (1966) with Equation (5.72).

Note: Values for $\alpha > 0$ include the Lighthill term which requires, in principle,that α is greater than the indicated levels. Values for α less than these levels are nonetheless seen to be generally good.

(Symmetrical profiles: $\alpha_i = 0$)

Present results added to those of Brockett, from Brockett, T. (1966). *Minimum Pressure Envelopes for Modified NACA-66 Sections with NACA a = 0.8 Camber and BuShips Type I and Type II Sections*, Hydromechanics Laboratory, Research and Development Report 1790. Washington, D.C.: David Taylor Model Basin — Department of the Navy. By courtesy of David Taylor Research Center, USA.

6 Design of Hydrofoil Sections

Criteria for the design of blade sections may be selected to include:

i. Minimum thickness and chord to meet strength requirements;

ii. Sufficient camber to generate the design lift;

iii. Distribution of thickness and camber to yield the least negative pressure coefficient to avoid or mitigate cavitation;

iv. Thickness- and loading-pressure distributions to avoid boundary layer separation with least chord to yield minimum drag consistent with requirements i. and iii.;

v. Leading and trailing edges to satisfy strength and manufacturing requirements.

The first part of this chapter follows from linearized theories developed by aerodynamicists more than 50 years ago, placing emphasis on the use of existing camber and thickness distributions yielding least negative minimum pressure coefficients, $C_{p_{min}}$ at ideal angle of attack. At non-ideal angles (which always occur in operation in the spatially and temporally varying hull wake flows) we are required to seek sections having greatest tolerance to angle deviations and at the same time having negative minimum pressure coefficients exceeding the level that indicates occurence of cavitation. This tolerance depends critically upon the leading edge radius and the forebody shape as well as upon the extent of the flat part of the pressure distribution. Thus we are led to the more recent findings of researchers who have developed profiles having greater tolerance to angle of attack. When cavitation is unavoidable the latest approach is to use blunter leading edges to generate shorter, more stable cavities thereby avoiding "cloud" cavitation which causes highly deleterious erosion or pitting of the blades. The older procedures are treated next under the heading of Applications of Linearized Theory and the modern developments are described in the section entitled Application of Non-Linear Theory, p. 103 and sequel.

Since propeller blade sections operate in real water, cavitation and friction must be taken into account in an actual design. The authors have chosen, however, to highlight the application of the theories developed in the preceding chapters without letting the real fluid effects obscure the development. Such effects are pursued later in Chapter 7 (Real Fluid Effects) and Chapter 8 (Cavitation), but are dealt with briefly in the present chapter in the form of comments in the text. This applies in particular to the section on Application of Non-Linear Theory since such applications have been devised to include more real fluid effects to provide superior designs. Readers who find this order awkward may prefer to read Chapters 7 and 8 prior to reading this chapter.

APPLICATION OF LINEARIZED THEORY

The Camber Design Problem

The Design Problem: To find the shape of the camber curve or mean line which yields a required lift coefficient when the distribution of pressure jump is prescribed.

We have seen that from the linearized Bernoulli equation (cf. Equation (1.26))

$$C_p = 2\frac{u}{U} \tag{6.1}$$

and that $u_+ = -\gamma/2$; $u_- = \gamma/2$. Hence the jump in pressure coefficient across a loaded curve is

$$\Delta C_p = 2\frac{\gamma}{U} \tag{6.2}$$

We have also seen that for vorticity distributions along the x-axis, we have the first order boundary condition on γ given by (cf. Equation (5.12))

$$\frac{\partial\phi}{\partial y} = \frac{1}{2\pi}\int_{-a}^{a}\frac{\gamma(x')}{x - x'}\,dx' = -U(\alpha + y_f') \; ; \; |x| \leq a \tag{6.3}$$

Eliminating the vorticity density γ between (6.2) and (6.3) yields

$$\alpha + y_f'(x) = -\frac{1}{4\pi}\int_{-a}^{a}\frac{\Delta C_p(x')}{x - x'}\,dx' \tag{6.4}$$

If we now specify $\Delta C_p(x')$ as a function of x', we see that (6.4) is a first order linear differential equation for the camber distribution y_f. As the integral operator is linear we can treat α and y_f' separately and write

$$\int_{-a}^{x} dy_f = -\frac{1}{4\pi}\int_{-a}^{x} dx'' ⨍_{-a}^{a} \frac{\Delta C_{pf}}{x'' - x'} dx' \tag{6.5}$$

where ΔC_{pf} is the loading associated only with camber.

We now wish to solve for y_f taking a $\Delta C_p(x')$ such that a specified lift coefficient (due only to camber) is met, i.e.

$$\frac{1}{c}\int_{-a}^{a} \Delta C_{pf} dx = C_{L_i} = \frac{1}{2}\int_{-1}^{1} \Delta C_{pf}(\check{x})\, d\check{x} \tag{6.6}$$

where $\check{x} = x/a$, $a = c/2$.

The simplest ΔC_p is independent of $\check{x}$ and we have from the constraint (6.6)

$$\Delta C_{pf} = C_{L_i} \tag{6.7}$$

and carrying out the x' integration in (6.5) with this pressure coefficient gives

$$y_f(x) - y_f(-a) = -\frac{C_{L_i}}{4\pi}\int_{-a}^{x} [\ln|x'' - a| - \ln|x'' + a|]\, dx''$$

$$y_f(\check{x}) - y_f(-1) = \frac{aC_{L_i}}{4\pi}\Big[(1-\check{x})\ln(1-\check{x}) + (1+\check{x})\ln(1+\check{x}) - 2\ln 2\Big] \tag{6.8}$$

We see that for $\check{x} = 1$ and $\check{x} = -1$

$$y_f(1) - y_f(-1) = 0$$

$$y_f(-1) - y_f(-1) = 0$$

Hence we meet requirements that $y_f = 0$ at $x = \pm a$ and the equation of the camber line which will yield the specified C_{L_i} is

$$\frac{y_f(\check{x})}{c} = \frac{C_{L_i}}{8\pi}\Big[(1 - \check{x})\ln(1 - \check{x}) + (1 + \check{x})\ln(1 + \check{x}) - 2\ln 2\Big] \tag{6.9}$$

This is called the *logarithmic camber line.*

We may note at once that $y_f(x) = y_f(-x)$, i.e., the camber distribution is symmetric or an even function about $x = 0$. Hence y_f' is odd and it follows from Equation (5.36) that the ideal angle α_i is zero. The pressure distribution is shown in Figure 6.1.

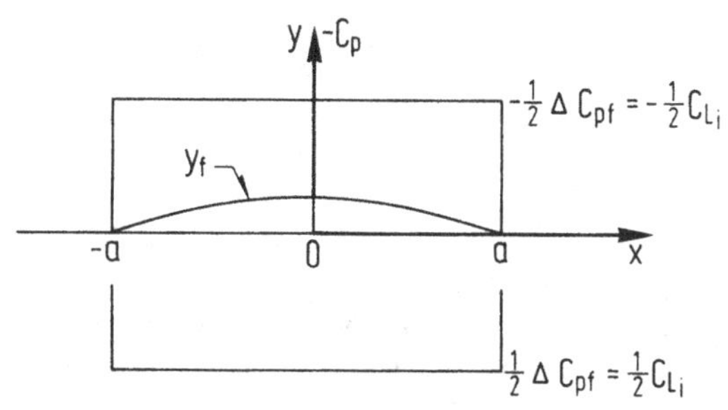

Figure 6.1 Pressure distribution for the logarithmic camber line (rectangular loading).

This is the best possible loading distribution since it meets the required lift coefficient with the least minimum pressure on the suction or upper side. For example for an elliptical distribution $\Delta C_P = \Delta C_{Pmin}\sqrt{1-(x/a)^2}$ yields, upon substitution in the constraint equation (6.6), a minimum pressure of

$$C_{Pmin} = -\frac{2}{\pi}C_{L_i} = -0.64\ C_{L_i} \tag{6.10}$$

which is 28 per cent more negative than for the rectangular loading.

However, the rectangular loading has an infinite adverse pressure gradient at the trailing edge which results in separation and high drag as has been shown by experiment. It was also found by Loftin (1948), that the experimental angle of zero lift of this camber was only 74 per cent of that given by theory resulting in $\partial C_P/\partial C_{L_i} = 0.67$[8] in lieu of 0.50. He also noted that the camber curve gave rise to cusped trailing edges which were structurally impracticable. Hence one must abandon this loading and camber distribution, adopting one which has a sufficiently gentle adverse pressure gradient in the region of the trailing edge. Highly useful distributions are the so-called "roof-tops" developed at NACA–Langley, Va., USA (now NASA) which have the general shape shown in Figure 6.2.

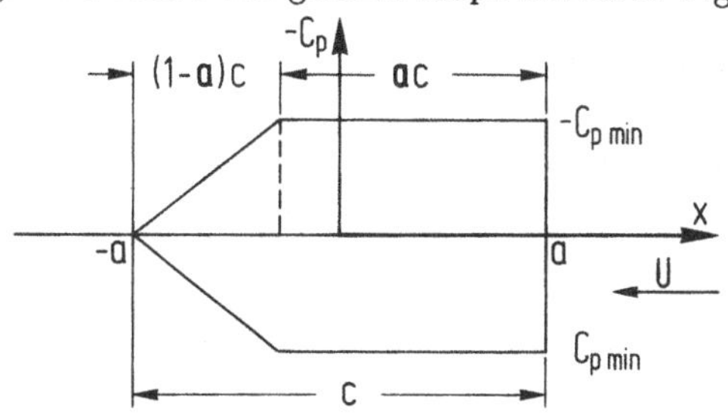

Figure 6.2 General shape of the roof-top pressure distribution.

[8] in order to overcome the deficiency in the angle of zero lift and achieve the design lift coefficient.

The minimum pressure is, by a simple calculation,

$$C_{Pmin} = -\frac{C_{Li}}{1 + \mathbf{a}} \tag{6.11}$$

where $\mathbf{a}$ is the fraction of the total chord c over which the pressure is uniform (over $c(1 - \mathbf{a})$ the pressure varies linearly to zero). Thus the value for the rectangular loading is of course recovered for $\mathbf{a} = 1.0$. Experiments with families of airfoils have shown that excellent lift and drag performance is obtained for $\mathbf{a} = 0.80$ (NACA notation for $\mathbf{a}$ is a), see Abbott & von Doenhoff, (1959). A deficiency in the simple $\mathbf{a} = 0.80$ mean line is that it was found to provide trailing edges which were still too thin for practical use when used with thickness distributions. Loftin, (1948) designed a modification of the descending portion of the pressure distribution which gave thicker trailing edges and showed experimental angles of zero lift somewhat greater (more negative) than the theory. This is designated the NACA $\mathbf{a} = 0.80$ (modified) camber or mean line. It is shown together with other data in Figure 6.4. This loading and associated camber have been widely adopted by aeronautical engineers and later by naval architects and propeller designers. Thus for $\mathbf{a} = 0.80$ (modified)

$$C_{Pmin} = -C_{L_i}/1.8 = -0.556\, C_{L_i} \approx -0.56\, C_{L_i} \tag{6.12}$$

The camber line can be calculated through (6.5) using

$$\Delta C_P(x) = \begin{cases} \dfrac{2C_{Li}}{1 + \mathbf{a}} & ;\ a(1 - 2\mathbf{a}) \le x \le a \\[2ex] \dfrac{(a + x)}{a(1 - \mathbf{a}^2)}\, C_{Li} & ;\ -a \le x \le a(1 - 2\mathbf{a}) \end{cases} \qquad 0 \le \mathbf{a} \le 1 \tag{6.13}$$

While the calculation is straightforward it is algebraically cumbersome. The result in Abbott & von Doenhoff, (1959) (p.74) when converted to our notation and axis with origin at mid-chord is

$$\frac{y_f}{c} = \frac{C_{Li}}{2\pi(1 + \mathbf{a})} \left\{ \frac{1}{1 - \mathbf{a}} \left[\frac{1}{2}\left[\frac{2\mathbf{a} - 1 + \tilde{x}}{2}\right]^2 \ln\frac{|2\mathbf{a} - 1 + \tilde{x}|}{2} \right.\right.$$
$$\left. - \frac{1}{2}\left[\frac{1 + \tilde{x}}{2}\right]^2 \ln\frac{1 + \tilde{x}}{2} + \left[\frac{1 + \tilde{x}}{4}\right]^2 - \left[\frac{2\mathbf{a} - 1 + \tilde{x}}{4}\right]^2 \right]$$
$$\left. - \left[\frac{1 - \tilde{x}}{2}\right] \ln\frac{1 - \tilde{x}}{2} + g - h\left[\frac{1 - \tilde{x}}{2}\right] \right\} \tag{6.14}$$

where $\check{x} = x/a$; $a = c/2$, ($\check{x} = 1$, leading edge; $\check{x} = -1$, trailing edge) and

$$g = \frac{-1}{(1 - \mathbf{a})}\left[\mathbf{a}^2\left[\frac{1}{2}\ln \mathbf{a} - \frac{1}{4}\right] + \frac{1}{4}\right]$$

$$h = \frac{-1}{(1 - \mathbf{a})}\left[\frac{1}{2}(1 - \mathbf{a})^2 \ln(1 - \mathbf{a}) - \frac{1}{4}(1 - \mathbf{a})^2\right] + g$$

The ideal angle for these mean lines is

$$\alpha_i = -\frac{h}{2\pi(1 + \mathbf{a})} C_{L_i} \qquad (6.15)^9$$

The equation for the camber curve is indeed cumbersome. It is not possible to write down an analytical expression for the camber $f = y_{f\max}$ since the calculation of the position of the maximum from $y_f' = 0$ involves a transcendental equation. This is best left to a computation of the camber curve via a computer for various **a**. It can be seen that $y_f(\pm 1) = 0$. On Figure 6.4 the ordinates of the camber lines for $\mathbf{a} = 0.80$ and $\mathbf{a} = 0.80$ (modified) are given together with a graphical presentation from which at 0.52 chord from the leading edge

$$\frac{y_{f\max}}{c} = \frac{f}{c} = \begin{cases} 0.067\ C_{L_i} & (\mathbf{a} = 0.80 \text{ modified}) \\ 0.068\ C_{L_i} & (\mathbf{a} = 0.80) \end{cases} \qquad (6.16)$$

Figure 6.3 displays the camber curves for the family of roof-top pressure distributions for an entire range of the parameter **a**.

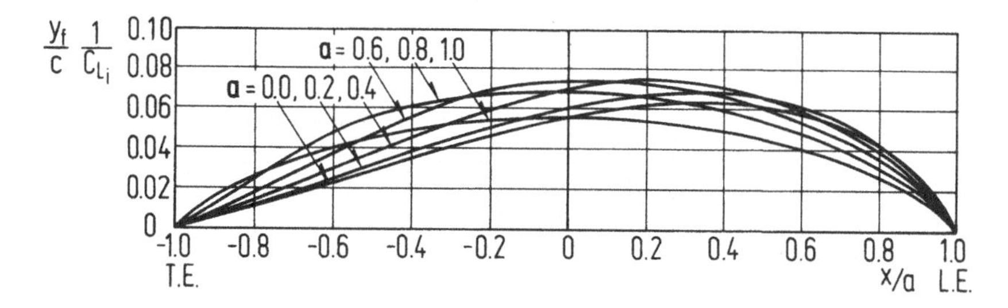

Figure 6.3 Family of camber curves for roof-top pressure distributions.

[9] There seems to be an error in sign in this formula as given on p. 74 of Abbott & von Doenhoff, (1959).

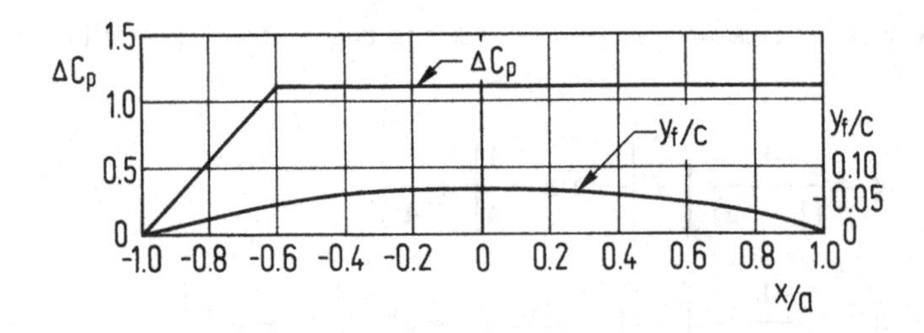

$C_{L_i} = 1.0$	$\alpha_i = 1.54^\circ$	$C_{M_{c/4}} = -0.202$		
x (per cent c)	y_f (per cent c)	dy_f/dx	ΔC_p	$q/U = \Delta C_p/4$
50	0			
49.5	0.287	−0.48535		
49.25	0.404	−0.44925		
48.75	0.616	−0.40359		
47.5	1.077	−0.34104		
45.0	1.841	−0.27718		
42.5	2.483	−0.23868		
40	3.043	−0.21050		
35	3.985	−0.16892		
30	4.748	−0.13734		
25	5.367	−0.11101		
20	5.863	−0.08775	1.111	0.278
15	6.248	−0.06634		
10	6.528	−0.04601		
5	6.709	−0.02613		
0	6.790	−0.00620		
−5	6.770	0.01433		
−10	6.644	0.03611		
−15	6.405	0.06010		
−20	6.037	0.08790		
−25	5.514	0.12311		
−30	4.771	0.18412		
−35	3.683	0.23921	0.833	0.208
−40	2.435	0.25583	0.556	0.139
−45	1.163	0.24904	0.278	0.069
−50	0	0.20385	0	0

Data for NACA Mean Line **a** = 0.8

x : per cent c from mid–chord. L.E.: x = 50 per cent
T.E.: x = −50 per cent

$$C_{L_i} = \frac{L_i}{\frac{1}{2}\rho U^2 c} \quad ; \quad C_{M_{c/4}} = \frac{M_{c/4}}{\frac{1}{2}\rho U^2 c^2} \quad ;$$

$M_{c/4}$: moment about quarter–chord point (from L.E.)

Figure 6.4a Pressure distribution for NACA mean line (camber curve **a** = 0.80).

From: Abbott, I.H. & von Doenhoff, A.E. (1959). *Theory of Wing Sections*, New York: Dover Publications.
By courtesy of Dover Publications, USA.

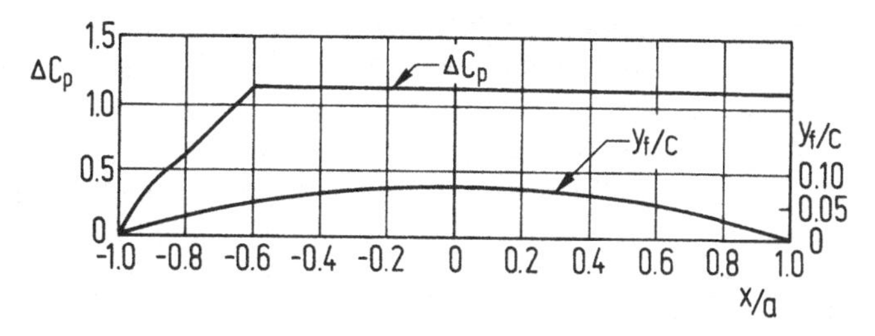

$C_{L_i} = 1.0$	$\alpha_i = 1.40^\circ$		$C_{M_{c/4}} = -0.219$	
x (per cent c)	y_f (per cent c)	dy_f/dx	ΔC_p	$q\_/U = \Delta C_p/4$
50	0			
49.5	0.281	−0.47539		
49.25	0.396	−0.44004		
48.75	0.603	−0.39531		
47.5	1.055	−0.33404	1.092	0.273
45.0	1.803	−0.27149		
42.5	2.432	−0.23378		
40	2.981	−0.20618		
35	3.903	−0.16546		
30	4.651	−0.13452	1.096	0.274
25	5.257	−0.10873		
20	5.742	−0.08595		
15	6.120	−0.06498		
10	6.394	−0.04507	1.100	0.275
5	6.571	−0.02559		
0	6.651	−0.00607		
− 5	6.631	0.01404	1.104	0.276
−10	6.508	0.03537		
−15	6.274	0.05887	1.108	0.277
−20	5.913	0.08610	1.108	0.277
−25	5.401	0.12058	1.112	0.278
−30	4.673	0.18034	1.112	0.278
−35	3.607	0.23430	0.840	0.210
−40	2.452	0.24521	0.588	0.147
−45	1.226	0.24521	0.368	0.092
−50	0	0.24521	0	0

Data for NACA Mean Line **a** = 0.8 (modified)

x : per cent c from mid–chord. L.E.: x = 50 per cent
T.E.: x = −50 per cent

$$C_{L_i} = \frac{L_i}{\frac{1}{2}\rho U^2 c} \quad ; \quad C_{M_{c/4}} = \frac{M_{c/4}}{\frac{1}{2}\rho U^2 c^2};$$

$M_{c/4}$: moment about quarter–chord point (from L.E.)

Figure 6.4b Pressure distribution for NACA mean line (camber curve **a** = 0.80 (modified)) in which the descending pressure is modified as compared to the linear descent of the simple **a** = 0.80 distribution.

From: Abbott, I.H. & von Doenhoff, A.E. (1959). *Theory of Wing Sections*, New York: Dover Publications.
By courtesy of Dover Publications, USA.

Addition of Thickness and Camber

The design or selection of the thickness distribution obviously cannot be made solely from hydrodynamic considerations as indicated by the first criterion i., p. 86. For conventional propellers beam bending stress theory forms the basis for ship classification society rules (as applied by Det Norske Veritas, American Bureau of Shipping, Lloyd's Register of Shipping etc.). The specification of thickness t_r at (or near) the root of the propeller blade then varies as

$$t_r \propto \frac{\text{coefficient}}{\sqrt{c_r}}$$

where c_r is the expanded chord at the root and the coefficient depends on the steady and fluctuating loading, allowable stress, fatigue factors, diameter and number of blades. The thickness at other radii decreases approximately linearly from the root to the tip.

It is clear from this that the chord must be known in order to determine the required thickness. A criterion for minimum expanded-blade-area ratio for avoidance or mitigation of cavitation as a function of thrust loading (given in Chapter 12) may be used together with a selection of blade outline to determine the expanded chord c_r.

It should be noted that the skin friction drag of the sections increases with chord and the viscous pressure drag as $c(t/c)^2$ as may be deduced from Equations (7.4) and (7.5). Hence increase of blade area or chord length is attended by a decrease in propeller efficiency.

Having determined the thickness and chord distribution from the foregoing we can proceed with the design process involving the addition of camber and thickness.

We are now in a position to complete the formulation of the approximate total minimum pressure on thin sections as arising from thickness, camber and angle of attack.

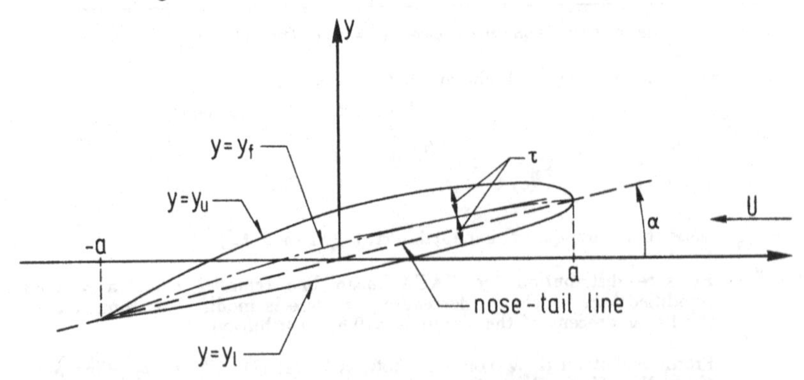

Figure 6.5 Hydrofoil section with thickness, camber and angle of attack.

Let the equation of the upper surface (in two dimensions) be designated by

$$y(x) = y_u(x) \tag{6.17}$$

and that of the lower surface

$$y(x) = y_l(x)$$

Then, by definition, the mean line or camber curve is

$$\frac{y_u + y_l}{2} = y_f(x) \tag{6.18}$$

and the semi-thickness distribution is

$$\frac{y_u - y_l}{2} = \tau(x) \tag{6.19}$$

If $y_f(x)$ and $\tau(x)$ are selected arbitrarily as independent functions then by adding (6.18) and (6.19) we have

$$y_u = y_f + \tau \tag{6.20}$$

and by subtracting

$$y_l = y_f - \tau \tag{6.21}$$

If however, the shape of the lower side, for example, is fixed for some reason then the thickness and the camber are constrained or coupled geometrically. Upon selection of both y_l and thickness, the camber is constrained to be

$$y_f = y_l + \tau \tag{6.22}$$

and

$$y_u = y_l + 2\tau \tag{6.23}$$

An example of this is the semi-ogival sections used in earlier times, where the lower side was taken to be flat, i.e. $y_l \equiv 0$. Clearly then

$$y_f = \tau \tag{6.24}$$

$$y_u = 2\tau \tag{6.25}$$

and the camber and semi-thickness are coupled in a special and obvious way.

Upon inclining the nose-tail line to any small angle α we then have the following simplified, first order kinematic conditions on the source and vorticity densities on the upper and lower sides (Equations (2.7), (2.8) for the thickness and (2.38) for the loading)

$$\frac{m(x)}{2} + \frac{1}{2\pi}\int_{-a}^{a}\frac{\gamma(x')}{x-x'}dx' = -U\frac{d}{dx}(y_u+\alpha x) = -U(y_f'+\tau'+\alpha) \tag{6.26}$$

$$-\frac{m(x)}{2} + \frac{1}{2\pi}\int_{-a}^{a}\frac{\gamma(x')}{x-x'}dx' = -U\frac{d}{dx}(y_l+\alpha x) = -U(y_f'-\tau'+\alpha) \tag{6.27}$$

We can see that the source and vorticity densities are hydrodynamically uncoupled when y_f and τ are independent of each other by adding and then subtracting (6.26) and (6.27) to get

$$\int_{-a}^{a}\frac{\gamma(x')}{x-x'}\,dx' = -2\pi U(y_f' + \alpha) \tag{6.28}$$

and

$$m(x) = -2U\tau' \tag{6.29}$$

which are the relations imposed by the kinematic conditions found earlier when we considered the symmetrical and asymmetrical flows about sections separately (cf. Equations (5.34) and (4.14)). This is not surprising as we have neglected the axial components of the perturbation velocities due to thickness, camber and angle of attack in the kinematic condition which if retained would couple the source and vorticity densities through the inclusion of terms which are of second order except in the neighborhood of the leading and trailing edges. It is important to realize that the boundary or kinematic condition is *linear* in all the velocity components and in the slopes y_f' and τ' [10]. This is not true of the full Bernoulli equation. We must not think of the first order kinematic condition as a linearization since the exact kinematic condition is always linear in u, v, U, y_f' and τ'. It is a first order approximation where terms of order of the square of the thickness and camber ratios and their products are neglected. Indeed, as we have seen, the approximations are not uniformly valid and give rise to unrealistic results at the leading edge (due to loading and thickness) and at the trailing edge (due to thickness).

As the pressure equation has been linearized, to obtain the combined pressures due to loading and thickness we can add their separate contributions to get, at ideal angles of attack, on the *upper side*

[10] Cf. Equations (3.14) and (3.15), p. 43 and 44.

$$C_{p\min} = -\frac{\partial C_p}{\partial(t/c)}\left[\frac{t}{c}\right] - \frac{\partial C_p}{\partial C_{L_i}} C_{L_i} \qquad \begin{matrix} y = 0_+ \\ \alpha = \alpha_i \end{matrix} \tag{6.30}$$

and on the *lower side*

$$C_{p\min} = -\frac{\partial C_p}{\partial(t/c)}\left[\frac{t}{c}\right] + \frac{\partial C_p}{\partial C_{L_i}} C_{L_i} \qquad \begin{matrix} y = 0_- \\ \alpha = \alpha_i \end{matrix} \tag{6.31}$$

where we have seen that the coefficient $A = \partial C_p/\partial(t/c)$ (cf. p. 60) is a function of the shape of the thickness distribution, (A being a minimum for an elliptical section, namely 2.0, cf. Table 4.1) and the coefficient $\partial C_p/\partial C_{L_i}$ is a functional of the shape of the camber or mean line (having the lowest theoretical but not realistic value of 0.50 for **a** = 1.0 and the very useful value of 0.56 for the NACA **a** = 0.8 (modified) mean line, a highly practical camber).

The determination of the coefficient $\partial C_p/\partial C_{L_i}$ as a functional of the camber distribution requires that we return to Equation (5.66) and pick out that part depending upon camber which we can designate as

$$C_{pf} = \mp \frac{2\sqrt{a^2 - x^2}}{\pi} \int_{-a}^{a} \frac{y_f'}{\sqrt{a^2 - x'^2}\,(x - x')}\,dx'$$

Replace x by $a\check{x}$, x' by $a\check{x}'$ and y_f by $f\check{y}_f$. Then

$$C_{pf} = \mp \frac{f}{c} \frac{4\sqrt{1 - \check{x}^2}}{\pi} \int_{-1}^{1} \frac{\check{y}_f'(\check{x}')}{\sqrt{1 - \check{x}'}\,(\check{x} - \check{x}')}\,d\check{x}' \tag{6.32}$$

Here we see that this contribution is proportional to the camber ratio and is a functional of $\check{y}_f'$. We have seen in Equation (5.43) that the lift coefficient is also a functional of the camber slope. Eliminating f/c between (5.43) and (6.32) gives upon differentiating with respect to C_{L_i}

$$\frac{\partial C_p}{\partial C_{L_i}} = \frac{C_{pf}}{C_{L_i}} = \pm \frac{\sqrt{1 - \check{x}^2} \displaystyle\int_{-1}^{1} \frac{\check{y}_f'}{\sqrt{1 - \check{x}'^2}\,(\check{x} - \check{x}')}\,d\check{x}'}{\pi \displaystyle\int_{-1}^{1} \frac{\check{x}\,\check{y}_f'}{\sqrt{1 - \check{x}'^2}}\,d\check{x}'} \tag{6.33}$$

a rather complicated functional of $\check{y}_f'$. Evaluation is to be made at that value of $\check{x}$ for which C_{pf} has an extreme value. For the camber curves de–

rived from roof-top loadings, $\partial C_P/\partial C_{L_i}$ is constant from the leading edge to the point where the "roof" slopes downward. For these cases

$$\frac{\partial C_P}{\partial C_{L_i}} = \mp \frac{1}{1 + a} \tag{6.34}$$

Although one could equally well write the contribution of camber as $[\partial C_P/\partial(f/c)]\,(f/c)$, the design C_{L_i} is usually specified and not the camber. Hence the form used in (6.30) and (6.31) is preferred here.

For operation at angles other than the ideal angle α_i we may again take the Lighthill term, Equation (5.61) plus the linearized thickness and camber contributions (6.30) and (6.31) to obtain an approximation to the minimum pressure envelopes in the form

$$C_{pmin} = -2\frac{c}{r_l}(\alpha-\alpha_i)^2 - \frac{\partial C_P}{\partial(t/c)}\left[\frac{t}{c}\right] \mp \frac{\partial C_P}{\partial C_{L_i}} C_{L_i} \tag{6.35}$$

where the upper sign applies to the upper side (normally the "suction" side) and the lower sign to the lower side (normally the "pressure" side or face) and where as before r_l is the radius of curvature of the nose or leading edge.

As observed in Chapter 5, the leading-edge radius is, as a fraction of the chord, nearly proportional to the square of the thickness ratio, cf. Equation (5.62). For example, in Abbott & von Doenhoff, (1959) the leading edge radius is specified for each basic thickness form as in Figure 6.6 for NACA 65-006. Here we see that $r_l/c = 0.00240$ for a thickness ratio $t/c = 0.06$ (the last two digits in the designation number). This yields

$$k = \frac{0.00240}{(0.06)^2} = 0.667 \tag{6.36}$$

This may be compared to the corresponding value for an ellipse (as stated following Equation (5.63)) which is 0.50.

This indicates that this family has leading edges which are blunter than an ellipse (of equal t/c) and hence the computed curve of $(q/U)^2$ in Figure 6.6 for $C_{L_i} = 0$ is erroneous in the immediate vicinity of the leading edge. This behavior has been indicated in Figures 4.12 and 4.13 and has been rediscovered in Brockett, (1966) as displayed in Figure 6.7.

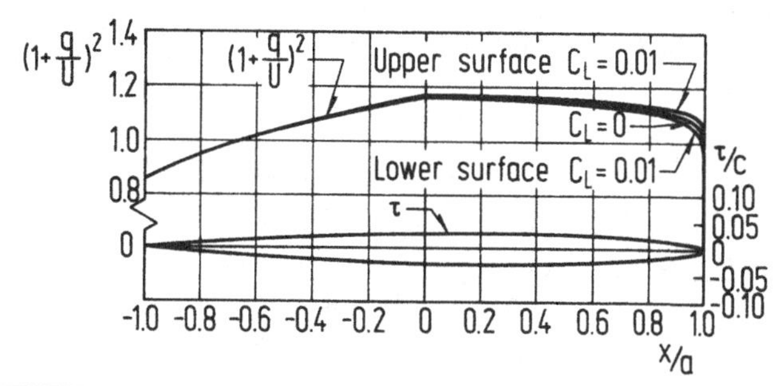

x (per cent c)	τ (per cent c)	$(1+\frac{q_t}{U})^2$	$1+\frac{q_t}{U}$	q_-^α/U
50	0	0	0	4.815
49.5	0.476	1.044	1.022	2.110
49.25	0.574	1.055	1.027	1.780
48.75	0.717	1.063	1.031	1.390
47.5	0.956	1.081	1.040	0.965
45.0	1.310	1.100	1.049	0.695
42.5	1.589	1.112	1.055	0.560
40	1.824	1.120	1.058	0.474
35	2.197	1.134	1.065	0.381
30	2.482	1.143	1.069	0.322
25	2.697	1.149	1.072	0.281
20	2.852	1.155	1.075	0.247
15	2.952	1.159	1.077	0.220
10	2.998	1.163	1.078	0.198
5	2.983	1.166	1.080	0.178
0	2.900	1.165	1.079	0.160
−5	2.741	1.145	1.070	0.144
−10	2.518	1.124	1.060	0.128
−15	2.246	1.100	1.049	0.114
−20	1.935	1.073	1.036	0.100
−25	1.594	1.044	1.022	0.086
−30	1.233	1.013	1.006	0.074
−35	0.865	0.981	0.990	0.060
−40	0.510	0.944	0.972	0.046
−45	0.195	0.902	0.950	0.031
−50	0	0.858	0.926	0
L.E. radius: 0.240 per cent c				

NACA 65–006 Basic Thickness Form

x : per cent c from mid–chord. L.E.: x = 50 per cent
T.E.: x = −50 per cent

q_t : velocity induced by thickness (values are tabulated for t/c = 0.06)

q_-^α : velocity (on lower surface) induced by angle of incidence (flat plate) (values are tabulated for $C_L = 1.0$)

Figure 6.6 Variation of surface velocity along NACA 65–006. For this section: $r_l/c = k(t/c)^2$ with k = 0.667.

From: Abbott, I.H. & von Doenhoff, A.E. (1959). *Theory of Wing Sections*. New York: Dover Publications.
By courtesy of Dover Publications, USA.

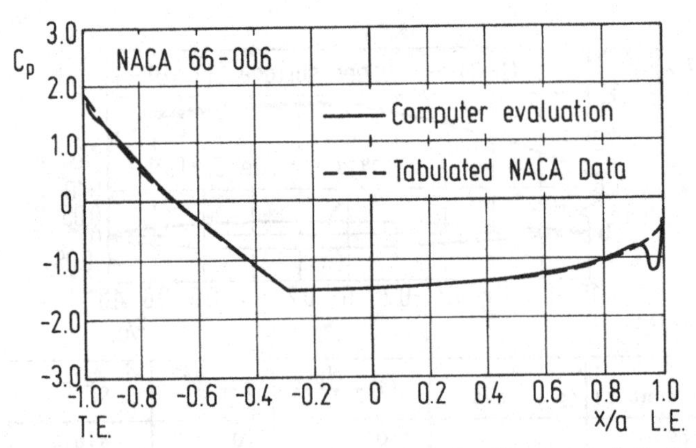

Figure 6.7 Pressure distributions along NACA 66–006 (Symmetrical Section, 6 per cent thick) from tabulation in Abbott & von Doenhoff, (1959), p. 359 and from computer evaluation given in Figure 1, Brockett, (1966)

From Brockett, T. (1966). *Minimum Pressure Envelopes for Modified NACA-66 Sections with NACA a = 0.8 Camber and BuShips Type I and Type II Sections.* Hydromechanics Laboratory, Research and Development Report 1790. Washington, D.C.: David Taylor Model Basin — Department of the Navy.
By courtesy of David Taylor Research Center, USA.

To ascertain the utility of Equation (6.35) evaluations are made for the NACA-66 sections with the **a** = 0.8 camber for the camber ratio of 0.01 for which computer-generated envelopes are given in Figure 6 of Brockett, (1966). In order to determine the C_{L_i} corresponding to a specified camber ratio it is necessary to use (6.16) which yields

$$C_{L_i} = 14.92\ f/c \quad (\text{for } \mathbf{a} = 0.8 \text{ (mod.) camber}) \tag{6.37}$$

As (6.35) requires the value of α_i, we may calculate this from Equation (6.15) or look up the value given in Abbott & von Doenhoff, (1959) as indicated in Figure 6.4b where we see that $\alpha_i = 1.40$ deg. at $C_{L_i} = 1.0$.

At $f/c = 0.01$, $C_{L_i} = 0.149$ (from (6.37)) and $\alpha_i = (1.40)(0.149) = 0.21$ deg. Thus the general formula, evaluated for $f/c = 0.01$ is

$$C_{p\min} \approx -0.001358\,\frac{(\alpha - 0.21)^2}{(t/c)^2} - 2.42\ t/c \mp 0.084 \tag{6.38}$$

Values are shown in Figure 6.8 where it can be seen that this approximation fits the computer generated envelopes closely for both positive and negative angles. It is believed that this type of approximation is useful for applicaton to other sections (when r_l/c, α_i and f/c are specified) although further correlations should be made.

To facilitate evaluations for other sections values of the leading-edge parameter k (in $r_l/c = k(t/c)^2$) are provided in Table 6.1.

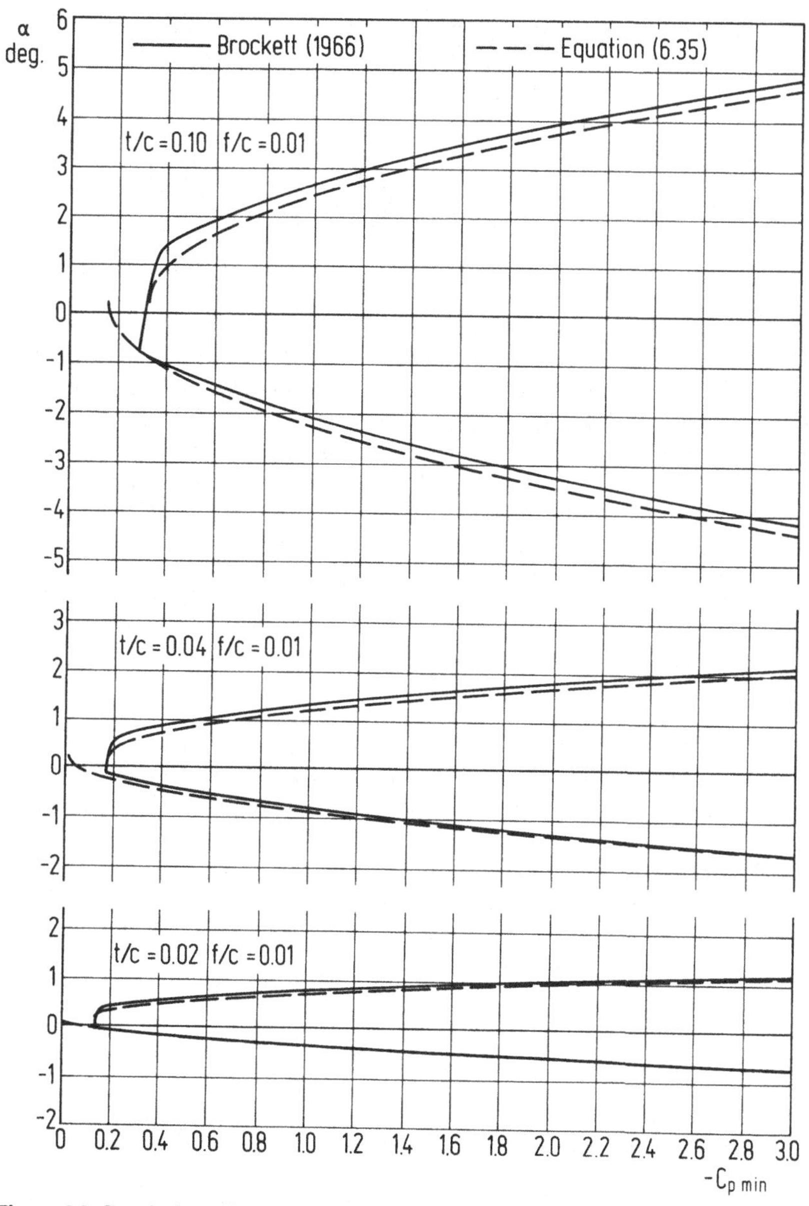

Figure 6.8 Correlation of minimum pressure coefficients from Equation (6.35) with computer-evaluated curves from Brockett, (1966).

Present results added to those of Brockett, from Brockett, T. (1966). *Minimum Pressure Envelopes for Modified NACA-66 Sections with NACA a = 0.8 Camber and BuShips Type I and Type II Sections.* Hydromechanics Laboratory, Research and Development Report 1790. Washington, D.C.: David Taylor Model Basin – Department of the Navy. By courtesy of David Taylor Research Center, USA.

NACA family	k
4-digit	1.10
16-0( )	0.489
63A-	0.742
64A	0.686
65	0.667
66-	0.662
67,1-	0.677

Table 6.1 Values of the leading-edge parameter k ($r_l/c = k(t/c)^2$) for $t/c \lesssim 0.10$.

It is clear that the tolerance of a section to deviations in angle of attack from the ideal angle is greater for larger values of the leading-edge parameter and larger values of the thickness-chord ratio. This is true provided the larger values of leading-edge radius of curvature do not generate pressure minima at $\alpha = \alpha_i$ which are in excess of the value at the maximum thickness (i.e. more negative than $-A(t/c)$). The NACA-16 series appears to be most favorable for which from Table 4.1 yields $A = \partial C_p/(\partial(t/c)) = 2.28$ and from Table 6.1, $k = 0.489$. For this family with the $\mathbf{a} = 0.8$ (mod.) camber

$$C_{pmin} = -\,0.00125\,\frac{(\alpha - \alpha_i)^2}{(t/c)^2} - 2.28\;t/c \mp 0.56\;C_{L_i} \tag{6.39}$$

with $C_{L_i} = 14.92\;(f/c)$ and $\alpha_i = 1.40°C_{L_i}$ where $(\alpha - \alpha_i)$ is in degrees.

Comparison with Equation (6.38) shows that the coefficients of the thickness and angle terms are significantly smaller for the 16-series than the 66-series.

It is important to note that the contributions of thickness and camber terms are nearly equal to one another for practical values of t/c and C_{L_i}. For example, a 5 per cent thick 16-series section at a $C_{L_i} = 0.20$ gives at $\alpha = \alpha_i$

$$\begin{aligned} C_{pmin} &= -(2.28(0.05) + 0.56(0.20)) \\ &= -(0.114 + 0.112) = -0.226 \end{aligned} \tag{6.40}$$

This near equality of thickness and loading contributions arises from the use of thin sections operated at light loading to avoid cavitation at ideal incidence.

The expectation that this parity also holds in certain near field regions led the first author to include the contribution of blade thickness along with those due to loading in the calculation of propeller-induced vibratory pressures in the vicinity of propellers operating in uniform flow circa 1957. Only then did theory and experiment agree!

APPLICATION OF NON-LINEAR THEORY

With the advent of digital computers and the development of versatile design theory, two-dimensional blade sections can now be "tailored" to specific applications. These procedures incorporate the effective camber arising from the three-dimensional inductions over the blade and produce total section shapes (camber and thickness) which have greater tolerance to angle of attack variations encountered in the hull wake.

The design method most commonly used by members of the International Towing Tank Conference (ITTC) as reported by the Propulsor Committee in the Proceedings of the 19th ITTC (1990b) is that developed by Eppler (1960, 1963, 1969). Applications of the Eppler procedure were made to design of sections for hydrofoil craft by Eppler & Shen (1979) (symmetrical sections) and by Shen & Eppler (1981). Wide utilization of Eppler's method is due to the readily available computer programs provided by Eppler & Somers (1980).

This procedure is based on a conformal mapping of the flow about a unit circle in a ζ-plane to generate a profile shape in the physical plane through a complex variable function $Z(\zeta)$. The mapping technique has been explained, for example, by von Kármán & Burgers (1935). Eppler's application permits specification of a velocity distribution over sub-regions along the profile in which the velocity is constant at some angle of attack, α, one for each such sub-region. These constant-velocity regions together with specification of velocity in the pressure-recovery region and in a closure region near the trailing edge provide through the mapping function an initial profile definition. Then a boundary-layer calculation is made to determine the margin (if any) against separation in the pressure-recovery region on the aft end of the suction side. Iterations are performed until acceptable cavitation buckets are determined consistent with no boundary-layer separation in the pressure recovery zone.

Generally at large angle α ($\alpha = \alpha_2$, cf. Figure 6.9) the aft suction side may be determined by a requirement that the forward suction side should have constant pressure (constant velocity) to suppress leading-edge cavitation and similarly at small $\alpha = \alpha_3$ on the pressure side. The adverse pressure gradient effect on the boundary layer in the aft region of the suction side must then be examined against empirical criteria for separation by a complete boundary-layer calculation over the entire profile. After the initial profile design, iterations are typically required to correct non-optimum features.

The sections developed for hydrofoil craft (much larger t/c than for propeller sections in outer radii) by Eppler & Shen (1979) were shown theoretically to have wider buckets of minimum pressure coefficient allowing larger cavitation-free angle-of-attack variation than the NACA-16 and NACA-66(mod.) sections as displayed in Figure 6.10. The shape of

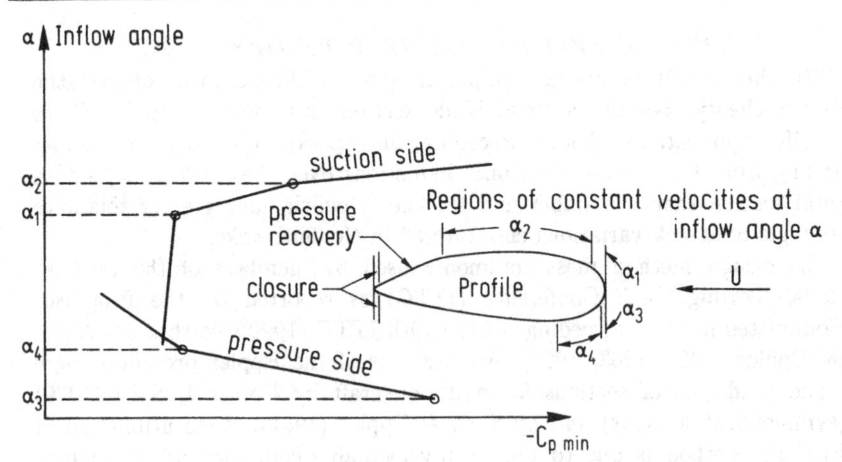

Figure 6.9 Regions on profile as designed by the method of Eppler.

Figure 16 from ITTC (1990b). Report of the Propulsor Committee. In *Proc. 19th International Towing Tank Conference*, vol. 1, pp. 109–60. El Pardo: Canal de Experiencias Hidrodinamicas.
By courtesy of ITTC, Spain.

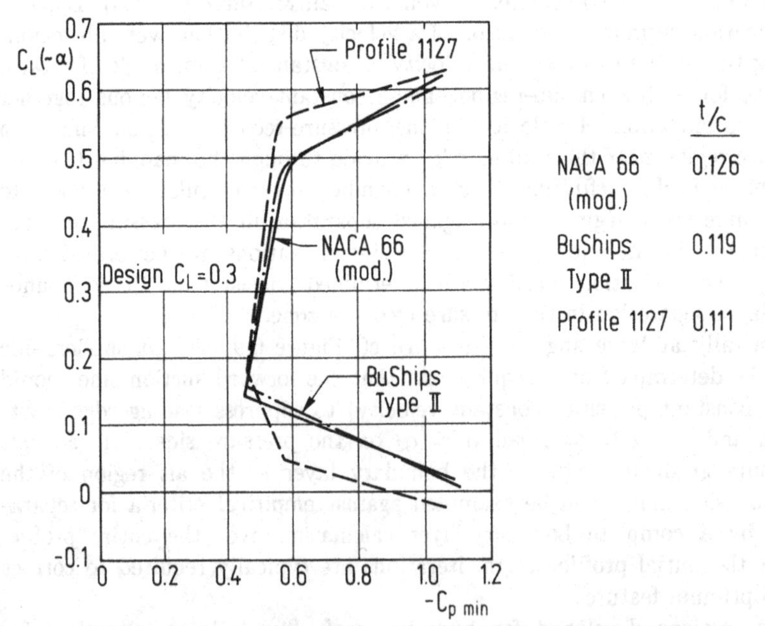

Figure 6.10 Minimum pressure envelopes of NACA–66 (mod.), NACA–16, and Eppler–Shen Profile 1127.

Figure 1 from Shen, Y.T. & Eppler, R. (1981). Wing sections for hydrofoils – part 2: nonsymmetrical profiles. *Journal of Ship Research*, vol. 25, no. 3, 191–200.

By courtesy of SNAME, USA.

this section and the velocity distributions obtained at three angles are shown in Figure 6.11.

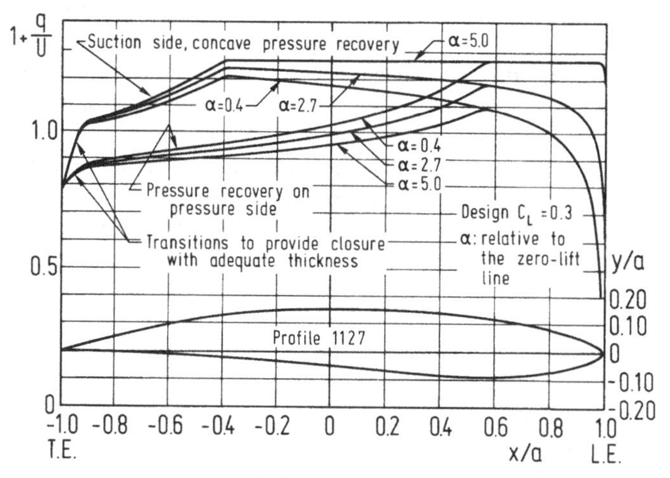

Figure 6.11 Shape of Eppler–Shen Profile 1127 and distributions of tangential velocity at three angles of attack.

From: Shen, Y.T. & Eppler, R. (1981). Wing sections for hydrofoils — part 2: nonsymmetrical profiles. *Journal of Ship Research*, vol. 25, no. 3, 191–200.

Experimental verification was conducted at DTRC by Shen (1985). Excellent agreement between measurements and predicted cavitation inception was obtained with leading-edge roughened models to overcome scale effects associated with boundary-layer transition pressure variations.

The sections developed in the foregoing are applicable to high-speed ships and naval vessels for which the mean blade loadings are low and the focus is upon raising the cavitation-inception speed. However, on large single-screw, slow-speed ships with consequent high mean loadings and strong wake variations cavitation is inevitable, most generally occurring intermittently on each blade during every revolution between 10 and 2 o'clock blade positions when viewing the propeller disc as the face of a watch.

The use of sections with small leading-edge radii and extensive, flat pressure distributions (at ideal angle) on such ships has been observed to generate long partial cavities which are very unstable, giving rise to "cloud" or fine bubble cavitation causing severe erosion of the blades. In addition, as Johnsson (1980) has pointed out, the NACA-type sections, originally designed for subsonic-aircraft wings, have leading-edge radii which are much too small (especially for the thin sections in the outer radii of ship propellers) to satisfy strength and manufacturing require ments. In actual constructions, the leading-edge radii are some 10 to 20 times those of the NACA tables. This requires modification of a substantial portion of the forebody. We have seen in Chapter 4, p. 59, that sections blunter than an ellipse produce sharp minimum pressures hard upon the leading edge at ideal incidence and that the minimum pressure at non-ideal incidence is dominated by the leading-edge contribution which varies inversely to the leading-edge radius of curvature. It is clear that to accommodate the large excursion in incidence angles in the wakes of full form ships it is necessary to employ sections with larger leading-edge radii and hence maximum thickness locations much closer to the nose. This and other factors lead Johnsson (ibid.) to design a new class of sections composed of

i. The NACA a=0.80 camber, and

ii. The blade symmetrical thickness distribution consisting of two semi-ellipses of semi-major axes of a_t and a_l joined by a fairing curve of axial extent a_f as shown in Figure 6.12.

The thickness ratio of the leading-edge ellipse is τ_l/a_l and that of the after ellipse is τ_0/a_t. The leading-edge radius of curvature is $r_l = \tau_l^2/a_l$ or for a selected r_l, the semi-major axis of the forward ellipse is

$$a_l = \frac{\tau_l^2}{r_l} \quad \text{or} \quad \frac{a_l}{c} = \frac{c}{4r_l}\left[\frac{t}{c}\right]^2$$

where t/c is the section thickness ratio.

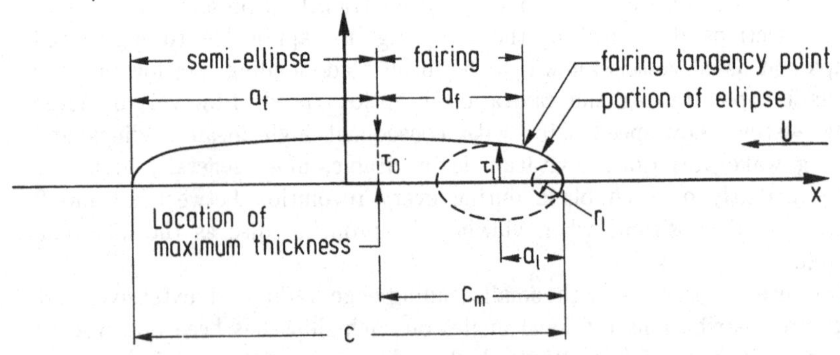

Figure 6.12 Semi-thickness distribution of Johnsson's Section.

Johnsson has determined the fairing graphically and has computed the pressure distribution by a surface-panel computer program by the Hess & Smith (1962) approach. For fixed thickness ratio $2\tau_0/a_t = t/c = 0.0346$ and a fixed leading-edge radius $r_l = 0.008c$ (13.3 times larger than that of an ellipse of this t/c!) he calculated the pressure coefficients for three positions of maximum thickness from the leading edge, $c_m/c = 0.50$, 0.137 and 0.0314. This corresponds to joining the nose ellipse to the after-body ellipse with $a_f = 0$ and $c_m = a_l$. The results are shown in Figure 6.13 where we see very sharp minimum C_p for $c_m = 0.50c$ and $0.157c$ because of their blunt shape.

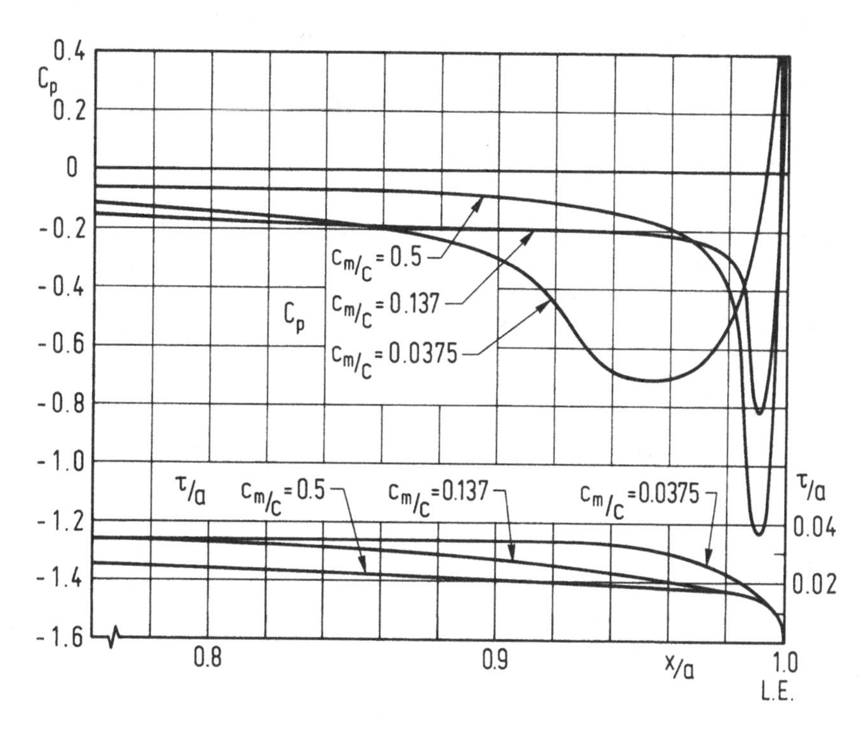

Figure 6.13 Influence of position of maximum thickness on pressure distribution for symmetrical profiles of elliptic type. $2\tau_0/c = 0.0346$. Leading-edge radius 0.8 per cent of c.

Figure 3 from Johnsson, C.-A. (1980). On the reduction of propeller excitation by modifying the blade section shape. *The Naval Architect*, no. 3, 113–5.

While a less negative C_{pmin} is obtained from the joining of two semi-ellipses giving the maximum thickness at $c_m = 0.0375c$ from the leading edge its value of −0.70 is more than 10 times that of a complete ellipse for which $C_{pmin} = -2(t/c) = -0.0692$! All of these surely giving low cavitation inception speeds at ideal incidence! Johnsson's measurements with a model propeller (built with this type of sections having radial distributions of position of maximum thicknesses as shown in Figure 6.14) operated in a model-produced wake showed the following as compared with a propeller composed of NACA-16 series sections:

- Intermittent cavitation observed to be markedly reduced (except for loading C, an extreme case);
- First and second harmonics of blade frequency pressure amplitudes greatly reduced (except for loading C);
- High-frequency noise reduced by 60 to 15 per cent;
- Improved efficiency.

The results are displayed in Figure 6.15 and 6.16.

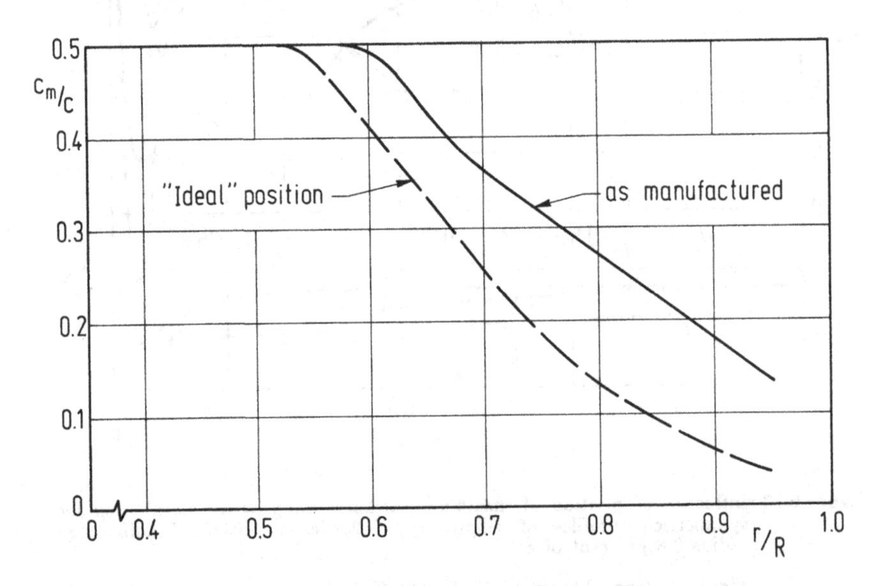

Figure 6.14 Chordwise position of maximum thickness for different blade sections of model propeller. "Ideal" condition is with respect to avoiding sharp pressure minima.

Figure 4 from Johnsson, C.-A. (1980). On the reduction of propeller excitation by modifying the blade section shape. *The Naval Architect*, no. 3, 113-5.

The pressure amplitudes are given in terms of the mean values (long term average) and the mean of 5 per cent of the largest amplitudes (model wakes vary temporally as well as spatially and hence the cavitation varies with time as well as with blade angle).

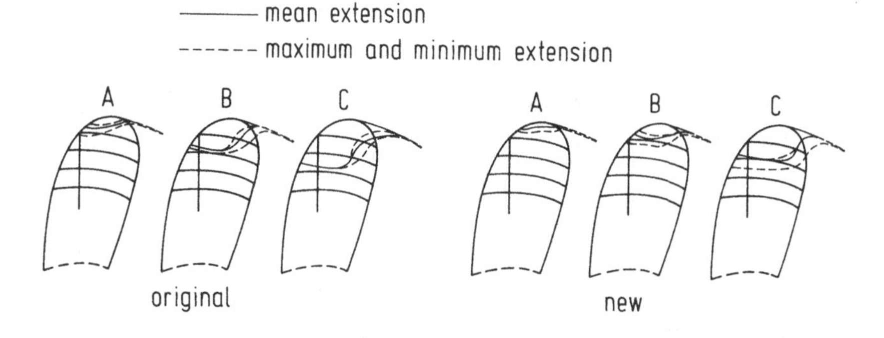

Figure 6.15 Cavitation extension observed at different loading cases. Behind condition. Blade position 30 degrees from upright.

Figure 6 from Johnsson, C.-A. (1980). On the reduction of propeller excitation by modifying the blade section shape. *The Naval Architect*, no. 3, 113–5.

By courtesy of RINA, Great Britain.

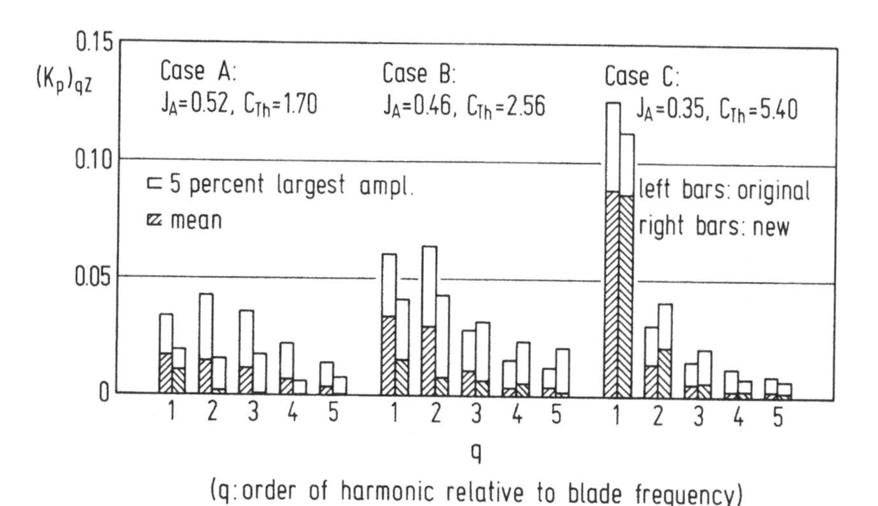

Figure 6.16 Amplitudes of pressure fluctuations measured above the propeller.

Figure 7 from Johnsson, C.-A. (1980). On the reduction of propeller excitation by modifying the blade section shape. *The Naval Architect*, no. 3, 113–5.

By courtesy of RINA, Great Britain.

Our own interpretation of the superior performance of Johnsson's sectional design relies on two mechanisms:

i. The rising pressure downstream of his minimum C_p produces short cavities as compared to the flat pressure distribution over an extensive portion of the chord provided by the NACA-16 sections.

ii. The far blunter leading edge dramatically reduces cavity length and volume.

These mechanisms are revealed by the theories elaborated in Chapter 8.

Similar studies of blunter blade-section design have been underway in Japan, see for example Yamaguchi *et al.* (1988). The subject of blade section design is well summarized in the ITTC Propulsor Committee Report, ITTC (1990b), which should be consulted for further information and references.

7 Real Fluid Effects and Comparisons of Theoretically and Experimentally Determined Characteristics

We have this far completely neglected the fact that all fluids possess viscosity. This property gives rise to tangential frictional forces at the boundaries of a moving fluid and to dissipation within the fluid as the "lumps" of fluid shear against one another. The regions where viscosity significantly alters the flow from that given by inviscid irrotational theory are confined to narrow or thin domains termed boundary layers along the surfaces moving through the fluid or along those held fixed in an onset flow. The tangential component of the relative velocity is zero at the surface held fixed in a moving stream and for the moving body in still fluid all particles *on* the moving boundary adhere to the body.

The resulting detailed motions in the thin shearing layer are complicated, passing from the laminar state in the extreme forebody through a transitional regime (due to basic instability of laminar flow) to a chaotic state referred to as turbulent. We do not calculate these flows.

In what follows we show that viscous effects are a function of a dimensionless grouping of factors known as the Reynolds number and review the significant influences of viscosity in terms of the magnitude of this number upon the properties of foils as determined by measurements in wind-tunnels at low subsonic speeds.

PHENOMENOLOGICAL ASPECTS OF VISCOUS FLOWS

The equations of motion for an incompressible but viscous fluid can be derived in the same way as for a non-viscous fluid, cf. Chapter 1, p. 3 and sequel, but now with inclusion of terms to account for the viscous shear stresses. The assumption that the fluid is Newtonian yields a simple relationship between stresses and rates of deformation, cf. for example Yih (1988). The equations of motion or ***Navier-Stokes Equations*** for an incompressible, Newtonian fluid are then, in vector notation, and in the presence of gravity

$$\rho \frac{D}{Dt} \mathbf{q} = \rho \mathbf{g} - \nabla p + \mu \nabla^2 \mathbf{q} \tag{7.1}$$

where

$\mathbf{q} = u\mathbf{i} + v\mathbf{j} + w\mathbf{k}$ is the fluid particle velocity in rectangular coordinates

$\mathbf{g} = -g\mathbf{k}$ (for positive z-axis vertical upwards)

μ is the dynamic viscosity (kg/m/sec)

$$\frac{D}{Dt} = \frac{\partial}{\partial t} + \mathbf{q}\cdot\nabla$$

The presence of gravity has no effect on the velocity field unless there is a free surface (air-water interface or a long cavity-water interface). We retain it here for it is highly significant in the flow about surface ships but has no contribution to the flow about sections well submerged.

In flows of interest to naval architects there are, of course, obvious reference velocities, i.e., the speed of the ship and the resultant velocity of the propeller blade sections. There are reference lengths, L the ship length and the propeller section chords c and the propeller diameter D. If we take generic reference quantities U_r and L_r then the foregoing equation of motion can be written in dimensionless variables by the substitutions

$$\mathbf{q} = U_r\check{\mathbf{q}} \quad ; \quad t = \frac{L_r}{U_r}\check{t} \quad ; \quad p = \rho U_r^2\check{p}$$

and by noting that $\nabla = (1/L_r)\,\check{\nabla}$ we find that in dimensionless variables

$$\left[\frac{\partial}{\partial\check{t}} + \check{\mathbf{q}}\cdot\check{\nabla}\right]\check{\mathbf{q}} = \mathbf{g}\frac{L_r}{U_r^2} - \check{\nabla}\check{p} + \frac{\mu}{\rho}\frac{\check{\nabla}^2\check{\mathbf{q}}}{U_rL_r} \tag{7.2}$$

The dimensionless quantity $g\,L_r/U_r^2$ ($L_r = L$) is the inverse of the square of the ***Froude number*** F_r and the dimensionless quantity

$$\frac{U_rL_r}{\mu/\rho} = \frac{U_rL_r}{\nu} = \text{the } \textbf{\textit{Reynolds number}}\ R_e$$

where $\nu = \mu/\rho$ is termed the kinematic viscosity.

Our primary interest in what follows is in the effects of viscosity or more properly the effects of Reynolds number (based on the properties of blade sections as deduced from extensive experimentation conducted in wind tunnels on airfoil sections). The Reynolds number can be seen to be the ratio of inertial to viscous forces and what our dimensionless equation tells us is that regardless of scale or fluid media (all being incompressible) the motion will be identical for geometrically similar forms if both the Froude number $F_r = U_r/\sqrt{gL_r}$ and the Reynolds number $R_e = U_rL_r/\nu$ are respectively the same in both scales.

One can immediately see the incompatibility in regard to model–to–ship scaling, but we shall not dwell on this here and consequently ignore F_r, setting $F_r = 0$.

We see that as $R_e \to \infty$ the dimensionless equation of motion reduces to that to which we have limited our attention thus far. Unfortunately in the real world R_e is never infinite. We can also see that as $R_e \to 0$ the viscous term dominates as it does for flows about micro–organisms, flow through capillaries, oil films etc.

It is also to be noted that the acceleration of a material element due to viscous stresses arising from a given rate of strain is evidently determined by the ratio μ/ρ and not by the viscosity μ alone. In as much as the equation of motion (in dimensional variables) has the form

$$\frac{\partial \mathbf{q}}{\partial t} = \ldots.. + \nu\nabla^2\mathbf{q} \tag{7.3}$$

$\nu(= \mu/\rho)$ is effectively a diffusivity coefficient for the velocity $\mathbf{q}$ having the dimension (length · velocity). It is like all diffusivities and playing the same rôle relative to the "dynamic" viscosity μ as the thermal diffusivity $\kappa = k/\rho c_p$ plays relative to the thermal conductivity k. Values for some common fluids at 15°C and one atmosphere pressure are listed in Table 7.1[11] below.

	$\mu \cdot 10$ kg/m/sec	$\nu \cdot 10^4$ m²/sec
Air	0.00018	0.15
Water	0.011	0.011
Mercury	0.016	0.0012
Olive oil	0.99	1.08
Glycerine	23.3	18.5

Table 7.1 Values of dynamic viscosity μ and kinematic viscosity $\nu = \mu/\rho$.

From Batchelor, G.K. (1967). *An Introduction to Fluid Dynamics.* Cambridge: Cambridge University Press.
By courtesy of Cambridge University Press, Great Britain.

We see that values of μ are in increasing order but that the order is reversed with respect to ν for air, water and mercury. Thus air is more viscous than water as measured by the kinematic viscosity by a factor of $0.15/0.011 = 13.64$. Hence to achieve the same Reynolds number in a wind tunnel with the same model used in a water tunnel requires the air speed to be 13.6 times the water speed.

Although the values of μ and ν are small for air and water it is not possible to state what these values should be compared with to ascertain the importance or insignificance of viscosity. We do know that in a nar-

[11] Above discussion and Table 7.1 taken from Batchelor, (1967), by courtesy of Cambridge University Press, Great Britain.

row region about a body in a flow, strong shearing and the development of strong vorticity takes place because of the clinging of the innermost molecules to the surface of the body. This region was termed the boundary layer by L. Prandtl. There the effects of viscosity dramatically alter the local flow from that predicted by inviscid theory.

The regions of the boundary layer and wake are provided in schematic fashion in Figure 7.1 for a flow without boundary-layer separation. The more general case with separated flow is shown in Figure 7.2. Here AB is the upstream dividing stream line, B the forward stagnation point, BC the region of the upper surface along which the pressure is decreasing and CT the region where it is increasing. The "edge" of the boundary layer is displayed by the dotted line which can be thought of as a locus along which the vorticity takes on some arbitrary small value.

Almost always there is a region Bt wherein the flow is laminar extending shortly abaft the point C beyond which the adverse pressure gradient begins. When the Reynolds number is sufficiently high, intermittent turbulent fluctuations appear in the boundary layer and immediately downstream of this region of transition the entire boundary layer becomes turbulent. Velocity profiles for these different regions are depicted in Figures 7.3 and 7.4. Here we can see that as shear stress at the wall shear is proportional to $\partial u/\partial z$ at $z = 0$ (z being distance normal to the wall) it is markedly different in laminar, transitional and turbulent flow and vanishes at separation points.

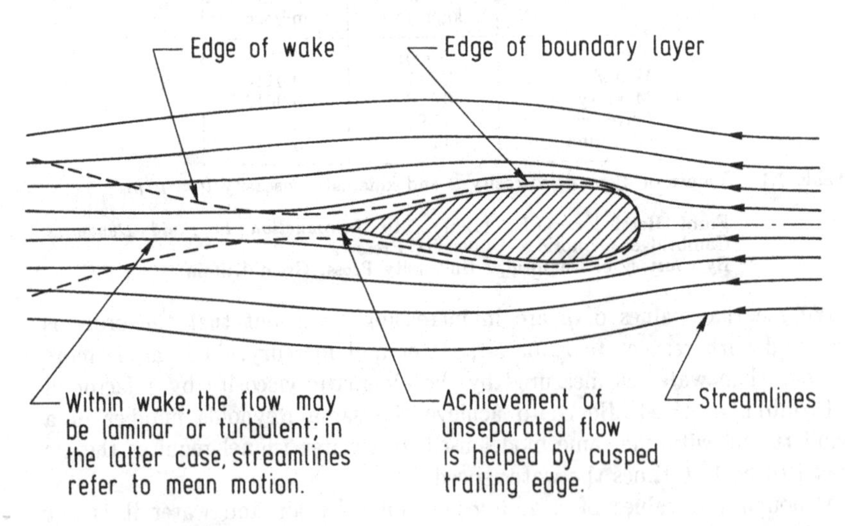

Figure 7.1 A sketch (not to scale) illustrating the unseparated flow of a uniform stream past a thin aerofoil at a low incidence.

Figure I.5 from Thwaites, B. (1960). *Incompressible Aerodynamics.* Oxford: Clarendon Press.
By permission of Oxford University Press, Great Britain.

When the adverse pressure gradient is sufficiently large the boundary layer will be deflected or separated away from the surface as shown in Figure 7.2 giving rise to a region of back flow, thwarting pressure recovery and engendering an increase in pressure drag beyond that for non-separated flows. The location of the onset of separation on smooth surfaces is dependent upon the Reynolds number and is one of the significant contributions to a lack of scaling of flow about model propellers and ship models to that of their respective prototypes.

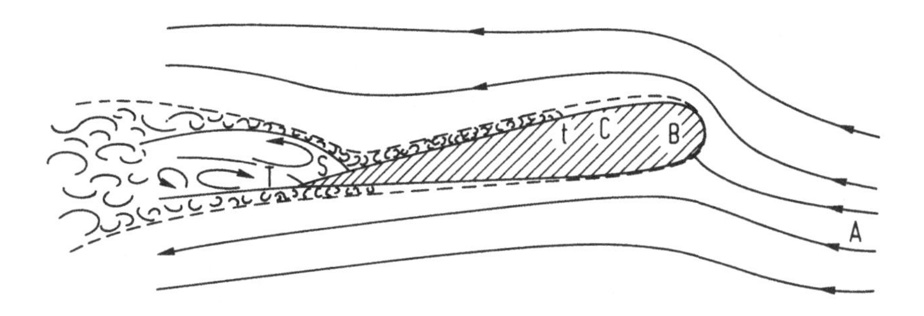

Figure 7.2 A sketch (not to scale) illustrating the nature of the flow of a uniform stream past an aerofoil when separation occurs near the trailling edge.

Figure I.10 from Thwaites, B. (1960). *Incompressible Aerodynamics*. Oxford: Clarendon Press.
By permission of Oxford University Press, Great Britain.

At elevated angles of attack, separation occurs just abaft the leading edge particularly at lower R_e whenever the flow in that region is laminar as compared to turbulent layers which are less prone to separation than laminar layers (because turbulent motion convects energy from the outer stream into the boundary layer). Leading-edge separation or "burble" leads directly to lift degeneration or "stall" as termed by aerodynamic engineers, although transition to turbulence in the detached layer and reattachment can ensue for a limited range of angle of attack.

This aspect is detailed schematically in Figure 7.5. It can be appreciated intuitively that laminar-turbulent transition is sensitive to leading-edge geometry and roughness.

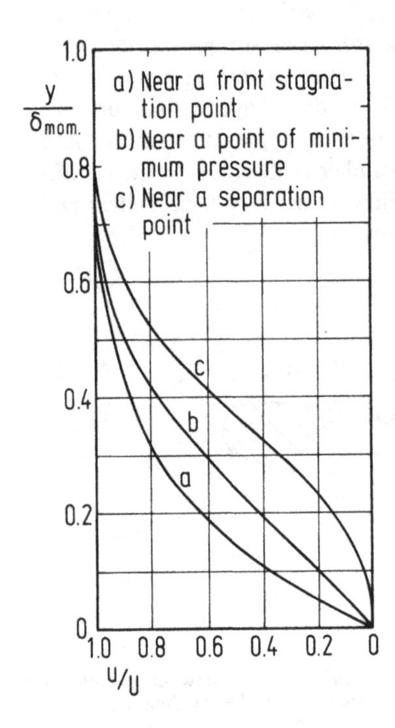

Figure 7.3 Some typical distributions of velocity in a laminar bound ary layer.

Figure 7.4 Some typical distributions of mean velocity in a turbulent boundary layer.

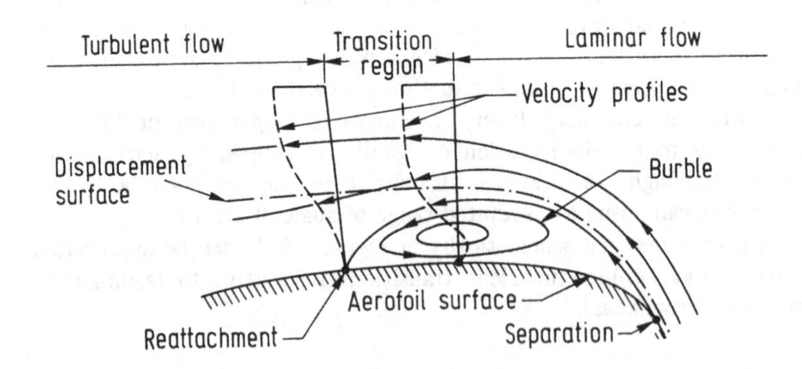

Figure 7.5 A sketch illustrating free transition and turbulent reattachment after laminar separation on an aerofoil.

Figure 7.3, 7.4 and 7.5, Figure I.11, I.12 and I.17, from Thwaites, B. (1960). *Incompressible Aerodynamics*. Oxford: Clarendon Press.
By permission of Oxford University Press, Great Britain

As turbulence is achieved in ship-hull and propeller-blade boundary layers hard upon the leading edge, the non-scaled persistence of laminar boundary layers on ship and propeller models downstream of their leading edges is a basic cause of lack of flow similarity which affects drag and lift or thrust at high angles of attack. For this reason trip wires and studs are used on hull models and fine leading-edge roughness is frequently applied to model propellers to artificially induce early transition to turbulence thereby securing a more stable boundary layer.

The foregoing overview of the phenomenological aspect of viscous flows about sections enables us to interpret, in a qualitative sense, the experimental characteristics obtained from wind tunnel measurements of airfoil sections of interest to us. We may now turn to an inspection of selected experimental results.

EXPERIMENTAL CHARACTERISTICS OF WING SECTIONS[12] AND COMPARISONS WITH THEORY

To overcome the deficiencies of earlier tests of wing models of finite aspect ratio (whose history predated the successful flight of aircraft) the U.S. National Committee for Aeronautics (NACA) built a two-dimensional, low turbulence wind tunnel. This enabled measurements of lift, drag, moments and pressure distributions to be made at elevated R_e using models of 0.61 m (2 ft.) chord length which completely spanned the 0.91 m (3 ft.) width of the test section. Lift was measured by integration of pressures arising from reactions on the floor and ceiling of the tunnel. Drag was obtained from wake survey measurements (momentum defect) and pitching moments directly by a balance.

Usual tests were made over a Reynolds number range of $3\text{-}9 \cdot 10^6$ and at Mach numbers (the ratio of the velocity of the flow to the velocity of sound) less than 0.17. Free stream turbulence intensities the order of a few hundredths of one per cent of the speed were maintained. With this understanding of the model test conditions we may turn to an inspection of the significant characteristics and their comparison with theory.

Reactions due to the hydrodynamic pressures and shearing stresses on a foil section can be specified by two force components, perpendicular and parallel to the direction of the ambient flow, (the lift and drag respectively) and by a moment perpendicular to their plane (the pitching moment). These forces and moments are primarily a function of the angle of attack and camber and secondarily of the thickness distribution. We may now examine several significant characteristics.

---

[12] Partly abstracted from Abbott & von Doenhoff, (1959), by courtesy of Dover Publications, USA.

Angle of Zero Lift

This parameter has been shown (in the foregoing linearized theory) to depend only on the camber. The **a** = 1.0 (uniform loading) camber is found to be deficient in angle of zero lift being about 0.74 times the theoretical value but for all other cambers the ratios are found from experiment to be quite close to unity. Independence of thickness is shown in Figure 7.6.

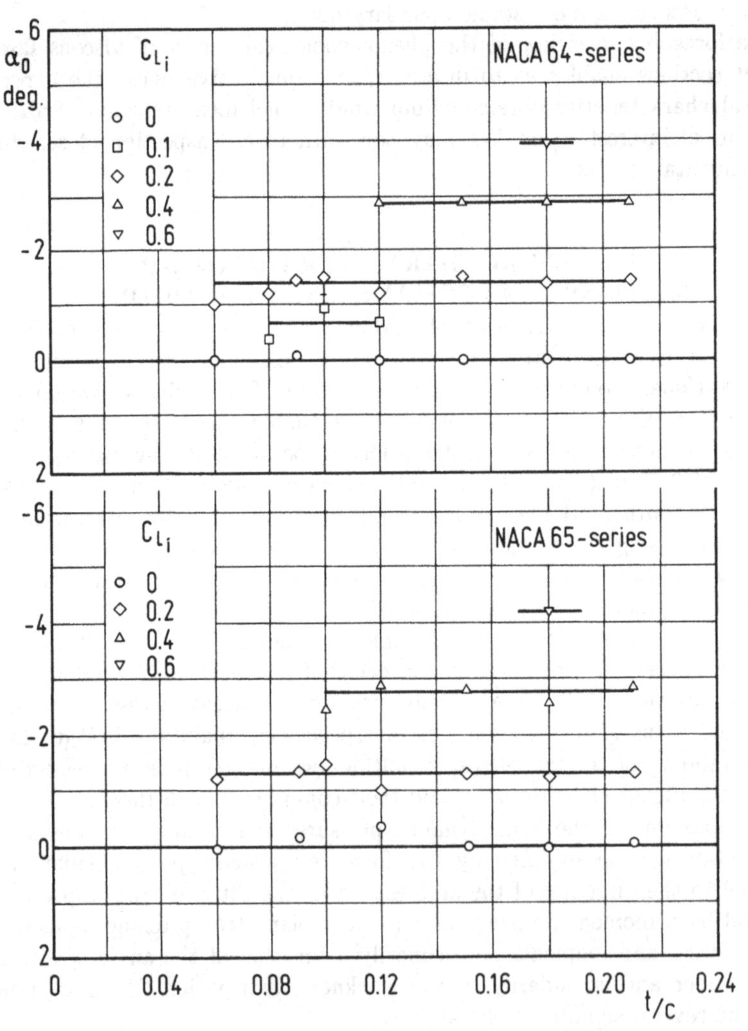

Figure 7.6 Values of the angle of zero lift for various cambers and thicknesses at Reynolds Number $R_e = 6 \cdot 10^6$.

Figure 56 from Abbott, I.H. & von Doenhoff, A.E. (1959). *Theory of Wing Sections*. New York: Dover Publications.
By courtesy of Dover Publications, USA.

Lift-Curve Slope ($dC_L/d\alpha$)

Variations of R_e from 3 to $9 \cdot 10^6$ and variations in camber up to 0.04 c are found to have no systematic effect on $dC_L/d\alpha$. Thickness has some effect. For the NACA-65 and 66 Series, as shown in Figure 7.7, the values increase slightly with increasing t/c for the smooth-surface condition and degrade beyond a t/c = 0.15 when roughened by sand. On these plots the theoretical value of $dC_L/d\alpha$ per degree $= 2\pi/57.3 = 0.1097 \approx 0.11$ is shown. Thus at t/c = 0.06 the mean experimental value is 0.96 of the theoretical and at t/c = 0.21 it is 1.05 (theoretical) for the 65 series and 0.99 (theoretical) for the 66 series for the smooth condition. The degradation with roughness is considerable for t/c > 0.20 indicating that propeller blade sections near the root will, when roughened by corrosion etc., have degraded lift (hence thrust) effectiveness.

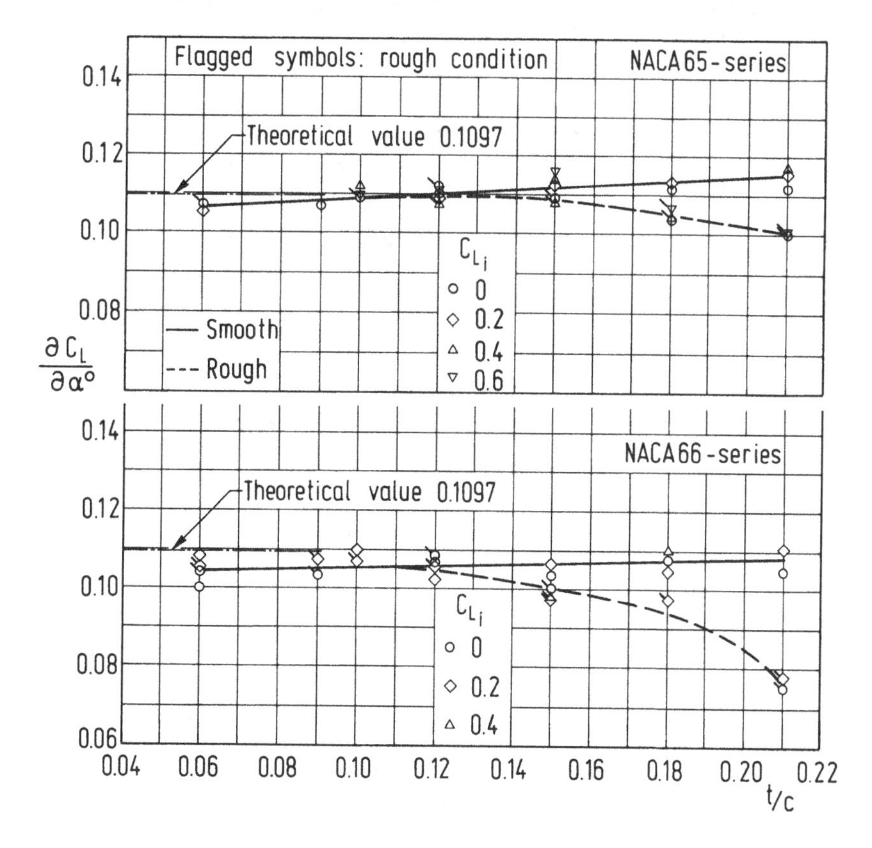

Figure 7.7 Variation of lift-curve slope with thickness ratio for NACA-65 and 66 series at $R_e = 6 \cdot 10^6$ in smooth and rough conditions.

Figure 57 from Abbott, I.H. & von Doenhoff, A.E. (1959). *Theory of Wing Sections*. New York: Dover Publications.
By courtesy of Dover Publications, USA.

Lift and Drag Variation with Angle of Attack

The variations of lift coefficient with angle of attack and five surface conditions are displayed in Figure 7.8. Here we see that although the lift-slope is unaffected at small angles, the maximum C_L is dependent on the surface condition of the model at the elevated $R_e = 26 \cdot 10^6$.

A typical ship propeller blade-section R_e (at the 0.7 radius at 90 RPM, a speed of advance of 14 knots, a chord of 1.50 m and diameter, 8 m) is $34.55 \cdot 10^6$, thus this wind tunnel test R_e is not much below that of typical merchant ship sections.

These results indicate that some degradation of propeller performance well off the design advance ratio may be expected as the blades become rough from corrosion and/or fouling.

The variation of $C_{L\max}$ with R_e and roughness is displayed for the same section in Figures 7.8 and 7.9. This portends scale effect (between model and ship) for operation well off the design point particularly for the lower Reynolds numbers ($\approx 5 \cdot 10^5$) at which ship model propellers are tested.

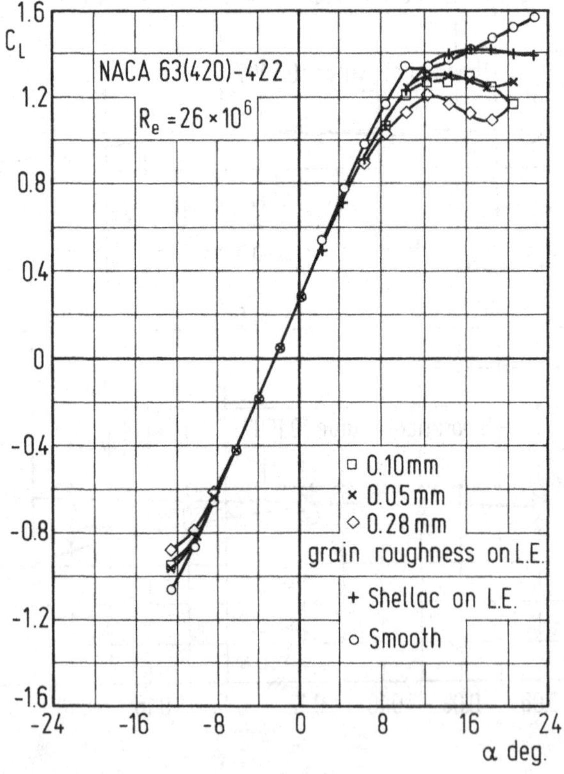

Figure 7.8 Lift characteristics of a NACA-63 (420)-422 airfoil with various degrees of roughness at the leading edge.

Figure 62 from Abbott, I.H. & von Doenhoff, A.E. (1959). *Theory of Wing Sections*. New York: Dover Publications.
By courtesy of Dover Publications, USA.

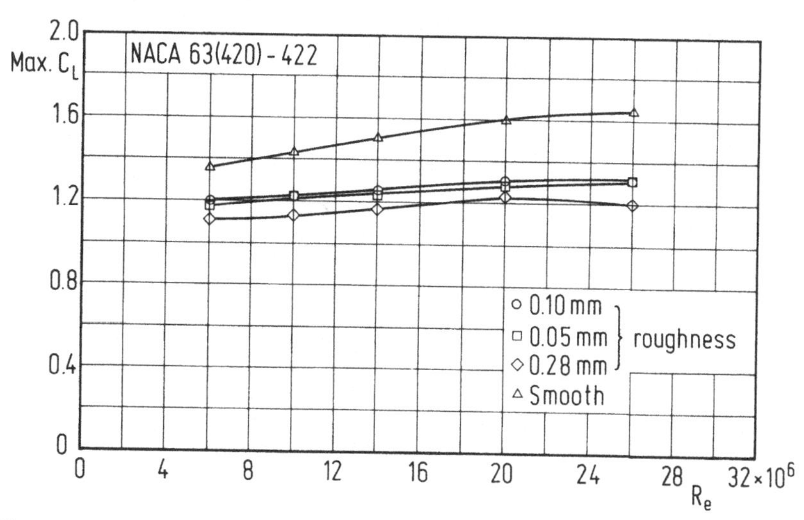

Figure 7.9 Effects of Reynolds number on maximum section lift coefficient C_L of the NACA-63 (420)-422 airfoil with roughness and smooth leading edge.

Figure 63 from Abbott, I.H. & von Doenhoff, A.E. (1959). *Theory of Wing Sections*. New York: Dover Publications.
By courtesy of Dover Publications, USA.

The variation of C_L with angle and with zero and standard roughness for four sections at $R_e = 9 \cdot 10^6$ is given in Figure 7.10. Here we see that the older section NACA 23012 (zero camber, 12 per cent thick) exhibits abrupt stall at 18 degrees. It also shows a small negative angle of zero lift which since it is a symmetrical section should be zero according to theory.

The dependence of drag coefficient on the operating C_L is also shown in Figure 7.10. Here, in smooth condition, it is important to note the low drag obtained at and to either side of the design $C_{L_i} = 0.40$, for NACA-64_2-415, the so-called drag bucket, and the steep rise in C_D below C_L = 0.2 and above C_L = 0.6. It is also to be noted that for moments taken about the aerodynamic center, the moment coefficients are independent of C_L. The listed values of the position, x_{ac}/a, of the aerodynamic center (about which point the moment is essentially constant) are quite close to the theoretical value of 0.50 (an aspect we did not examine in our theoretical development but it is of structural significance).

The drag or resistance of a section is made up of the tangential stress arising from skin friction and a form drag from the lack of full pressure recovery (attained in inviscid flow) over the afterbody. The skin-friction drag is comparable to that of a flat plate of the same length in that the shearing stresses in the region of the forebody where the pressure is falling (negative pressure gradient) are greater than that of a flat plate (both having turbulent or both having laminar flows) and in the afterbody the

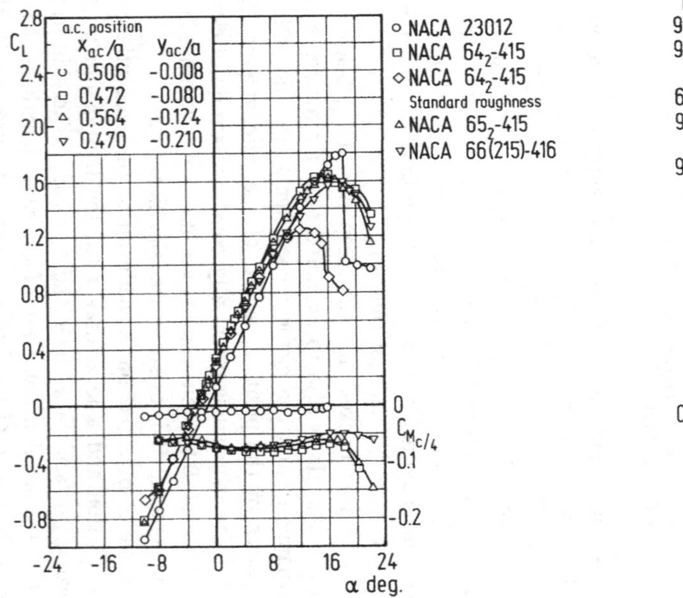

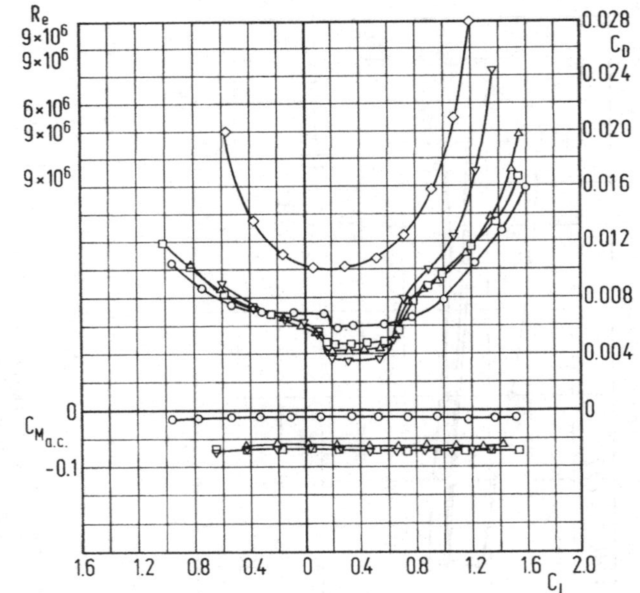

Figure 7.10 Comparison of the aerodynamic characteristics of some NACA airfoils from tests in the Langley two-dimensional low-turbulence pressure tunnel.

Figure 73 from Abbott, I.H. & von Doenhoff, A.E. (1959). *Theory of Wing Sections*. New York: Dover Publications.
By courtesy of Dover Publications, USA.

shearing stresses are less than that of the plate because of the greater thickening of the boundary layer on the foil arising from the rising pressure (adverse or positive pressure gradient). The viscous form drag is dependent on thickness ratio for symmetrical forms at zero angle of attack. Figure 7.11 displays the variation of drag coefficient $C_D = D/\frac{1}{2}\rho c U^2$, of a number of NACA sections with Reynolds number together with laminar and turbulent lines for flat plates.

It is important to note that the drag coefficients are based on the chord and not on the "wetted" surface as is customary in naval architecture.

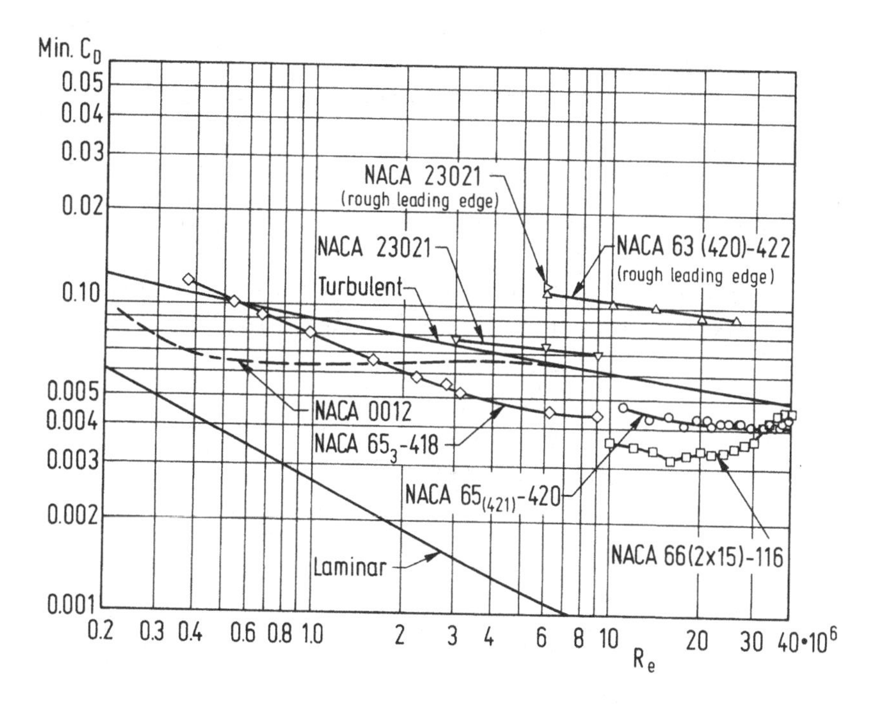

Figure 7.11 Variation of minimum drag coefficient with Reynolds number for several airfoils, together with laminar and turbulent skin-friction coefficients for a flat plate.

Figure 66 from Abbott, I.H. & von Doenhoff, A.E. (1959). *Theory of Wing Sections*. New York: Dover Publications.
By courtesy of Dover Publications, USA.

It seems clear that the smooth sections almost all exhibit the effect of partial laminar flow, their results lying below the turbulent line for the flat plate. Those with roughened leading edge are well above the flat plate. It is therefore difficult to ascertain the form-drag portion from these data in contrast to ship model hulls tested with turbulence stimulators.

Figure 7.12 shows that for a constant thickness ratio the minimum drag coefficient at design C_{L_i} is virtually independent of C_{L_i}. That C_D increases with increasing thickness ratio is apparent from Figure 7.13. The NACA-66 series C_D in smooth condition, can be fitted by

$$C_D = 0.00300 + 0.0272\ (t/c)^2 \tag{7.4}$$

showing that the drag increment of thickness rises quadratically with thickness. For the rough condition the variation for this section can be fitted by

$$C_D = 0.00850 + 0.0864\ (t/c)^2 \tag{7.5}$$

displaying a much more pronounced increment due to t/c.

In theoretical propeller calculations an empirical allowance for section drag commonly used is to take a constant $C_D = 0.0080$. This appears to be non-conservative although at full scale R_e both the coefficients in the last equation will be reduced. For correlations of theory with model propeller data it would appear much more reasonable to use Equation (7.5).

Again it should be noted from Figure 7.12 that the drag coefficient for different thickness ratios is insensitive to the design or ideal C_{L_i}.

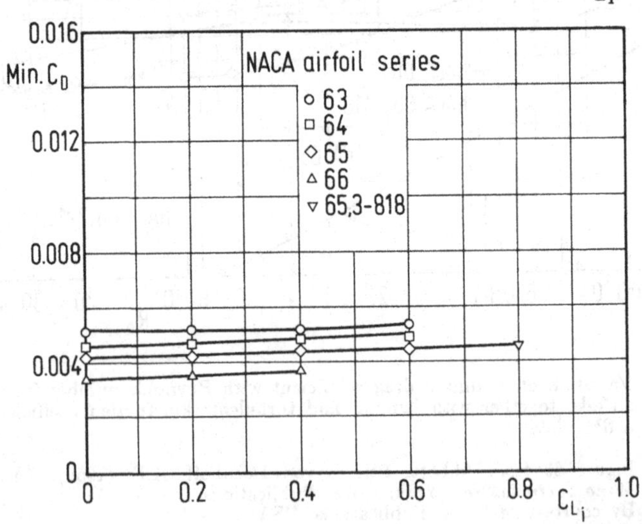

Figure 7.12 Variation of section minimum drag coefficient with camber for several NACA-6 series airfoil sections of 18 per cent thickness ratio. $R_e = 6 \cdot 10^6$.

Figure 67 from Abbott, I.H. & von Doenhoff, A.E. (1959). *Theory of Wing Sections*. New York: Dover Publications.
By courtesy of Dover Publications, USA.

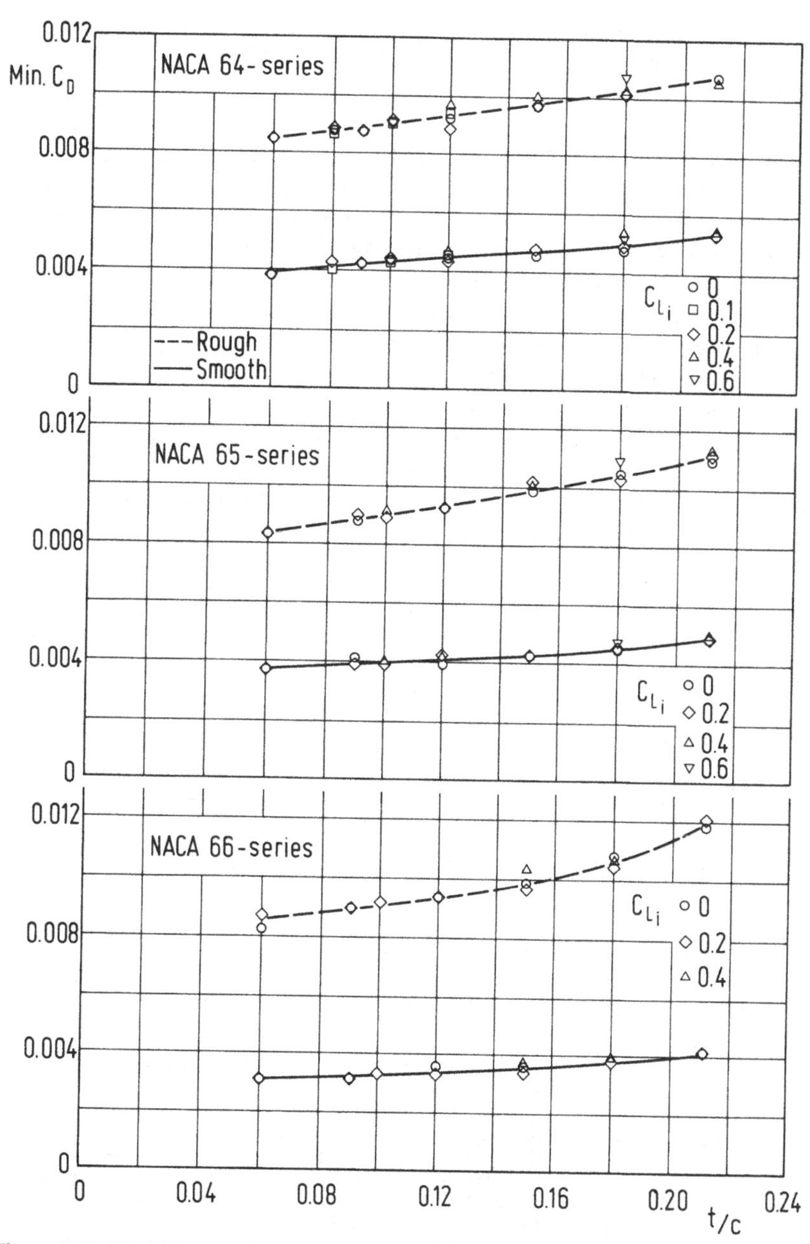

Figure 7.13 Variation of section minimum drag coefficient with airfoil thickness ratio for several NACA sections of different cambers in both smooth and roughened conditions at $R_e = 6 \cdot 10^6$.

Figure 68 from Abbott, I.H. & von Doenhoff, A.E. (1959). *Theory of Wing Sections*. New York: Dover Publications.
By courtesy of Dover Publications, USA.

Pressure Distributions

A single comparison of calculated and measured pressure coefficient distributions on a section is provided in Figure 7.14 showing excellent agreement on both lower and upper sides, especially in the region of most negative C_p. This comparison is based on a superposition-of-velocities procedure described in Abbott & von Doenhoff, (1959) (pp. 75 - 79) and cannot be regarded as representative of all theoretical processes which include non-linear and precise representation of the geometry. Nonetheless other correlations between measured and calculated pressure distributions have shown comparable agreement so the reader may rest assured that two-dimensional theoretical estimates can be used with confidence and that they will generally be conservative when applied to propeller blade sections when the angle of flow incidence is known to be accurate. Both the unsteady and three-dimensional effects present on ship propeller sections serve to reduce the negativeness of C_{pmin} as determined from quasi-steady, two-dimensional values as deduced from the theory presented herein. This is in keeping with the traditional conservatism of the naval architect!

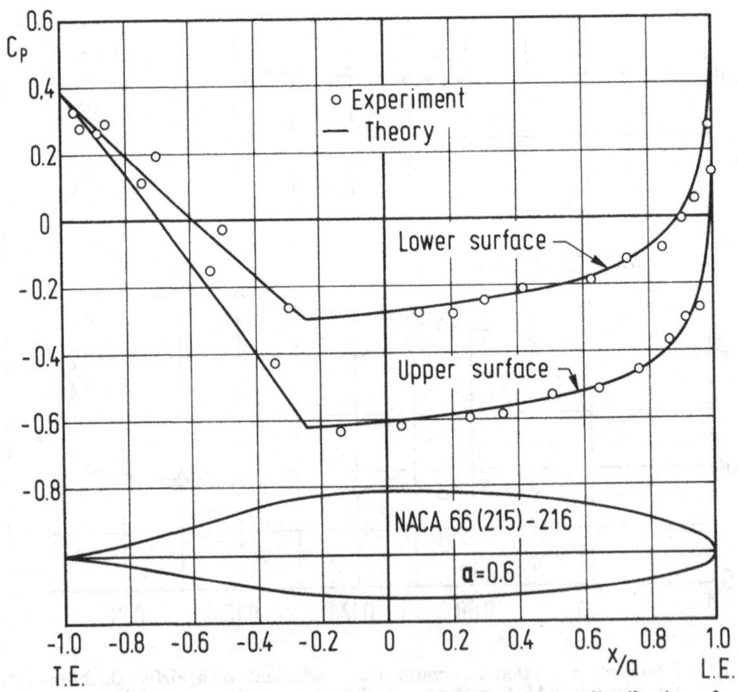

Figure 7.14 Comparison of theoretical and experimental pressure distributions for the NACA-66 (215)-216 airfoil; $C_L = 0.23$.

Figure 39 from Abbott, I.H. & von Doenhoff, A.E. (1959). *Theory of Wing Sections*. New York: Dover Publications.
By courtesy of Dover Publications, USA.

In summary, the foregoing overview of experimental characteristics reveals that in a comfortable range about the design lift coefficient, the lift associated characteristics are very well predicted by theory (with the exception of the rectangular ($\mathbf{a} = 1.0$) camber loading). As inviscid theory predicts zero drag and no stalling characteristics we must employ empiricisms for these aspects. Thus relatively simple theory is found to be of great utility in the understanding of the performance of blade sections and in the pragmatic design of propellers.

We may now turn with increased confidence to the prediction of condition of cavitation inception and thereafter to the theory of cavitating hydrofoils.

8 Cavitation

Here, following a brief account of early observations of the effects of cavitation on ship propellers, we present methods of estimation of conditions at inception of cavitation followed by an outline of the development of linearized theory of cavitating sections. Application of this theory is made to partially cavitating sections, employing the rarely used method of coupled integral equations. The chapter concludes with important corrections to linear theory and a brief consideration of unsteady cavitation.

HISTORICAL OVERVIEW

Cavitation or vaporization of a fluid is a phase change observed in high speed flows wherein the local absolute pressure in the liquid reaches the vicinity of the vapor pressure at the ambient temperature. This phenomenon is of vital importance because of the damage (pitting and erosion) of metal surfaces produced by vapor bubble collapse and degradation of performance of lifting surfaces with extensive cavitation. It is also a source of high-frequency noise and hence of paramount interest in connection with acoustic detection of ships and submarines. Both "sheet" and "bubble" forms of cavitation are shown in Figure 8.1.

One of the earliest observations of the effects of extensive cavitation on marine propellers was made by Osborne Reynolds (1873) when investigating the causes of the "racing" or over-speeding of propellers. The first fully recorded account of cavitation effects on a ship was given by Barnaby (1897) in connection with the operation of the British destroyer *Daring* in 1894. About that time, Sir Charles Parsons (inventor of the steam turbine) obtained very disappointing results from the initial trials (1894) of his vessel *Turbinia*, fitted with a single, two-bladed propeller 0.75 m in diameter. He concluded from the trials of the *Daring*, that the limiting thrust because of the formation of large cavities corresponded to an average pressure on the blades of 77.6 kN/m^2 (11.25 lbs/in^2). After experimentation with three tandem propellers on a single shaft, he finally fitted the *Turbinia* with three shafts, each with three tandem propellers of 0.46 m (1.50 ft) diameter and blade-area ratio of 0.60. His vessel then achieved the very remarkable speed of 32.75 knots at 1491 kW (2000 hp) and later was said to have reached 34 knots as detailed by Burrill (1951).

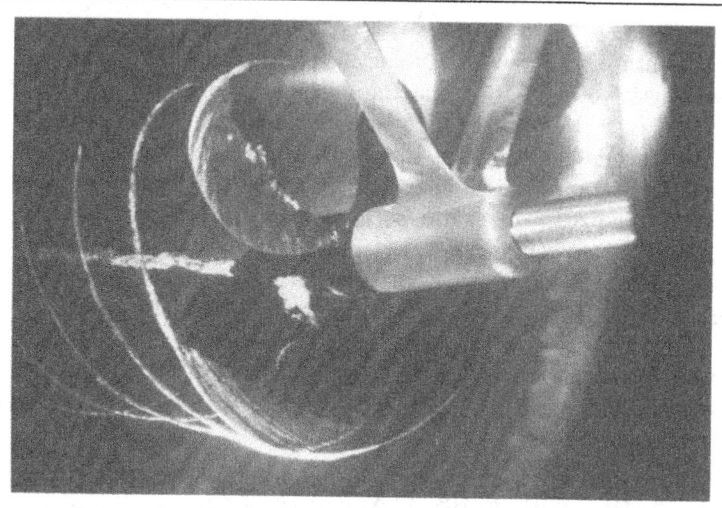

High-speed photograph of blade surface, tip vortex and hub vortex cavitation on a model ship propeller fitted to an inclined shaft in the 36-inch Cavitation Tunnel at the David Taylor Research Center, USA.
By courtesy of David Taylor Research Center, USA.

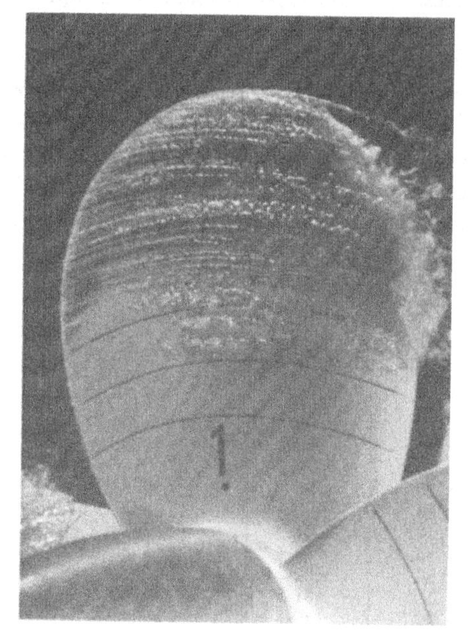

High-speed photograph of propeller cavitation. The sheet cavitation breaks up in a cloud. On the inner radii bubble cavitation can be seen. P/D = 0.8, J = 0.4, σ_N = 0.92, $R_{e,N} = 2.11 \cdot 10^6$
From Kuiper, G. (1981). *Cavitation Inception on Ship Propeller Models.* Publ. no. 655. Wageningen: Netherlands Ship Model Basin.
By courtesy of Maritime Research Insititute Netherlands, The Netherlands.

Figure 8.1 Cavitation on ship propeller models.

Since the turn of this century an enormous literature has grown dealing with the physics of cavitation and the damage to pumps, propellers, hydrofoils etc. as well as the effect on their hydrodynamic performance. It will be far beyond the limits of this book to give even a review of this literature. Instead we refer interested readers to the limited number of books on the subject where also such reviews of the literature can be found. They give comprehensive treatments of the phenomena of cavitation such as its formation, the dynamics of bubbles, their collapse and the erosion. Knapp, Dailey & Hammitt (1970) also describe the effects on flow over hydrofoils and treat cavitation scaling which is of importance in connection with model testing of hydrofoils and propellers. We also mention Young (1989), and Isay (1981) who includes cavitation on hydrofoils and propellers.

PREDICTION OF CAVITATION INCEPTION

To predict the inception of cavitation we are interested in finding the conditions, in particular the position on the body, where the local pressure drops to the vapor pressure. The vapor pressure of a liquid is a fundamental characteristic (analogous to density, surface tension, viscosity etc.) which depends on the temperature. Volatile liquids such as benzine have high vapor pressures relative to water for which at 10°C (50°F) the vapor pressure p_v = 1227.1 N/m^2 (0.178 lbs/in^2). We may immediately note that this vapor pressure at sea temperature is very small compared to the atmospheric pressure which is 101325 N/m^2 (14.696 lbs/in^2). The variation of p_v with temperature for water is given in Table 8.1.

Temperature °C	Temperature °F	Vapor Pressure, p_v N/m^2
0	32	610.8
5	41	871.8
10	50	1227.1
15	59	1704.0
20	68	2336.9
25	77	3166.6
30	86	4241.4
35	95	5622.2
40	104	7374.6
45	113	9582.1
50	122	12334.8
55	131	15740.7
60	140	19917.3
65	149	25007.0
70	158	31155.7
75	167	38549.9
80	176	47356.3
85	185	57800.4
90	194	70107.7
95	203	84523.5
100	212	101325.3

Table 8.1 Vapor pressure of water for various temperatures.

Although cavitation according to theory should take place when the pressure drops to the vapor pressure it has been observed to occur at pressures above and below the vapor pressure depending on the amount and distribution of nuclei or particles to which minute pockets of undissolved gas or air are attached. These act as interfaces on which vaporization or boiling initiates. Indeed, Harvey, the great researcher of human blood, showed long ago that distilled water requires an enormous negative pressure, some –60 atmospheres before rupture and vaporization ensued. However in our application to propellers, sea water is profused with undissolved air and we may take the condition for the onset of cavitation to occur where the local total pressure is close to the vapor pressure.

In this connection it is important to realize that in model test facilities such as variable pressure tunnels and towing tanks it is necessary to ensure a sufficient supply of nuclei; otherwise cavitation inception and its subsequent extent will not comport to full scale. There are many other factors which affect the scaling of such test results, cf. for instance Knapp, Daily & Hammitt (1970).

In flowing liquids the tendency of the flow to cavitate is indicated by the so-called cavitation index or ***vapor cavitation number*** which (by the ITTC Standard Symbols (1976)) is expressed by

$$\sigma_V = \frac{p - p_V}{\frac{1}{2}\rho U^2} \tag{8.1}$$

where

p	is defined as the absolute ambient pressure
p_V	is the vapor pressure
$\frac{1}{2}\rho U^2$	is the dynamic stagnation pressure
U	is some reference velocity.

It has been observed that the pressure within the cavity, p_c, is usually a little higher than the vapor pressure, p_V. Consequently the ("normal") cavitation number σ is defined in analogy with (8.1) as

$$\sigma = \frac{p - p_c}{\frac{1}{2}\rho U^2} \tag{8.2}$$

Here we prefer to replace U by V and V is the total relative resultant velocity between the body and quiescent fluid.

If a section is considered on a blade rotating at N revs/sec in a wake where the local speed of advance (axial component) is $U_a(r)$, then the dynamic pressure to be used in the cavitation index is

$$\tfrac{1}{2}\rho(U_a^2(r) + r^2(2\pi N)^2) = \tfrac{1}{2}\rho V^2(r) \tag{8.3}$$

where V(r) is the resultant "free stream" velocity. Returning to (8.1) we note that the ITTC designation of p as the absolute ambient pressure conflicts with our use of p as a general pressure. To obviate possible confusion we define their p by

$$p = p_{at} + \rho gh \tag{8.4}$$

where p_{at} is the atmospheric pressure acting on the surface of the fluid; ρ, the fluid mass density; g, the acceleration due to gravity and h the submergence of the point of interest below the fluid surface. Use of (8.4) and (8.3) in (8.1) gives

$$\sigma_V = \frac{p_{at} + \rho gh - p_v}{\frac{1}{2}\rho(U_a^2(r) + (2\pi Nr)^2)} = \frac{p_{at} + \rho gh - p_v}{\frac{1}{2}\rho V^2} \tag{8.5}$$

Thus the vapor cavitation number is the ratio of the net pressure inhibiting cavitation to the relevant dynamic pressure. It is a measure of the tendency of the moving fluid to cavitate. When σ_V is large, cavitation is unlikely; when σ_V is small the likelihood is great. It is a combined property of the ambient flow analogous to the Mach number in the flow of compressible gases. It has nothing to do with the local flow – it is best thought of as characterizing the relative free stream in regard to the tendency to cavitate. The local flow and hence the geometry and loading enter the picture through the minimum Euler pressure coefficient (cf. p. 6) peculiar to the specific section and its operating condition.

We may write the minimum pressure coefficient of a body as

$$\frac{p_{min} - p_\infty}{\frac{1}{2}\rho V^2} = C_{pmin} \tag{8.6}$$

where p_{min} is the minimum pressure at some point on the body. In the absence of free surface (Froude number) effects and viscous (Reynolds number) effects the above pressure coefficient is independent of speed or body size. The pressure p_∞ is the ambient pressure which is $p_{at} + \rho gh$ and so we can express the minimum pressure as

$$p_{min} = p_{at} + \rho gh + \tfrac{1}{2}\rho V^2 C_{pmin} \tag{8.7}$$

Now as we have seen, $C_{pmin} < 0$ so we can replace it by $-|C_{pmin}|$ to put the minus character in evidence. As the relative resultant velocity V becomes sufficiently great, there will be a critical or inception value V_i such that

$$p_{min} = p_{at} + \rho gh - \tfrac{1}{2}\rho V_i^2 |C_{pmin}| = p_v \tag{8.8}$$

that is, the available pressure is reduced to the vapor pressure at the point on the section or body where the pressure coefficient is a minimum.

To compare this to the cavitation number, recast (8.8) to obtain

$$|C_{pmin}| = \frac{p_{at} + \rho gh - p_v}{\frac{1}{2}\rho V_i^2} \tag{8.9}$$

We see that the right side is that value achieved by the cavitation number when the reference velocity V achieves the critical value V_i.

Thus the *nominal* condition for cavitation inception can be stated as

$$\sigma_{vi} = |C_{pmin}| = \frac{p_{at} + \rho gh - p_v}{\frac{1}{2}\rho V_i^2} \tag{8.10}$$

where C_{pmin} is the most negative value of the pressure coefficient on the body.

Whenever the cavitation number is larger than $\sigma_i = |C_{pmin}|$ then no cavitation will be present. Whenever $\sigma < \sigma_i$ then cavitation will exist and as $\sigma \to 0$, the cavity lengths will, in first order theory, become infinitely long.

Equation (8.10) provides a means for finding the inception speed V_i by solving for that quantity, i.e.,

$$V_i = \sqrt{\frac{2\left[\frac{p_{at} - p_v}{\rho} + gh\right]}{|C_{pmin}|}} \tag{8.11}$$

In metric units, say for 10°C, $p_v = 1227.1$ N/m² (from Table 8.1), $p_{at} =$ 101324.4 N/m², $g = 9.81$ m/sec² and $\rho = 1025$ kg/m³ (for salt water) we have

$$V_i = \sqrt{\frac{2(97.655 + 9.81\ h)}{|C_{pmin}|}}\ \text{m/sec} \tag{8.12}$$

For negligible submergence, i.e. h < 1 m, we have the simple formula

$$V_i = \frac{14}{\sqrt{|C_{pmin}|}}\ \text{m/sec} \approx \frac{27.2}{\sqrt{|C_{pmin}|}}\ \text{knots} \tag{8.13}$$

As noted earlier the vapor pressure at normal sea temperature is extremely small in comparison to the atmospheric pressure, i.e., at 10°C $p_v/p_{at} =$ 0.012. If neglected, the factor 14 (actually 13.975) in (8.13) becomes 14.06. Hence the vapor pressure may be neglected when dealing with full scale or ship conditions. Thus formula (8.13) is accurate enough and is easy to carry in mind (*when both* p_v *and* h *can be neglected*).

As an example, we may consider the favorable NACA-16 section having $t/c = 0.05$ and $C_{L_i} = 0.20$ which from Equation (6.39) at $\alpha = \alpha_i$ gives $C_{pmin} = -0.225$ and evaluate (8.11) at h = 0, 1, 5 and 10 meters. The results for the nominal inception speeds are

h	m	0	1	5	10
V_i	m/s	29.4	30.8	36.0	41.6

We compare these inception speeds with the resultant velocity at the 0.8 radius of a propeller having a diameter of 8 meters turning at 90 RPM = 1.50 rps at a speed of advance U_A = 15 knots = 7.72 m/s. Then

$$V = \sqrt{(7.72)^2 + (2\pi(1.50)(0.8)(4))^2} = 31.13 \text{ m/s} \tag{8.14}$$

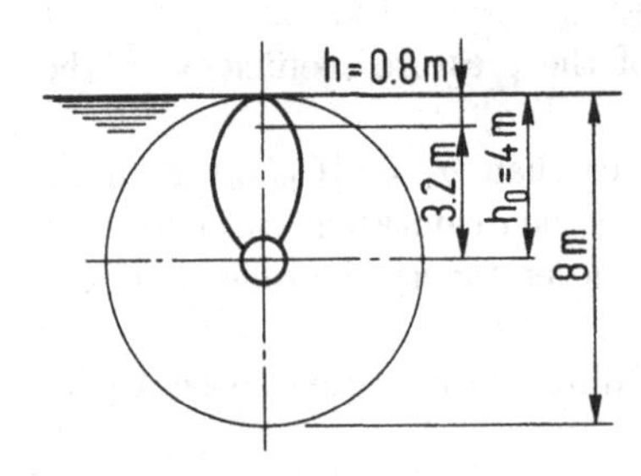

Figure 8.2

If this propeller operates with axis at say, h_0 = 4 m (one radius), a minimum submergence for which at 12 o'clock our 0.8 section, cf. Figure 8.2, is at h = 0.8 m then the inception speed is 30.55 m/s and hence the inception speed is exceeded slightly. Indeed the tip speed of this propeller is 38.5 m/s and it is virtually impossible to avoid cavitation at small submergences especially when we always have to remember that the sections do not enjoy angles of incidence at the most favorable angle of $\alpha = \alpha_i$. Indeed, the spatial variation of the wake gives rise to angles of attack of the order of 4° to 6° (in the 12 o'clock region) above the circumferential mean. Allowing for induced effects then $\alpha - \alpha_i \approx 2^\circ$ to 3° and C_{pmin} from (6.39) with k = 0.489 (for the highly favorable 16 series from Table 6.1) yields for $t/c = 0.05$, $C_{L_i} = 0.20$, a = 0.8 camber for which $\alpha_i = 1.40^\circ(0.2) = 0.28^\circ$

$$C_{pmin} = -\left[2.28(0.05) + 0.56(0.20) + \frac{0.00125}{(0.05)^2}\left[\begin{Bmatrix}2^\circ\\3^\circ\end{Bmatrix} - 0.28\right]^2\right]$$

$$= \begin{cases} -1.71\ ;\ \alpha - \alpha_i = 2^\circ \\ -3.93\ ;\ \alpha - \alpha_i = 3^\circ \end{cases} \tag{8.15}$$

Then the inception speeds at h = 0.8 m using (8.11) are found to be V_i = 11.12 m/s and 7.33 m/s which are far below the operating local speed of 31.13 m/s. Hence extensive transient cavitation will occur. These estimates are slightly pessimistic because of neglect of the relief of the dynamic pressures provided by the presence of the water surface and three-dimensional effects alluded to earlier which would provide for somewhat less negative C_{pmin}.

For negligible h and p_v an easy formula to commit to memory is

$$V_i = \frac{14}{\sqrt{2.28(t/c) + 0.56\ C_{L_i} + 0.00125(\alpha-\alpha_i)^2/(t/c)^2}} \text{ (m/s)} \tag{8.16}$$

where $\alpha_i = 1.4\ C_{L_i}$, for the 16 series, **a** = 0.8 mean line, $(\alpha - \alpha_i)$ in degrees, cf. (6.39) p. 102.

An upper bound of possible cavitation speeds at small submergence (neglecting the water surface effect on the sectional pressures) is

$$V_i = \frac{14}{\sqrt{2\ (t/c) + 0.5\ C_{L_i}}}\ (m/s) \tag{8.17}$$

where the coefficient 2 applies to the ellipse and the 0.5 for the uniform loading or logarithmic camber line (**a** = 1.0). If inception speeds at small submergences are found to be in excess of the values from (8.17) then the results must be considered suspect (when $h < 1$ m). To account for intermediate submergences

$$V_i = \frac{14(1 + 0.050\ h - 0.00126\ h^2 + \ldots)}{\sqrt{2.28(t/c) + 0.56C_{L_i} + 0.00125(\alpha-\alpha_i)^2/(t/c)^2}}\ (m/s) \tag{8.18}$$

provided $\rho gh/p_{at} < 0.60$ or $h < 6$ m; $(p_{at}/\rho g = 10.09$ m); for an error less than one per cent.

In general, the cavitation inception speed formula is (8.11) with (6.35) giving

$$V_i = \sqrt{\frac{2(97.66 + 9.81\ h)}{A(t/c) + (\partial C_p/\partial C_{L_i})C_{L_i} + 2(\alpha-\alpha_i)^2/((57.3)^2 k(t/c)^2)}}$$
$$A = \partial C_p/\partial(t/c) \tag{8.19}$$

with $\partial C_{pmin}/\partial C_{L_i} = 0.56$ for **a** = 0.8 roof top pressure distribution; A and k are appropriately found from Tables 4.1 (p. 62) and 6.1 (p. 102) for the various thickness families. The factor $k(t/c)^2$ may be replaced by r_l/c, r_l being the leading-edge radius. The angle deviation $(\alpha - \alpha_i)$ is in degrees and the ideal angle α_i found from Equation (6.15) for any roof-top loading.

It must be appreciated that the foregoing formulas involve estimates of the minimum pressure coefficients for foils as provided by the first order theory with the Lighthill correction. They give a good indication of what rôles are played by the parameters and can be used for preliminary estimates. However they do not give precise results in the vicinity of very small $(\alpha - \alpha_i)$ as has been shown in Chapters 5 and 6 (e.g. Figure 5.8). The locus of the exact (but two-dimensional!) C_{pmin} as a function of angle of attack for a specified camber ratio and thickness distribution has the form shown in Figure 8.3 where we have also indicated the types of cavitation to be expected if $|C_{pmin}| < \sigma_V$.

As a convenience in evaluating (8.19) the nomograph in Figure 8.4 is provided. One need only connect the depth to the value $\sigma_{vi} = |C_{pmin}|$ with a straight line and its intercept on the speed scale is the answer.

For comparative purposes the cavitation inception speeds for a variety of two-dimensional and axially symmetric forms are graphed in Figure 8.6 for the forms displayed in Figure 8.5. Two features should be noted. Firstly, axially symmetric forms have much higher speeds of inception than two-dimensional sections because the flow is "relieved" by three-dimensionality. Secondly, decrease in cavitation speed is more rapid with decrease in fineness ratio (length/diameter) for axially symmetric forms than for two-dimensional sections of increasing thickness ratio.

The minimum pressure coefficient for spheroids at zero angle of attack has been found by Breslin & Landweber (1961) to be expressed simply by

$$C_{pmin} = -\frac{4}{3}\left[\frac{d}{\ell}\right]^{3/2} \qquad 0 < \frac{d}{\ell} \leq 0.25 \tag{8.20}$$

where d is the diameter and ℓ the length. Thus we see that C_{pmin} varies more sharply with "thickness ratio" d/ℓ than two-dimensional forms.

The variation of minimum pressure with nose radius for various bodies of revolution calculated from axially symmetric source-sink distributions are shown in Figure 8.7. Again it is seen that forms blunter than ellipses of revolution yield relatively large negative C_{pmin}.

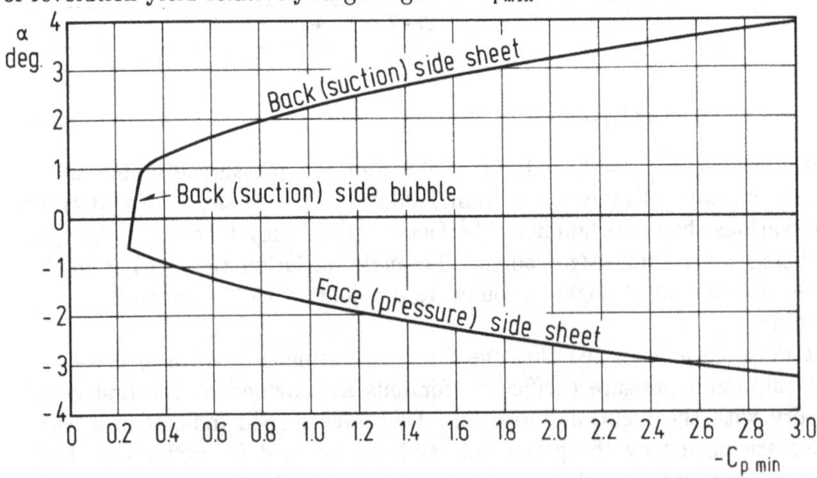

Figure 8.3 Variation of C_{pmin} with angle of attack and cavitation types to be expected if $|C_{pmin}| < \sigma_v$.

Adapted from Brockett, T. (1966). *Minimum Pressure Envelopes for Modified NACA-66 Sections with NACA a = 0.8 Camber and BuShips Type I and Type II Sections.* Hydromechanics Laboratory, Research and Development Report 1790. Washington, D.C.: David Taylor Model Basin — Department of the Navy.
By courtesy of David Taylor Research Center, USA.

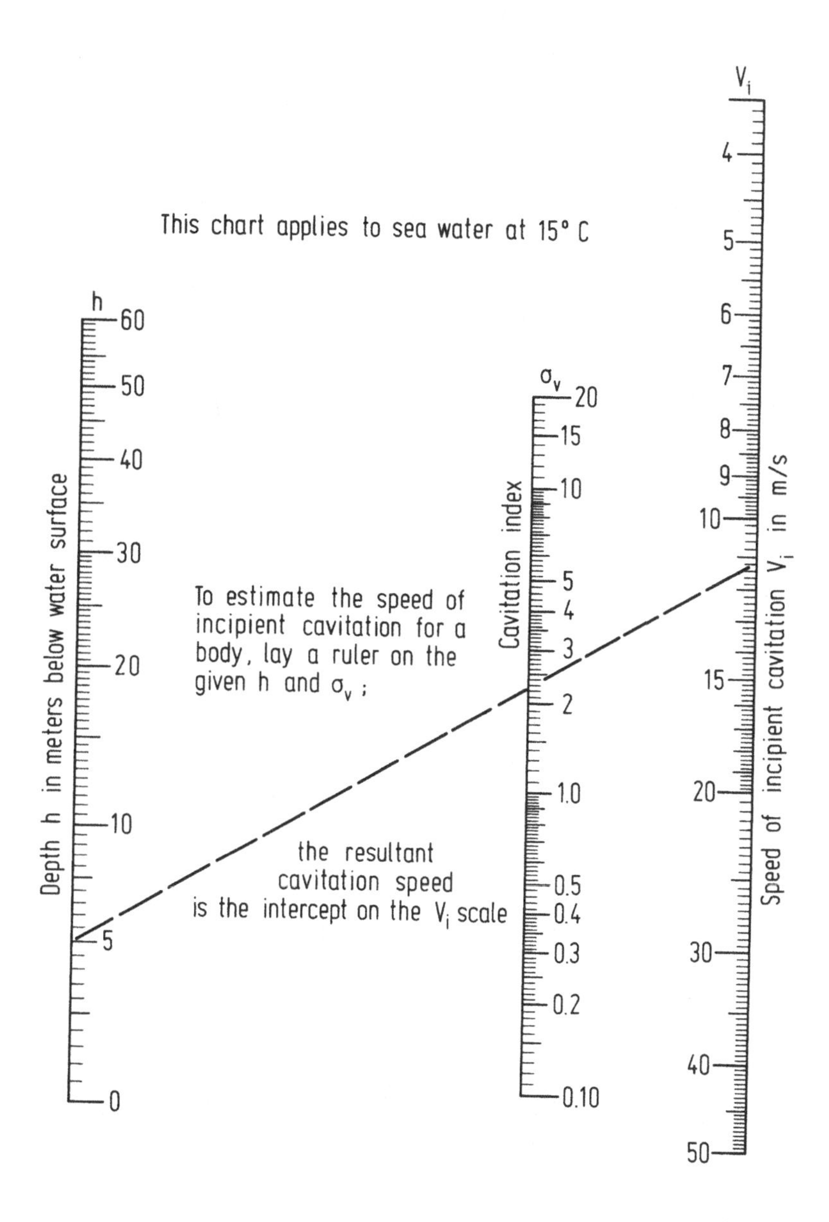

Figure 8.4 Nomograph showing the relationship of depth of submergence, cavitation index, and speed of incipient cavitation for any body.

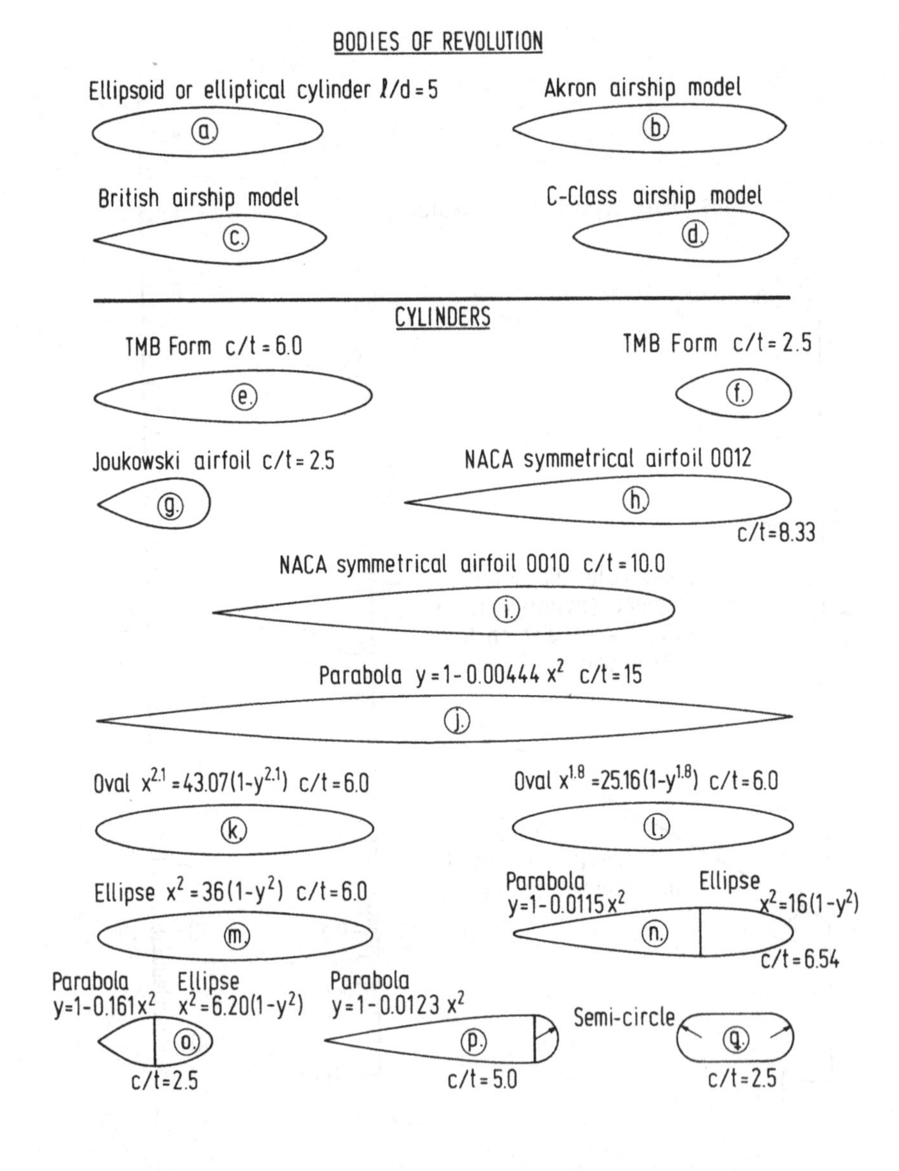

Figure 8.5 Sections of cylinders and bodies of revolution for which cavitation speeds were calculated.

Figure 1 from Landweber, L. (1951). *The Axially Symmetric Potential Flow about Elongated Bodies of Revolution.* Report No. 761. Washington, D.C.: Navy Department, The David W. Taylor Model Basin.
By courtesy of David Taylor Research Center, USA.

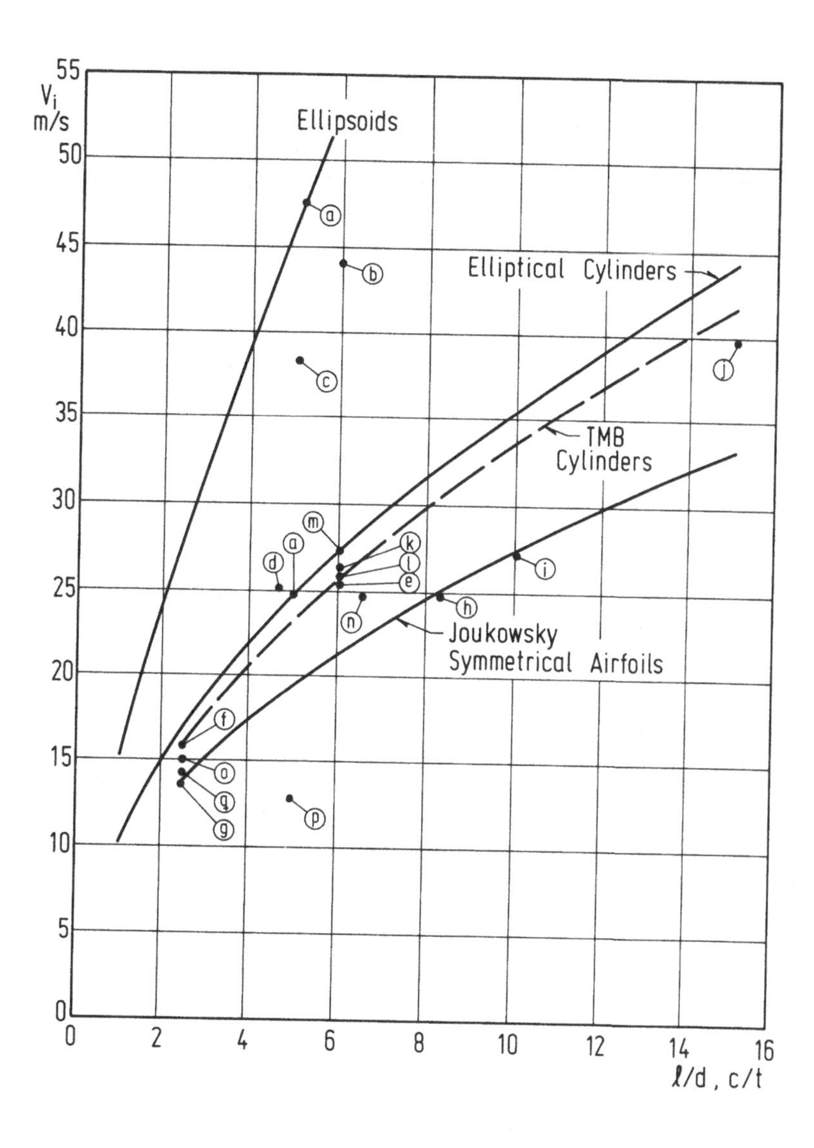

Figure 8.6 Calculated cavitation speeds for various bodies of revolution and cylinders. The lower case letters in circles refer to the body shapes shown in Figure 8.5.

Figure 2 from Landweber, L. (1951). *The Axially Symmetric Potential Flow about Elongated Bodies of Revolution.* Report No. 761. Washington, D.C.: Navy Department, The David W. Taylor Model Basin.
By courtesy of David Taylor Research Center, USA.

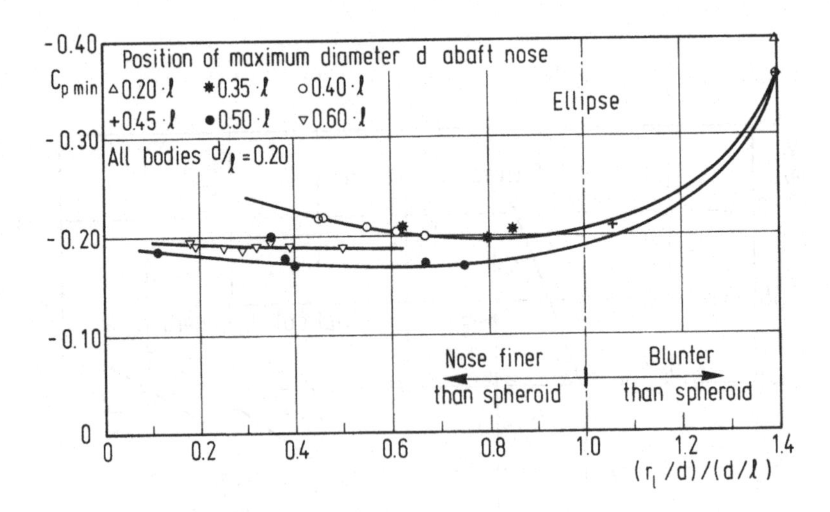

Figure 8.7 Variation of C_{pmin} with nose radius index for a series of bodies developed from source–sink distributions.

Figure 14, from Breslin J.P. & Landweber, L. (1961). A manual for calculation of inception of cavitation on two– and three–dimensional forms. *SNAME T&R Bulletin*, no. 1–21. New York, N.Y.: SNAME.

By courtesy of SNAME, USA.

CAVITATING SECTIONS

The dominant feature of steady high-speed flows of inviscid fluids about blunt forms, as for example, displayed in Figure 8.8, is the development of free-stream surfaces or streamlines. Along these the pressure is the constant vapor pressure and hence, by Bernoulli's equation, the magnitude of the tangential velocity is also constant. For the pressure in the cavity equal to the ambient *total* pressure, the tangential velocity component along the stream surface is equal to the speed of the incident flow far upstream, yielding an open cavity extending (theoretically) to infinity downstream.

Mathematical solutions of such flows about two-dimensional forms were derived more than 120 years ago by Helmholtz (1868) and Kirchhoff (1868) using complex-variable theory and mapping procedures, see Milne-Thomson (1955) for a detailed account. An excellent summary of the extensive literature about cavity flows has been given by Wu (1972) who has made basic contributions. However, as Tulin (1964) pointed out, there were no important applications of free-streamline theory to flows about forms of engineering interest for more than 60 years until Betz & Petersohn (1931) explained the stalled operation of pumps and compressors by extension of the theory to an infinite cascade of flat foils.

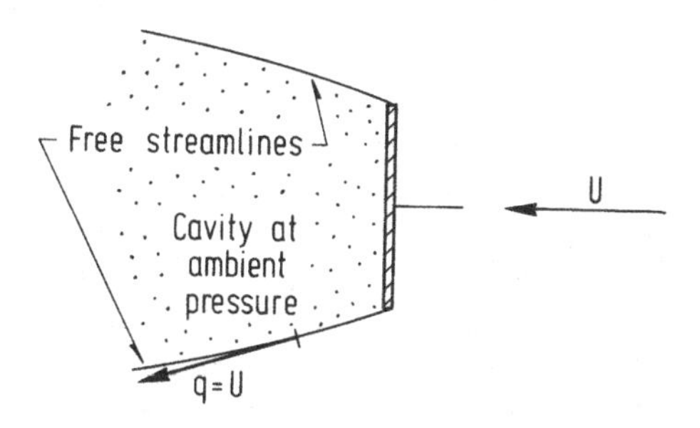

Figure 8.8 Flow past a flat plate with a cavity at ambient pressure corresponding to a cavitation number of zero.

During World War II the impetus of naval problems lead to the development of approximate theories for cavities of finite length by assuming the existence of re-entrant flow at the aft end of the cavity and alternatively by use of a flat plate to permit the cavity to close and the pressure to rise on the flat plate to the stagnation pressure. The assumption was that for cavities of sufficient length, the flow at their termini has little effect on the flow about the body. This allowed the use of mapping procedures to cavities of finite length. Posdunin (1944, 1945) pioneered research on propellers designed to operate with large trailing cavities which he referred to as "supercavitating" propellers. This appellation has been in common use thereafter for all flows with cavities which extend beyond the trailing edge of the body.

The highly significant application of theory to supercavitating flows of interest to naval architects began in the early 1950's by the dramatic breakthroughs made by Tulin (1953) at DTRC who introduced linearization of the equations of motion. As an aeronautical engineer fully familiar with the application of linearized theory to thin wing sections, Tulin quickly perceived that the same procedure (with some modifications) could be applied to cavity flows generated by thin wedges and hydrofoils. In contrast, and prior to that time most professors of naval architecture as well as mathematicians regarded the non-linearity of the Bernoulli equation as sacrosant and were uninformed of the wide use of linear theory by aerodynamicists.

Tulin's first analysis in 1953, clearly displayed the adaption of small perturbation, aerodynamic-type procedures to the flow about slender, blunt-based strut sections (with port-starboard symmetry) with trailing cavities of finite length at small cavitation indices. His result for the limiting case of $\sigma = 0$ for which the cavity is theoretically open and of infinite length agreed with the exact results of Kirchhoff for wedges of small apex angles. He next demonstrated (with Burkart), Tulin & Burkart

(1955), by an ingenious procedure, that the flow about supercavitating lifting sections at $\sigma = 0$ is directly related to a shortened aerodynamic section and gave simple formulas relating the force coefficients of supercavitating sections to those of the related airfoil.

Thereafter, Tulin and others extended the theory to a variety of boundary-value problems and developed high-efficiency sections as illustrated in Tulin (1956), Johnson & Starley (1962) and Auslaender (1962); see also Tulin (1964) for a second order theory and for an extensive citation of the literature. More recently, interesting and useful extensions have been made by Tulin & Hsu (1977) and by Kinnas (1985), and Kinnas & Fine (1989, 1991) at MIT to which reference is made later.

PARTIALLY CAVITATING HYDROFOILS

A linearized theory of a flat plate with partial, sheet cavitation extending from the leading edge has been given by Acosta (1955). Later Geurst (1959, 1960) solved the problems of partial and supercavitating cambered sections. Geurst represented the foil by the camber curve alone and solved the problem using complex-variable theory in which the physical plane with boundary conditions on the upper and lower banks of a cut in way of the foil are mapped on the entire real axes of an auxiliary plane. This gives a Hilbert problem with mixed conditions for which a known integral plus two simple functions with unknown coefficients is the solution. The unknown constants are (laboriously) computed from conditions at infinity. Here we shall proceed in the way with which we have become familiar from the previous chapters and address this problem (including section thickness) with distributions of sources for the blade and cavity thicknesses and vortices for the loading. This is analogous to the analysis of supercavitating sections by Davies[13] (1970) done independently of the work of Hanaoka (1964), also later by Kinnas (1985).

When we attempt to regard in exact theory a cavity as closed and also as a surface of constant pressure (the vapor pressure) in an inviscid fluid we are confronted with a dilemma because the pressure cannot be constant at the terminus rising as it must in the ideal fluid to the stagnation pressure. This led to formulations by earlier researchers to provide flow terminations on flat plates, normal to the surface. Wu (1975) and Tulin (1964) give extensive discussions of cavity termination models. The basic assumption is that generally the character of the cavity termination affects only the local behavior for long-enough cavities, but the cavity shape and length can depend critically upon the termination model and the detachment point (cf. Franc & Michel (1985) and Kinnas & Fine (1989)).

[13] Accomplished as Visiting Scientist at Davidson Laboratory, USA during 1969–70.

In contrast to this, the cavity thickness in linearized theory is found to be of higher order in the vicinity of the terminus, allowing one to impose closure, or zero ordinate there. However, as both Acosta (1955) and Geurst (1959, 1960) found, linearized theory becomes invalid for cavity terminations approaching the foil trailing edge. Experimentally, cavities on hydrofoil sections are found to be unstable for lengths in the region 3c/4 to 5c/4.

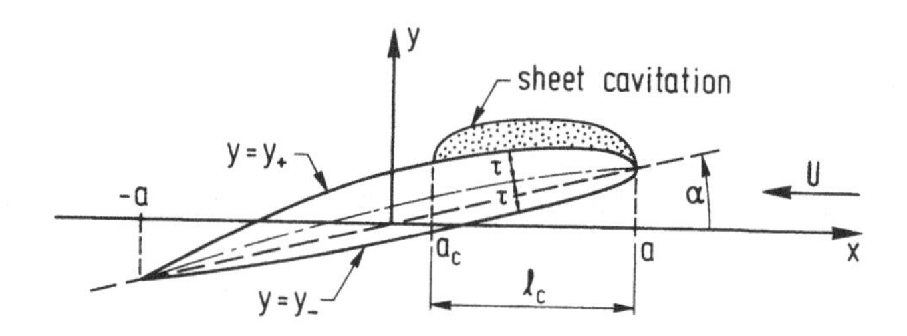

Figure 8.9 Schematic of a partially cavitating foil.

The geometry connected to the partially cavitating foil problem is outlined in Figure 8.9. The following six conditions must be met:

i. The velocity potential for the fluid disturbance flow, ϕ, must be a solution of Laplace's Equation, (1.17);

ii. The velocity components must vanish for $\sqrt{x^2 + y^2} \to \infty$;

iii. The kinematic condition on the wetted positions of the foil, viz. $\partial(\phi - Ux)/\partial n = 0$, Equation (3.9);

iv. The pressure on the cavity surface is constant, being equal to the pressure p_c (cf. p. 131);

v. The cavity is closed;

vi. A Kutta condition at the trailing edge, $x = -a$, i.e., $\nabla\phi$ finite.

The pressure at any field point is provided by the Bernoulli equation (see Equation (1.23) reduced to two dimensions and dropping of subscript b)

$$\frac{(u - U)^2 + v^2}{U^2} = 1 - \frac{p - p_\infty}{\frac{1}{2}\rho U^2} = \frac{q^2}{U^2} \tag{8.21}$$

where p_∞ is the *total* ambient pressure. Condition iv. requires that $p = p_c$ on the cavity which yields for the *tangential* velocity along the cavity q_c

$$q_c = -\sqrt{1 + \sigma}\, U \tag{8.22}$$

in which σ is the cavitation number given by (8.2).

Linearizing for thin sections gives $q_c \approx u - U$, so the horizontal component in way of the cavity is approximately

$$u = -\sigma U/2 \quad ; \quad a_c \leq x \leq a \tag{8.23}$$

This is the linearized dynamic condition on the *total* horizontal component due to all disturbances.

As in the treatment of fully wetted sections in Chapter 4, we convey the physical conditions applicable to the flow indicated in Figure 8.9 to the banks of a slit along the x-axis and evaluate the relevant functions along $y = 0$ as displayed in Figure 8.10. We can satisfy conditions i. and ii. by constructing our foil with the partial cavity by distributions of sources and vortices in analogy to the process for fully wetted foils as indicated in the figure. $m(x)$ and $\gamma(x)$ are the source and vortex densities which are required to generate the foil in the absence of cavitation.

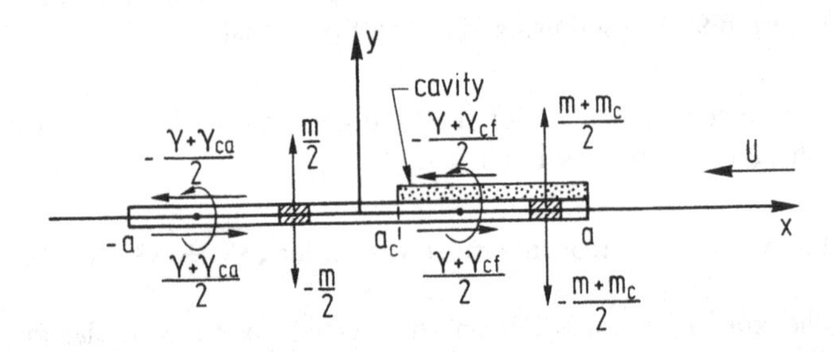

Figure 8.10 Schematic indicating vertical components from source distributions m over the entire foil and m_c in way of the cavity. Vortex distribution γ applies to the fully wetted flow, γ_{cf} is the additional vorticity in way of the cavity and γ_{ca}, that required aft of the cavity.

As the Laplace equation is linear we can install an additional source distribution of strength $m_c(x)$ in way of the cavity to "look after" the cavity thickness and add the vortex distributions γ_{cf} (forward, in way of the cavity) and γ_{ca} (aft of the cavity) to account for the alteration in circulation arising from the presence of the cavity.

The requirement ii. when simplified is of the form given by (5.3) and (5.4) p. 67 upon neglect of $u_\pm$ there. The vertical components from the sources are given by (2.7) and (2.8) and those from the vortex distributions are of the form of (2.38) with the applicable limits of integration.

Consequently, the kinematic condition iii. on the lower side in way of the cavity is explicitly given (via the thin-section approximations) by

$$-\frac{m}{2}-\frac{m_c}{2}+\frac{1}{2\pi}\fint_{a_c}^{a}\frac{\gamma+\gamma_{cf}}{x-x'}\,dx'+\frac{1}{2\pi}\int_{-a}^{a_c}\frac{\gamma+\gamma_{ca}}{x-x'}\,dx'$$
$$=-U\frac{d}{dx}y_{-}\quad;\quad a_c\le x\le a \tag{8.24}$$

and upon the upper (+) and lower (−) sides abaft the cavity, condition iii. requires that

$$\pm\frac{m(x)}{2}+\frac{1}{2\pi}\int_{a_c}^{a}\frac{\gamma+\gamma_{cf}}{x-x'}\,dx'+\frac{1}{2\pi}\fint_{-a}^{a_c}\frac{\gamma+\gamma_{ca}}{x-x'}\,dx'$$
$$=-U\frac{d}{dx}y_{\pm}\quad;\quad -a\le x\le a_c \tag{8.25}$$

The pressure condition iv. on the cavity yields the approximate condition on the total induced horizontal component given by (8.23). The horizontal component from sources is of the form given by (2.4) (with appropriate limits) and by (2.36) for vortex densities. Hence the explicit equation arising from the linearized condition of constant pressure on the cavity is

$$\frac{1}{2\pi}\fint_{-a}^{a}\frac{m(x')}{x-x'}\,dx'+\frac{1}{2\pi}\fint_{a_c}^{a}\frac{m_c(x')}{x-x'}\,dx'-\frac{\gamma+\gamma_{cf}}{2}=-\frac{\sigma U}{2}$$
$$;\quad a_c\le x\le a \tag{8.26}$$

As m(x) and $\gamma(x)$ satisfy the kinematic conditions on the wetted portion, these quantities drop out of Equation (8.24) and (8.25) together with their right-hand sides, leaving the following coupled integral equations

$$\fint_{a_c}^{a}\frac{\gamma_{cf}}{x-x'}\,dx'+\int_{-a}^{a_c}\frac{\gamma_{ca}}{x-x'}\,dx'-\pi m_c=0\;;\;a_c\le x\le a;\;y=0_{-} \tag{8.27}$$

$$\int_{a_c}^{a}\frac{\gamma_{cf}}{x-x'}\,dx'+\fint_{-a}^{a_c}\frac{\gamma_{ca}}{x-x'}\,dx'=0\qquad;\;-a\le x\le a_c;\;y=0_{\pm} \tag{8.28}$$

$$-\pi\,\gamma_{cf}+\int_{a_c}^{a}\frac{m_c}{x-x'}\,dx'=-g(x)\qquad ;\ a_c\le x\le a;\ y=0_+\tag{8.29}$$

where

$$g(x)=\pi(\sigma U+2u_0(x,0_+))\tag{8.30}$$

and

$$u_0(x,0_+)=\frac{1}{2\pi}\int_{-a}^{a}\frac{m}{x-x'}\,dx'-\frac{\gamma(x)}{2}\qquad ;\ a_c\le x\le a\tag{8.31}$$

$u_0(x,0_+)$ is the horizontal component of the velocity induced along the cavity by the section thickness and the non-cavitating vortex distribution. We can see that the cavity is "driven" by the right side of (8.29) which is composed of the cavitation number, the speed U and the horizontal induction of the fully wetted flow, u_0 on the cavity.

We now have three integral equations involving four unknown functions, γ_{cf}, γ_{ca}, m_c and the cavity length $\ell_c = a-a_c$ for a given cavitation index σ and foil geometry. The fourth relationship is given by enforcing requirement v. above. This condition for closure implies

$$\int_{a_c}^{a} m_c(x)dx=0\tag{8.32}$$

It is expeditious to fix the cavity length and solve for the required cavitation number. We can solve (8.28) for γ_{ca} in terms of γ_{cf}, using that result in (8.27). This last equation can then be inverted to give γ_{cf} as an operation on m_c which is used in (8.29) to obtain an integral equation for m_c alone.

For the inversion of (8.28) we use the general form in Appendix A (p. 484 and sequel) in which the function h(x) is now the integral of γ_{cf} in (8.28). The solution is given in (A.20) p. 489, with $a_1 = -a$ and $a_2 = a_c$. The constant K in (A.20) is found by imposing the condition vi. that $\gamma_{ca}(-a)$ is finite (the same procedure as explained on p. 71 for the flat plate). This yields

$$\gamma_{ca}=\frac{1}{\pi^2}\sqrt{\frac{a+x}{a_c-x}}\int_{-a}^{a_c}dx''\sqrt{\frac{a_c-x''}{a+x''}}\int_{a_c}^{a}\frac{\gamma_{cf}}{(x-x'')(x''-x')}\,dx'\tag{8.33}$$

Expansion of the inner kernel in partial fractions gives

$$\frac{1}{(x-x'')(x''-x')} \equiv \frac{1}{x-x'}\left[\frac{1}{x-x''} - \frac{1}{x'-x''}\right] \tag{8.34}$$

and upon interchange of the order of integration, (8.33) becomes

$$\gamma_{ca} = \frac{1}{\pi^2}\sqrt{\frac{a+x}{a_c-x}}\int_{a_c}^{a} dx'\,\frac{\gamma_{cf}}{x-x'}\int_{-a}^{a_c}\sqrt{\frac{a_c-x''}{a+x''}}\left[\frac{1}{x-x''} - \frac{1}{x'-x''}\right]dx''$$

$$;\quad -a \leq x \leq a_c \tag{8.35}$$

The x''-integrals are evaluated by (M8.21), p. 525 noting that x is inside the integration limits and x' is outside, giving

$$\gamma_{ca}(x) = \frac{1}{\pi}\sqrt{\frac{a+x}{a_c-x}}\int_{a_c}^{a}\sqrt{\frac{x'-a_c}{a+x'}}\,\frac{\gamma_{cf}}{x-x'}\,dx'\ ;\ -a \leq x \leq a_c \tag{8.36}$$

Insertion of this into (8.27), noting that x then becomes x'', yields

$$\int_{a_c}^{a}\frac{\gamma_{cf}}{x-x'}\,dx' + \frac{1}{\pi}\int_{-a}^{a_c}dx''\sqrt{\frac{a+x''}{a_c-x''}}\int_{a_c}^{a}\sqrt{\frac{x'-a_c}{a+x'}}\,\frac{\gamma_{cf}(x')}{(x-x'')(x''-x')}\,dx'$$

$$= \pi\, m_c(x)\ ;\quad a_c \leq x \leq a$$

We again expand in partial fractions, interchange orders of integration and evaluate the x''-integrals by use of (M8.22), p. 525 selectively to obtain

$$\int_{a_c}^{a}\sqrt{\frac{x'-a_c}{a+x'}}\,\frac{\gamma_{cf}}{x-x'}\,dx' = \pi\, m_c(x)\sqrt{\frac{x-a_c}{a+x}} \tag{8.37}$$

Now $h(x)$ in Appendix A is the right-hand side of (8.37) and the unknown function is $\gamma_{cf}\sqrt{(x'-a_c)/(a+x')}$. Applying the general inversion formula in Appendix A, taking the result which gives regularity at $x = a_c$, we find

$$\gamma_{cf} = -\frac{1}{\pi}\sqrt{\frac{a+x}{a-x}}\int_{a_c}^{a}\sqrt{\frac{a-x'}{a+x'}}\,\frac{m_c(x')}{x-x'}\,dx' \tag{8.38}$$

Use of (8.38) in (8.29) yields the final integral equation for m_c

$$\int_{a_c}^{a}\left[\sqrt{\frac{a-x}{a+x}}+\sqrt{\frac{a-x'}{a+x'}}\right]\frac{m_c(x')}{x-x'}\,dx' = -\,g(x)\sqrt{\frac{a-x}{a+x}} \tag{8.39}$$

The forms in the kernel suggest the substitutions

$$z = \sqrt{\frac{a-x}{a+x}}\ ;\ z' = \sqrt{\frac{a-x'}{a+x'}} \tag{8.40}$$

With these transformations (8.39) (quite miraculously!) transforms to

$$\int_0^{s}\frac{z'\,m_c(z')}{(1+z'^2)(z-z')}\,dz' = \frac{z\,g(z)}{2(1+z^2)} = G(z) \tag{8.41}$$

where $s = \sqrt{(a-a_c)/(a+a_c)} = \sqrt{\ell_c/(c-\ell_c)}$; $c = 2a$ is the chord, ℓ_c the cavity length.

This equation is of the form

$$\int_0^{s}\frac{M(z')}{z-z'}\,dz' = G(z)$$

with $M(z') = z'\,m_c(z')/(1+z'^2)$.

We may now use the general inversion formula (A.20) from Appendix A to solve for $M(z)$ and hence for $m_c(z)$ to obtain

$$m_c(z) = -\frac{1+z^2}{\pi^2 z^{3/2}\sqrt{s-z}}\left\{\int_0^{s}\frac{\sqrt{z'(s-z')}}{z-z'}\,G(z')dz' + K\right\} \tag{8.42}$$

where K is the arbitrary constant which must be determined by imposing a side condition somewhat analogous to the Kutta condition.

The behaviour of the flow at a sharp leading edge was examined by Geurst (1959) who analyzed the local velocity there from the exact solution of Helmholtz for a cavitating plate. He found the singularity in the vertical component at the leading edge to vary as $(a-x)^{-1/4}$ for $x \to a$. As $z = \sqrt{(a-x)/(a+x)}$, this corresponds to $1/\sqrt{z}$. We find K in a similar way to that in Chapter 5, p. 71 by requiring $\lim_{z\to 0} z^{3/2}m_c(z) = 0$ yielding

$$K = \int_0^s \sqrt{\frac{s - z'}{z'}}\, G(z')dz'$$

and with this (8.42) becomes upon substitution for G and z

$$m_c(z) = -\frac{1 + z^2}{2\pi\sqrt{z(s-z)}} \int_0^s \frac{\sqrt{z'(s-z')}}{1 + z'^2} \frac{\sigma U + 2u_0(x,0_+)}{z - z'} dz' \tag{8.43}$$

This gives the quarter root singularity at $x = a$ and a square-root singularity at the cavity end $x = a_c$, both of which are integrable.

Cavitation Number – Cavity Length Relation

The cavitation number, σ, can be found as a function of the cavity length, $\ell_c = a - a_c$, by evaluation of Equation (8.32) using the transformation (8.40) and the source density (8.43) and solving for σ. This relation is found to be

$$\sigma = -\frac{2}{U} \frac{\displaystyle\int_0^s dz' \frac{z'}{1+z'^2} \sqrt{\frac{s-z'}{z'}}\, u_0(z',0_+) \int_0^s \frac{1}{z-z'} \frac{1}{1+z^2} \sqrt{\frac{z}{s-z}}\, dz}{\displaystyle\int_0^s dz \frac{1}{1+z^2} \sqrt{\frac{z}{s-z}} \int_0^s \frac{1}{z-z''} \frac{z''}{1+z''^2} \sqrt{\frac{s-z''}{z''}}\, dz''}; \tag{8.44}$$

The z- and z''- integrals can be evaluated by application of relevant integrals (M8.23)-(M8.32) at the end of Section 8, Table of Airfoil Integrals of the Mathematical Compendium. This gives

$$\sigma = \frac{4\sqrt{2}}{\pi U} \frac{1}{1-\sqrt{1-\tilde{\ell}_c}} \cdot \int_0^s \frac{z}{(1-\tilde{\ell})^{1/4}} \frac{\left[1+\sqrt{1-\tilde{\ell}_c}\right]^{1/4} + z\left[1-\sqrt{1-\tilde{\ell}_c}\right]}{(1+z^2)^2} \sqrt{\frac{s-z}{z}}\, u_0(z,0_+)dz \tag{8.45}$$

where $\tilde{\ell}_c$ is the cavity length in fraction of the chord.

Evaluating for the case of a flat plate for which $m \equiv 0$ and from (5.29) $u_0(x,0_+) = -\gamma/2 = -U\alpha\sqrt{(a+x)/(a-x)} = -U\alpha/z$, we find (with assistance from the Table of Airfoil Integrals) the simple result

$$\frac{\alpha}{\sigma} = \frac{1}{2}\sqrt{\frac{1-\tilde{\ell}_c}{\tilde{\ell}_c}} \left[\frac{1 - \sqrt{1-\tilde{\ell}_c}}{1 + \sqrt{1-\tilde{\ell}_c}}\right] \tag{8.46}$$

This result can be shown to agree with that of Acosta (1955) and Geurst (1959). We observe that, in general, the cavity length is an implicit function of σ and a functional of the horizontal component of the flow induced by the camber, thickness and angle of attack of the fully wetted section.

A graph of (8.46) given in Figure 8.11 shows that two cavity lengths are given for a single value of the parameter α/σ in the range $\alpha/\sigma < 0.0962$ and none for larger values as found by Acosta (ibid.). Observations of tests of two-dimensional sections in a water tunnel by Wade & Acosta (1966) confirmed that stable short cavities exist for $\alpha/\sigma < 0.10$, giving good agreement with the theory. For $0.10 < \alpha/\sigma < 0.20$, the flow was found to give cavity oscillations having low reduced frequencies $\omega c/U$ in the range 0.10 to 0.20. For $\alpha/\sigma > 0.20$ quite steady, supercavitating flow ensued. Hence the theory is useful only for cavity/chord length, $\tilde{l}_c < 0.75$ or $\alpha/\sigma < 0.10$. Unrealistic lift and cavity cross-sectional areas are obtained for $\tilde{l}_c > 0.75$.

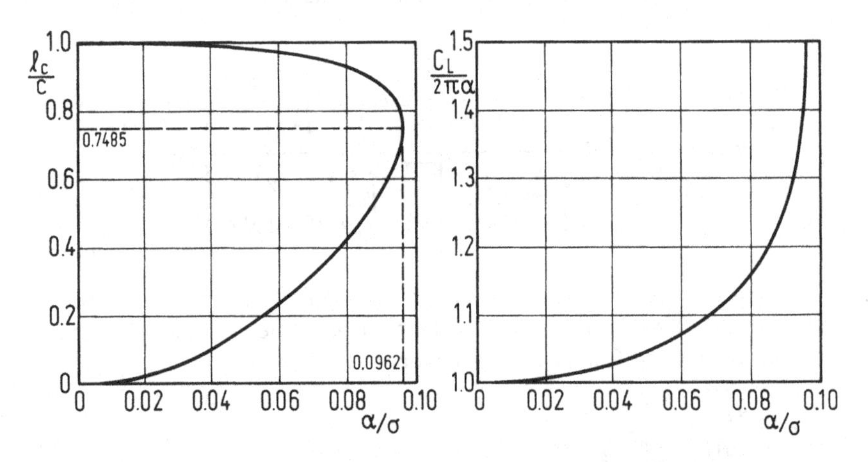

Figure 8.11 Cavity-length dependence upon α/σ for a partially cavitating flat plate.

Figure 8.12 Lift coefficient due to cavity on a flat plate.

Lift

The total sectional lift may be calculated by the Kutta-Joukowsky law

$$L = \rho U \int_{-a}^{a} \gamma(x)\, dx + L_c \tag{8.47}$$

where L_c is the alteration arising from cavitation given by

$$L_c = \frac{4a\rho U}{\pi^2}\int_0^s dz \sqrt{\frac{z}{s-z}}\,\frac{z}{1+z^2}\int_0^s \frac{1}{z-z'}\sqrt{\frac{s-z'}{z'}}\,\frac{z'}{1+z'^2}\,g(z')\,dz' \qquad (8.48)$$

where we use (8.36), (8.38) and (8.42).

For a flat plat, the lift coefficient is

$$C_L = \pi\,\frac{1+\sqrt{1-\ell_c}}{\sqrt{1-\ell_c}}\,\alpha \qquad (8.49)$$

which for $\ell_c = 0$ ($\sigma \to \infty$) gives agreement with the fully wetted result. However as $\ell_c \to 1$, $C_L \to \infty$! Again we see that the theory applies only to short cavities. It is to be noted in Figure 8.12 that the lift coefficient is composed of the fully wetted linear variation with α plus a non-linear increment which grows rapidly with α/σ to give $C_L/2\pi = 1.4$ at the largest value for which the theory may be valid. The increase in lift may be thought of as due to the effective curvature or camber provided by the cavity.

The fact that cavity length and area are highly non-linear in α/σ means that we cannot identify contributions of individual harmonics of the inflow velocity to propellers as cavity area etc. depends on the total inflow angle α to the sections.

MODIFICATION OF LINEAR THEORY

Linear theory is, of course, approximate because of the several simplifications made to secure tractable mathematics. Primary motivation to improve prediction of cavity size has arisen from the need to predict fluctuating pressures from intermittent sheet cavitation on propeller blades. Numerical experience with representations of three-dimensional unsteady cavities on blades revealed that the extent and volume of the cavities were overly large in comparison with ship and model observations.

Tulin & Hsu (1977) were the first to show by an ingenious modification of the exact flow about hydrofoils that the effect of forebody thickness is to markedly reduce cavity length and sectional area. This follows from the fact that linear theory gives a poor prediction of the horizontal velocity of the non-cavitating section $u_0(x,0_+)$ at and near the leading edge. We have in Chapters 5 and 6 and the beginning of this chapter seen that the Lighthill correction is essential to give predictions of inception of cavitation. A most adroit correction to linearized cavity flow analogous to that of Lighthill for the non-cavitating case has been developed by Kinnas (1985, 1991). Although Kinnas' procedure involves only the leading-edge radius of curvature of the section and not the entire forebody thickness distribution as does that of Tulin & Hsu (1977), it is more readily understood in light of the prior developments recounted in this text. An outline of Kinnas' correction follows.

In analogy to Lighthill (see Equation (5.51)), Kinnas (*ibid.*) takes the surface velocity on the cavity to be (in our notation),

$$q_c = (u(x,0_+) - U)\sqrt{\frac{a - x}{a - x + r_c/2}}; \; a_{ct} \lesssim x \lesssim a_{cl} \tag{8.50}$$

where a_{cl} and a_{ct} are the location of the cavity leading and trailing edges, respectively and r_c is the radius of curvature of the cavity and the foil at its leading edge. He shows that this is equivalent to the radius of curvature of the body at $x = a$.

The surface velocity q_c should also satisfy the exact condition imposed by the constancy of pressure along the cavity given by (8.22). Then, eliminating q_c between (8.50) and (8.22), yields

$$u(x,0_+) = \left[1 - \sqrt{1 + \sigma}\sqrt{\frac{a - x + r_c/2}{a - x}}\right]U \; ; \; a_{ct} \lesssim x \lesssim a_{cl} \tag{8.51}$$

This is the modified horizontal component which replaces $-\sigma U/2$ on the right side of Equation (8.26).

Equations (8.27) and (8.28) remain the same for cavities detaching from the leading edge, otherwise the upper limit a is replaced by a_{cl} for detachment at $x = a_{cl}$. Equation (8.29) is modified by replacing g by

$$g' = 2\pi\left\{\left[\sqrt{1 + \sigma}\sqrt{\frac{a - x + r_c/2}{a - x}} - 1\right]U + u_0(x,0_+)\right\}$$

and the limit a replaced by a_{cl} as noted above except in (8.31) where a is unchanged. The solution of the integral equation is identical in form as before with g replaced by g'.

Evaluations of cavity geometries without and with corrections carried out for three foils of increasing thicknesses are displayed in Figure 8.13. Here we see that linear theory shows rapid growth of cavity size with thickness in contrast to the corrected theory which shows *decreasing* cavity size with *increasing section thickness.*

Cavity areas from four calculational methods are directly compared in Figures 8.14 & 8.15 for thickness ratios of 0.06 and 0.12. Here we see that both the Kinnas and Tulin-Hsu results are close to that of the totally non-linear evaluation of Uhlman (1987) for the 6 per cent thick section.

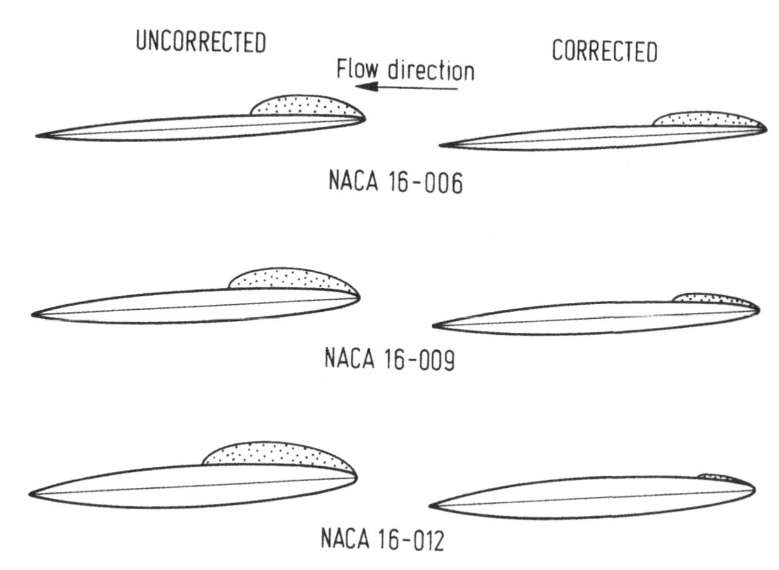

Figure 8.13 Cavity shapes given by uncorrected and corrected theory for foils of 6, 9 and 12 per cent thickness ratios at $\alpha = 4^{\mathrm{o}}$ inflow angle and a cavitation number, $\sigma = 1.07$.

Figure 6 and 10 from Kinnas, S.A. (1991). Leading–edge corrections to the linear theory of partially cavitating hydrofoils. *Journal of Ship Research*, vol. 35, no. 1, 15–27.

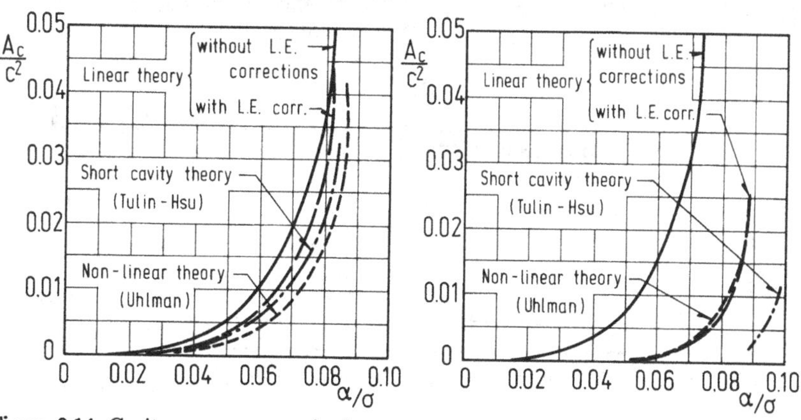

Figure 8.14 Cavity area versus α/σ for NACA–16–006 section at $\alpha = 4^{\mathrm{o}}$ as computed from various theories

Figure 8.15 Cavity area versus α/σ for NACA–16–012 section at $\alpha = 4^{\mathrm{o}}$ as computed from various theories.

Figure 8.14 and 8.15 compiled from Figures 8, 12, 14 & 15 of Kinnas, S.A. (1991). Leading–edge corrections to the linear theory of partially cavitating hydrofoils. *Journal of Ship Research*, vol. 35, no. 1, 15–27.

However for the thicker (12 per cent) section there are dramatic differences with Kinnas's answers being surprisingly close to that of Uhlman. It is clear that the correction is much smaller for thinner sections which are located at the outer radii where most of the cavitation ensues on propeller blades. Hence only the results for the 6 per cent and thinner sections are germaine to our interest.

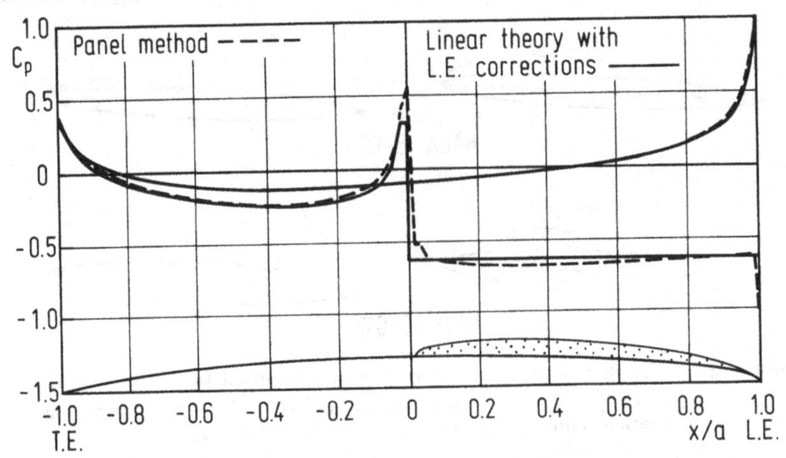

Figure 8.16 Pressure distributions on an NACA-16-009 section with a 50 per cent cavity at $\alpha = 3^{\circ}$ from panel method and linear theory with leading-edge corrections.

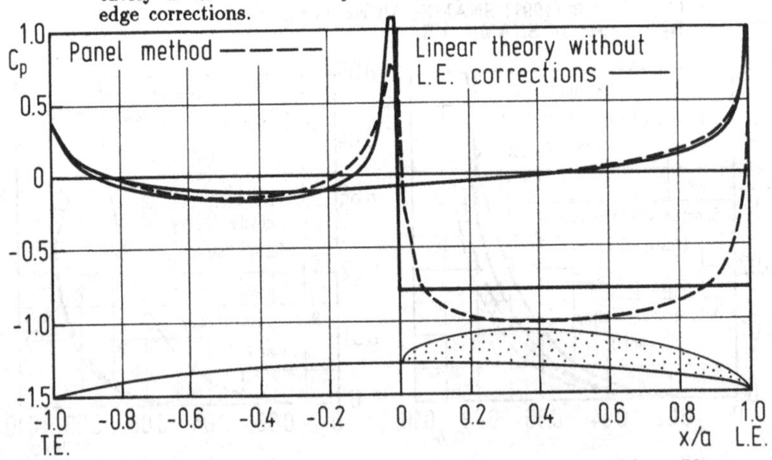

Figure 8.17 Pressure distributions on an NACA-16-009 section with a 50 per cent cavity at $\alpha = 3^{\circ}$ from panel method and linear theory without leading-edge corrections.

Figure 8.16 and 8.17 from Figures 16 and 17 of Kinnas, S.A. (1991). Leading-edge corrections to the linear theory of partially cavitating hydrofoils. *Journal of Ship Research*, vol. 35, no. 1, 15–27.

As a check of his theory Kinnas added the cavity offsets obtained from the corrected theory for an NACA-16-009 section at $\alpha = 3°$ to the foil ordinates and computed the flow about the total form using a potential-based panel method, enforcing the kinematic conditions everywhere. The pressure distributions computed this way are compared in Figure 8.16 where we see that the pressure is very nearly constant over the region where the cavity ordinates were used and agrees elsewhere except in the immediate vicinity of the end of the cavity as expected. In contrast, repeating the calculation using cavity ordinates without leading-edge corrections gave the results shown in Figure 8.17 with consequent non-constant pressure over the extent of the cavity.

The leading-edge correction procedure has been incorporated into a large program for prediction of intermittent propeller cavitation as described by Kerwin, Kinnas, Wilson & McHugh (1986). The remarkable reduction in cavity volume for the case of a propeller operating in the wake of a navy oiler is displayed in Figure 8.18, showing slower volume increase and more precipitous collapse.

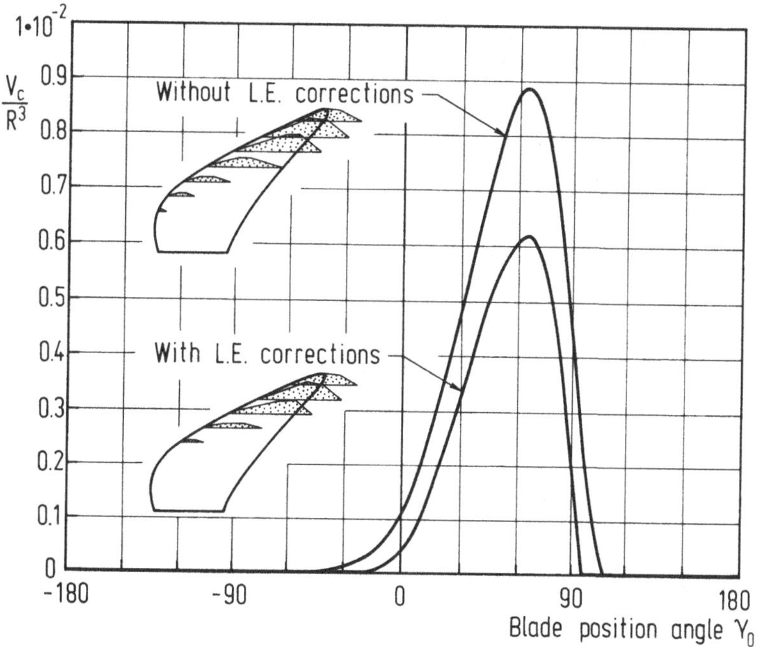

Figure 8.18 Cavity volume per blade as predicted by the propeller unsteady force program PUF–3A at MIT.

Figure 39 from Kerwin, J., Kinnas, S., Wilson, M.B. & McHugh, J. (1986). Experimental and analytical techniques for the study of unsteady propeller sheet cavitation. In *Proc. Sixteenth Symposium on Naval Hydrodynamics*, ed. W.C. Webster, pp. 387–414. Washington D.C.: National Academy Press.
By courtesy of Office of Naval Research, USA.

SUPERCAVITATING SECTIONS

At sufficiently low values of the cavitation number (about one-half the σ at inception) sheet cavitation on the back of sections extends beyond the trailing edge. These flows are of interest to designers of high-speed hydrofoils and propellers for operation at speeds in excess of 40 knots. As linear theory for supercavitating sections such as shown schematically in Figure 8.19 is analogous to that exhibited for partial cavitation we omit details, giving formulas only for the flat plate which are equivalent to those obtained by Geurst (1960) but are expressed in physical parameters.

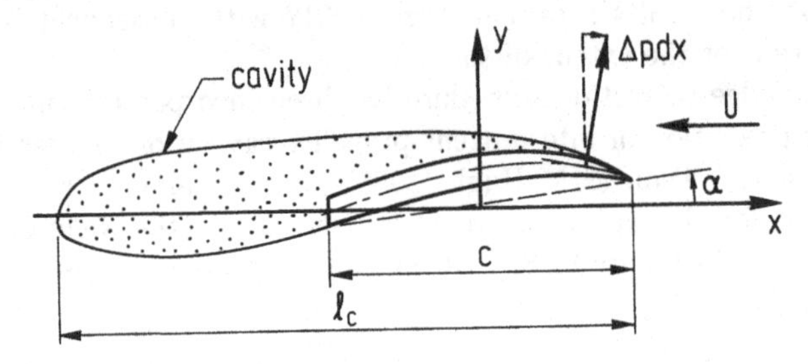

Figure 8.19 Schematic of idealized supercavitating section (cavity thickness exaggerated).

The relation of the dimensionless cavity length $\tilde{\ell}_c$ to α/σ for the supercavitating flat plate is extremely simple, being

$$\tilde{\ell}_c = 1 + \left[\frac{2\alpha}{\sigma}\right]^2 \tag{8.52}$$

Hence for $\sigma \to 0$, $\tilde{\ell}_c \to \infty$, giving an open cavity. For $\sigma \to \infty$, $\tilde{\ell}_c \to 1$ which is an unrealistic result to be discussed later. The lift coefficient is

$$C_L = \frac{\pi \tilde{\ell}_c \left[\sqrt{\tilde{\ell}_c} - \sqrt{\tilde{\ell}_c - 1}\,\right]}{\sqrt{\tilde{\ell}_c - 1}}\,\alpha \tag{8.53}$$

or

$$C_L = \frac{\pi\sigma(1 + (2\alpha/\sigma)^2)}{2}\left[\sqrt{1 + (2\alpha/\sigma)^2} - 2\alpha/\sigma\right] \tag{8.54}$$

This subsumes Tulin's (1956) result for $\sigma = 0$ as seen next. For $\sigma \to 0$, using the binomial expansion of (8.54) we find

$$C_L \to \frac{2\;\pi\;\alpha^2}{\sigma}\;\;\frac{1}{2}\left[\frac{\sigma}{2\alpha}\right] = \frac{\pi}{2}\,\alpha \tag{8.55}$$

Hence for a flat plate with a long cavity the lift rate $dC_L/d\alpha = \pi/2$ which is one fourth of that of non-cavitating sections (cf. (5.48) p. 77). This says that supercavitating propellers will produce much less vibratory forces from the non-uniform flow in the hull wake.

For $\sigma \to \infty$, $C_L \to \infty$ which is again a non-physical result because the theory becomes invalid as the end of the cavity approaches the trailing edge of the foil. Tulin (1979) has pointed out that the cavity lengths for $0.096 < \alpha/\sigma < 0.20$ as shown in Figure 8.20 given by the linearized steady theory for partial and supercavitation have not been observed experimentally because of flow instability. He developed a qualitative unsteady theory for partial cavitation on thin foils of finite aspect ratio at reduced frequences $\omega c/U$ of order unity which explains the occurence of stable flows for $C_L/2\pi\sigma < 0.096$ and instability for $C_L/2\pi\sigma > 0.096$ as well as flow hysteresis.

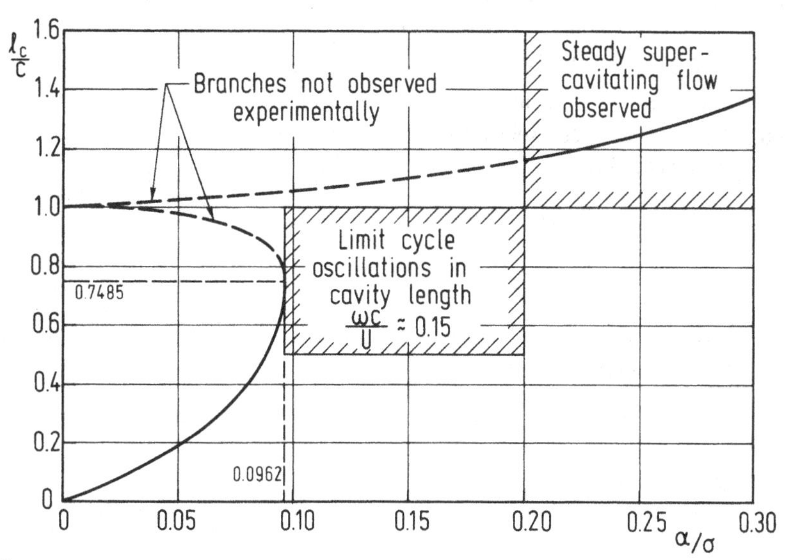

Figure 8.20 Cavity length as function of inflow angle/cavitation number for flat plate.

Figure 9 from Tulin, M.P. & Hsu, C.C. (1980). New applications of cavity flow theory. In *Proc. Thirteenth Symp. on Naval Hydrodynamics*, ed. T. Inui, pp. 107–31. Tokyo: The Shipbuilding Research Association of Japan.
By courtesy of Office of Naval Research, USA.

Cavity-caused drag is simply the horizontal component of the local pressure jump integrated over the length of the section, i.e.

$$D = \int_{-a}^{a} \Delta p(x) \frac{dy_-}{dx} dx \tag{8.56}$$

since there is no leading-edge suction force with leading-edge cavitation. For the flat plate, $dy_-/dx = \alpha$ yielding an aft directed force or $C_D = \alpha C_L$, giving a lift-drag ratio of $1/\alpha$. However, for a foil with a curved lower side, forward components can be obtained over part of the surface, as indicated in Figure 8.19. This gives the possibility of designing supercavitating sections of high lift-drag ratios. Figure 8.21 displays designs of supercavitating sections having large lift-drag ratio as developed by Posdunin (1945), by Tulin, by Newton & Rader (1961) and at Hydronautics Inc.. Those labeled TMB-Tulin and Hydronautics were designed by the use of linearized theory by Tulin & Burkart (1955) and by Johnson (1957).

Tulin & Burkart (*ibid.*) elegantly showed that the flow at $\sigma = 0$ about supercavitating sections of chord c is equivalent to that about aerodynamic sections of chord length equal to $\sqrt{c}$, and deduced the equivalence relations between the coefficients

$$C_L = \breve{C}_M \;\; ; \;\; C_D = \breve{C}_L{}^2/8\pi \tag{8.57}$$

where C_L and C_D are the hydrofoil coefficients and $\breve{C}_M$ and $\breve{C}_L$ are the moment (about the leading edge) and lift coefficient respectively of the related airfoil.

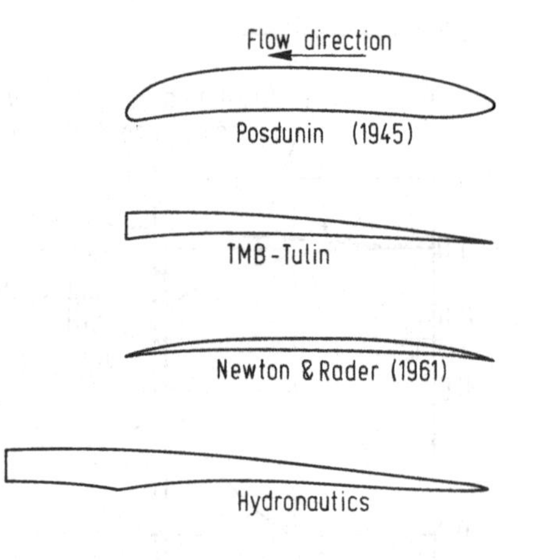

Figure 8.21 Supercavitating propeller profiles.

Figure 13 from Tulin, M. (1964). Supercavitating flows — small perturbation theory. *Journal of Ship Research*, vol. 7, no. 3, 16–37.

As a consequence of the equivalences, Tulin pointed out that the supercavitating foil of minimum drag for a fixed C_L is given by the shape of the related airfoil with all its lift concentrated at its trailing edge, i.e.,

$$\breve{C}_M = \frac{cL}{\frac{1}{2}\rho U^2 c^2} = C_L \tag{8.58}$$

giving from (8.57)

$$\left[C_L/C_D\right]_{opt} = 8\pi/C_L \tag{8.59}$$

This ideal upper bound cannot be attained in practice for as he found, the flow in the physical plane has negative cavity thickness in the region of the hydrofoil. It is mandatory to account for the relative shape of foil and cavity and, of course, to provide for sufficient foil thickness for strength. Auslaender (1962) has taken this geometry into account, yielding remarkably high lift–drag ratios.

Application of supercavitating sections to propellers must take into account the effects of the thick, trailing cavity shed by the preceding blade. Early ignorance of this proximity of a "free surface" has, in one case, resulted in a 50 per cent underdevelopment of design thrust! An approximate theory has been outlined by Tulin (1964) to correct for the presence of a nearby cavity on sectional performance. Nowadays, computer–effected representations in three dimensions include this interference effect.

Use of supercavitating sections for propellers is necessary for operation at low cavitation numbers according to a criterion devised by Tachmindji & Morgan (1958) as indicated in Figure 8.22.

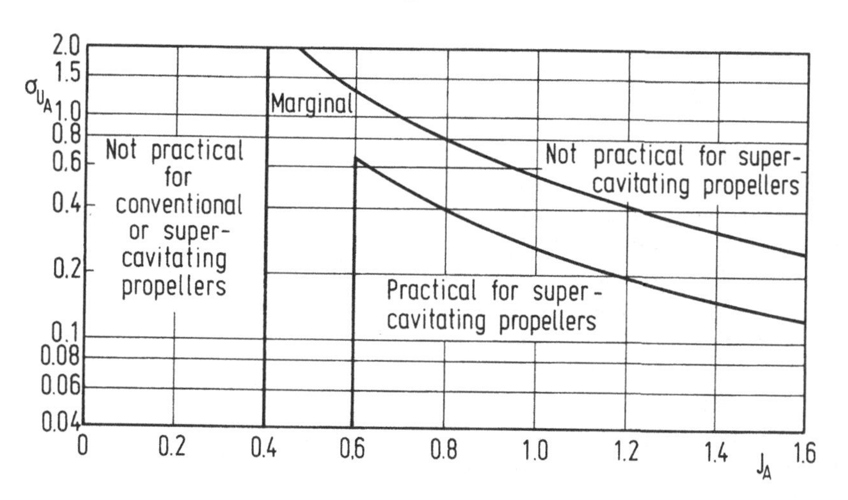

Figure 8.22 Areas for practical use of supercavitating propellers.

Figure 1 from Tachmindji, A.J. & Morgan, W.B. (1958). The design and estimated performance of a series of supercavitating propellers. In *Proc. Second Symp. on Naval Hydrodynamics*, ed. R.D. Cooper, pp. 489–532. Washington, D.C.: Office of Naval Research — Department of the Navy. By courtesy of Office of Naval Research, USA.

UNSTEADY CAVITATION

Most ship screws do not cavitate continuously as do those on very high speed craft but are prone to intermittent cavitation, generally in the case of centerline propellers, in the angular region from 10 to 2 o'clock. This could be modelled in two dimensions by considering the case of the section moving at constant speed through a stationary cyclic variation in cross flow. This is known as a gust problem and is here treated in detail in Chapter 17 for the non-cavitating section. For the cavitating sections in such a travelling transverse gust the varying cavity is modelled by a source distribution which varies with time. This yields the untenable result that at large distance the perturbation pressure becomes infinite. This obtains because the potential for the source distribution at large distance R takes the form

$$\phi = \frac{1}{2\pi}\frac{d}{dt}\int_{a_c}^{a} \eta(x',t)\ \ln R\ dx' = \frac{1}{2\pi}\frac{\partial A_c}{\partial t}\ln R \tag{8.60}$$

where $R = \sqrt{x^2 + y^2}$, η is the time-dependent cavity ordinate and $A_c(t)$ is the cavity area. We can neglect the convective part of the pressure (cf. Equation (1.30); a justification of this is given on p. 411) and the dominant pressure is then

$$p = \rho\frac{\partial\phi}{\partial t} = \rho\frac{1}{2\pi}\frac{\partial^2 A_c}{\partial t^2}\ln R \tag{8.61}$$

Hence p becomes infinite as $R \to \infty$, unless $\partial^2 A_c/\partial t^2 = 0$, which in turn requires $\partial A_c/\partial t$ to be a constant. This is the consequence of the incompressibility of the fluid and the absence of any pressure-relieving boundaries in this two-dimensional boundless domain.

In actuality there is always a free water surface and none of the bodies of interest are two-dimensional in character. Introduction of the water surface with a boundary condition appropriate to high Froude number (Appendix F) (consistent with low σ) would provide a velocity potential from the sources and their negative image in the water surface on $y \equiv 0$, cf. Figure 8.23

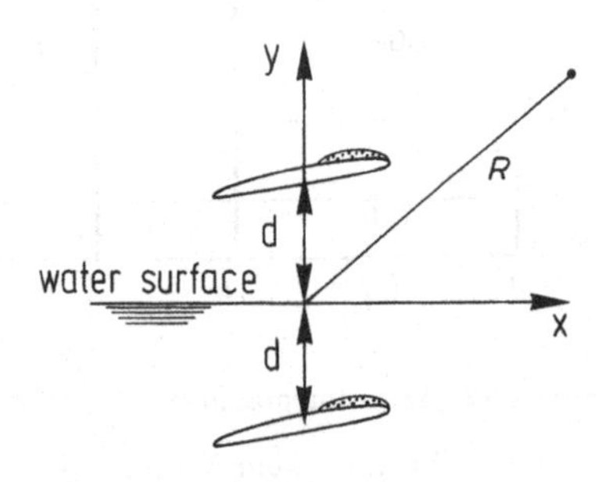

Figure 8.23

$$\phi = \frac{1}{2\pi}\frac{d}{dt}\int_{a_c(t)}^{a} \eta(x',t)\ \ln\sqrt{\frac{1 + \dfrac{d^2 + x'^2 + 2(yd - x'x)}{R^2}}{1 + \dfrac{d^2 + x'^2 - 2(yd + x'x)}{R^2}}}\,dx' \tag{8.62}$$

where d is the depth to the line source below the water surface which as $R \to \infty$ becomes

$$\phi \to -\frac{1}{\pi}\frac{\partial A_c}{\partial t}\frac{yd}{R^2} \qquad \text{or} \qquad p = -\frac{1}{\pi}\frac{\partial^2 A_c}{\partial t^2}\frac{yd}{R^2} \to 0 \quad \text{as } R \to \infty \tag{8.63}$$

This representation (together with vertical dipoles over the chord and onto the "wake" and their negative images) would be amenable to solution by computer as there is no inversion formula which applies to the coupled integral equations. (Surprisingly, this problem has not been solved to the authors' knowledge).

Intermittent cavitation on propeller blades is, in any event, very three-dimensional and must be represented by three-dimensional singularity distributions which yield convergent pressure fields. This topic is outlined in Chapter 20.

9 Actuator Disc Theory

After the foregoing excursion into basic hydrodynamics and hydrofoil theory we can now at long last begin on the subject we set out to examine: propellers. As a start we shall consider the propeller in its simplest form as an actuator disc, a circular slab of negligible thickness of diameter equal to that of the propeller. Through the disc the flow passes continuously while subjected to an axial pressure jump across the disc. For the time being we shall not bother ourselves with details, such as how this pressure jump is created.

Since we are dealing with rotational symmetry it is obviously advantageous to use a cylindrical coordinate system, with the x-axis along the axis of the disc (or propeller) and the angle γ measured from the vertical axis (z- or r-axis), cf. Figure 9.1.

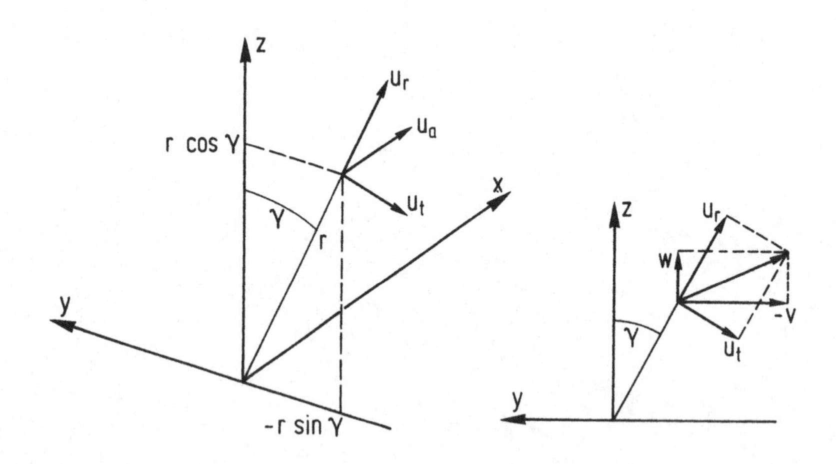

Figure 9.1 Definition of cylindrical coordinate system.

We now have for the coordinates (x remains unchanged)

$$y = -r \sin\gamma$$
$$z = r \cos\gamma \tag{9.1}$$

and for the velocities (u, v, w are x–, y– and z–components)

$$u = u_a$$
$$v = -u_t \cos\gamma - u_r \sin\gamma \tag{9.2}$$
$$w = -u_t \sin\gamma + u_r \cos\gamma$$

Application of the chain rule gives

$$\frac{\partial}{\partial y} = \frac{\partial}{\partial r}\frac{\partial r}{\partial y} + \frac{\partial}{\partial \gamma}\frac{\partial \gamma}{\partial y}$$
$$\frac{\partial}{\partial z} = \frac{\partial}{\partial r}\frac{\partial r}{\partial z} + \frac{\partial}{\partial \gamma}\frac{\partial \gamma}{\partial z} \tag{9.3}$$

which, for example by insertion of (9.1) and solving for $\partial r/\partial y$, $\partial\gamma/\partial y$ etc. yields

$$\frac{\partial r}{\partial y} = -\sin\gamma \quad ; \quad \frac{\partial r}{\partial z} = \cos\gamma$$
$$\frac{\partial \gamma}{\partial y} = -\frac{1}{r}\cos\gamma \quad ; \quad \frac{\partial \gamma}{\partial z} = -\frac{1}{r}\sin\gamma \tag{9.4}$$

We shall also need the body–mass forces which are transformed like the velocities ($F'_x = F'_a$ remains unchanged)

$$F'_y = -F'_t \cos\gamma - F'_r \sin\gamma$$
$$F'_z = -F'_t \sin\gamma + F'_r \cos\gamma \tag{9.5}$$

By using (9.2), (9.3) and (9.4) on the continuity equation (1.4) we obtain

$$\frac{\partial u_a}{\partial x} + \left[-\sin\gamma\frac{\partial}{\partial r} - \frac{1}{r}\cos\gamma\frac{\partial}{\partial \gamma}\right]\left[-u_t \cos\gamma - u_r \sin\gamma\right]$$
$$+ \left[-\cos\gamma\frac{\partial}{\partial r} - \frac{1}{r}\sin\gamma\frac{\partial}{\partial \gamma}\right]\left[-u_t \sin\gamma + u_r \cos\gamma\right] = 0 \tag{9.6}$$

which can be reduced to give the ***continuity equation in the cylindrical coordinate system*** (for an incompressible fluid)

$$\frac{\partial u_a}{\partial x} + \frac{\partial u_r}{\partial r} + \frac{1}{r}u_r + \frac{1}{r}\frac{\partial u_t}{\partial \gamma} = 0 \tag{9.7}$$

By applying a similar technique to the Euler Equations (1.8) we arrive at the ***Euler Equations in the cylindrical coordinate system*** (this is left to the reader as an exercise)

$$
\begin{aligned}
&\frac{\partial u_a}{\partial t} + u_a \frac{\partial u_a}{\partial x} + u_r \frac{\partial u_a}{\partial r} + u_t \frac{1}{r}\frac{\partial u_a}{\partial \gamma} + \frac{1}{\rho}\frac{\partial p}{\partial x} = \frac{1}{\rho} F'_a \\
&\frac{\partial u_r}{\partial t} + u_a \frac{\partial u_r}{\partial x} + u_r \frac{\partial u_r}{\partial r} + u_t \frac{1}{r}\frac{\partial u_r}{\partial \gamma} - \frac{1}{r} u_t^2 + \frac{1}{\rho}\frac{\partial p}{\partial r} = \frac{1}{\rho} F'_r \qquad (9.8) \\
&\frac{\partial u_t}{\partial t} + u_a \frac{\partial u_t}{\partial x} + u_r \frac{\partial u_t}{\partial r} + u_t \frac{1}{r}\frac{\partial u_t}{\partial \gamma} + \frac{1}{r} u_r u_t + \frac{1}{\rho r}\frac{\partial p}{\partial \gamma} = \frac{1}{\rho} F'_t
\end{aligned}
$$

To be able to use (9.8) for describing the flow around lifting surfaces we must include the effect of the surfaces. This is done by transforming the extraneous forces (F'_a, F'_r, F'_t) acting on the fluid into concentrated loads over the surfaces.

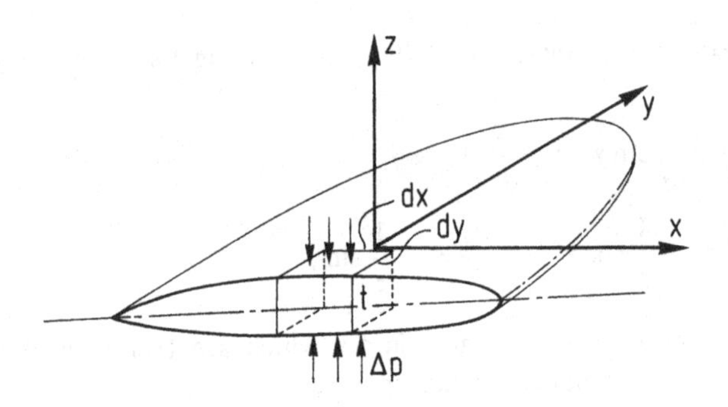

Figure 9.2

Concentrated Loads

It is obvious that when we ignore the effects of gravity the extraneous forces must be zero everywhere in the fluid outside the bodies disturbing the flow. But since our mathematical description of the flow also includes the interior of the bodies we see that the extraneous forces are non-zero in these regions even if we generally do not concern ourselves with the interior flow. For a wing of finite thickness t(x,y) placed in the x–y plane, Figure 9.2, the extraneous force term must then be

$$
\begin{aligned}
&F'_x = 0 \\
&F'_y = 0 \qquad (9.9) \\
&F'_z = -\frac{\Delta p(x,y)\ dx\ dy}{t(x,y)\ dx\ dy} = \frac{\text{Force}}{\text{volume}}
\end{aligned}
$$

since the pressure difference over the wing at the particular position is Δp. Note that we define $\Delta p = p_- - p_+$ ($= p(z = 0_-) - p(z = 0_+)$), cf. p. 68, so Δp is the net upward pressure difference which when integrated produces an upward-directed force or lift on the wing. The force on the fluid is equal but opposite, hence the minus.

We can lend credence to (9.9) by using Equation (1.8). For the wing we can assume that in the point (x,y) we have the vertical velocity $w = 0$ by the boundary condition. Then the z-component simply gives

$$\frac{\partial p}{\partial z} = F'_z$$

which can be integrated to

$$p\left[x,y,\frac{t}{2}\right] - p\left[x,y,-\frac{t}{2}\right] = \int_{-t/2}^{t/2} F'_z \, dz \tag{9.10}$$

If we adopt the concept of a wing of zero thickness, cf. Chapter 5, by letting t approach zero F'_z will become infinite according to (9.9) but the pressure difference over the wing must still be finite $\Delta p(x,y) = p(x,y,0_-) - p(x,y,0_+)$. This difficulty is overcome by use of Dirac's δ-function, cf. Section 2 of the Mathematical Compendium, which makes it possible to write (9.10) as

$$p(x,y,0_+) - p(x,y,0_-) = -\int_{-\infty}^{\infty} \Delta p(x,y) \; \delta(z')dz' \tag{9.11}$$

Since the pressure difference acts along the normal we must take this into account when specifying the components of the force. However, to find the influence of the entire surface, we must integrate over it and at the same time make sure that at the point (x,y,z) we only obtain the pressure difference at this point. This is achieved by additional application of δ-functions. The extraneous forces are then expressed as

$$\mathbf{F}' = -\,\mathbf{n}\;\Delta p(x',y',z')\;\delta(x - x')\;\delta(y - y')\;\delta(z - z')dS' \tag{9.12}$$

where

$\mathbf{n}$	is the unit normal to the lifting surface
dS'	is the surface element
(x',y',z')	are the surface coordinates over which the integration is done.

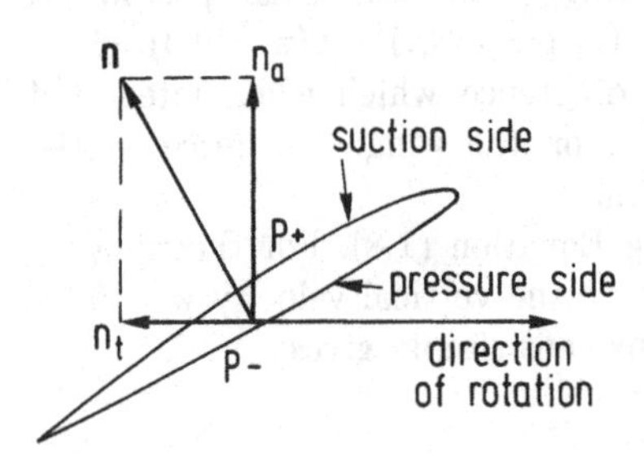

Figure 9.3

The force on an element of a propeller blade is then $\mathbf{n}\Delta p dS$ where $\mathbf{n}$ is defined in the same direction as Δp, i.e. from the pressure to the suction side. The axial component n_a will be numerically positive while the tangential component n_t for a right-handed propeller will be numerically negative, cf. Figure 9.3.

In our cylindrical system the force components of (9.8) will then become

$$F_a' = -\,n_a \frac{\Delta p(x',r',\gamma')}{r}\,\delta(x-x')\,\delta(r-r')\,\delta(\gamma-\gamma')\,r'dr'd\gamma'$$

$$F_r' = -\,n_r \frac{\Delta p(x',r',\gamma')}{r}\,\delta(x-x')\,\delta(r-r')\,\delta(\gamma-\gamma')\,r'dr'd\gamma'$$

$$F_t' = -\,n_t \frac{\Delta p(x',r',\gamma')}{r}\,\delta(x-x')\,\delta(r-r')\,\delta(\gamma-\gamma')\,r'dr'd\gamma' \tag{9.13}$$

where we have used $\delta(r(\gamma-\gamma')) = 1/r\ \delta(\gamma-\gamma')$.

HEAVILY LOADED DISC

In all textbooks of naval architecture the flow around an actuator disc is described by a partly non-linear theory. We shall try to obtain additional insight by avoiding this simplification as this allows us to deal with a disc of heavy loading. We follow the elegant theory by Wu (1962) (a landmark paper) with some alternatives and with greater exposition of mathematical details.

In the previous paragraphs of this chapter all velocities were total velocities. Since in our applications we generally consider propellers in an axial inflow of velocity $-U$ we shall distinguish between the total velocity and the perturbation velocity like we have done with wing sections in the previous chapters. We shall then use u_a for the perturbation velocity and u_a^* for the total velocity $u_a^* = u_a - U$. Since the inflow contains neither swirl nor contraction, u_r and u_t represent both total and perturbation velocities.

As a start we make a further, general modification of (9.8) by defining the velocity vector

$$
\begin{aligned}
&\mathbf{q} = (u_a^*, u_r, u_t) \\
&q = \sqrt{u_a^{*2} + u_r^2 + u_t^2} \\
&u_a^* = -U + u_a
\end{aligned}
\tag{9.14}
$$

We then add and subtract terms in each equation, respectively

$$
\frac{\partial}{\partial x}\left[\frac{1}{2}q^2\right], \frac{\partial}{\partial r}\left[\frac{1}{2}q^2\right], \frac{1}{r}\frac{\partial}{\partial \gamma}\left[\frac{1}{2}q^2\right]
$$

to obtain

$$
\begin{aligned}
&\frac{\partial u_a^*}{\partial t} + \frac{\partial}{\partial x}\left[\frac{1}{2}q^2\right] + u_t\left[\frac{1}{r}\frac{\partial u_a^*}{\partial \gamma} - \frac{\partial u_t}{\partial x}\right] - u_r\left[\frac{\partial u_r}{\partial x} - \frac{\partial u_a^*}{\partial r}\right] + \frac{1}{\rho}\frac{\partial p}{\partial x} = \frac{1}{\rho}F_a' \\
&\frac{\partial u_r}{\partial t} + \frac{\partial}{\partial r}\left[\frac{1}{2}q^2\right] + u_a^*\left[\frac{\partial u_r}{\partial x} - \frac{\partial u_a^*}{\partial r}\right] - u_t\left[\frac{\partial u_t}{\partial r} - \frac{1}{r}\frac{\partial u_r}{\partial \gamma} + \frac{1}{r}u_t\right] + \frac{1}{\rho}\frac{\partial p}{\partial r} = \frac{1}{\rho}F_r' \\
&\frac{\partial u_t}{\partial t} + \frac{1}{r}\frac{\partial}{\partial \gamma}\left[\frac{1}{2}q^2\right] + u_r\left[\frac{\partial u_t}{\partial r} - \frac{1}{r}\frac{\partial u_r}{\partial \gamma} + \frac{1}{r}u_t\right] - u_a^*\left[\frac{1}{r}\frac{\partial u_a^*}{\partial \gamma} - \frac{\partial u_t}{\partial x}\right] + \frac{1}{\rho}\frac{1}{r}\frac{\partial p}{\partial \gamma} = \frac{1}{\rho}F_t'
\end{aligned}
\tag{9.15}
$$

It can easily be verified that the third and fourth terms in each equation in (9.15) are the components of the product $\mathbf{q} \times \zeta$ where $\zeta = (\xi,\eta,\zeta)$ is given by

$$
\begin{aligned}
&\xi = \frac{\partial u_t}{\partial r} - \frac{1}{r}\frac{\partial u_r}{\partial \gamma} + \frac{1}{r}u_t \\
&\eta = \frac{1}{r}\frac{\partial u_a^*}{\partial \gamma} - \frac{\partial u_t}{\partial x} \\
&\zeta = \frac{\partial u_r}{\partial x} - \frac{\partial u_a^*}{\partial r}
\end{aligned}
\tag{9.16}
$$

The procedure above is exactly the same procedure followed in Chapter 1, p. 2-4. We then recognize ζ as the vorticity vector with components in our cylindrical system given by (9.16) (and which we shall use shortly). We also see that the gradient operator can be written in the cylindrical system as

$$
\nabla = \left[\frac{\partial}{\partial x}, \frac{\partial}{\partial r}, \frac{1}{r}\frac{\partial}{\partial \gamma}\right]
\tag{9.17}
$$

The equation of motion is then in vector notation

$$\frac{\partial \mathbf{q}}{\partial t} + \nabla\left[\frac{p}{\rho} + \frac{1}{2}q^2\right] - \mathbf{q} \times \zeta = \frac{\mathbf{F}'}{\rho} \tag{9.18}$$

(which is the same as Equation (1.14)).

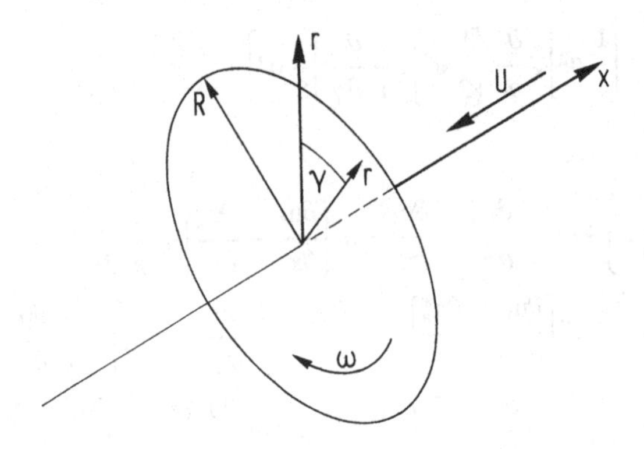

Figure 9.4 Actuator disc.

The disc of radius R is located in the plane $x = 0$ in a uniform inflow $-U$, see Figure 9.4. It applies axial, radial and tangential forces to the fluid having the components F_a, F_r, F_t, cf. (9.13) with $x' = 0$. These forces only apply on the disc. The disc is rotating with angular velocity ω.

We now impose that the flow is stationary and that it is axisymmetric so that

$$u_a^* = u_a^*(x,r)$$
$$u_r = u_r(x,r)$$
$$u_t = u_t(x,r)$$

and that $u_t = 0$ upstream of the disc and outside the race but is non-zero in the race because of the swirl imposed by the tangential blade forces on the fluid. This will be seen later.

For axisymmetric flow we can employ the Stokes' stream function ψ, cf. Milne-Thomson (1955). The axial and radial velocities are then given by

$$u_a^* = \frac{1}{r}\frac{\partial \psi}{\partial r} \quad ; \quad u_r = -\frac{1}{r}\frac{\partial \psi}{\partial x} \tag{9.19}$$

We see that ψ is constructed to satisfy the continuity equation (9.7)

$$\frac{\partial}{\partial x}\left[\frac{1}{r}\frac{\partial \psi}{\partial r}\right] + \frac{\partial}{\partial r}\left[-\frac{1}{r}\frac{\partial \psi}{\partial x}\right] + \frac{1}{r}\left[-\frac{1}{r}\frac{\partial \psi}{\partial x}\right] + 0 \equiv 0$$

The stream function is constant along stream tubes (Milne–Thomson (*ibid.*)). For the uniform stream the stream function is

$$\psi_0 = -U\frac{r^2}{2} \tag{9.20}$$

or

$$u_0 = \frac{1}{r}\frac{\partial \psi_0}{\partial r} = -\frac{1}{r}\frac{U}{2}\frac{\partial r^2}{\partial r} = -U$$

i.e., far upstream of the disc.

The kinematics of the flow is completely specified by ψ and the tangential velocity component u_t which we must now find. The geometry of the flow in any meridional plane is sketched in Figure 9.5.

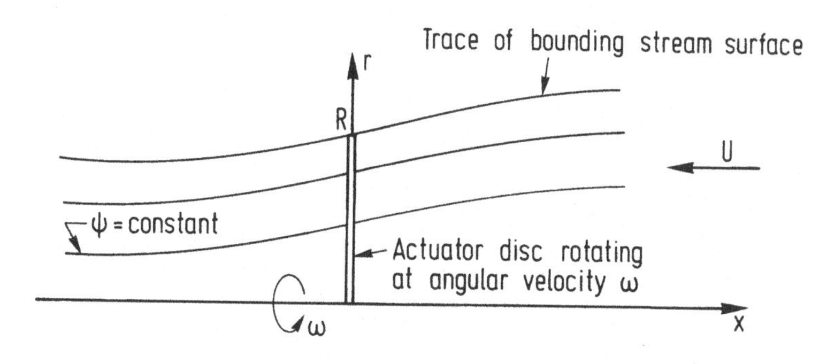

Figure 9.5 Upper half of meridional plane.

As the disc is heavily loaded the equations of motion cannot be linearized. The only simplification in (9.18) is that the derivative with respect to time vanishes because of stationary flow.

With the force terms given in (9.13), (9.18) is valid for an infinitesimal force element. To obtain the influence from the entire disc we must integrate. For the axial component

$$F'_a(r) = \int_0^R dr' \int_0^{2\pi} -n_a \frac{\Delta p(r')}{r} \delta(x)\delta(r-r')\delta(\gamma-\gamma')r'd\gamma'$$

$$= \begin{cases} -n_a\,\Delta p(r)\delta(s) & 0 < r < R \\ 0 & R < r \end{cases} \tag{9.21}$$

where the longitudinal coordinate x has been substituted by s which is the curvilinear coordinate along a trace of stream surface, cf. Figure 9.6.

For the other force components we replace n_a with the relevant component of the normal. Note that F'_a is still a force density or force per unit volume ($\Delta p \approx$ force/unit area, $\delta \approx 1$/unit length) for which reason we have kept the ' on F'_a. Note also that **n** is not the geometrical normal to the disc since this normal is axially directed. But **n** defines the direction of the forces and we must therefore keep all components. The force can now be written as

$$\mathbf{F}' = \begin{cases} -\mathbf{n}\,\Delta p(r)\delta(s) & 0 < r < R \\ 0 & R < r \end{cases} \tag{9.22}$$

From (9.22) we see that outside the disc the force term is everywhere zero leaving

$$\nabla\left[\frac{p}{\rho} + \frac{1}{2}q^2\right] = \mathbf{q} \times \boldsymbol{\zeta} \tag{9.23}$$

Taking the scalar products

$$\mathbf{q} \cdot \nabla\left[\frac{p}{\rho} + \frac{1}{2}q^2\right] = \mathbf{q} \cdot (\mathbf{q} \times \boldsymbol{\zeta}) = 0$$

$$\boldsymbol{\zeta} \cdot \nabla\left[\frac{p}{\rho} + \frac{1}{2}q^2\right] = \boldsymbol{\zeta} \cdot (\mathbf{q} \times \boldsymbol{\zeta}) = 0$$

since the vector $\mathbf{q} \times \boldsymbol{\zeta}$ is orthogonal to both $\boldsymbol{\zeta}$ and $\mathbf{q}$. Therefore the surfaces

$$\frac{p}{\rho} + \frac{1}{2}q^2 = \text{constant}$$

outside the blade row (the disc) coincide with the axisymmetric stream and vortex tubes, the latter being generally in the race of the propeller. (In the case of incident flow with shear or vorticity there will also be vortex tubes in the entire region of the ship hull wake).

Wu (1962) has pointed out that the relations of dynamical characteristics can be most clearly put in evidence when expressed in a system of curvilinear coordinates (s,ψ) in any meridional plane, where s is along a trace of a stream surface where ψ is constant and ψ varies along the normal.

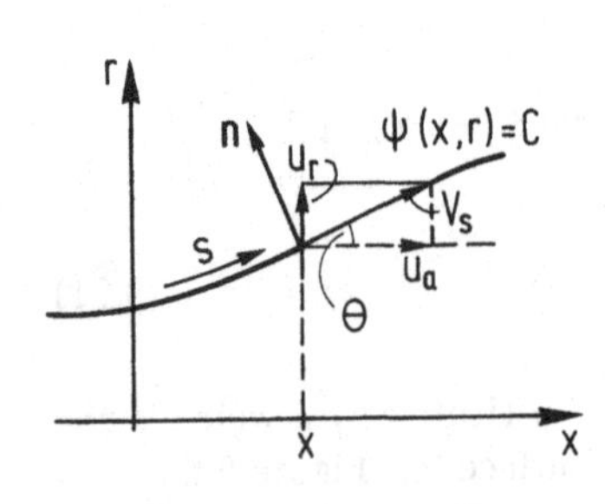

Figure 9.6

To determine the relationships between the differential operators we use Equation (3.6) or similar relations from the differential geometry of surfaces to calculate the components of the normal

$$n'_x = \frac{\partial\psi/\partial x}{\sqrt{(\partial\psi/\partial x)^2+(\partial\psi/\partial r)^2}}$$
$$n'_r = \frac{\partial\psi/\partial r}{\sqrt{(\partial\psi/\partial x)^2+(\partial\psi/\partial r)^2}} \tag{9.24}$$

By way of (9.19) we also have (as we should since $\psi(x,r) = c$ is a streamline)

$$n'_x = -\frac{u_r}{V_s} = -\sin\theta$$
$$n'_r = \frac{u_a^*}{V_s} = \cos\theta \tag{9.25}$$

where $V_s = \sqrt{u_a^{*2} + u_r^2}$, the velocity along the stream surface.

Using the chain rule and from geometrical considerations (Figure 9.6)

$$\frac{\partial}{\partial s} = \frac{\partial x}{\partial s}\frac{\partial}{\partial x} + \frac{\partial r}{\partial s}\frac{\partial}{\partial r} = \cos\theta\frac{\partial}{\partial x} + \sin\theta\frac{\partial}{\partial r}$$
$$\frac{\partial}{\partial n'} = \frac{\partial x}{\partial n'}\frac{\partial}{\partial x} + \frac{\partial r}{\partial n'}\frac{\partial}{\partial r} = -\sin\theta\frac{\partial}{\partial x} + \cos\theta\frac{\partial}{\partial r}$$

The derivatives can now be expressed by the velocities using (9.25)

$$\frac{\partial}{\partial s} = \frac{1}{V_s}\left[u_a^*\frac{\partial}{\partial x} + u_r\frac{\partial}{\partial r}\right]$$
$$\frac{\partial}{\partial n'} = \frac{1}{V_s}\left[u_a^*\frac{\partial}{\partial r} - u_r\frac{\partial}{\partial x}\right] \tag{9.26}$$

Noting that $\delta\psi/\delta n' = rV_s$ and

$$\frac{\partial}{\partial\psi} = \frac{\partial n'}{\partial\psi}\frac{\partial}{\partial n'} = \frac{1}{rV_s}\frac{\partial}{\partial n'}$$

then from (9.26)

$$\frac{\partial}{\partial\psi} = \frac{1}{rV_s^2}\left[u_a^*\frac{\partial}{\partial r} - u_r\frac{\partial}{\partial x}\right] \tag{9.27}$$

We now examine aspects of the equation of motion using (9.26) and (9.27). Consider the tangential component of (9.15) and remember that we have stationary, axisymmetric flow. We then have

$$u_r \frac{1}{r} \frac{\partial (r u_t)}{\partial r} + u_a^* \frac{\partial u_t}{\partial x} = - n_t \frac{\Delta p(r)}{\rho} \delta(s) \tag{9.28}$$

Multiplying the second term on the left side above and below with r gives for the left side

$$\frac{1}{r} \left[u_r \frac{\partial}{\partial r} + u_a^* \frac{\partial}{\partial x} \right] (r u_t)$$

whereby (9.28) becomes, by use of (9.26)

$$\frac{\partial}{\partial s} (r u_t) = - n_t r \frac{\Delta p(r)}{\rho V_s} \delta(s) \tag{9.29}$$

We integrate (9.29) along a stream tube from far upstream, $s(\infty)$, where $r u_t = 0$ to position x and note that for an accelerating disc $s(\infty) \approx -\infty$ (as can be seen from (9.26) since $u_a^* < 0$)

$$r u_t(x) = \int_{s(\infty)}^{s(x)} -n_t \frac{r \Delta p(r)}{\rho V_s} \delta(s')\, ds'$$

We carry out the integration

$$u_t(x, r(s)) = \begin{cases} -n_t \dfrac{\Delta p(r(0))}{\rho V_s} & \text{in the race abaft the propeller; } x < 0 \\ -\dfrac{1}{2} n_t \dfrac{\Delta p(r(0))}{\rho V_s} & \text{at } x = 0,\ r < R \\ 0 & \text{everywhere else} \end{cases} \tag{9.30}$$

where r(s) indicates that r varies along a stream tube. This shows that the angular momentum $r u_t$ varies along a stream tube at a rate proportional to the moment density on the blades for $x < 0$.

Taking the scalar product of (9.18) and **q** we find

$$\mathbf{q} \cdot \nabla \left[\frac{p}{\rho} + \frac{1}{2} q^2 \right] = \left[u_a^* \frac{\partial}{\partial x} + u_r \frac{\partial}{\partial r} \right] \left[\frac{p}{\rho} + \frac{1}{2} q^2 \right] = \mathbf{q} \cdot \frac{\mathbf{F}'}{\rho}$$

Using (9.26) we have

$$V_s \frac{\partial}{\partial s}\left[\frac{p}{\rho} + \frac{1}{2} q^2\right] = \mathbf{q} \cdot \frac{\mathbf{F}'}{\rho} \tag{9.31}$$

This equation shows that the rate of change of total head $p/\rho + 1/2\ q^2$ within the blade row (where $\mathbf{F}'$ is non-zero) is directly proportional to the rate of working of the blade forces and this rate is zero elsewhere.

However, the blade-force components must be such in an ideal fluid that their direction is perpendicular to the *relative* velocity. For rotation at angular velocity ω this condition is

$$(\mathbf{q} - \mathbf{i}_t \omega r) \cdot \mathbf{F}' = 0 \qquad \text{at } x = 0 \tag{9.32}$$

where $\mathbf{i}_t$ is the unit vector in the tangential direction. Using (9.31) in (9.32) gives

$$\frac{\partial}{\partial s}\left[\frac{p}{\rho} + \frac{1}{2} q^2\right] = \frac{\omega r}{V_s}\left[- n_t \frac{\Delta p(r)}{\rho}\, \delta(s)\right] \tag{9.33}$$

Eliminating terms between (9.29) and (9.33)

$$\frac{\partial}{\partial s}\left[\frac{p}{\rho} + \frac{1}{2} q^2\right] = \omega \frac{\partial}{\partial s}(r u_t) \tag{9.34}$$

and integrating along the stream tube

$$\frac{p}{\rho} + \frac{1}{2} q^2 = \begin{cases} \dfrac{p_\infty}{\rho} + \dfrac{1}{2} U^2 + \omega r u_t & \text{in the race abaft the propeller} \\ \dfrac{p_\infty}{\rho} + \dfrac{1}{2} U^2 & \text{everywhere else} \end{cases} \tag{9.35}$$

To secure relations giving the derivative of the total head normal to the stream surface, i.e. $\partial(p/\rho + 1/2\ q^2)/\partial\psi$, Wu (*ibid.*) examined the tangential component of the vector product of $\mathbf{q}$ and (9.18), i.e.

$$\left[\mathbf{q} \times \nabla\left[\frac{p}{\rho} + \frac{1}{2} q^2\right]\right]_t - \frac{1}{\rho}(\mathbf{q} \times \mathbf{F}')_t = (\mathbf{q} \times (\mathbf{q} \times \boldsymbol{\zeta}))_t \tag{9.36}$$

From vector analysis the triple vector product formula is

$$\mathbf{A} \times (\mathbf{B} \times \mathbf{C}) = (\mathbf{A} \cdot \mathbf{C})\mathbf{B} - (\mathbf{A} \cdot \mathbf{B})\mathbf{C}$$

hence

$$\mathbf{q} \times (\mathbf{q} \times \boldsymbol{\zeta}) = (\mathbf{q} \cdot \boldsymbol{\zeta})\mathbf{q} - q^2 \boldsymbol{\zeta}$$

As $\boldsymbol{\zeta} = (\xi, \eta, \zeta)$ we obtain for the tangential component of the right side of (9.36)

$$(u_a^*\xi + u_r\eta + u_t\zeta)u_t - (u_a^{*2} + u_r^2 + u_t^2)\zeta$$

$$= u_t\left[u_a^*\frac{1}{r}\frac{\partial(ru_t)}{\partial r} - u_r\frac{1}{r}\frac{\partial(ru_t)}{\partial x}\right] - V_s^2\zeta$$

$$= u_tV_s^2\frac{\partial(ru_t)}{\partial\psi} - V_s^2\zeta$$

using (9.16) and (9.27) and $V_s^2 = u_a^{*2} + u_r^2$.

By manipulation of the left side of (9.36) (including use of (9.27)) we obtain

$$rV_s^2\frac{\partial}{\partial\psi}\left[\frac{p}{\rho} + \frac{1}{2}q^2\right] + (u_a^*n_r - u_rn_a)\frac{\Delta p(r)}{\rho}\delta(s)$$
$$= u_tV_s^2\frac{\partial(u_tr)}{\partial\psi} - V_s^2\zeta \tag{9.37}$$

From (9.16) and (9.19) we have

$$\zeta = -\frac{1}{r}\frac{\partial^2\psi}{\partial x^2} + \frac{1}{r^2}\frac{\partial\psi}{\partial r} - \frac{1}{r}\frac{\partial^2\psi}{\partial r^2}$$

and using this and (9.35) we can obtain for (9.37)

$$\frac{\partial^2\psi}{\partial x^2} - \frac{1}{r}\frac{\partial\psi}{\partial r} + \frac{\partial^2\psi}{\partial r^2}$$
$$= (\omega r^2 - u_tr)\frac{\partial(u_tr)}{\partial\psi} + r(u_a^*n_r - u_rn_a)\frac{\Delta p(r)}{\rho V_s^2}\delta(s) \tag{9.38}$$

The non-linearity is two-fold. The right side is clearly a non-linear function of ψ since we have $u_tr = f(\psi)$ and the boundary of the slipstream or race wherein $f = u_tr \neq 0$ is unknown, i.e. the locus of the bounding stream surface is unknown.

The radiation conditions are

$$\frac{\partial\psi(+\infty,r)}{\partial r} = -U$$
$$\frac{\partial\psi(+\infty,r)}{\partial x} = 0 \tag{9.39}$$
$$\frac{\partial\psi(-\infty,r)}{\partial x} = 0$$

An analytic solution of (9.38) appears to be impossible. Instead we now convert (9.38) to an integral equation as such equations are more amenable to iterative solution.

The Integral Equation for the Perturbation Stream Function

We introduce the perturbation stream function $\psi'(x,r)$ and the free stream function $\psi_0(r)$ defined by

$$\psi(x,r) = \psi_0(r) + \psi'(x,r) \tag{9.40}$$

where

$$\psi_0(r) = -U \frac{r^2}{2}$$

In order to obtain $+ 1/r \cdot \partial/\partial r$ in the operator yielding a Bessel equation we can now rewrite (9.38) in a form similar to Wu's (1962)

$$\left[\frac{\partial^2}{\partial x^2} + \frac{\partial^2}{\partial r^2} + \frac{1}{r}\frac{\partial}{\partial r} - \frac{1}{r^2}\right]\left[\frac{\psi'}{r}\right] = -\frac{g(x,r,\psi)}{r} \tag{9.41}$$

where

$$g(x,r,\psi) = -(\omega r^2 - f(\psi))\frac{\partial f(\psi)}{\partial \psi} - r(u_a^* n_r - u_r n_a)\frac{\Delta p(r)}{\rho V_s^2}\,\delta(s)$$

$$f(\psi) = u_t r \qquad (9.42)^{14}$$

Note that f is only non-zero inside the slipstream region.

From (9.39) and (9.40) it is obvious that the boundary conditions on the perturbation stream function are homogeneous

$$\frac{\partial \psi'(+\infty,r)}{\partial r} = 0 \;;\quad \frac{\partial \psi'(+\infty,r)}{\partial x} = 0 \;;\quad \frac{\partial \psi'(-\infty,r)}{\partial x} = 0 \tag{9.43}$$

The bounding streamline $r = R_b(x)$ will pass through the outer edge of the actuator disc at $x = 0$, $r = R_b(0) = R$. Then the bounding stream tube, ψ = constant, is given by

$$\psi(x,r) = \psi(0,R) = \psi_0(R) + \psi'(0,R)$$

$$= -U\frac{R^2}{2} + \psi'(0,R)$$

or

$$\psi'(x,r) = \frac{U}{2}(r^2 - R^2) + \psi'(0,R) \tag{9.44}$$

[14] Wu does not retain the last term in g.

To convert (9.41) to an integral equation we use Green function technique, see Mathematical Compendium, section 3. The Green function in our case must satisfy the same boundary condition as ψ'/r and the field equation (cf. (M3.5))

$$\left[\frac{\partial^2}{\partial x^2}+\frac{\partial^2}{\partial r^2}+\frac{1}{r}\frac{\partial}{\partial r}-\frac{1}{r^2}\right]G=-\frac{\delta(x-x')\delta(r-r')}{r} \tag{9.45}$$

We then express ψ' in terms of G making use of Green's second identity (M3.3). In here we take $\phi \approx G$, $\psi \approx \psi'/r$, the Laplace operator $\nabla^2 = \partial^2/\partial x^2 + \partial^2/\partial r^2 + 1/r\cdot\partial/\partial r$, the volume element $dV = r'dr'dx'$ and let S stretch to infinity. We also use the sifting properties of the δ-function to obtain

$$\psi' = r\int_{-\infty}^{\infty}dx'\int_0^{\infty} g(x',r')\ G(x,r;x',r')\ dr' \tag{9.46}$$

The derivation of G (omitted by Wu (*ibid.*)) can be done by using the Hankel transform and its inversion which are, cf. Sneddon (1951)

$$\begin{aligned}\hat{H}_n(x,k) &= \int_0^{\infty} rJ_n(kr)\ H(x,r)dr\\ H(x,r) &= \int_0^{\infty} kJ_n(kr)\ \hat{H}_n(x,k)dk\end{aligned} \tag{9.47}$$

The presence of the term $-1/r^2$ in the operator of (9.45) signals that one should use the Hankel transform of order unity ($n = 1$). Multiplying (9.45) by $rJ_1(kr)$ and integrating, we have

$$\int_0^{\infty}\left\{\frac{d^2G}{dr^2}rJ_1(kr)+\frac{1}{r}\frac{dG}{dr}rJ_1(kr)-\frac{G}{r^2}rJ_1(kr)\right\}dr+\frac{d^2\hat{G}_1}{dx^2}$$

$$=-\int_0^{\infty}\frac{\delta(x-x')\delta(r-r')}{r}rJ_1(kr)dr \tag{9.48}$$

where $\hat{G}_1$ is the Hankel transform of G. The terms on the left side of (9.48) containing derivatives of G must integrated by parts (d^2G/dr^2 integrated twice). During this process we must also use that G must satisfy the same boundary conditions as ψ' and the behaviour of J_1 at 0 and ∞. We then have

$$\int_0^{\infty}\frac{1}{r}\left[r^2k^2J_1''(rk)+rkJ_1'(rk)-J_1\right]Gdr+\frac{d^2\hat{G}_1}{dx^2}=-J_1(kr')\delta(x-x') \tag{9.49}$$

By remembering the differential equation fulfilled by the Bessel function (cf. Abramowitz & Stegun (1972)) the integral can be transformed into

$$\int_0^{\infty}\frac{1}{r}(-(rk)^2J_1(rk))Gdr$$

We recognize the Hankel transform of G, (9.47), and can now write (9.49) as

$$\left[\frac{d^2}{dx^2} - k^2\right]\hat{G}_1 = -\,J_1(kr')\delta(x-x') \tag{9.50}$$

As the right side of (9.50) is zero for $x > x'$ and $x < x'$ we have the following solutions of the homogeneous equations in these regions

$$\begin{aligned}\hat{G}_{1+} &= A_1\, e^{-k(x-x')} \qquad && x' \lesssim x \\ \hat{G}_{1-} &= A_2\, e^{k(x-x')} \qquad && x \lesssim x'\end{aligned} \tag{9.51}$$

having discarded the solutions which are unbounded at $\pm\infty$. There are two conditions that $\hat{G}_1$ must satisfy

$$\begin{aligned}&\text{i.} \quad \hat{G}_{1+}(x') = \hat{G}_{1-}(x') \quad ; \quad \text{continuity at } x' \\ &\text{ii.} \quad \frac{d\hat{G}_{1+}}{dx} - \frac{d\hat{G}_{1-}}{dx} = -\,J_1(kr')\end{aligned} \tag{9.52}$$

(This last condition follows from integration of (9.50) over x from $x' - \varepsilon$ to $x' + \varepsilon$ and passing to the limit as $\varepsilon \to 0$. The integral of $k^2\hat{G}_1$ contributes nothing as $\hat{G}_1$ is continuous). Then $A_1 = A_2$ from condition i. and ii. yields

$$A_1 = \frac{J_1(kr')}{2k}$$

whence

$$\hat{G}_1 = \frac{J_1(kr')}{2k}\, e^{-k|x-x'|} \qquad \text{for all } x \tag{9.53}$$

and the inversion formula (9.47) gives

$$G(x,r;x',r') = \frac{1}{2}\int_0^\infty J_1(kr)J_1(kr')e^{-k|x-x'|}\,dk \tag{9.54}$$

We shall now go a little step further and express the k–integrals by the Legendre function of degree 1/2, see Mathematical Compendium, Section 5, Equation (M5.15)

$$\int_0^\infty J_1(kr)J_1(kr')e^{-k|x-x'|}\,dk = \frac{1}{\pi\sqrt{rr'}}\,Q_{\frac{1}{2}}(Z) \tag{9.55}$$

and we then have for G

$$G(x,r;x',r') = \frac{1}{2\pi\sqrt{rr'}}\,Q_{\frac{1}{2}}(Z) \tag{9.56}$$

where

$$Z = \frac{(x - x')^2 + r^2 + r'^2}{2rr'}$$

We now insert (9.56) together with (9.42) into (9.46) and obtain the solution for ψ'

$$\psi' = -\frac{1}{2\pi}\int_{-\infty}^{0} dx' \int_{0}^{R_b(x')} (\omega r' - u_t)\,\frac{\partial(u_t r')}{\partial\psi}\sqrt{rr'}\;Q_{\frac{1}{2}}(Z)dr'$$

$$-\frac{1}{2\pi}\int_{0}^{n(R)} (u_a^* n_r - u_r n_a)\,\frac{\Delta p(r')}{\rho V_s^2}\sqrt{rr'}\;Q_{\frac{1}{2}}(Z)dn' \qquad (9.57)$$

where in the first term we have taken account of the fact that u_t is non-zero only in the race and the sifting by $\delta(s')$ has been included in the last term.

The last term in the right hand side of (9.57) does not appear in Wu's integral equation as he contends that the term is small compared to the first term and moreover argues that in the regions of interest outside of the blade row, this term does not contribute. Greenberg (1972) does not include the term either, with the argumentation that it would produce a jump in the radial velocity over the disc. We shall see this later.

Whereas Wu (*ibid.*) did not include numerical results, Greenberg (1972) calculated streamlines and velocities by solving a non-linear integral equation similar to (9.57), (without the last term on the right side of the equation). Results similar to his are shown in Figure 9.7 for three different disc loadings[15]. It is clearly seen that with increasing disc loading C_{Th} the effects of non-linearities become more pronounced. The distribution of loading is representative of propellers (Greenberg (*ibid.*)).

From Figure 9.7a, the heavily loaded disc, it is seen that the flow at the tip is nearly in the plane of the disc. Analyses and calculations by Schmidt & Sparenberg (1977) (for a disc of constant loading) indicate that the flow in this region is even more complex. Their results show that in the neighborhood of the edge there is a region where the fluid particles cross the disc more than once! In the larger part of this region the particles come in from behind, cross the disc twice, turn around the edge and then pass again aft through the front of the disc. For further details the reader should consult the references.

An identification of the physical significance of Equation (9.57) (in the form of Wu's expressions) has been provided by Greenberg & Powers (1970). The kernel function $1/2\pi\sqrt{rr'}\;Q_{\frac{1}{2}}(Z)$ is the stream function of a ring vortex of unit strength located in general at $x = x'$, $r = r'$ in a plane perpendicular to the x-axis as shown in Figure 9.8.

[15] The results presented here are obtained by the computer program listed by Greenberg & Powers (1970) and re-implemented by the authors. The authors are grateful not only for the permission to use this efficient code but also for the help and advice of Professor Greenberg both in connection with the implementation of the code and with various aspects of disc theory.

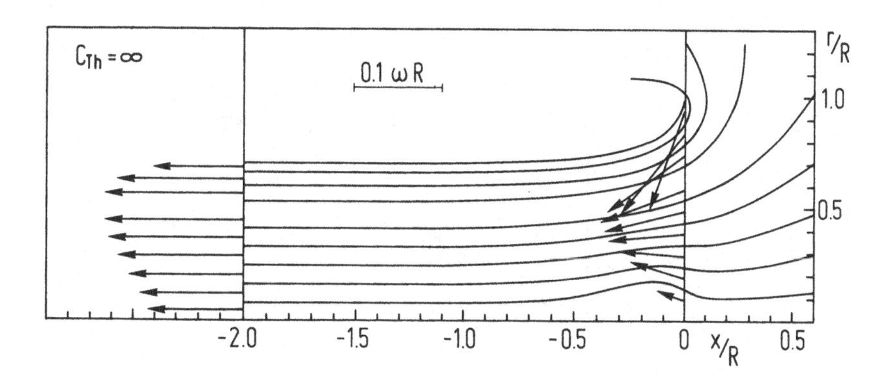

Figure 9.7a Stream tubes for non-linear actuator disc in uniform inflow, $C_{Th} = \infty$.

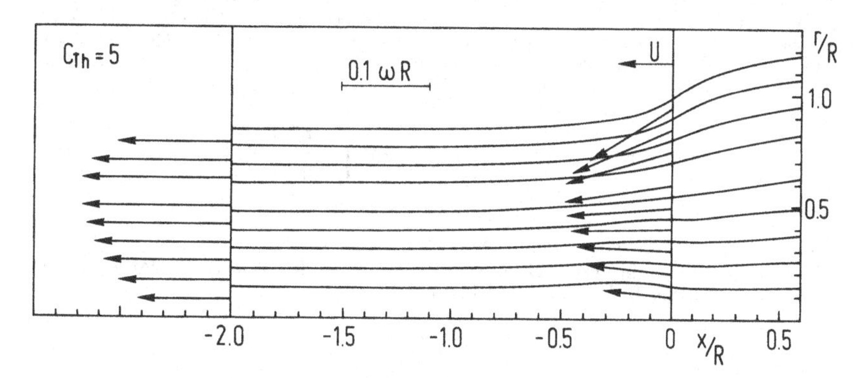

Figure 9.7b Stream tubes for non-linear actuator disc in uniform inflow, $C_{Th} = 5$.

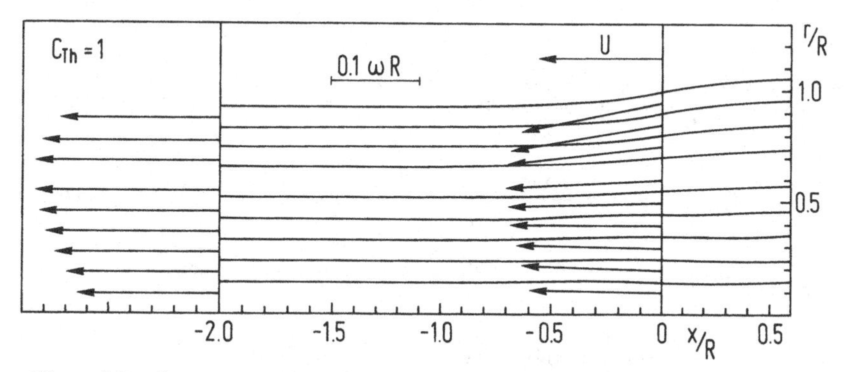

Figure 9.7c Stream tubes for non-linear actuator disc in uniform inflow, $C_{Th} = 1$.

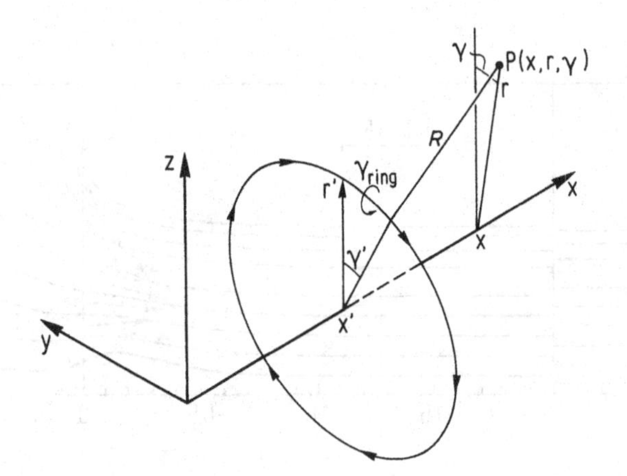

Figure 9.8 Geometry of a ring vortex at x = x', r = r'.

As Greenberg & Powers (1970) made this assertion without proof we give a demonstration of this here for sake of completeness. We have seen in Chapter 1 (p. 23), that the velocity potential of a ring vortex of strength γ_{ring} is given by a uniform distribution of dipoles normal to the disc bounded by the ring. Thus for a ring at x = x' and radius r = r' we have from Equation (1.90)

$$\phi_{ring}(x,r,x',r') = -\frac{\gamma_{ring}}{4\pi}\frac{\partial}{\partial x''}\int_0^{2\pi} d\gamma' \int_0^{r'} \frac{r''}{R}\bigg|_{x''=x'} dr'' \tag{9.58}$$

where $R = \sqrt{(x - x'')^2 + r^2 + r''^2 - 2rr''\cos(\gamma - \gamma')}$.

We now express R in terms of integrals over Bessel functions, cf. Mathematical Compendium, Section 5, Equation (M5.14), p. 516

$$\phi_{ring}(x,r,x',r') = -\frac{\gamma_{ring}}{4\pi}\frac{\partial}{\partial x''}\int_0^{2\pi} d\gamma' \int_0^{r'} dr''r'' \cdot \sum_{m=0}^{\infty} \varepsilon_m \int_0^{\infty} J_m(kr)J_m(kr'')e^{-k|x-x''|}\cos m(\gamma-\gamma')\bigg|_{x''=x'} dk \tag{9.59}$$

In this way we have separated out the angular dependence and the integration over γ' can easily be carried out, picking out the m = 0 term yielding 2π. Using that in this context $\partial/\partial x'' = -\partial/\partial x$ gives

$$\phi_{ring}(x,r,x',r') = \frac{\gamma_{ring}}{2}\frac{\partial}{\partial x}\int_0^{r'} dr''r'' \int_0^{\infty} J_0(kr)J_0(kr'')e^{-k|x-x'|}dk \tag{9.60}$$

Now, to calculate the stream function ψ_{ring} we can integrate the radial velocity over x (cf. Equation (9.19))

$$\psi_{ring} = -r\int u_r(x)dx + \Psi(r) = -r\int \frac{\partial\phi_{ring}}{\partial r}dx + \Psi(r)$$

where $\Psi(r)$ is to be determined later. Since ϕ_{ring} is expressed as a derivative in x we find easily

$$\psi_{ring} = -\frac{\gamma_{ring}}{2} r \int_0^{r'} dr'' r'' \int_0^{\infty} k J_1(kr) J_0(kr'') e^{-k|x-x'|} dk \tag{9.61}$$

and where we have used that $\partial J_0(kr)/\partial r = -kJ_1(kr)$. From the Bessel function relations, cf. Abramowitz & Stegun (1972)

$$J_0(kr'') = \frac{\partial J_1(kr'')}{\partial(kr'')} + \frac{1}{kr''} J_1(kr'') = \frac{1}{kr''} \frac{\partial}{\partial r''} (r'' J_1(kr'')) \tag{9.62}$$

so (9.61) becomes, by integration

$$\psi_{ring} = \frac{\gamma_{ring}}{2} rr' \int_0^{\infty} J_1(kr) J_1(kr') e^{-k|x-x'|} dk + \Psi(r) \tag{9.63}$$

As ψ_{ring} must vanish for $|x - x'| \to \infty$ and also for $r \to \infty$ then $\Psi(r) = 0$. Finally setting $\gamma_{ring} = 1$ and using (9.55) gives the result.

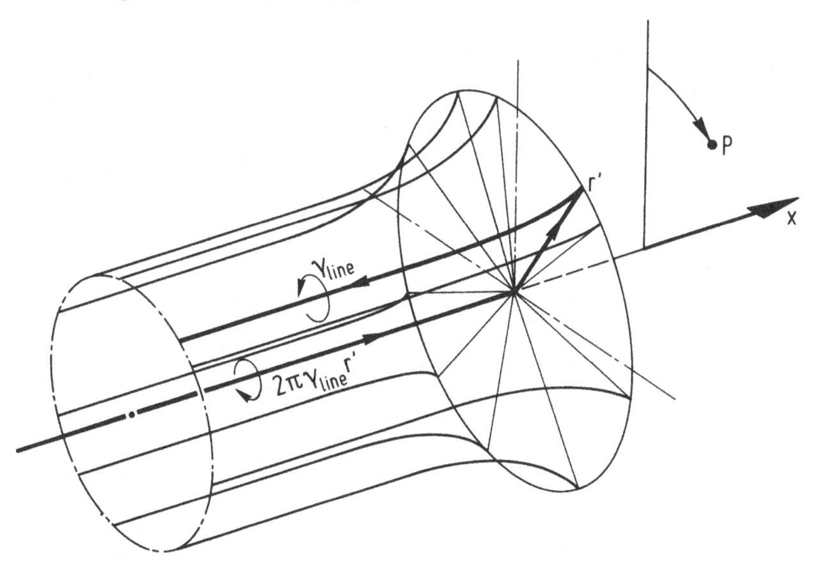

Figure 9.9 Non-tangentially directed vortices.

To understand the meaning of the tangential velocity u_t consider the system of vortices outlined in Figure 9.9. The system consists of a "horseshoe" vortex filament, cf. Figure 2.6, p. 37 of strength $\gamma_{line} d\gamma dn$ with one leg along the negative x-axis and the whole horseshoe rotated about the x-axis. The bounded parts lie in the disc plane x = 0 directed radially but along the normal to the stream surfaces like (curved) spokes in a wheel. The free parts not on the negative x-axis are assumed to lie on the stream tube described by ψ = constant in the previous derivations. There is rotational symmetry and the strength of the vortex on the x-axis is therefore $d\Gamma_{line} = 2\pi r'\gamma_{line}dn'$. Instead of going through a series of lengthy manipulations we shall use the following argumentation to find the velocities:

First, because of symmetry, one can always find vortex filaments whose axial and radial velocity contributions separately annul.

Secondly, also because of symmetry, the tangential velocity u_t is independent of angle γ. We can then find the circulation by integrating along any circular path, lying in a plane perpendicular to the x-axis and with the axis as center

$$d\Gamma_{line} = \int_0^{2\pi} du_t(x,r)d\gamma = 2\pi r\, du_t(x,r) \tag{9.64}$$

If we apply this procedure upstream of the disc or outside the vortex tube $r > r'(x)$ we do not encompass any net circulation, hence $d\Gamma_{line} = 0$. But inside, i.e. $r < r'(x)$, $x < 0$, we have

$$2\pi r\, du_t(x,r) = 2\pi r'\gamma_{line}dn' \tag{9.65}$$

since we encompass the vortex along the negative x-axis. We can then write

$$du_t(x,r) = \frac{\gamma_{line}r'}{r}(1 - H(n - n'))dn' \quad \text{for } x < 0 \tag{9.66}$$

where H(n) is the Heaviside step function, cf. Mathematical Compendium, Section 2, p. 504.

To obtain the entire contribution from all vortex tubes we integrate along the normal to the vortex tubes or streamlines (cf. Figure 9.10)

$$u_t(x,r) = \int_0^{n'_R} \frac{\gamma_{line}(n'')r'(n'')}{r} \cdot (1 - H(n' - n''))dn'' \tag{9.67}$$

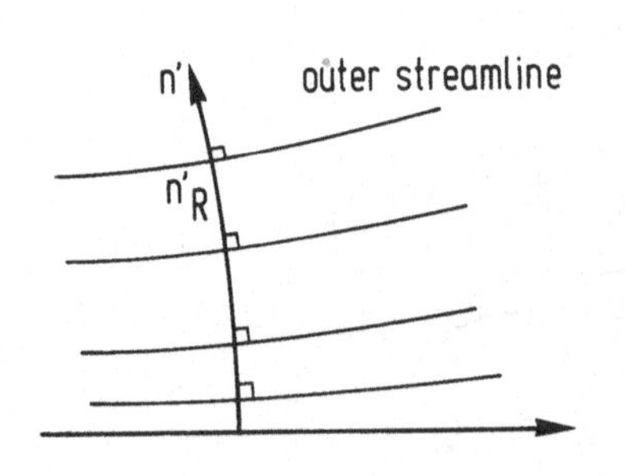

Figure 9.10 Integration path along normal.

where (x,r) corresponds to (s,n) in the streamline system. By use of (9.64) and (9.65) we can introduce $d\Gamma_{line}/dn'$ in (9.67) which we can integrate by parts

$$u_t(x,r) = \frac{1}{2\pi r}\int_0^{n'_R} \Gamma_{line}(n'')\delta(n'-n'')dn'' = -\frac{\Gamma_{line}(n')}{2\pi r} \tag{9.68}$$

and using (9.64) and (9.65) on this result gives

$$\gamma_{line}(n) = -\frac{1}{r}\frac{\partial(u_t r)}{\partial n'} \tag{9.69}$$

which is the strength of the vortices. Note that γ_{line} (as well as γ_{ring} below) has the dimension 1/sec since the vortices are distributed in the race, i.e. both in the radial and circumferential direction.

We saw previously that the induced flow described by the first term on the right side of (9.57) was created by ring vortices lying in the wake abaft the disc. From the derivation it is clear that rings of constant vorticity are aligned on stream surfaces and from (9.57) that the vorticity is

$$\gamma_{ring} = -(\omega r - u_t)\frac{\partial(u_t r)}{\partial\psi} \tag{9.70}$$

The additional vortex lines inducing the tangential velocities are also aligned on the stream surfaces having vortex strength given by (9.69) as just described. The resulting vortex system, obtained by adding the ring and line vortices, as outlined in Figure 9.11, will still lie on the stream tubes. The pitch angle, defined in the figure is, by use of (9.26) and (9.27)

$$\mathrm{tg}\beta_i = \frac{\gamma_{line}}{\gamma_{ring}} = \frac{-\frac{1}{r}\frac{\partial(u_t r)}{\partial n'}}{-(\omega r - u_t)\frac{\partial(u_t r)}{\partial\psi}} = \frac{V_s}{\omega r - u_t} \tag{9.71}$$

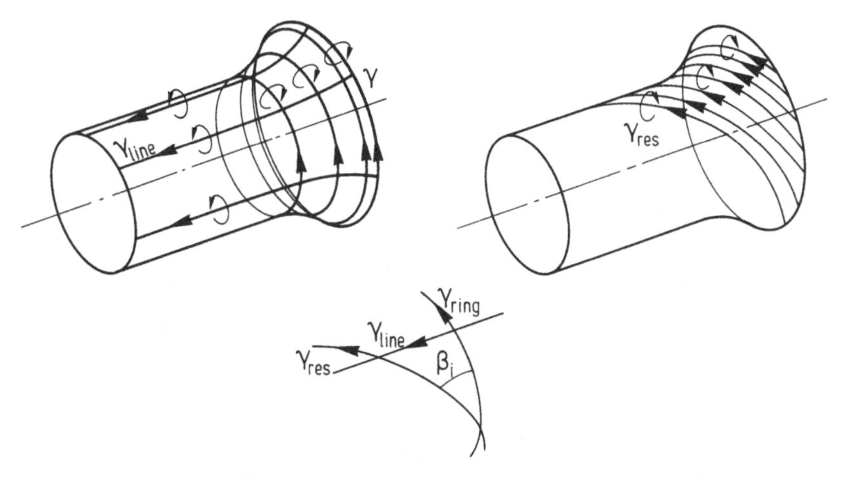

Figure 9.11 Vortex system behind disc.

Now we see that the flow abaft the disc is described by vortices that trail after the rotating disc along the streamlines, as we should have expected since the trailing vortices must be force free. However, since the vortices are spread over the entire wake volume, we conclude that the actuator disc models a propeller with infinitely many blades.

We have now just seen a demonstration of the theorem of ***Helmholtz*** which says that

- ***The vortex lines in an inviscid fluid move with the fluid.***

We shall later (Chapter 11) use Helmholtz's theorem to model a similar geometry of the trailing vortices of a single blade, when dealing with propellers with a finite number of blades.

The Local Term

The only term which we now need to explain is the second term of (9.57) or the local term. From the previous analysis we now see that this flow can be represented by a distribution of ring vortices confined to the plane of the disc. The strength of these vortices is

$$\gamma_l = -(u_a^* n_r - u_r n_a) \frac{\Delta p}{\rho V_s^2} \tag{9.72}$$

where all quantities are functions of n. To interpret this we use (9.25) but note that for most cases $u_a^* < 0$ and $u_r < 0$. We then take $\theta_l = \theta - \pi$, see Figure 9.12, and use

$$\cos\theta_l = -\frac{u_a}{V_s} \quad ; \quad \sin\theta_l = \frac{u_r}{V_s}$$

in (9.72) to obtain

$$\gamma_l = (\cos\theta_l \, n_r - \sin\theta_l \, n_a) \frac{\Delta p}{\rho V_s}$$

$$= -n_l \frac{\Delta p}{\rho V_s} \tag{9.73}$$

Figure 9.12

where n_l is the projection of the force direction on the outward normal of the stream tubes, Figure 9.12.

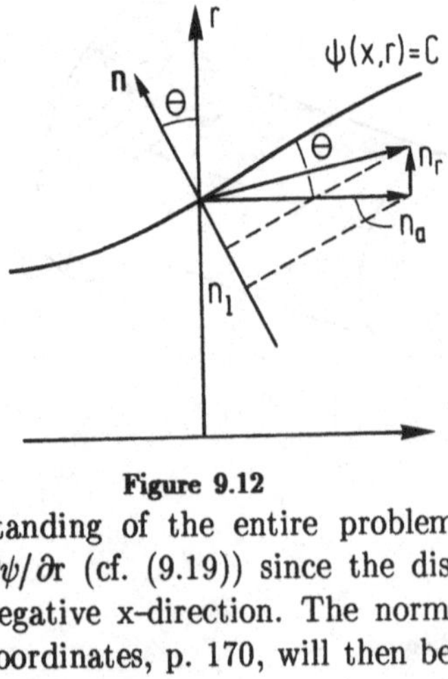

It complicates the physical understanding of the entire problem that ψ is generally negative and so is $\partial\psi/\partial r$ (cf. (9.19)) since the disc accelerates the flow downstream, in the negative x-direction. The normal connected to this system of curvilinear coordinates, p. 170, will then be negative outwards.

We now see that the force on the disc can be resolved into three mutually perpendicular components F_s , F_n and F_t, respectively tangent to the stream surface and with no tangential component, along the normal and in the tangential direction, cf. Figure 9.13. By use of the Kutta-Joukowsky law (Appendix B, p. 490) we can identify the corresponding vortex systems (Figure 9.13) as

$$dF_s = \rho\,(\omega r - u_t)\;\gamma_{line}\;d\gamma\;dn$$
$$dF_n = \rho\;V_s\;\gamma_l\;d\gamma\;dn$$
$$dF_t = \rho\;V_s\;\gamma_{line}\;d\gamma\;dn$$

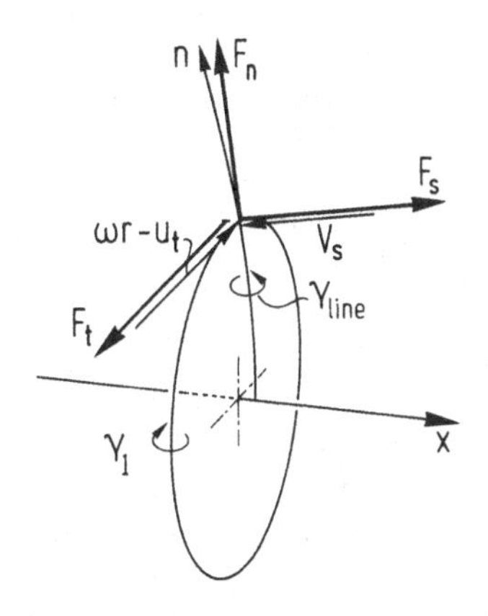

Figure 9.13

The velocities of the local term are obtained by using (9.19) on the second term of (9.57). We moreover place the disc at x = 0 instead of along the normal. This is consistent with the approximation of airfoil theory, p. 69 and sequel, where the singularities were placed on the axis and not on the cambered mean line. By this operation we then have $\cos\theta_l\; dn' = -dr'$. For $Z \to 1$ the Legendre function of (9.57) will be singular, according to Equation (M6.10) of the Mathematical Compendium, Section 6. The contribution to the velocities from this singular term is then

$$\Delta u_a = \frac{1}{4\pi r}\int_0^R \frac{n_l}{\cos\theta_l}\frac{\Delta p}{\rho V_s}\frac{\partial}{\partial r}\left[\sqrt{rr'}\,\ln(Z-1)\right]dr'$$

$$\Delta u_r = -\frac{1}{4\pi r}\int_0^R \frac{n_l}{\cos\theta_l}\frac{\Delta p}{\rho V_s}\sqrt{rr'}\,\frac{\partial}{\partial x}\left[\ln(Z-1)\right]dr' \tag{9.74}$$

With $Z - 1 = ((x^2 + r^2 + r'^2)/2rr') - 1 = (x^2 + (r - r')^2)/2rr'$ we have, after a series of reductions that for $x \to 0$

$$\Delta u_a = \frac{1}{4\pi r^2}\int_0^R \frac{n_l}{\cos\theta_l}\frac{\Delta p}{\rho V_s}\sqrt{rr'}\left[\frac{1}{2}\ln\frac{r^2 - r'^2}{2rr'} - \frac{r + r'}{r - r'}\right]dr'$$

$$\Delta u_r = -\frac{x}{4\pi r^2}\int_0^R \frac{n_l}{\cos\theta_l}\frac{\Delta p}{\rho V_s}\sqrt{rr'}\,\frac{1}{x^2 + (r - r')^2}\,dr' \tag{9.75}$$

The ln-term in the expression for Δu_a can be integrated without difficulty. The second term must be treated as a Cauchy principal-value integral, and we recognize this term and the integral for u_r as the expressions for two-dimensional line vortices, cf. p. 33, in particular Equations (2.34) and (2.35). This should be expected since close to the ring vortex the dominating terms will be that of the line. Using the results from (2.34) and (2.36) we have for the radial component

$$\Delta u_r(r) = \pm \frac{1}{2} \frac{n_l}{\cos\theta_l} \frac{\Delta p(r)}{\rho V_s(r)} \quad , \quad \begin{array}{l} + \text{ for } x = 0_+ \\ - \text{ for } x = 0_- \end{array} \tag{9.76}$$

We observe that the local term gives a jump in the radial velocity over the disc. By comparison with Equation (9.30) we see that this jump corresponds to the jump in the tangential velocity component. Both jumps are due to the forces from the blades acting on the fluid particle when it travels along the streamline from immediately upstream to immediately downstream of the disc. But if the propeller forces are so directed that the projection of the blade force in the x–r plane is parallel to the stream tube (ψ = c, cf. Figure 9.12) the local term contribution from this point will vanish.

Thrust, Torque and Efficiency

The thrust and torque on the disc are calculated by integrating the pressure over the disc. We prefer, as is customary, to calculate the torque as positive for right-handed "propellers". This torque is actually the torque provided by the shaft on the disc and is equal and opposite to that exerted on the disc by the fluid. The thrust and torque can then be expressed as

$$T = 2\pi \int_0^R \Delta p(r)\, n_a(r)\, r\, dr \tag{9.77}$$

$$Q = -2\pi \int_0^R \Delta p(r)\, n_t(r)\, r^2\, dr \tag{9.78}$$

From these the usual thrust and torque propeller coefficients can easily be calculated.

The amount of useful work transferred to the fluid per unit time is

$$P_{out} = TU \tag{9.79}$$

while the power absorbed by the disc is

$$P_{in} = Q\omega$$

and hence the efficiency

$$\eta = \frac{TU}{Q\omega} = \frac{J}{2\pi}\frac{K_T}{K_Q} \tag{9.80}$$

in terms of propeller coefficients $J = U/ND$, $K_T = T/\rho N^2D^4$, $K_Q = Q/\rho N^2D^5$, N being the rate of revolutions ($= 2\pi\omega$) and D the diameter of the disc.

The designer of a propeller (or in this case a disc) is usually faced with the problem of finding the geometry which for a given set of conditions, in this case U, ω, R produces a given thrust T (9.77) while having a high efficiency, i.e. absorbing as little power as possible. The numerical complications of this problem are severe, both when solving the non-linear integral equation (9.57) and when carrying out the optimization procedure. But on the basis of the linearized integral equation the velocities can be calculated directly and a simple optimum criterion can be derived. This will be considered next.

LIGHTLY LOADED DISC

From Figure 9.7 we saw that with decreasing blade loading the radial velocity components become smaller relative to the axial components and the stream tubes look more like concentric circular cylinders. In other words, the angle $\theta \approx 0$ in Figure 9.6, and Equation (9.27) becomes, since $u_a^* \approx -U$

$$\frac{\partial}{\partial\psi} \approx -\frac{1}{rU}\frac{\partial}{\partial r} \tag{9.81}$$

Assuming furthermore, that the normal component of the disc force is small, $n_l \approx 0$, in (9.74) and the local term of (9.57) vanishes whereby this equation becomes

$$\psi' = \frac{1}{2\pi}\int_{-\infty}^{0} dx' \int_{0}^{R} \omega r' \frac{1}{r'U}\frac{\partial(u_t r')}{\partial r'}\sqrt{rr'}\, Q_{\frac{1}{2}}(Z)dr' \tag{9.82}$$

where the stream-tube radius over which we integrate is now constant equal to the disc radius. By use of (9.30), (9.82) becomes

$$\psi' = \frac{1}{2\pi}\int_{-\infty}^{0} dx' \int_{0}^{R} \frac{\omega}{U}\frac{\partial}{\partial r'}\left[-\,n_t\,\frac{\Delta p(r')r'}{\rho V_s}\right]\sqrt{rr'}\, Q_{\frac{1}{2}}(Z)dr' \tag{9.83}$$

The induced velocities can now be expressed directly, using (9.19)

$$u_a = \frac{1}{2\pi}\int_{-\infty}^{0} dx' \int_0^R \frac{\omega}{U}\frac{\partial}{\partial r'}\left[-n_t\frac{\Delta p(r')r'}{\rho V_s}\right]\frac{1}{r}\frac{\partial}{\partial r}\left[\sqrt{rr'}\,Q_{\frac{1}{2}}(Z)\right]dr'$$

$$u_r = -\frac{1}{2\pi}\int_{-\infty}^{0} dx' \int_0^R \frac{\omega}{U}\frac{\partial}{\partial r'}\left[-n_t\frac{\Delta p(r')r'}{\rho V_s}\right]\frac{1}{r}\frac{\partial}{\partial x}\left[\sqrt{rr'}\,Q_{\frac{1}{2}}(Z)\right]dr' \tag{9.84}$$

For the radial velocity we can very easily carry out the integration by observing that in here $\partial/\partial x = -\partial/\partial x'$ (definition of Z in (9.56)) and that the lower limit will contribute nothing since the Legendre function vanishes. Hence

$$u_r = \frac{1}{2\pi}\int_0^R \frac{\omega r'}{U}\frac{\partial}{\partial r'}\left[-n_t\frac{\Delta p(r')r'}{V_s}\right]\frac{1}{\sqrt{rr'}}\,Q_{\frac{1}{2}}\left[\frac{x^2+r^2+r'^2}{2rr'}\right]dr' \tag{9.85}$$

For the axial velocity a series of manipulations will be necessary. For the sake of notation we designate

$$\Gamma(r) = -n_t\frac{\Delta p(r)}{\rho V_s}2\pi r \tag{9.86}$$

and use (9.55) to re-introduce Bessel functions whereby (9.84) can be written

$$u_a = \frac{\omega}{4\pi U}\int_{-\infty}^{0} dx' \int_0^R dr'\frac{\partial\Gamma}{\partial r'}\frac{\partial}{\partial r}\left[rr'\int_0^\infty J_1(kr)J_1(kr')e^{-k|x-x'|}dk\right] \tag{9.87}$$

Integration over x' gives

$$\int_{-\infty}^{0} e^{-k|x-x'|}dx' = \begin{cases}\dfrac{1}{k}e^{-k|x|} & x \geq 0\\[2ex] \dfrac{2}{k}-\dfrac{1}{k}e^{-k|x|} & x < 0\end{cases} \tag{9.88}$$

which we can insert back into (9.87), and by use of the step function

$$u_a = \pm\frac{\omega}{4\pi U}\int_0^R dr'\frac{\partial\Gamma}{\partial r'}\frac{1}{r}\frac{\partial}{\partial r}\left[rr'\int_0^\infty J_1(kr)J_1(kr')\frac{1}{k}e^{-k|x|}dk\right]$$

$$+\frac{\omega}{2\pi U}\int_0^R dr'\frac{\partial\Gamma}{\partial r'}\frac{1}{r}\frac{\partial}{\partial r}\left[rr'\int_0^\infty J_1(kr)J_1(kr')\frac{1}{k}dk\right](1-H(x))$$

$$\pm \text{ for } x \gtrless 0 \tag{9.89}$$

Now use (9.62) for the r-variable and in connection with integration by parts over r' to obtain

$$u_a = \mp \frac{\omega}{4\pi U} \int_0^R dr' \, \Gamma(r')r' \int_0^\infty kJ_0(kr)J_0(kr') \, e^{-k|x|} dk$$

$$+ \frac{\omega}{2\pi U} \int_0^R dr' \, \frac{\partial\Gamma(r')}{\partial r'} \, r' \int_0^\infty J_0(kr)J_1(kr') \, dk \, (1 - H(x))$$

$$\mp \text{ for } x \gtrless 0 \tag{9.90}$$

The inner integral of the last expression can be found for example in Watson (1966) (p.406)

$$\int_0^\infty J_0(kr)J_1(kr') \, dk = \begin{cases} 0 & r' < r \\ 1/2r' & r' = r \\ 1/r' & r < r' \end{cases} \tag{9.91}$$

so we have

$$u_a = \frac{\omega}{4\pi U} \int_0^\infty dr'\Gamma(r')r' \, \frac{\partial}{\partial x} \int_0^\infty J_0(kr)J_0(kr')e^{-k|x|}dk$$

$$+ \frac{\omega}{2\pi U} (1 - H(x)) \int_0^R \frac{\partial\Gamma(r')}{\partial r'} \, H(r' - r)dr' \tag{9.92}$$

where the last integral vanishes if $r > R$. This integral can now be carried out and by re-introducing the Legendre function in the first term we obtain

$$u_a = \frac{x}{4\pi^2 r\sqrt{r}} \, \frac{\omega}{U} \int_0^R \Gamma(r') \, \frac{1}{\sqrt{r'}} \, Q'_{-\frac{1}{2}}(Z)dr' - (1-H(x))\Gamma(r) \, \frac{\omega}{2\pi U} \tag{9.93}$$

where

$$Q'_{-\frac{1}{2}}(Z) = \partial Q_{-\frac{1}{2}}/\partial Z \quad ; \quad Z = (x^2 + r^2 + r'^2)/2rr'$$

When integrating by parts we have used that $\Gamma(0) = \Gamma(R) = 0$, which is justified in the definition of Γ in (9.86) since we cannot have a pressure difference over the propeller blade at the tip and presumably at the root.

We can now outline the behaviour of u_a with x by considering (9.93). The integral over r' is asymmetric with respect to x, it is zero for $x = 0$ and the Q-function will make it vanish like $x/|x|^3$ for large x. When we combine this with the behaviour of the step function we see that the velocity is zero (as we must expect) far upstream and increases to a finite value $-\Gamma(r)\omega/4\pi U$ at the disc. Aft of the disc it increases further until far downstream it obtains a value of $-\Gamma(r)\omega/2\pi U$ or twice the value at the disc. All this is valid for a particle which crosses the disc. Outside the disc the axial velocity will only display the asymmetric behaviour of the xQ'-term and it will go to zero again far downstream.

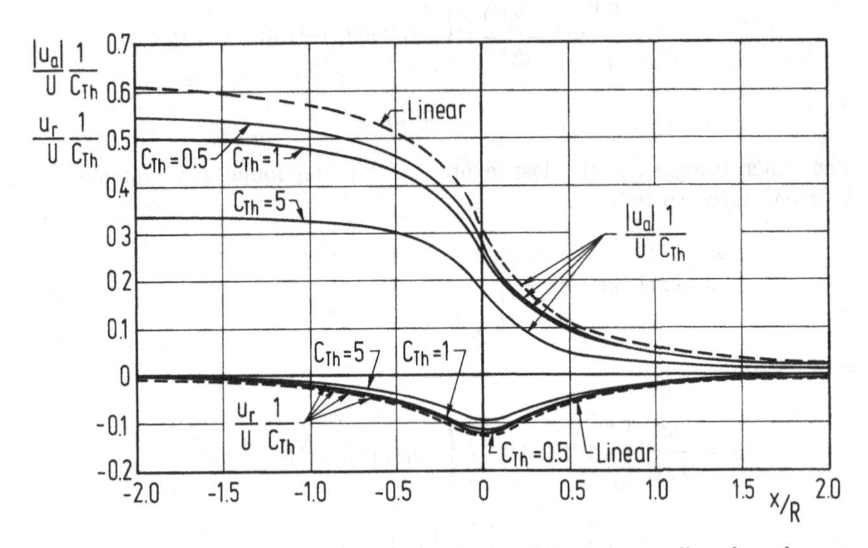

Figure 9.14 Axially and radially induced velocities along a streamline through $r = 0.7\ R$ at disc. Additional results for the non-linear case are given in Figure 9.7.

The expression for the velocities derived here are similar to those obtained by Hough & Ordway (1965). Results of calculations for the linear case are presented in Figure 9.14 and compared with results of the non-linear case. The symmetry in the behaviour of the linear results is apparent.

Optimum Efficiency

We shall now try to solve, within the limits of linearized theory, the problem of finding the load distribution which gives the highest possible efficiency while producing a specified thrust. We consider the velocities on the disc and assume no radial velocity so the condition on the disc (9.32) becomes

$$\frac{n_a}{n_t} = \frac{\omega r - u_t}{-U + u_a} \qquad \text{at } x = 0 \tag{9.94}$$

where u_a is the axial perturbation velocity. Using the circulation as defined in (9.86) (with $V_s \approx U - u_a$) we can write the thrust and torque (9.77) and (9.78) as

$$
\begin{aligned}
T &= \rho \int_0^R \Gamma(\omega r - u_t)dr \\
Q &= \rho \int_0^R \Gamma(U - u_a)rdr
\end{aligned}
\tag{9.95}
$$

Our problem is now, that T is specified and Q must be minimized. This problem can be solved by the Calculus of Variations, as outlined in Section 7 of the Mathematical Compendium, p. 522.

With δT and δQ the first variations and ℓ the inverse of the Lagrange multiplier we set the first variation of the functional $T - \ell Q$ to zero or

$$\delta T - \ell \delta Q = 0 \tag{9.96}$$

Suppose that $\Gamma(r)$ is the sought-for, or optimum distribution and $\Gamma(r) + \delta\Gamma(r)$ is taken to be any neighboring distribution which yields the same thrust and vanishes at $r = 0$ and R. Then

$$
\begin{aligned}
\delta T &= \rho \int_0^R (\delta\Gamma(\omega r - u_t) - \Gamma \frac{\partial u_t}{\partial \Gamma} \delta\Gamma)dr \\
\delta Q &= \rho \int_0^R (\delta\Gamma(U - u_a) - \Gamma \frac{\partial u_a}{\partial \Gamma} \delta\Gamma)rdr
\end{aligned}
\tag{9.97}
$$

Now we have from (9.93) that on the disc

$$u_a(r) = -\frac{\omega}{4\pi U}\Gamma(r) \quad , \quad x = 0 \tag{9.98}$$

and similarly, from (9.30) and (9.86)

$$u_t(r) = \frac{\Gamma(r)}{4\pi r} \quad , \quad x = 0 \tag{9.99}$$

Using these expressions in (9.97) which is then inserted into (9.96) gives the following equation for the optimum

$$\rho \int_0^R \delta\Gamma((\omega r - 2u_t) - \ell(U - 2u_a)r)dr = 0 \tag{9.100}$$

The only way this condition can be fulfilled for any variation $\delta\Gamma(r)$ is when the expression is parenthesis is zero, or when

$$\frac{\omega r - 2u_t}{r(U - 2u_a)} = \ell \tag{9.101}$$

Since we have, from (9.93) and (9.30) that

$$\begin{aligned} u_a(\infty,r) &= 2u_a(0,r) \\ u_t(\infty,r) &= 2u_t(0,r) \end{aligned} \tag{9.102}$$

(9.101) can be reformulated into

$$\tan \beta_i(\infty,r) = \frac{U - u_a(\infty\,,\,r)}{\omega r - u_t\,(\,\infty,r)} = \frac{1}{r\ell} = \frac{\tan\beta}{\ell^*} \tag{9.103}$$

where β_i is the hydrodynamic pitch angle, cf. Figure 9.15. We shall later see (Chapter 12) that this simple optimum criterion holds approximately even for the more complicated case of a propeller with a finite number of blades. The physical significance of (9.103) is that the streamlines far downstream form a true helicoidal surface of pitch $2\pi/\ell$.

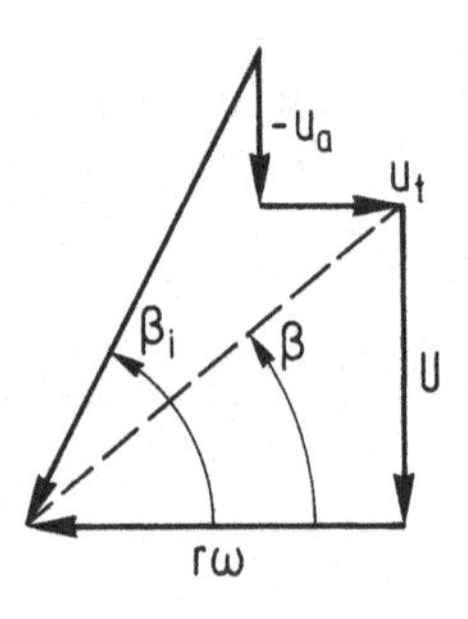

Figure 9.15 Definition of pitch angles β and β_i.

In a different approach Betz (1920) arrived at the criterion (in our notation)

$$\frac{\frac{1}{2}\,u_t(\infty\,,\,r)}{U - u_a\,(\,\infty,r)} + \frac{U - \frac{1}{2}\,u_a\,(\,\infty,r)}{\omega r - u_t(\infty\,,\,r)} = C\,\frac{U}{\omega r} \tag{9.104}$$

Now, because of (9.102), which also Betz assumed, (9.104) can be recast into

$$\frac{U - u_a(\infty,r)}{\omega r - u_t(\infty,r)} = C\,\frac{U}{\omega r} + \frac{1}{2}\left[\frac{-u_a(\infty,r\,)}{\omega r - u_t(\infty,r)} - \frac{u_t(\infty,r\,)}{U - u_a(\infty,r)}\right] \tag{9.105}$$

Since both $-u_a/\omega r$ and u_t/U are small and positive the difference between them will be even smaller and (9.105) when neglecting this difference is the same as (9.103).

On the basis on this criterion Betz (*ibid.*) gave curves of optimum distributions of induced velocities u_a and u_t. We shall only briefly outline such a procedure based on the simpler formulas presented here. From (9.103) together with (9.98), (9.99) and (9.100) we can obtain the optimum distribution of circulation as

$$\Gamma = 2\pi\,(\omega - \ell U)\,\frac{r^2}{1 + \ell \dfrac{\omega}{U} r^2} \tag{9.106}$$

We see that this distribution is not feasible from a practical point of view since we had to impose the boundary condition, that $\Gamma(R) = 0$ which (9.106) does not meet. This could have been remedied by multiplying the distribution by a factor, e.g. $\sqrt{1-r/R}$.

The circulation distribution can then be inserted into the expressions for thrust and torque (9.95). The integrations involved can be carried out analytically to obtain the efficiency (9.80) as function of the thrust loading coefficient and advance ratio. The results are presented in Figure 9.16.

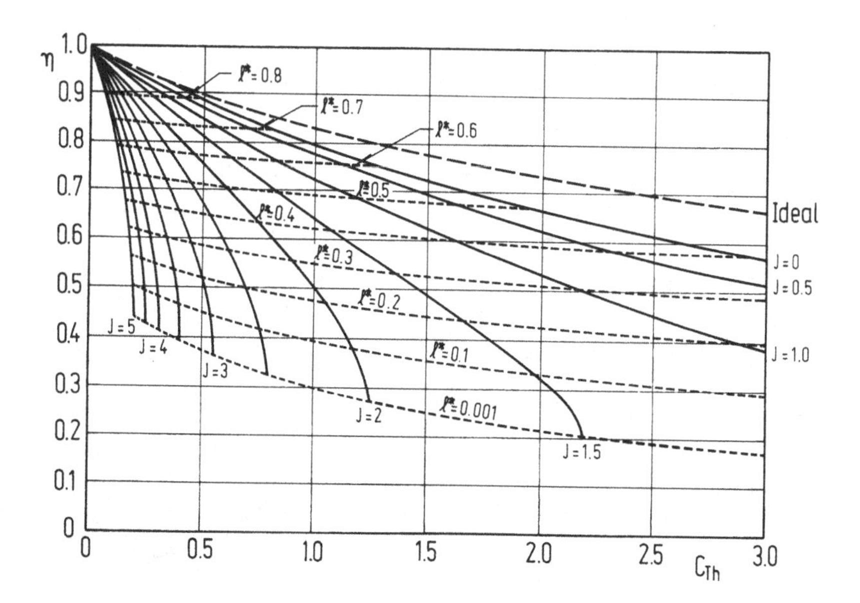

Figure 9.16 Efficiency as function of thrust loading, C_{Th}, advance ratio J and parameter ℓ^*, for optimum circulation distribution, for linearized flow.

The limiting case $J = 0$ corresponds to the case where the induced tangential velocity is negligible. This can be seen by inserting the expression for optimum Γ (9.106) into (9.98) and (9.99) and then let J → 0. The efficiency can be expressed by the simple formula

$$\eta_{\text{lin}} = \frac{1}{1 + C_{Th}/4} \tag{9.107}$$

This expression can also be obtained by linearizing the expression for the *ideal efficiency*

$$\eta_{\text{ideal}} = \frac{2}{1 + \sqrt{1 + C_{Th}}} \tag{9.108}$$

The derivation of this formula is given in many textbooks on naval architecture and will thus be omitted here. Parallel to the above procedure the ideal efficiency can also be derived out of Betz's (1920) expressions by neglecting rotation.

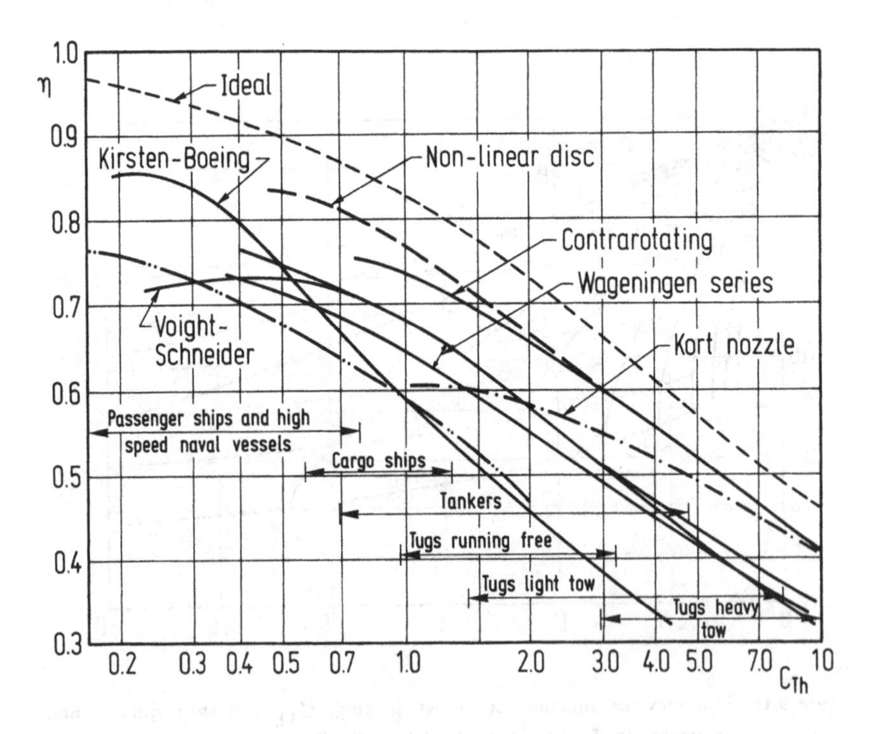

Figure 9.17 Efficiencies of various propulsion devices and C_{Th} ranges for ship classes. The figure is based on data from around 1960 but has been extended for tankers to 1985 by the authors. (For Kirsten–Boeing and Voight–Schneider vertical shaft propellers, see for example Lewis (1988)).

The ideal efficiency must then be interpreted as an unrealistically high efficiency that can never be achieved but which may be used when comparing two propellers for instance by comparing their efficiencies relative to the ideal efficiency. This can be seen in Figure 9.17 where η_{ideal} has been plotted over an enormous range of C_{Th} together with practical ranges of C_{Th} and efficiencies for various classes of vessels.

Despite the lengthy treatment of actuator-disc theory in this chapter we have so far omitted the case of the disc in a wake behind a ship where the disc not only sees a radially varying inflow but where also the disc loading must be increased relative to the open water situation to account for the effects of suction from the disc on the hull. Goodman (1979) has examined this case including the effects of shear. He calculated the optimum load distribution and presented curves for the induced velocities.

We shall not go deeper into details about the actuator disc, in particular because the aspects of wake in connection with propellers (still, however, without shear) will be treated later.

10 Wing Theory

The flows about wings of finite span are sufficiently analogous to those about propeller blades to warrant a brief examination before embarking on the construction of a mathematical model of propellers. A more detailed account of wing theory is given for example by von Kármán & Burgers (1935).

A basic feature of the flow about a straight upward-lifting wing of finite span and starboard-port symmetry in a uniform axial stream is that the velocity vectors on both the lower and upper surfaces are not parallel to the longitudinal plane of symmetry. On the lower side the vectors are inclined outboard and on the upper side they are inclined inwardly in any vertical plane or section parallel to the vertical centerplane. This is a consequence of the pressure relief at the wing tips and the largest increase in pressure being in the centerplane on the lower side and the greatest decrease in pressure being in the centerplane on the upper or suction side. Thus there are positive spanwise pressure gradients on the lower side and negative spanwise gradients on the upper side which give rise to spanwise flow components which are obviously not present in two-dimensional flows.

In many treatments of wing theory one finds figures purporting to show the flow about a wing of finite aspect ratio in which there is a continuous line of stagnation points near the leading edge, extending from one wing tip to the other. Wehausen, (1976) has claimed that this is not possible and has conjectured that continuous irrotational flow about a simply connected body of finite extent placed in a uniform stream will have just two stagnation points, one on the upstream side and one on the downstream side. Wehausen points out that the analog of this statement for two-dimensional flow is readily established via conformal mapping. Unfortunately, in three-dimensional flows the same procedure does not work (as he points out by referring to Kellogg, (1967) pp. 235-6). While he states that his conjectured theorem has not been found in any treatises on potential theory or fluid mechanics he points out that it is true for ellipsoids in general and in particular for those of dimensions comparable to those of a wing.

Assuming his conjecture to be true, Wehausen then provided the flow schematic shown in Figure 10.1. Here we see that streamlines emanating from the *single* stagnation point S on the lower side in the centerplane cover the lower and upper side but give rise to trajectories on the upper and lower sides at or near the trailing edge which are not co-linear. In a real fluid this behavior gives rise to a swirling or shearing flow directly downstream of the trailing edge. The opposing directions of spanwise components pointed out earlier are consonant with Wehausen's picture of the flow pattern on upper and lower sides and along the trailing edges. The presence of viscosity must then give rise to the sheet of vorticity downstream of the trailing edge driven by the difference in the transverse components on upper and lower surfaces.

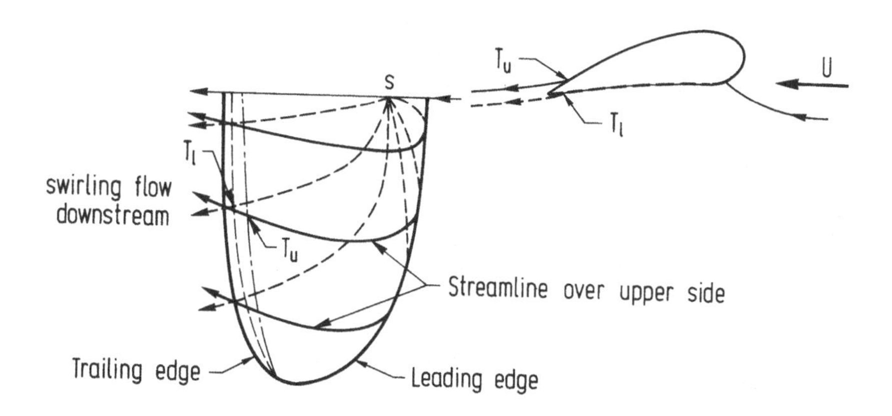

Figure 10.1 Schematic of flow about wings.

Figure 1 from Wehausen, J.V. (1976). A qualitative picture of the flow about a wing of finite aspect ratio. *Schiffstechnik*, 23. Band, 114. Heft, 215–7.
By courtesy of Schiffstechnik, Schiffahrts–Verlag "Hansa", Germany.

We owe it to the genius of L. Prandtl who from observations of the flow about wing models devised an inviscid flow model to mimic the actual flow about wings of large aspect ratio. Prandtl's model was built up of an array of horseshoe vortices as depicted in the sketches on pages 37 and 38 in which the chord of the foil is considered to be shrunk to a loaded line or bound vortex located at the quarter chord abaft the leading edge.

There are several approaches or representations of this lifting-line model of wings. We have seen that the velocity field of any vortex array can be calculated from the law of Biot and Savart, Equations (1.83) or (1.84).

Here we choose to write the velocity potential of the Prandtl lifting-line model of wings of finite, but large aspect ratios, by exploiting Stokes' theorem, p. 23. The result is given by Equation (2.50) and sequel. Let us start with Equation (2.52) and set $x' = 0$, i.e. placing the bound vortex on the y-axis. Then

$$\phi(x,y,z) = -\frac{z}{4\pi}\int_{-b}^{b}\frac{\Gamma(y')}{(y-y')^2+z^2}\left[\frac{x}{\sqrt{x^2+(y-y')^2+z^2}}-1\right]dy' \tag{10.1}$$

To achieve a general picture of the variation of the most important component, viz. the vertical velocity or downwash we may take the spanwise circulation $\Gamma(y')$ to be uniform say Γ_0.

The vertical velocity can be secured by differentiating (10.1) after effecting the y'-integrals. However, the result is cumbersome. The most adroit procedure is to return to Equation (2.50) and replace $\partial/\partial z'$ by $-\partial/\partial z$ and set $z' = 0$ (as well as x'). Then we see that

$$w(x,y,z) = \frac{\partial\phi}{\partial z} = -\frac{\Gamma_0}{4\pi}\int_{-\infty}^{0}dx''\int_{-b}^{b}\frac{\partial^2}{\partial z^2}\left[\frac{1}{R}\right]dy' \tag{10.2}$$

As $1/R$ $(= 1/\sqrt{(x-x'')^2+(y-y')^2+z^2}\,)$ satisfies Laplace's equation at all (x,y,z) except (x'',y',0) we may write (10.2) as

$$w = \frac{\Gamma_0}{4\pi}\int_{-\infty}^{0}dx'\int_{-b}^{b}\left[\frac{\partial^2}{\partial x^2}+\frac{\partial^2}{\partial y^2}\right]\frac{1}{R}\,dy' \tag{10.3}$$

Exploiting the fact that $\partial(1/R)/\partial x = -\,\partial(1/R)/\partial x''$ and likewise for y and y' we may integrate (10.3) by inspection to secure

$$w = -\frac{\Gamma_0}{4\pi}\left\{\int_{-b}^{b}\frac{\partial}{\partial x}\left[\frac{1}{R}\bigg|_{x'=0}-\frac{1}{R}\bigg|_{x'=-\infty}\right]dy' + \int_{-\infty}^{0}\frac{\partial}{\partial y}\left[\frac{1}{R}\bigg|_{y'=b}-\frac{1}{R}\bigg|_{y'=-b}\right]dx'\right\}$$

Carrying out the indicated operations we obtain

$$w = -\frac{\Gamma_0}{4\pi}\left\{\frac{x}{x^2+z^2}\left[\frac{y-b}{\sqrt{x^2+(y-b)^2+z^2}}-\frac{y+b}{\sqrt{x^2+(y+b)^2+z^2}}\right]\right.$$

$$+\frac{y-b}{(y-b)^2+z^2}\left[\frac{x}{\sqrt{x^2+(y-b)^2+z^2}}-1\right]$$

$$\left.-\frac{y+b}{(y+b)^2+z^2}\left[\frac{x}{\sqrt{x^2+(y+b)^2+z^2}}-1\right]\right\} \tag{10.4}$$

We can see that as $x \to +\infty$, $w \to 0$ and that as $x \to -\infty$, w becomes independent of x and

$$w(-\infty,y,z) = -\frac{\Gamma_0}{2\pi}\left[\frac{b-y}{(b-y)^2+z^2}+\frac{b+y}{(b+y)^2+z^2}\right] \tag{10.5}$$

Whereas at $x = 0$

$$w(0,y,z) = \frac{1}{2}\,w(-\infty,y,z) \tag{10.6}$$

We may also observe that if we first take $z = 0$ and then at $x \to 0_{\pm}$, the first term in (10.4) becomes either positively or negatively infinite because we are running into the bound vortex along $x = 0$, $z = 0$. However if we first set $x = 0$ and then let $z \to 0$ we get

$$\begin{aligned} w(0,y,z) &= -\frac{\Gamma_0}{4\pi}\left[\frac{b-y}{(b-y)^2+z^2}+\frac{b+y}{(b+y)^2+z^2}\right] \\ &\to -\frac{\Gamma_0}{4\pi}\left[\frac{1}{b-y}+\frac{1}{b+y}\right];\ x = 0,\ z \to 0 \end{aligned} \tag{10.7}$$

The singular behavior of the first term of (10.4) is shown in physical terms in Figure 10.2.

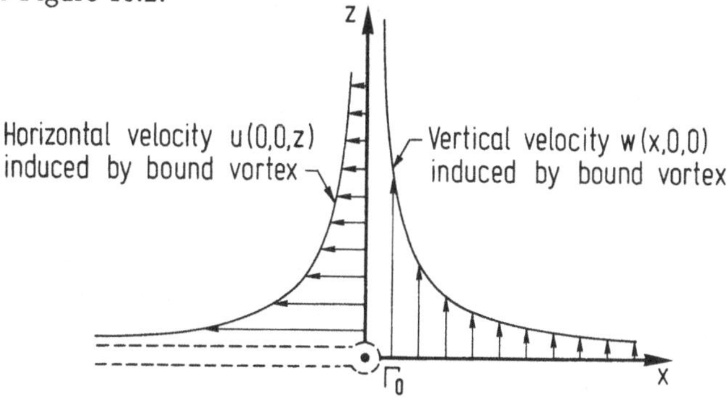

Figure 10.2 Velocity induced by bound vortex (first term of Equation (10.4)).

The finite limits of w(0,y,0) are given by the terms from the lateral edges which, via the Biot-Savart formulation, would be identified as the contribution of the two trailing vortices along $-\infty < x < 0$, $y = \pm b$, $z = 0$. Thus, in general, the flow components as derived from the dipole-sheet potential are found as the aggregate effect of the distribution and consequently the origins of these components are not usually as readily discernible as from the Biot-Savart law.

However, the velocity potential approach is essential for use in other contexts such as in the case of hydrofoils of finite span beneath the water surface in the presence of gravity (Breslin, (1957)) and in unsteady wing theory.

The distribution of vertical velocity is given in Figure 10.3. It is seen that w/U quickly becomes nearly constant as the distance from the bound vortex downstream increases and the asymptotic value of w/U downstream is attained for all practical purposes (with an error of 2 per cent) at $x/b = -3.4$.

The variation spanwise of the downwash along the lifting line and the upwash outboard of the tips is also shown in Figure 10.3. The singular behavior at the tips is physically untenable (arising from the non-feasible uniform loading) and we must turn back to the general case wherein $\Gamma(\pm b) = 0$ and hence Γ varies spanwise.

To calculate the downwash along the bound vortex we return to (10.1) set $x = 0$ and integrate by parts. Thus

$$\phi(0,y,z) = \frac{z}{4\pi}\left\{-\frac{\Gamma(y')}{z}\tan^{-1}\left[\frac{y-y'}{z}\right]\Bigg|_{-b}^{b} + \frac{1}{z}\int_{-b}^{b}\Gamma'(y')\tan^{-1}\left[\frac{y-y'}{z}\right]dy'\right\}$$

The integrated part vanishes by imposing $\Gamma(\pm b) = 0$ and then upon differentiation with respect to z

$$w(0,y,z) = -\frac{1}{4\pi}\int_{-b}^{b}\frac{(y-y')\Gamma'(y')}{(y-y')^2+z^2}dy'$$

and thus along $z = 0$ (on the bound vortex)

$$w(0,y,0) = -\frac{1}{4\pi}\int_{-b}^{b}\frac{\Gamma'(y')}{y-y'}dy' \tag{10.8}$$

We now consider an element of the wing at any y assumed to be acting as a two-dimensional section whose lift coefficient is expressed by

$$C_L = \frac{\rho U\Gamma(y)}{\frac{1}{2}\rho c(y)U^2} = 2\pi\left[\alpha - \alpha_0 + \frac{w(0,y,0)}{U}\right]$$

or

$$\frac{\Gamma(y)}{U} = \pi c(y)\left[\alpha - \alpha_0 - \frac{1}{4\pi U}\int_{-b}^{b}\frac{\Gamma'(y')}{y-y'}dy'\right] \tag{10.9}$$

where

α is the geometric angle of attack
α_0 is the angle of zero lift
$c(y)$ is the total chord at any y where the circulation is $\Gamma(y)$

and the last term is the *induced angle* $\alpha_{in} = w(0,y,0)/U < 0$, assuming $w \ll U$.

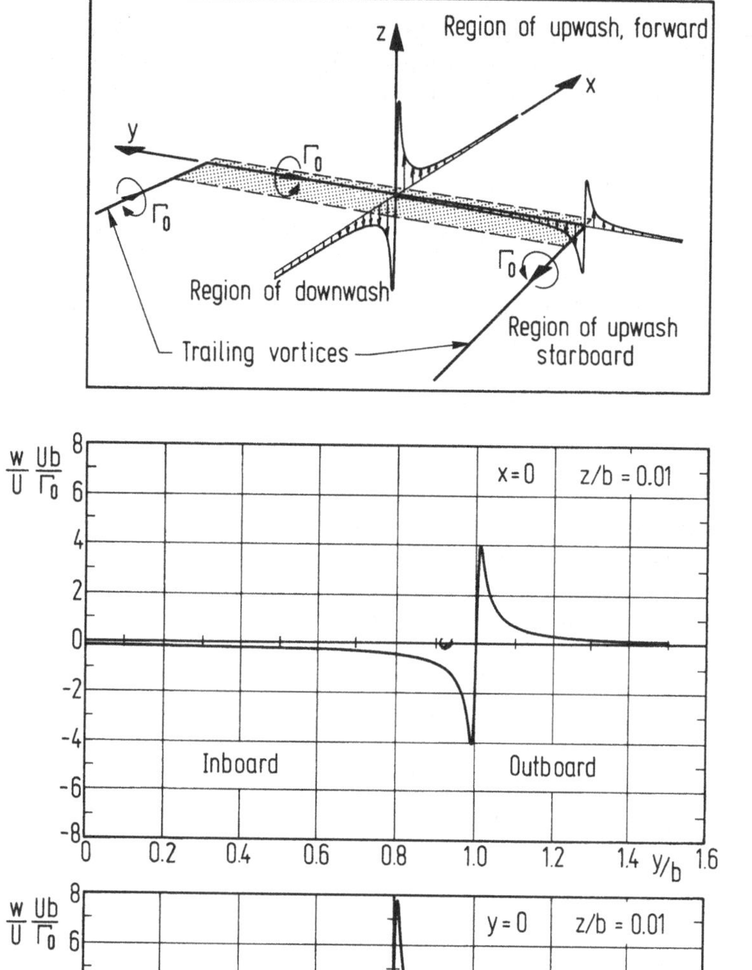

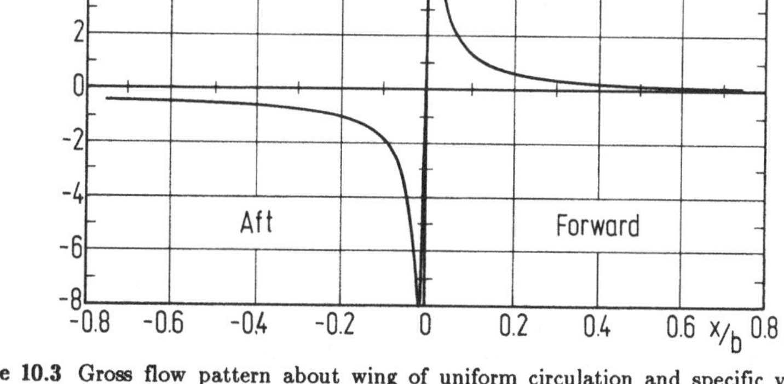

Figure 10.3 Gross flow pattern about wing of uniform circulation and specific values of vertical velocity distribution.

Equation (10.9) is an integro-differential equation for the circulation density Γ which must be solved numerically for a specified distribution of chord $c(y)$. The equation can be transformed by taking $y = -b\cos\theta$, $y' = -b\cos\theta'$ and assuming

$$\Gamma(y') = \Gamma(\theta') = \sum_{n=1}^{\infty} \Gamma_n \sin n\theta'$$

so that $\Gamma(\pm\pi)$ (at the wing tips) is zero. The coefficients can be solved for, once c is expressed as

$$c = \sum_{n=1}^{\infty} c_n \sin n\theta'$$

by dividing the span into a finite number of stations and thereby replacing the integro-differential equation by a set of simultaneous algebraic equations. We shall not attempt this detail here.

To secure a special but most useful result we assume that $\Gamma(y')$ is distributed elliptically as well as $c(y)$. Thus we take

$$\Gamma(y) = \Gamma_0 \sqrt{1 - \left[\frac{y}{b}\right]^2} \tag{10.10}$$

and

$$c(y) = c_0 \sqrt{1 - \left[\frac{y}{b}\right]^2} \tag{10.11}$$

Then (10.9) becomes

$$\frac{\Gamma_0}{c_0} = \pi U \left\{\alpha - \alpha_0 + \frac{\Gamma_0}{4\pi b U} \int_{-b}^{b} \frac{y'}{\sqrt{b^2 - y'^2}\,(y - y')} dy'\right\} \tag{10.12}$$

From the Table of Airfoil Integrals, Equation (M8.7), we see that the integral is $-\pi$. Upon solving for the mid-span circulation Γ_0 we obtain

$$\Gamma_0 = \frac{4\pi U b c_0 (\alpha - \alpha_0)}{4b + \pi c_0} \tag{10.13}$$

The lift on the entire wing is

$$L^w = \rho U \Gamma_0 \int_{-b}^{b} \sqrt{1 - \left[\frac{y}{b}\right]^2}\, dy = \frac{\pi}{2} \rho U \Gamma_0 b \tag{10.14}$$

Defining the lift coefficient based on the area of the elliptic wing $\pi c_0 b/2$ we obtain from (10.13) and (10.14)

$$C_L^w = \frac{2\pi(\alpha - \alpha_0)}{1 + \left[\frac{\pi}{4}\frac{c_0}{b}\right]} \tag{10.15}$$

The convention for defining the *aspect ratio* is

$$A_R = \frac{(\text{span})^2}{\text{area}} \tag{10.16}$$

so with the chord distribution given in (10.11) we have

$$A_R = \frac{4b^2}{\pi \frac{c_0}{2} b} = \frac{8}{\pi}\frac{b}{c_0} \tag{10.17}$$

Substituting in (10.15) for c_0/b from (10.16) yields finally

$$C_L^w = \frac{2\pi(\alpha - \alpha_0)}{1 + \frac{2}{A_R}} \quad \text{(for elliptic planform and loading)} \tag{10.18}$$

This is Prandtl's famous formula for the lift of wings of "high" aspect ratio which is found to agree well with data for aspect ratios from about 4 to infinity. Clearly the two-dimensional result is secured by inspection as the aspect ratio is made infinite. For wings of rectangular planform (10.18) requires only a small correction.

For delta wings of small aspect ratio (≈ 1) a cross flow analysis yields[16]

$$\frac{dC_L^w}{d\alpha} = \frac{\pi}{2} A_R \qquad 0 < A_R < 2 \tag{10.19}$$

A graph of $1/2\pi \; dC_L^w/d\alpha$ from (10.18) is presented in Figure 10.4. We see that the effect of aspect ratio persists to an effective value of 100!

We see that the induction of the trailing vortex system has the effect of reducing the lift-effectiveness (lift-rate or $dC_L^w/d\alpha$) of a wing as compared to the two-dimensional value.

---

[16] This result is derived in Chapter 17 in connection with a discussion of effects of aspect ratio on the unsteady force on the wing.

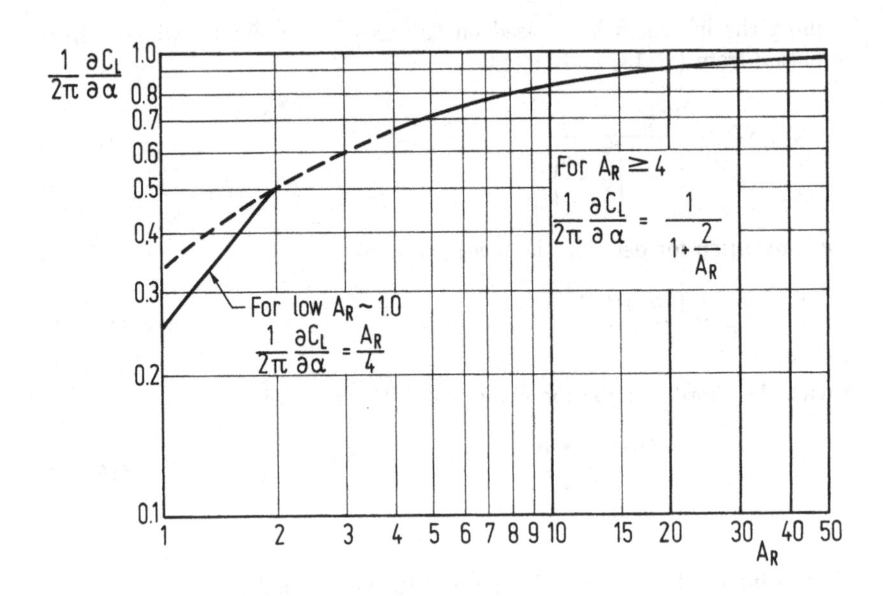

Figure 10.4 Lift–slope of wings as a fraction of the two–dimensional value, as a function af aspect ratio, from theories for high and low aspect ratios.

The trailing vortex system also carries away energy which is reflected in an *induced drag.* This can be seen from the direction of the resultant force acting on the bound vortex, cf. Figure 10.5.

The resulting force is perpendicular to the resultant velocity which is at the induced angle α_{in} (= w/U, cf. p. 200) and the total induced drag is

$$D_i^w = \rho U \int_{-b}^{b} \alpha_{in}(y)\Gamma(y)dy$$

$$= \rho \int_{-b}^{b} w(y)\Gamma(y)dy \qquad (10.20)$$

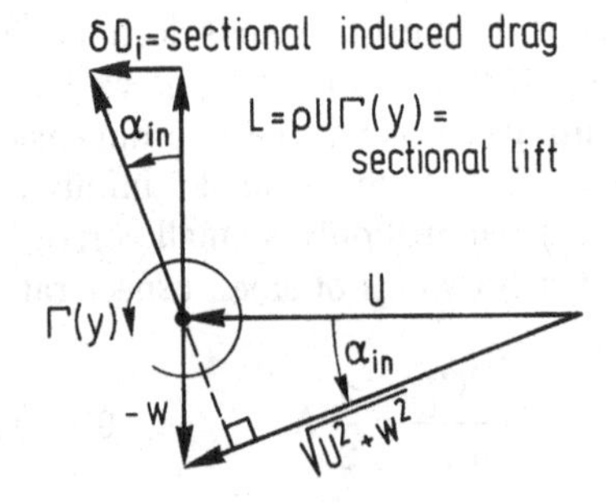

Figure 10.5 Resulting force acting on bound vortex.

From (10.8) we secure for an elliptic $\Gamma(y')$

$$w(0,y,0) = -\frac{\Gamma_0}{4b} = \text{a constant}$$

whence

$$D_i^w = -\frac{\pi\rho\Gamma_0^2}{8} \qquad (10.21)$$

Substituting for the mid-span circulation Γ_0 from (10.14) we can cast (10.21) into the form

$$C^w_{D_i} = \frac{C^{w\,2}_L}{\pi A_R} \tag{10.22}$$

where $C^w_{D_i}$ is the induced-drag coefficient defined by

$$C^w_{D_i} = \frac{|D^w_i|}{\frac{1}{2}\rho\pi \frac{c_0}{2} bU^2} \tag{10.23}$$

The total drag coefficient for a wing in a real fluid is then

$$C^w_{D_T} = C^w_{D_i} + C^w_{D_v} \tag{10.24}$$

where $C^w_{D_v}$ is the viscous-shear and form-drag coefficient and hence the lift-drag ratio of wings of aspect ratios of 4 or greater is

$$\frac{L^w}{D^w} = \frac{C^w_L}{C^w_{D_T}} = \frac{C^w_L}{\frac{C^{w\,2}_L}{\pi A_R} + C^w_{D_v}} \tag{10.25}$$

Then for $C^w_L = 0.20$, an aspect ratio of 6 and $C^w_{D_v} = 0.004$, $L^w/D^w = 32.7$ whereas for $C^w_L = 1.0$, $L^w/D^w = 17.5$. Hence at elevated C^w_L the induced drag is very high relative to the viscous drag.

Although the foregoing theory for wings having spanwise variation of lift or circulation and hence a planar wake of vorticity extending downstream in the plane of the wing has been shown to give excellent correlation with measurements, the actual vortex wake is unstable and in a short distance rolls up into two concentrated vortices as depicted in Figure 10.6.

The spanwise distance to the "wrapped-up" vortices from the center plane is (from Thwaites, (1960)) approximately,

$$b_r = \frac{\pi}{4} b = 0.78\ b$$

and the effective downstream distance to complete wrap-up is about

$$\frac{x_r}{b} \approx -0.14 \frac{A_R}{C^w_L}$$

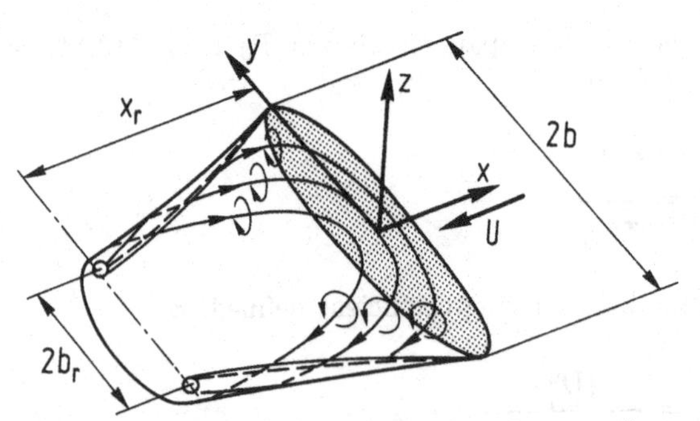

Figure 10.6 Schematic of trailing vortex sheet "roll" or "wrap-up".

Figure XII.5a from Thwaites, B. (1960). *Incompressible Aerodynamics.* Oxford: Clarendon Press.

Thus for an aspect ratio of 4 and a $C_L^w = 0.2$, $x_r/b = -2.8$, which is approaching the value of 3.4 where the theoretical asymptotic downwash is practically achieved. Thus the assumption of a planar wake of vorticity is fortunately adequate, i.e., the effective downwash at the wing is presumably accomplished by the portion of the assumed vortex wake forward of the wrap-up region. The effect of vortex wake instability is significant at low A_R and high C_L^w. The helical sheets of vorticity abaft each propeller blade also exhibit a similar instability and wrap-up.

In summary we see that the effects of finite aspect ratio are to degrade the lift-effectiveness of wings relative to the two-dimensional value and to incur induced drag which is directly proportional to the square of the lift and inversely proportional to the aspect ratio. These effects are a consequence of the downwash induced at the wing by the system of trailing vortices shed into the wake abaft the wing. We are now prepared to understand the analogous behavior encountered by propellers having a finite number of blades. It is to the application of lifting-line theory to propeller blades that we may now turn.

11 Lifting-Line Representation of Propellers

To construct a lifting-line model of propellers we can use the insight on wings obtained in the previous chapter, since a propeller blade is analogous to a rolling wing, pivoting about one tip while advancing. When the circulation distribution is taken to be constant along the bound vortex then an axial vortex is shed from the pivoting end (root) and a vortex is shed from the tip which, while preserving the history of the motion, forms a helical "chain" of vorticity extending downstream to the starting point where an initially shed radial vortex closes the array.

Thus at the outset, for sake of simplicity, we consider a simple vortex array as shown in the following Figure 11.1 which consists of hub, bound and helical vortices.

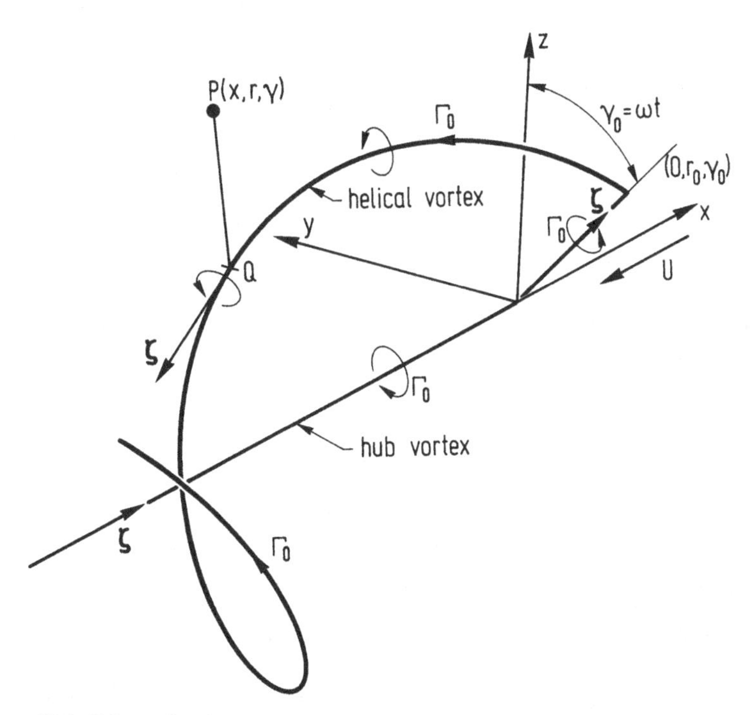

Figure 11.1 Schematic of vortex array for constant circulation density Γ_0 on bound vortex showing the vorticity vector ζ tangent to each of the vortex elements.

It is to be noted that the positive x-axis is taken forward (in conformity with ITTC notation which stems from the long tradition in naval architecture that the bow is to the right hand when viewed from the starboard. However our z-axis is positive upward in opposition to the ITTC because it is customary to measure the blade position angle from the vertical). This axis system is in opposition to that used by many who take +x downstream, a practice of aerodynamicists. Our objection to that system is that the angular velocity of a right-handed propeller is then negative and, worse, the thrust is negative and the torque is positive when the ship is driven ahead. We depart, then, from most writers and incur the trouble of comparison which requires x,y and ω (angular velocity) to be replaced by their negatives.

We have three ways of calculating the velocity field induced by this array:

i. By differentiating the velocity potential of dipoles of strength Γ_0 distributed normally to the helicoidal surface bounded by the hub, bound, and helical vortices.

ii. By differentiating the potential resulting from an integral of the pressure field arising from a concentrated load on the bound vortex.

iii. By the Biot-Savart formulation as given by Equation (1.85).

A fourth approach would be to find the solution to the Laplace equation directly, under the assumption that the vortex system was infinitely long in both directions of the x-axis. This was done by Kawada (1939), among others. His expressions are the basis of the numerical evaluation of the velocity field by Lerbs (1952).

We take option iii. as being most commonly employed in the literature. Our ultimate goal is to calculate the induction normal to the lifting line from all elements in analogy to the wing and thereby to secure an analogous integro-differential equation for the circulation density.

To this end we examine the induced components generated by a hub vortex, a single bound-to-the-blade vortex and a trailing "tip" vortex. As the flow *relative* to a blade which rotates at angular rate ω while advancing at speed U_a is of tangential velocity $-r\omega$ and axial velocity $-U_a$ the vorticity swept away from the blade lies on a distorted helical surface (as demonstrated in Chapter 9, p. 184) because by the theorem of Helmholtz (cf. for example Yih (1988)) vorticity is convected with the total flow components. Hence the trailing vortices are actually convected with component velocities of $u_a - U_a$, $u_t - r\omega$ and u_r where u_a, u_t and u_r are the axial, tangential and radial components induced by all the vortices and all other disturbances. For lightly loaded propellers the trajectories of the trailing vortices are virtually helices of pitch $2\pi U_a/\omega$ or pitch angle $\beta_a = \tan^{-1} U_a/r\omega$. Figure 11.2 is a photograph of a propeller in a water tunnel at reduced pressure in which the tip and hub vortices are made evident because the pressures on their physical cores are sufficiently low to produce local cavitation.

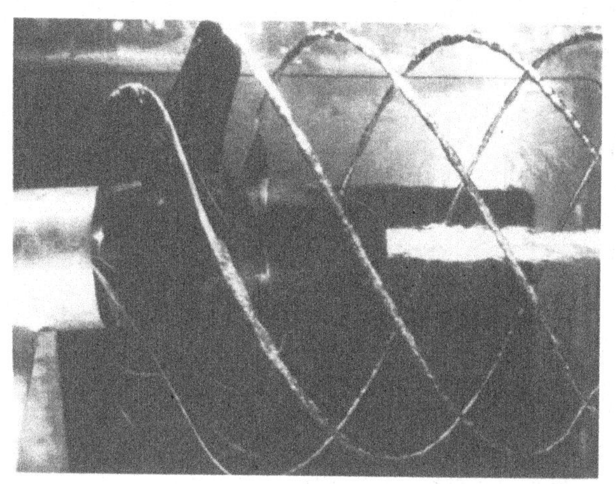

Figure 11.2 High–speed photograph of developed tip and hub vortex cavitation on a model ship propeller tested in the 36–inch Cavitation Tunnel at the David Taylor Research Center, USA.

By courtesy of David Taylor Research Center, USA.

INDUCED VELOCITIES FROM VORTEX ELEMENTS

The Biot–Savart formula, Equation (1.85) is now applied to the hub, bound and helical vortices.

Hub Vortex Inductions

Let $\mathbf{i}_a$, $\mathbf{i}_r$ and $\mathbf{i}_t$ be the unit vectors in the x–, r– and γ–(tangential) directions. Then the vorticity vector along the hub vortex is

$$\boldsymbol{\zeta} = \Gamma_0 \mathbf{i}_a + 0\mathbf{i}_r + 0\mathbf{i}_t \tag{11.1}$$

and the vector $\boldsymbol{R}$ is given by

$$\boldsymbol{R} = (x - x')\mathbf{i}_a + (r - 0)\mathbf{i}_r + 0\mathbf{i}_t \tag{11.2}$$

Along the hub vortex ds = dx', $-\infty \leq x' \leq 0$. Then

$$\mathbf{q}^h = u_a^h \mathbf{i}_a + u_r^h \mathbf{i}_r + u_t^h \mathbf{i}_t \tag{11.3}$$

$$= \frac{1}{4\pi} \int_{-\infty}^{0} \frac{\begin{vmatrix} \mathbf{i}_a & \mathbf{i}_r & \mathbf{i}_t \\ \Gamma_0 & 0 & 0 \\ x - x' & r & 0 \end{vmatrix}}{[(x - x')^2 + r^2]^{3/2}} dx' \tag{11.4}$$

Expanding by minors of the unit vectors we see that

$$u_a^h \equiv 0 \quad ; \quad u_r^h \equiv 0 \tag{11.5}$$

and

$$u_t^h = \frac{\Gamma_0 r}{4\pi}\int_{-\infty}^{0}\frac{1}{[(x-x')^2+r^2]^{3/2}}\,dx'$$
$$= \frac{\Gamma_0}{4\pi r}\left[1-\frac{x}{\sqrt{x^2+r^2}}\right] \tag{11.6}$$

It is seen that

$$u_t^h(\infty,r) = 0 \quad ; \quad u_t^h(-\infty,r) = \frac{\Gamma_0}{2\pi r} = 2u_t^h(0,r) \tag{11.7}$$

and that $u_t^h > 0$ as expected.

Bound Vortex Inductions

Here

$$\zeta = 0i_a + \Gamma_0 i_r + 0i_t \tag{11.8}$$

$$\boldsymbol{R} = (x-0)i_a + (r\cos(\gamma-\gamma_0)-r')i_r + (r\sin(\gamma-\gamma_0)-0)i_t \tag{11.9}$$

$$ds = dr' \quad ; \quad 0 \le r' \le r \tag{11.10}$$

Then

$$q^b = \frac{1}{4\pi}\int_0^{r_0}\frac{\begin{vmatrix} i_a & i_r & i_t \\ 0 & \Gamma_0 & 0 \\ x & r\cos(\gamma-\gamma_0)-r' & r\sin(\gamma-\gamma_0)\end{vmatrix}}{[x^2+(r\cos(\gamma-\gamma_0)-r')^2+(r\sin(\gamma-\gamma_0))^2]^{3/2}}\,dr' \tag{11.11}$$

It may be seen by inspection that $u_r^b \equiv 0$ as required physically.
The axial and tangential components are

$$u_a^b = \frac{\Gamma_0}{4\pi}\int_0^{r_0}\frac{r\sin(\gamma-\gamma_0)}{[x^2+r^2+r'^2-2rr'\cos(\gamma-\gamma_0)]^{3/2}}\,dr' \tag{11.12}$$

$$u_t^b = -\frac{x\Gamma_0}{4\pi}\int_0^{r_0}\frac{1}{R^3}\,dr' \tag{11.13}$$

To put these in different forms we may note that

$$\frac{\partial}{\partial\gamma}\frac{1}{R} = -\frac{rr'\sin(\gamma-\gamma_0)}{R^3} \quad ; \quad \frac{\partial}{\partial x}\frac{1}{R} = -\frac{x}{R^3}$$

so that

$$u_a^b = -\frac{\Gamma_0}{4\pi}\int_0^{r_0}\frac{1}{r'}\frac{\partial}{\partial\gamma}\left[\frac{1}{R}\right]dr' \quad ; \quad u_t^b = \frac{\Gamma_0}{4\pi}\int_0^{r_0}\frac{\partial}{\partial x}\left[\frac{1}{R}\right]dr' \qquad (11.14)$$

These manipulations allow us to express these components in terms of the harmonics of the relative angle $\gamma - \gamma_0$ by making use of one of the Fourier expansions of $1/R$, cf. Section 5 of the Mathematical Compendium. We use the form with the associated Legendre function of the second kind of half-integer degree, Equation (M5.23), and

$$\mathcal{Z} = \frac{x^2 + r^2 + r'^2}{2rr'} \qquad (x' = 0 \text{ here}) \qquad (11.15)$$

Then (11.14) becomes

$$u_a^b = -\frac{i\Gamma_0}{4\pi^2}\sum_{-\infty}^{\infty} m\int_0^{r_0}\frac{1}{r'\sqrt{rr'}}\,Q_{|m|-\frac{1}{2}}(\mathcal{Z})\,e^{im(\gamma-\gamma_0)}\,dr' \qquad (11.16)$$

$$u_t^b = +\frac{\Gamma_0}{4\pi^2}\sum_{-\infty}^{\infty}\int_0^{r_0}\frac{1}{\sqrt{rr'}}\frac{\partial \mathcal{Z}}{\partial x}\,Q'_{|m|-\frac{1}{2}}(\mathcal{Z})\,e^{im(\gamma-\gamma_0)}\,dr' \qquad (11.17)$$

As $\partial\mathcal{Z}/\partial x = x/rr'$ we observe that in the plane of the bound vortex, $x \equiv 0$, $u_t^b(0,r,(\gamma-\gamma_0)) \equiv 0$.

For a Z-bladed propeller (ITTC notation, not to be confused with (11.15)) we will have Z equally-spaced bound vortices with the ν-th bound vortex being located at $\gamma_0 + 2\pi\nu/Z$, the key blade being specified by $\nu = 0$. Then the total axial component due to all these "blades" involves us with the sum

$$e^{im\gamma}\sum_{\nu=0}^{Z-1} e^{-i(m\gamma_0+\frac{2\pi}{Z}\nu m)} = e^{im(\gamma-\gamma_0)}\sum_{\nu=0}^{Z-1} e^{-\frac{i2\pi\nu m}{Z}} = e^{im(\gamma-\gamma_0)}S$$

This sum can be recognized to be the well-known geometric series by writing

$$S = \sum_{\nu=0}^{Z-1} e^{-\frac{i2\pi\nu m}{Z}} = \sum_{\nu=0}^{Z-1} X^\nu \quad ; \quad X = e^{-\frac{i2\pi m}{Z}} \qquad (11.18)$$

and then the sum is

$$S = \sum_{0}^{Z-1} X^{\nu} = 1 + X + X^2 + \dots + X^{Z-1} \tag{11.19}$$

Multiplying both sides of (11.19) by 1–X gives

$$S = \frac{1 - X^Z}{1 - X} = \frac{1 - e^{-i2\pi m}}{1 - e^{-\frac{i2\pi m}{Z}}} \equiv 0 \qquad \text{for } m \neq qZ$$

with $q = 0, \pm1, \pm2, \dots$ (since m is an integer $e^{-i2\pi m} \equiv 1$). For $m = qZ$ the sum is indeterminate. Returning to (11.18) using $m = qZ$

$$\sum_{\nu=0}^{Z-1} e^{-i2\pi\nu q} = 1 + 1 + \dots = Z \ ; \ \text{all } q$$

and hence

$$S = \sum_{\nu=0}^{Z-1} e^{-\frac{i2\pi\nu m}{Z}} = \begin{cases} 0 \ ; \ m \neq qZ \\ Z \ ; \ m = qZ, \ q = 0, \pm1, \pm2, .. \end{cases} \tag{11.20}$$

Then the *total axial component* from Z bound vortices is

$$u_a^b = -\frac{i\Gamma_0 Z^2}{4\pi^2} \sum_{q=-\infty}^{\infty} q \int_0^{r_0} \frac{1}{r'\sqrt{rr'}} Q_{|qZ|-\frac{1}{2}}(Z)\, e^{iqZ(\gamma-\gamma_0)} dr' \tag{11.21}$$

The mean value $(q = 0)$ is zero for all r. Upon folding the series[17] the cosine terms in $e^{iqZ(\gamma-\gamma_0)}$ drop out and the sine terms double up (because q is an odd function) and we achieve

$$u_a^b = \frac{\Gamma_0 Z^2}{2\pi^2} \sum_{q=1}^{\infty} q \int_0^{r_0} \frac{1}{r'\sqrt{rr'}} Q_{|qZ|-\frac{1}{2}}(Z) \sin qZ(\gamma - \gamma_0) dr' \tag{11.22}$$

Evaluating this on the bound vortex at $x = 0$, $\gamma = \gamma_0$ clearly gives

$$u_a^b(0, r, \gamma_0) \equiv 0 \tag{11.23}$$

---

[17] Folding a series means $\sum_{-\infty}^{\infty} F(m) = \sum_{-\infty}^{-1} F(0) + F(0) + \sum_{1}^{\infty} F(m)$ and in first sum replace m by –m to get $F(0) + \sum_{1}^{\infty} (F(m) + F(-m))$.

Hence

The bound vortices in concert play no rôle in the induction on themselves.

Helical Vortex Inductions

Vortex elements shed by the propeller blade rotating about a fixed point at angular velocity ω (positive clockwise viewed from aft) in a stream $-U_a$ are in principle convected by the resultant relative velocity composed of $-U_a$, $-r\omega$ plus the axial, tangential and radial components induced at the shed element by all members of the vortex array. Thus the trajectory of vortices shed from any radial blade element is not a true helix as the induced velocities vary with distance from the propeller. In the ultimate wake of the screw propeller (some two – three diameters downstream) a true helical pattern is achieved as the axial inductions achieve their asymptotic values and the radial component vanishes. Moreover, as the vortices act on each other the sheet of vorticity shed from all blade elements quickly becomes unstable as in the flow abaft wings and wraps up into two concentrated vortices, a straight one streaming aft of the hub and one inboard of the tip radius in a quasi-helical pattern. At this point we shall be content to ignore these vortex interactions and moreover take U instead of U_a. We then examine the implications of a single tip vortex whose trajectory relative to the bound vortex can be specified in the following parametric way.

A vortex element shed t' seconds ago from the tip of the key blade which at present time is at angle γ_0 is assumed to be located by

$$\begin{aligned} x' &= Ut' \quad ; \quad -\infty < t' \leq 0 \\ r' &= r_0 \\ \gamma' &= \gamma_0 + \omega t' \\ y' &= -r_0 \sin(\gamma_0+\omega t') \\ z' &= r_0 \cos(\gamma_0+\omega t') \end{aligned} \tag{11.24}$$

The element of arc length along this trajectory is

$$ds = \sqrt{(dx')^2 + (dy')^2 + (dz')^2} = \sqrt{U^2 + (r_0\omega)^2}\, dt' \tag{11.25}$$

The vorticity vector along the helix is aligned with the resultant *relative* velocity and is perpendicular to the radius.

Then from Figure 11.3 we can see that the normalized resultant velocity relative to the bound vortex is, in rectangular components

$$\mathbf{V} = \frac{-U\mathbf{i} + (r_0\omega \cos\gamma')\mathbf{j} + (r_0\omega \sin\gamma')\mathbf{k}}{\sqrt{U^2 + (r_0\omega)^2}} \tag{11.26}$$

and upon introduction of the flow angle β, defined in Figure 11.3, so that

$$\frac{U}{\sqrt{U^2 + (r_0\omega)^2}} = \sin\beta \quad ; \quad \frac{r_0\omega}{\sqrt{U^2 + (r_0\omega)^2}} = \cos\beta \tag{11.27}$$

we may then express the vorticity vector as

$$\zeta = (-\Gamma_0 \sin\beta)\mathbf{i} + (\Gamma_0 \cos\beta \cos\gamma')\mathbf{j} + (\Gamma_0 \cos\beta \sin\gamma')\mathbf{k}$$

Then the velocity vector induced by the entire helical vortex is

$$\mathbf{q}^{he} = \frac{\sqrt{U^2+(r_0\omega)^2}}{4\pi} \int_{-\infty}^{0} \frac{\begin{vmatrix} \mathbf{i} & \mathbf{j} & \mathbf{k} \\ -\Gamma_0 \sin\beta & \Gamma_0 \cos\beta' \cos\gamma' & \Gamma_0 \cos\beta' \sin\gamma' \\ x-Ut' & -(r\sin\gamma - r_0\sin\gamma') & r\cos\gamma - r_0\cos\gamma' \end{vmatrix}}{[(x-Ut')^2 + r^2 + r_0^2 - 2rr_0\cos(\gamma-\gamma')]^{3/2}} dt' \tag{11.28}$$

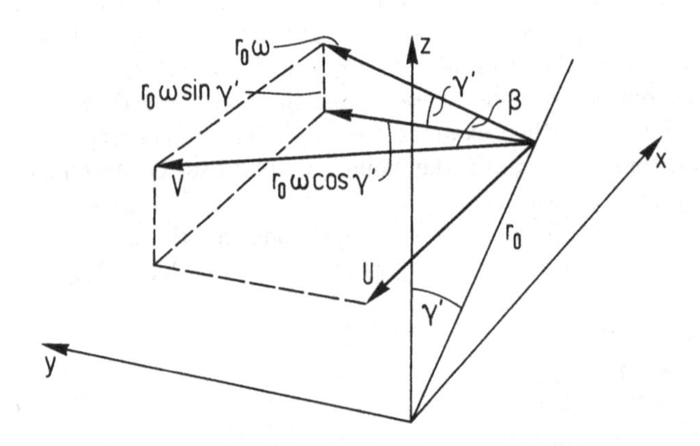

Figure 11.3

The axial component can be seen, through the use of (11.27), to be

$$u_a^{he} = -\frac{r_0\omega\Gamma_0}{4\pi} \int_{-\infty}^{0} \frac{r_0 - r\cos(\gamma-\gamma_0-\omega t')}{R^3} dt'$$

$$= \frac{r_0\omega\Gamma_0}{4\pi} \frac{\partial}{\partial r_0} \int_{-\infty}^{0} \frac{dt'}{R} \tag{11.29}$$

We may now manipulate (11.29) by taking $x' = x - Ut'$ to obtain

$$u_a^{he} = \frac{r_0\omega\Gamma_0}{4\pi U}\frac{\partial}{\partial r_0}\int_x^\infty \frac{1}{\sqrt{x'^2+r^2+r_0^2-2rr_0\cos\left[\gamma-\gamma_0-\frac{\omega(x-x')}{U}\right]}}dx' \tag{11.30}$$

We may observe immediately that $u_a^{he}(\infty,r,\gamma) = 0$ (limits come together). As $x \to -\infty$ the evaluation of the integral in (11.30) is not straightforward. To determine this behavior we make use of an alternate form of $1/R$ given by Equation (M5.24) of Section 5 of the Mathematical Compendium. The integrand $1/R$ of (11.30) can then be expressed as

$$\frac{1}{R} = \frac{1}{\pi}\sum_{m=-\infty}^{\infty}\int_{-\infty}^{\infty} I_{|m|}(|k|r)\ K_{|m|}(|k|r_0)\ e^{ikx'}\ e^{im(\gamma-\gamma_0-\frac{\omega(x-x')}{U})}dk \tag{11.31}$$

where I_m and K_m are the modified Bessel functions of the first and second kind.

Now the x'-integral of (11.30) can be seen to involve (for $x \to -\infty$) from (11.31)

$$\int_{-\infty}^{\infty} e^{i(k+\frac{m\omega}{U})x'}dx' = 2\pi\delta\left[k+\frac{m\omega}{U}\right] \tag{11.32}$$

where δ is the Dirac delta-function. This stroke allows us to evaluate the k-integral in (11.31) at once, to yield from (11.31) and (11.32) into (11.30)

$$u_a^{he}(-\infty,r,(\gamma-\gamma')) = \frac{r_0\omega\Gamma_0}{2\pi U}\frac{\partial}{\partial r_0}\sum_{m=-\infty}^{\infty} I_{|m|}\left[|m|\frac{\omega r}{U}\right] \cdot K_{|m|}\left[|m|\frac{\omega r_0}{U}\right]\ e^{im(\gamma-\gamma_0-\frac{\omega x}{U})} \tag{11.33}$$

We may now carry out the differentiation which, upon use of one of the Bessel function identities[18], *summing for* Z *blades* by the procedure of p. 212, and folding the series, produces

$$u_a^{he} = -\frac{\pi Z\Gamma_0}{4r_0J^2}\sum_{q=0}^{\infty}\varepsilon_q q I_{qZ}\left[\kappa\frac{r}{r_0}\right]\ \left[K_{qZ-1}(\kappa)+K_{qZ+1}(\kappa)\right] \cdot\cos qZ\left[\gamma-\gamma_0-\frac{\pi x}{Jr_0}\right] \tag{11.34}$$

[18] From Abramowitz and Stegun (1972), $dK_m(z)/dz = \frac{1}{2}(K_{m-1} + K_{m+1})$

where the parameter $\kappa = \pi qZ/J$; $J = U/ND$ (here $D = 2r_0$), the advance ratio and $\varepsilon_q = 1$, $q = 0$, otherwise $= 2$. This result is valid for $r/r_0 < 1$; for $r/r_0 > 1$ interchange the arguments of the Bessel functions.

This result shows that the axial component varies harmonically with x and hence at $q = 1$ (blade frequency) the components will exhibit a cyclic variation x-wise of length $\tilde{x}$

$$\frac{\pi Z}{J}\frac{\tilde{x}}{r_0} = 2\pi \quad \text{or} \quad \tilde{x} = \frac{U}{ZN} \tag{11.35}$$

Defining the undisturbed-flow pitch angle β and the undisturbed-flow or fluid pitch P_f by Figure 11.4 then

$$\frac{U}{r_0\omega} = \frac{P_f}{2\pi r_0} \tag{11.36}$$

and upon eliminating U/ω between (11.35) and (11.36) we see that

$$\tilde{x} = \frac{P_f}{Z} \quad \text{or} \quad \tilde{x} = P_f \text{ for } Z = 1 \tag{11.37}$$

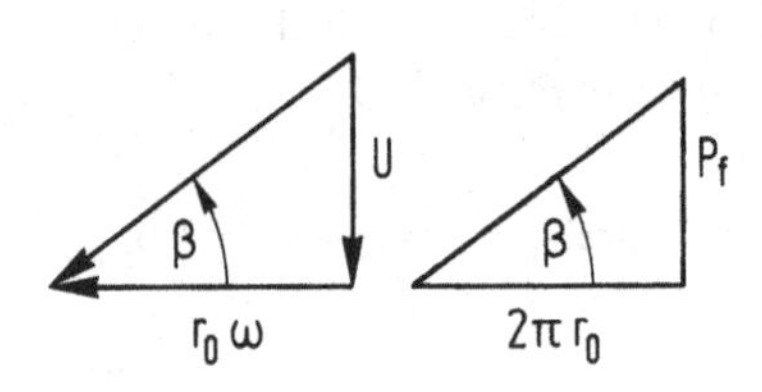

Figure 11.4 Undisturbed-flow pitch (fluid pitch) P_f and pitch angle β.

This result confirms our physically based expectation. When we account for the axial and tangential induction at the propeller the pitch of the helices will be a function of the hydrodynamic pitch angle β_i (cf. Figure 9.15, p. 192).

Returning to (11.34) we inquire as to the mean value $\overline{u}_a^{he}$, that is, the value at frequency order $q = 0$. At first sight this appears to be zero because of the factor $|qZ|$. However, from the properties of the Bessel functions we can find (see for example Abramowitz and Stegun, (1972)) that

$$I_{qZ}\left[\frac{\pi qZ}{J}\frac{r}{r_0}\right] \longrightarrow \frac{1}{(qZ)!}\left[\frac{\pi qZ}{2J}\frac{r}{r_0}\right]^{qZ} \quad \text{as } q \to 0 \tag{11.38}$$ [19]

and, in general

$$K_m(\kappa) \longrightarrow \tfrac{1}{2}|m-1|!\left[\frac{2}{\kappa}\right]^m \quad \text{as } \kappa = \frac{\pi qZ}{J} \longrightarrow 0 \tag{11.39}$$ [19]

Consequently,

[19] $I_{-m} = I_m$; $K_{-m} = K_m$. The arguments of I_m and K_m should contain $|q|$. We omit the modulus symbol for simplicity but we must understand that the arguments are always positive.

$$\lim_{q \to 0} qZ I_{qZ}\left[\kappa \frac{r}{r_0}\right] K_{qZ-1}(\kappa) = 0$$

and (11.40)[19]

$$\lim_{q \to 0} qZ I_{qZ}\left[\kappa \frac{r}{r_0}\right] K_{qZ+1}(\kappa) = \frac{J}{\pi}$$

$$\therefore \bar{u}_a^{he} = -\frac{Z\Gamma_0}{4 r_0 J} \quad ; \quad 0 < r < r_0, \quad x \to -\infty \tag{11.41}$$

We now wish to compare (11.41) and (11.34) with corresponding values in the plane x = 0. Returning to (11.30), setting x = 0 and using (11.31) we are confronted by

$$u_a^{he}(0,r,\psi) = \frac{r_0 \omega \Gamma_0}{4\pi^2 U} \frac{\partial}{\partial r_0} \sum_{m=-\infty}^{\infty} \int_0^{\infty} dx' \int_{-\infty}^{\infty} I_m(kr)\, K_m(kr_0)\, e^{i(k+mp_f^*)x'} e^{im\psi} dk \tag{11.42}$$

where for brevity $\psi = \gamma - \gamma_0$; $p_f^* = \omega/U$ (= $2\pi/P_f$, P_f the fluid pitch).
The x'-integral can be written as

$$I = \int_0^{\infty} \cos((k+mp_f^*)x')dx' + i\int_0^{\infty} \sin((k+mp_f^*)x')dx' \tag{11.43}$$

These integrals do not exist in the ordinary sense. The first is known to be $\pi\delta(k + mp_f^*)$, cf. (11.32). To evaluate the second we assume that

$$\int_0^{\infty} \sin((k+mp_f^*)x')dx' = Im\left\{\lim_{\mu \to 0} \int_0^{\infty} e^{(i(k+mp_f^*)-\mu)x'} dx'\right\}$$

(*Im* indicates to take imaginary part), where μ is a small positive "damping" parameter.
This integral exists in the ordinary sense and we carry out the integration to obtain

$$\lim_{\mu \to 0}\left[Im\left\{\frac{1}{\mu - i(k+mp_f^*)}\right\}\right] = \lim_{\mu \to 0}\left[\frac{k + mp_f^*}{\mu^2 + (k+mp_f^*)^2}\right]$$

$$= \frac{1}{k + mp_f^*} \tag{11.44}$$

Thus (11.43) becomes

$$I = \pi\delta(k + mp') + \frac{i}{k + mp_f^*} \tag{11.45}$$

and hence (11.42) is reduced to

$$u_a^{he}(0) = \frac{r_0\omega\Gamma_0}{4\pi^2 U}\frac{\partial}{\partial r_0}\sum_{m=-\infty}^{\infty}\int_{-\infty}^{\infty} I_m(kr)\ K_m(kr_0)\left[\pi\delta(k+mp')+\frac{i}{k+mp_f^*}\right]\cdot e^{im\psi}dk \tag{11.46}$$

The term involving the δ-function is clearly one-half of what we found at $x = -\infty$, provided we take the modulus of that result which by (11.34) is seen to vary harmonically with x. The terms involving $i/(k + mp_f^*)$ requires further manipulations. Folding the m-summation (cf. footnote on p. 212) as I_m and K_m are even in m

$$\sum_{-\infty}^{\infty} F(m) = \frac{iI_0K_0}{k} + i\sum_{m=1}^{\infty}\left[\frac{e^{im\psi}}{k + mp_f^*} + \frac{e^{-im\psi}}{k - mp_f^*}\right] I_m(kr)\ K_m(kr_0)$$

$$= \frac{iI_0K_0}{k} + i\sum_{m=1}^{\infty}\left[\frac{2k\ \cos m\psi\ -\ 2\,i\,mp_f^*\ \sin m\psi}{k^2\ -\ (mp_f^*)^2}\right] I_m(kr)\ K_m(kr_0) \tag{11.47}$$

We may observe that the terms involving 2k in (11.47) will integrate to zero as k is odd and $I_m\ K_m$ is even in k (recall that arguments of Bessel functions involve $|k|$). Thus (11.46) is reduced to

$$u_a^{he}(0) =$$

$$\frac{r_0\omega\Gamma_0}{8\pi U}\sum_{m=-\infty}^{\infty}|mp_f^*|\,I_m(mp_f^*r)\left[K_{|m-1|}(mp_f^*r_0)+K_{|m+1|}(mp_f^*r_0)\right]e^{im\psi}$$

$$+\frac{ir_0\omega\Gamma_0}{4\pi^2U}\left\{-\int_{-\infty}^{\infty}\frac{|k|}{k}\ I_0(kr)\ K_1(kr_0)\ dk + p_f^*\sum_{m=1}^{\infty} m\int_{-\infty}^{\infty}\frac{|k|I_m(kr)[K_{m-1}(kr_0)+K_{m+1}(kr_0)]}{k^2 - (mp_f^*)^2}\sin m\psi\ dk\right\} \tag{11.48}$$

where $\partial/\partial r_0$ has been carried out.

The first sum is one-half the amplitude of $u_a^{he}(-\infty)$ (where we sum the effects of Z blades by the procedure of p. 212 replacing m by qZ, the m-summation by the q-summation, and multiply by Z). The first integral (involving I_0K_1) vanishes because $|k|/k$ is an odd function. The last summation contains no zero harmonic or mean term as $m \geq 1$, but it vanishes for $\psi = \gamma - \gamma_0 = 0$, i.e. when evaluating on the bound vortex.

Thus we conclude that the mean values (from all blades) obey

$$\bar{u}_a^{he}(0) = \frac{1}{2}\,\bar{u}_a^{he}(-\infty) \tag{11.49}$$

which corresponds to the result obtained in Chapter 9 for the actuator disc (linearized flow), Equation (9.102). But we also have that unless $\gamma = \gamma_0$

$$(u_a^{he}(0,r,\gamma))_{qZ} \neq \frac{1}{2}((u_a^{he}(-\infty,r,\gamma))_{qZ} \tag{11.50}$$

GENERALIZATION TO A CONTINUOUS RADIAL VARIATION OF CIRCULATION

As we saw in the previous chapter on wing theory, the simple model of a wing or propeller blade of radially uniform loading with trailing vortices from the blade root and tip gave physically untenable results. We then make our model of the vortex system more realistic by using a distribution of vorticity over the radius. To generalize from the single helical vortex to a continuum of shed vorticity we return to Equation (11.29) and replace r_0 by a dummy radial variable r', and Γ_0 by $\Gamma(r')$ and form the difference in the axial velocity arising from a pair of adjacent helical vortices at $r = r' + \delta r'$ and $r = r'$. The increase in bound vorticity is $-d\Gamma/dr'$, where the minus is due to the vortex leaving the radial line (ζ directed downstream) as shown in Figure 11.1. Then schematically

$$u_a(x,(r'+\delta r'),.....) = \omega r'\left[\Gamma(r') - \frac{d\Gamma}{dr'}\delta r'\right]\int_{-\infty}^{0}....$$

$$u_a(x,r',.....) = \omega r'\Gamma(r')\int_{-\infty}^{0}....$$

and the difference divided by $\delta r'$ is

$$\frac{\delta u_a}{\delta r'} = -r'\frac{d\Gamma}{dr'}\int_{-\infty}^{0}.... \tag{11.51}$$

Thus the generalization of Equation (11.29) is, upon integrating the form (11.51) over the dummy variable, r', and writing $\partial\Gamma/\partial r' = \Gamma'$

$$u_a = \frac{\omega}{4\pi}\int_{R_h}^{R} dr'r'\Gamma'\int_{-\infty}^{0}\frac{r' - r\cos(\psi-\omega t')}{[(x-Ut')^2 + r^2 + r'^2 - 2rr'\cos(\psi-\omega t')]^{3/2}}\,dt' \tag{11.52}$$

where R is the propeller radius; R_h = hub radius; $\psi = \gamma - \gamma_0$ the angle between any radial line to a space point (x,r,γ) and a point $(0,r',\gamma_0)$ on the bound vortex on the key blade.

To cast this into a form generally used we effect two transformations. First let $x' = x - Ut'$, $dt' = -dx'/U$ yielding

$$u_a = \frac{\omega}{4\pi U}\int_{R_h}^{R} dr' r' \Gamma' \int_{x}^{\infty} \frac{r' - r\cos(\psi - p_f^*(x-x'))}{[x'^2 + r^2 + r'^2 - 2rr'\cos(\psi - p_f^*(x-x'))]^{3/2}} dx' \tag{11.53}$$

and as our interest is to evaluate the induction on the bound vortex we set $x = 0$, $\psi = 0$. The peripherally mean axial component of the wake flow abaft a ship is not uniform radially so we replace U by $U_a(r)$. Furthermore the hydrodynamic (fluid) pitch $2\pi U/\omega$ is not obtained because of the propeller induction. To allow for this we perform the following replacements by writing

$$p_f^* = \frac{\omega}{U} \longrightarrow \left[\frac{\omega r'}{U_a}\right]\frac{1}{r'} = \frac{1}{r'\tan\beta} \longrightarrow \frac{1}{r'\tan\beta_i(r')} \tag{11.54}$$

that is, we replace the undisturbed flow angle by the induced angle $\beta_i(r')$ which varies radially. The angle β_i is defined by Figure 11.5. We next change the dummy variable x' which locates a point on any "helical" vortex element by the transformation

$$x' = r'\psi' \tan\beta_i(r') \tag{11.55}$$

which is geometrically defined in Figure 11.5. We see that

$$\beta_i(r') = \tan^{-1}\left[\frac{u_a(0,r',0) - U_a(r')}{u_t(0,r',0) - r'\omega}\right] = \tan^{-1}\left[\frac{U_a(r') + |u_a(0,r',0)|}{r'\omega - u_t(0,r',0)}\right] \tag{11.56}$$

as $u_a(0,r',0) < 0$, $u_t > 0$.

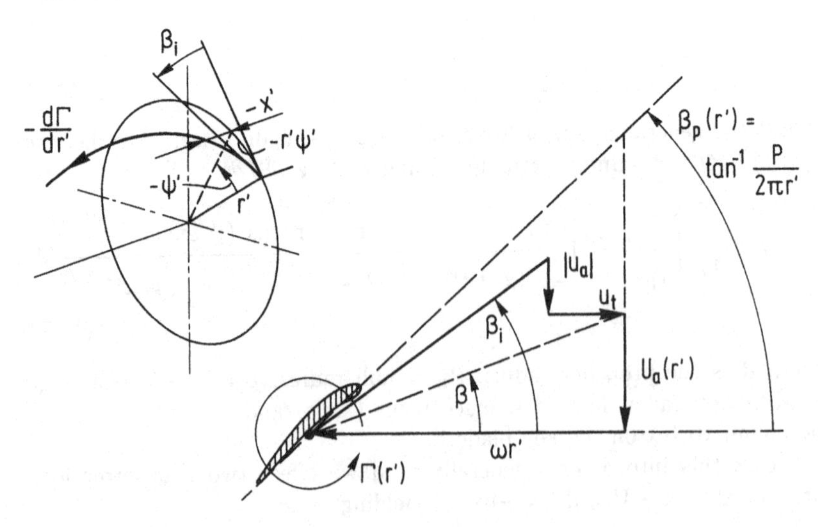

Figure 11.5 Helical vortex element and inflow velocities.

With the replacement (11.54) and the transformation (11.55) employed in (11.53), the axial induction from the vortex wake of a Z-bladed propeller (via this modified representation) is, on the key blade

$$u_a(0,r,0) = \frac{1}{4\pi}\sum_{\nu=0}^{Z-1}\int_{R_h}^{R} dr'\Gamma'\int_0^{\infty}\frac{(r'^2-r\,r'\,\cos(\psi'+\gamma_\nu))\,d\psi'}{[(r'\psi'\tan\beta_i)^2+r^2+r'^2-2\,r\,r'\cos(\psi'+\gamma_\nu)]^{3/2}} \tag{11.57}$$

with $\gamma_\nu = 2\pi\nu/Z$, the inter-blade angle being $2\pi/Z$, $\nu = 0$ the key blade.

It is noted that the modification of the trajectories of the shed vorticity or "trailers" is based on the induced axial and tangential velocities u_a, u_t evaluated at the bound vortex whereas, physically, the vortex elements are convected by the particle velocities which vary spatially downstream until they achieve their asymptotic values. An expression similar to (11.57) for the tangential component will be omitted here.

The foregoing modifications provide us with the induced velocities of a *moderately* loaded propeller whose blades are of large aspect ratio. If the propeller is lightly loaded the free, or trailing vortices, can be considered to be convected by the undisturbed flow relative to the blades and consequently β_i can be replaced by

$$\beta_a(r') = \tan^{-1}\left[\frac{U_a(r')}{r'\omega}\right] \tag{11.58}$$

which removes the implicitness of (11.57) (and similarly for $u_t(0,r,0)$) since β_i depends on u_a and u_t. Thus the lightly loaded propeller is analogous to the wing where the trajectories of the trailing vortices are assumed to be independent of the loading on the wing.

However, unlike the wing the inductions from the free vortices are not easy to evaluate. In the linear (lightly loaded) theory a special case arises when the pitch of the free vortices is constant as always obtains when the inflow is uniform. This special case was first solved analytically by Goldstein (1929) by considering the flow generated by a translating, rigid helicoidal surface.

The case of radially non-uniform pitch of the trailing vortices, as occurs in operation abaft a hull, was first solved by Lerbs (1952) whose method will be outlined herein as it forms the basis of many existing lifting-line computer programs (for example, Caster, Diskin & La Fone (1975) and van Oossanen, (1975)).

The integrations required in (11.57) cannot be effected analytically, a major obstacle in the pre-computer era. An approximate development by Lerbs (1952) for the induction of single helical vortex elements resulted in a formulation involving an infinite series of products of modified Bessel functions of the first and second kind (analogous to the development leading to Equation (11.34) which is more adroit than that of Lerbs (*ibid.*)). The advantage of this procedure is that it allows the use of asymptotic approximations of the Bessel functions and then the sums are relatively easy to compute. An improved version of Lerbs' development was evaluated by Wrench (1957) via computer with an algorithm currently in use.

Kerwin (1981) relates that his subsequent direct numerical integration of expressions for all components of the induced velocities confirmed the accuracy of the Lerbs-Wrench procedure (and vice versa) and proceeds to remark that "while the computational effort of direct numerical integration is now trivial by current standards the Wrench formulation is still more efficient".

Once the ψ'-integration is performed by either procedure there remains the tasks of evaluating the singular integral over the radius. Lerbs (1952) evidently exploited the work of Moriya (1933) who factored out the singular part thereby introducing "induction factors" which have finite limits as $r' \to r$ ($R_h \leq r \leq R$). We now consider this process in some detail.

INDUCTION FACTORS

The singular behavior of the velocity components when the field point is brought to a helical (or bound) vortex is asserted by various writers to be of the Cauchy type, i.e., proportional to $1/(r-r')$ on the basis that all curved vortices appear straight and two-dimensional when one is close enough, and that is the singularity of straight vortices (cf. Newman, (1977)). For sake of completeness we offer the following demonstration to show that this is indeed the character of the singularity.

Returning to the axial component induced by a single helical vortex (11.29) we rewrite as follows replacing r_0 by r'

$$u_a(x,r,\gamma) = -\frac{r'\omega\Gamma_0}{4\pi}\left\{\int_{-\infty}^{\frac{x-\varepsilon}{U}} + \int_{\frac{x-\varepsilon}{U}}^{\frac{x+\varepsilon}{U}} + \int_{\frac{x+\varepsilon}{U}}^{0} \frac{r'-r\cos(\gamma-\gamma_0-\omega t')}{[(x-Ut')^2+r^2+r'^2-2rr'\cos(\gamma-\gamma_0-\omega t')]^{3/2}}\,dt'\right\} \tag{11.59}$$

where $-\infty < x < 0$, and ε is a small positive parameter.

It is apparent that the singularity occurs in the second integral in which we substitute $x' = x-Ut'$ and obtain (after reversing the new limits)

$$I = -\frac{r'\omega\Gamma_0}{4\pi U}\int_{-\varepsilon}^{\varepsilon} \frac{r'-r\cos(\gamma-\gamma_0-p_f^*(x-x'))}{[x'^2+r^2+r'^2-2rr'\cos(\gamma-\gamma_0-p_f^*(x-x'))]^{3/2}}\,dx' \tag{11.60}$$

We now position our field point in the radial plane $\gamma = \gamma_0 + p_f^*x$, noting that the cosines become cos p_f^*x' and as the range of x' is 2ε we can take cos $p_f^*x' = 1$ in this interval. The x'-integral can now be effected to give

$$I = -\frac{r'\omega\Gamma_0}{4\pi U}\left\{\frac{r - r'}{(r - r')^2}\left[\frac{2\varepsilon}{\sqrt{\varepsilon^2 + (r - r')^2}}\right]\right\} \tag{11.61}$$

Holding ε fixed we see that as we close on to the vortex by letting $r \to r'$ we find

$$u_a^{he} \to -\frac{r'\omega\Gamma_0}{2\pi U(r - r')} \quad ; \quad -\infty < x < 0 \quad ; \quad \gamma = \gamma_0 + p_f^*x \tag{11.62}$$

as the other integrals provide bounded contributions. The tangential and radial components can also be shown to have the same type of singularity.

This suggests that in the generalized forms (as in (11.57)) we can place the singularity in evidence by multiplying above and below by (r–r') to achieve the following expressions for the axial, u_a, radial, u_r, and tangential, u_t, components

$$u_{a,r,t} = \frac{1}{4\pi}\int_{R_h}^{R} \frac{d\Gamma}{dr'}\frac{i_{a,r,t}}{r - r'}\,dr' \tag{11.63}$$

where the i's are the corresponding *induction factors* which are defined by

$$i_a = \sum_{\nu=0}^{Z-1}\int_0^\infty \frac{(r-r')(r'^2-rr'\cos(\psi'+\gamma_\nu))}{R^3}\,d\psi' \tag{11.64}$$

$$i_r = -\sum_{\nu=0}^{Z-1}\int_0^\infty \frac{(r-r')r'^2[\sin(\psi'+\gamma_\nu)-\psi'\cos(\psi'+\gamma_\nu)]\tan\beta_i}{R^3}\,d\psi' \tag{11.65}$$

$$i_t = -\sum_{\nu=0}^{Z-1}\int_0^\infty \frac{(r-r')r'[r'\psi'\sin(\psi'+\gamma_\nu)-(r-r'\cos(\psi'+\gamma_\nu))]\tan\beta_i}{R^3}\,d\psi' \tag{11.66}$$

where

$$R = \sqrt{(r'\tan\beta_i\psi')^2 + r^2 + r'^2 - 2rr'\cos(\psi'+\gamma_\nu)} \tag{11.67}$$

Note that by factoring out r^3,

$$i_{a,r,t} = i_{a,r,t}\left[\frac{r'}{r}\right] \tag{11.68}$$

We observe that the induction factors are dimensionless velocities induced by helical vortices of unit strength. Since they hence depend only of the geometry of the vortex element they can be calculated when the geometry is known, i.e. on the basis of pitch angle β_i, number of blades Z and radius ratio r'/r. Curves of the axial and tangential factors are given for example by Lerbs (1952), based on an approximate procedure. This is described in more detail in the next chapter (p. 239 and sequel) where it is also shown that as $r \to r'$, $|i_a| \to \cos\beta_i$ and $i_t \to \sin\beta_i$. We shall have no application here for the radial component as we will not treat raked blades. This requires one to start the trailers from a raked bound vortex. The radial component must be included in the lifting-surface treatment of skewed and raked blades.

The radial component at all $x > 0$ can easily be seen to vanish as $x \to \infty$ (upstream)[20]. However as $x \to -\infty$, $u_r(x,r,\gamma)$ can be shown to vanish in the planes of the local helices, i.e., for

$$\gamma = \gamma_0 + \frac{2\pi\nu}{Z} + \frac{x\ \cot\beta_i}{r'}$$

but otherwise its determination is difficult via the form given by the Biot–Savart law. The proof which shows that the mean or zero harmonic of $u_r \equiv 0$ at $x = -\infty$ is presented in Appendix C.

FORCES ACTING ON THE BLADES AND THE EQUATION FOR THE CIRCULATION DENSITY

The differential force acting at any radius r is, by the Kutta-Joukowsky law (cf. Appendix B) applied locally at the bound vortex

$$dF = \rho V\Gamma(r) \tag{11.69}$$

This force is at right angles to the resultant velocity V which includes the induced velocities as well as the relative axial and tangential components $-U_a$ and $-r\omega$ as indicated in Figure 11.6.

The axial and tangential components are then

$$dF_a = \rho V\Gamma\ \cos\beta_i \tag{11.70}$$

$$dF_t = -\rho V\Gamma\ \sin\beta_i \tag{11.71}$$

20 The x'-integral will have a form comparable to (11.53) where we can see that as $x \to \infty$ the x'-limits come together.

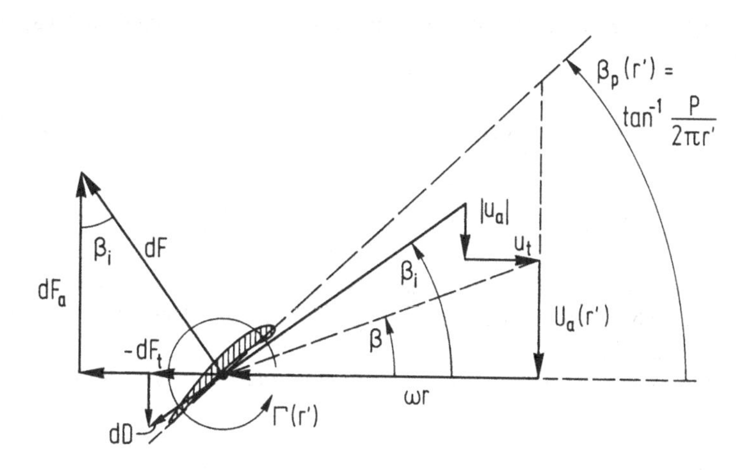

Figure 11.6 Force elements on lifting line.

Analogous to the wing we may write this local lift force as

$$\rho V\Gamma = 2\pi(\beta_P - \alpha_0 - \beta_i)\ \tfrac{1}{2}\rho c(r)\ V^2 \tag{11.72}$$

i.e. the force per unit length as though the blade element of chord c(r) were a two-dimensional section at the effective angle $\beta_P - \alpha_0 - \beta_i$, α_0 is the angle of zero lift, < 0 (Equation (5.44)).

Upon insertion of the definition of β_i which is a functional of Γ' and β_i, (11.72) becomes

$$\Gamma(r) = \pi c(r)\left\{\beta_P - \alpha_0 - \tan^{-1}\left[\frac{U_a(r) + \frac{1}{4\pi}\displaystyle\int_{R_h}^{R}\left|i_a\left[\frac{r'}{r},\beta_i\right]\right|\left|\frac{\Gamma'(r')}{r - r'}\right| dr'}{\omega r - \frac{1}{4\pi}\displaystyle\int_{R_h}^{R} i_t\left[\frac{r'}{r},\beta_i\right]\frac{\Gamma'(r')}{r - r'}\, dr'}\right]\right\} V \tag{11.73}$$

This is a non-linear integro-differential equation for the circulation $\Gamma(r)$ which while analogous to that for the wing is much more complicated in several respects, particularly because the induction or "influence functions", i_a and i_t, are functions of β_i and, in addition, because the arc tangent is not a linear function. For the lightly loaded propeller β_i is replaced by β_a ($= \tan^{-1}(U_a(r')/\omega r')$) and then the calculation is simplified. For the moderately loaded propeller, an iterative scheme must be adopted until a desired (or specified) thrust is achieved.

The total thrust and torque for a Z-bladed propeller are obtained by integrating, summing and changing sign on the torque equation to define the torque in concordance with current practice, as explained in Chapter 9, p. 186. This gives

$$T = \rho Z \int_{R_h}^{R} [V\Gamma \cos\beta_i - \tfrac{1}{2}c(r)C_D \sin\beta_i V^2] dr \tag{11.74}$$

$$Q = \rho Z \int_{R_h}^{R} r[V\Gamma \sin\beta_i + \tfrac{1}{2}c(r)C_D \cos\beta_i V^2] dr \tag{11.75}$$

where the effect of viscosity is included via the viscous drag coefficient at each radius which depends on the thickness ratio and the local Reynolds number (cf. p. 121-125).

The implications of (11.73) and the evaluations to obtain designs are discussed next.

12 Propeller Design via Computer and Practical Considerations

Here we present the essential steps in the problems posed by the design and analysis of propellers. In design we are required to develop the diameter, pitch, camber and blade section to deliver a required thrust at maximum efficiency (minimum torque). There are other criteria such as to design a propeller to drive a given hull (of known or predicted resistance over a range of speeds) with a specified available shaft horsepower and to determine the ship speed.

The analysis procedure requires prediction of the thrust, torque and efficiency of propellers of specified geometry and inflow.

We begin with the development of the criteria for the radial distribution of thrust-density to achieve maximum efficiency in uniform and non-uniform inflows. This is followed by methods for determining optimum diameter for given solutions and optimum solutions for a given diameter.

The derivation and reason for the induction factors in the lifting-line theory of discrete number of blades, as displayed in the previous chapter, is followed by formulas for the thrust and torque coefficients in terms of the circulation amplitude function G_n.

Applications are then made to the design and analysis of propellers. Means for selection of blade sections to avoid or mitigate cavitation are followed by extensive discussion of practical aspects of tip unloading via camber and pitch variation. Effects of blade form and skew on efficiency and pressure fluctuations at blade frequency (number of blades times revolution per second) are presented.

CRITERIA FOR OPTIMUM DISTRIBUTIONS OF CIRCULATION

Before going into details we may remark (with Kerwin, Coney & Hsin (1986)) that lifting-line theory is the basis for propeller design since it provides the radial distribution of loading or circulation. This distribution is obtained by use of criteria for optimum efficiency (as will be derived shortly) or modifications of such a distribution, for example to reduce the tip loading, to avoid unacceptable unsteady cavitation, high vibratory forces and noise, as outlined at the end of this chapter.

In the design process the more elaborate lifting-surface methods are used to trace the blade geometry from data mainly generated by lifting-line procedures, viz. overall geometry and radial distribution of loading. Chordwise load distribution may be of the roof-top type.

A more numerically oriented version of the lifting line method has been developed by Kerwin *et al.* (*ibid.*). While it has been designed to treat multi-component propulsors such as contrarotating propellers, propellers with vane wheels etc., it can also be used for the single propeller. We postpone a description of this method until Chapter 23 on unconventional propulsion and take what one may call the classical approach to lifting-line propeller theory.

Propellers in Uniform Inflow

Formal analyses yielding criteria for optimum propeller efficiency when viscous-drag and cavity-drag effects are included have been carried out by Yim, (1976). His conclusion is that for non-cavitating propellers the criteria for optimum efficiency in both uniform and non-uniform inflow can be secured by ignoring viscous effects but that the optimum distribution of pitch for supercavitating propellers is highly dependent on the lift-drag ratio of the sections. Consequently we proceed to determine a criterion for optimum efficiency by setting $C_D = 0$ in (11.74) and (11.75). We follow the development by Wehausen, (1964) in part, considering at first the case of the propeller in open water (freely running).

As the forces on an element of the bound vortex are at right angles to the incident velocity, the thrust is generated by the net tangential velocity and the torque-producing force by the net axial component. Hence we have, cf. p. 226, (11.74) and (11.75), with $V \cos\beta_i = \omega r - u_t$ and $V \sin\beta_i = U + |u_a|$

$$T = \rho Z \int_{R_h}^{R} \Gamma(r) \left\{ \omega r - \frac{1}{4\pi} \int_{R_h}^{R} i_t \left[\frac{r'}{r}\right] \frac{\Gamma'(r')}{r - r'} dr' \right\} dr \tag{12.1}$$

$$Q = \rho Z \int_{R_h}^{R} r\Gamma(r) \left\{ U + \frac{1}{4\pi} \int_{R_h}^{R} \left| i_a \left[\frac{r'}{r}\right] \right| \left| \frac{\Gamma'(r')}{r - r'} dr' \right| \right\} dr \tag{12.2}$$

where we look after the fact that the axially induced velocity is in the same direction as U by taking the modulus of i_a. Here we see that T and Q are functionals of $\Gamma(r)$ and $\Gamma'(r')$. To secure expressions which are functionals of only Γ' we integrate by parts using the requirement that $\Gamma(R) = 0$ and the assumption that $\Gamma(R_h) = 0$, thereby obtaining

$$T = -\rho Z \int_{R_h}^{R} dr \left\{ \frac{\omega}{2}(r^2 - R_h^2) - \int_{R_h}^{R} dr'' \Gamma'(r'') \frac{1}{4\pi} \int_{R_h}^{r} i_t \left[\frac{r''}{r'}\right] \frac{dr'}{r' - r''} \right\} \Gamma'(r) \quad (12.3)$$

$$Q = -\rho Z \int_{R_h}^{R} dr \left\{ \frac{1}{2} U(r^2 - R_h^2) + \int_{R_h}^{R} dr'' \Gamma'(r'') \frac{1}{4\pi} \int_{R_h}^{r} \left| i_a \left[\frac{r''}{r'}\right] \right| r' \frac{dr'}{r' - r''} \right\} \Gamma'(r) \quad (12.4)$$

As in Chapter 9 we now seek to minimize the torque Q for a specified thrust T and thereby obtain maximum efficiency at the specified or fixed revolutions and a given distribution of the axial, peripheral-mean-inflow velocity, $U_a(r)$, although at first $U_a = U$, a constant.

This requirement can be met by setting the first variation of $T - \ell Q$ to zero or

$$\delta T - \ell \delta Q = 0 \quad (12.5)$$

where δT and δQ are the first variations and ℓ is the inverse of the Lagrange multiplier, see Mathematical Compendium, Section 7. We then proceed as in Chapter 9, p. 191 and sequel. Because of differences in expressions this parallel development is included here in detail.

Suppose that $\Gamma(r)$ is the sought-for, or optimum, distribution. Then $\Gamma(r) + \delta\Gamma(r)$ is taken to be any neighboring distribution which yields the same thrust and vanishes at R_h and R. If we designate δT as the part of $T(\Gamma + \delta\Gamma) - T(\Gamma)$ which is of *first* order in $\delta\Gamma'$, then from (12.3) we obtain as the first variation

$$\delta T = -\rho Z \int_{R_h}^{R} dr \left\{ \frac{\omega}{2}(r^2 - R_h^2) - \int_{R_h}^{R} dr'' \Gamma'(r'') \frac{1}{4\pi} \int_{R_h}^{r} i_t \left[\frac{r''}{r'}\right] \frac{dr'}{r' - r''} \right\} \delta\Gamma'(r)$$

$$+ \rho Z \int_{R_h}^{R} dr\, \Gamma'(r) \int_{R_h}^{R} dr'' \delta\Gamma'(r'') \frac{1}{4\pi} \int_{R_h}^{r} i_t \left[\frac{r''}{r'}\right] \frac{dr'}{r' - r''} \quad (12.6)$$

To gather these terms together as a factor of $\delta\Gamma'$ we interchange the r- and r''- dummy variables in the last line of (12.6) and collecting the coefficient of $\int_{R_h}^{R} \delta\Gamma'(r) dr'$ we secure

$$\delta T = -\rho Z \int_{R_h}^{R} \left\{ \frac{\omega}{2}(r^2 - R^2) - \int_{R_h}^{R} dr'' \Gamma'(r'') \frac{1}{4\pi} \left[\int_{R_h}^{r} i_t \left[\frac{r''}{r'}\right] \frac{dr'}{r' - r''} + \int_{R_h}^{r''} i_t \left[\frac{r}{r'}\right] \frac{dr'}{r' - r} \right] \right\} \delta\Gamma'(r) dr \quad (12.7)$$

Similarly the variation in Q can be expressed

$$\delta Q = -\rho Z \int_{R_h}^{R} \Bigg\{ U(r^2 - R_h^2) + \int_{R_h}^{R} dr''\Gamma'(r'')\frac{1}{4\pi}\left[\int_{R_h}^{r} \left| i_a\left[\frac{r''}{r'}\right] \right| \left|\frac{r'dr'}{r'-r''}\right| + \int_{R_h}^{r''} \left| i_a\left[\frac{r}{r'}\right] \right| \left|\frac{r'dr'}{r'-r}\right| \right] \Bigg\} \delta\Gamma'(r)dr \tag{12.8}$$

Combining (12.7), (12.8) into the condition (12.5) yields an integral over r of the form

$$\int_{R_h}^{R} \Big[\{....\} - \ell\{....\} \Big] \, \delta\Gamma'(r)dr = 0 \tag{12.9}$$

and as the variation $\delta\Gamma'$ is arbitrary the quantity in the square brackets must vanish. This requirement produces the following integral equation for the optimum circulation distribution Γ

$$\int_{R_h}^{R} dr''\Gamma'(r'') \frac{1}{4\pi} \Bigg\{ \int_{R_h}^{r} \left[i_t\left[\frac{r''}{r'}\right] - \ell\, r' \left| i_a\left[\frac{r''}{r'}\right] \right| \right] \frac{dr'}{r'-r''} + \int_{R_h}^{r''} \left[i_t\left[\frac{r}{r'}\right] - \ell\, r' \left| i_a\left[\frac{r}{r'}\right] \right| \right] \frac{dr'}{r'-r''} \Bigg\} = -\frac{\omega}{2}(r^2 - R_h^2) + \frac{\ell U}{2}(r^2 - R_h^2) \tag{12.10}$$

We observe that the solution of this equation is parametrically dependent upon the inverse of the Lagrange parameter, ℓ. The value of ℓ is found for a given thrust by inserting the solution of (12.10), say $\Gamma_{opt} = \Gamma(r;\ell)$ in (12.3) and solving for ℓ.

We now seek the analogue of a criterion which can be established for finite wings that the circulation for minimum induced drag is that for which the spanwise downwash is constant. To accomplish this we rearrange the condition (12.5) to $\delta T/\delta Q = \ell$ and write (12.10)

$$\frac{\dfrac{\omega}{2}(r^2-R_h^2) - \displaystyle\int_{R_h}^{R} dr''\Gamma'(r'')\frac{1}{4\pi}\left[\int_{R_h}^{r} i_t\left[\frac{r''}{r'}\right] \frac{dr'}{r'-r''} + \int_{R_h}^{r''} i_t\left[\frac{r}{r'}\right] \frac{dr'}{r'-r} \right]}{\dfrac{U}{2}(r^2-R_h^2) + \displaystyle\int_{R_h}^{R} dr''\Gamma'(r'')\frac{1}{4\pi}\left[\int_{R_h}^{r} \left| i_a\left[\frac{r''}{r'}\right] \right| \left|\frac{r'dr'}{r'-r''}\right| + \int_{R_h}^{r} \left| i_a\left[\frac{r}{r'}\right] \right| \left|\frac{r'dr'}{r'-r}\right| \right]} = \ell \tag{12.11}$$

In the numerator interchange the order of the r''- and the first r'-integrals to secure

$$-\int_{R_h}^{r} dr' \frac{1}{4\pi} \int_{R_h}^{R} \Gamma'(r'') \frac{i_t\left[\frac{r''}{r'}\right] dr''}{r' - r''} = -\int_{R_h}^{r} u_t(r')\, dr'$$

by (11.63). Next, integrate the second r''-, r'-integrals in the numerator of (12.11) by parts, i.e.,

$$\int_{R_h}^{R} dr''\Gamma'(r'') \int_{R_h}^{r''} i_t\left[\frac{r}{r'}\right] \frac{dr'}{r' - r}$$

$$= \Gamma(r'') \int_{R_h}^{r''} i_t\left[\frac{r}{r'}\right] \frac{dr'}{r' - r} \Bigg|_{r''=R_h}^{R} - \int_{R_h}^{R} \frac{\Gamma(r'')}{r'' - r} i_t\left[\frac{r}{r''}\right] dr''$$

where the integrated part vanishes because $\Gamma(R) = \Gamma(R_h) = 0$.

The denominator of (12.11) can be manipulated similarly so we may change (12.11) to

$$\frac{\displaystyle\int_{R_h}^{r} (\omega r' - u_t(r'))dr' + \frac{1}{4\pi} \int_{R_h}^{R} \frac{\Gamma(r'')}{r'' - r} i_t\left[\frac{r}{r''}\right] dr''}{\displaystyle\int_{R_h}^{r} r'(U + |u_a(r')|)dr' - \frac{1}{4\pi} \int_{R_h}^{R} \frac{r''\Gamma(r'')}{r'' - r} \left| i_a\left[\frac{r}{r''}\right] \right| dr''} = \ell = \frac{\delta T}{\delta Q} \tag{12.12}$$

We may now recall that as $\Gamma + \delta\Gamma$ is an allowable variation which keeps T constant then

$$T(\Gamma + \delta\Gamma) = T(\Gamma) \qquad \text{or} \qquad \delta T = 0$$

If Γ is to be the minimizing distribution we must also have $\delta Q = 0$.

Hence the ratio given by (12.12) is seen to be 0/0, i.e., it is indeterminate. To resolve this we apply l'Hospital's rule, differentiating the numerator and the denominator of (12.12) with respect to r. This operation yields,

$$\frac{\displaystyle \omega r - u_t(r) + \frac{\partial}{\partial r} \frac{1}{4\pi} \int_{R_h}^{R} \frac{\Gamma(r'')}{r'' - r} i_t\left[\frac{r}{r''}\right] dr''}{\displaystyle rU + r|u_a(r)| - \frac{\partial}{\partial r} \frac{1}{4\pi} \int_{R_h}^{R} \frac{r''\Gamma(r'')}{r'' - r} \left| i_a\left[\frac{r}{r''}\right] \right| dr''} = \ell, \text{ a constant} \tag{12.13}$$

As Yim, (1976) notes this equation is the same as those derived by Wehausen, (1964) or Morgan & Wrench, (1965). For the special case of a propeller with infinite number of blades, i.e. an actuator disc, Morgan & Wrench (*ibid.*) considered the integral terms and were able to show that these terms are equal to $-u_t$ and ru_a, respectively. A comparison of terms in (12.13) and (9.101) clearly confirms this finding. Andersen (1986) found by numerical calculations that this was also the case for a propeller with finite number of blades.

For the general case of a propeller with finite number of blades we shall not go through the complication of inverting (12.13). Instead let us as Yim (1976) seek an answer via another procedure in which one assumes that Munk's (1919) "displacement law" developed for wing theory applies to propellers. This was the basis of Betz's theory. Munk's law (as applied to propellers) states that the thrust and the loss of energy are not altered when the blade elements are displaced along the vortex lines of the free vortex sheet without altering the circulation of the blade elements. In other words, the Munk law indicates that the thrust and energy loss is only a function of the radial distribution of integrated chordwise pressures when the chord is considered to lie along a free vortex line of the sheet and does not depend upon the modes of chordwise distribution as long as the integrated lift is the same. Let us now return to (12.1) and write the variation in T as

$$\delta T = \rho Z\left\{-\int_{R_h}^{R} dr\Gamma(r)\frac{1}{4\pi}\fint_{R_h}^{R}\frac{\delta\Gamma'}{r-r'}i_t\left[\frac{r'}{r}\right]dr' + \int_{R_h}^{R}\delta\Gamma\left[\omega r - \frac{1}{4\pi}\fint_{R_h}^{R}\frac{\Gamma'}{r-r'}i_t\left[\frac{r'}{r}\right]dr'\right]\right\} \tag{12.14}$$

Applying Munk's law we replace $i_t(r'/r)$ by $i_t(x;r'/r)$ and regard the variation $\delta\Gamma'$ as an additional small circulation which is now attached to an "auxiliary" foil (as envisioned by Lerbs (1952)) which is shifted downstream. This induces a decreasing downwash at the blade as $x \to -\infty$ and hence vanishes whereas by (11.49) which applies similarly to the tangential component

$$\lim_{x\to-\infty}\frac{1}{4\pi}\fint_{R_h}^{R}\frac{\Gamma'}{r-r'}\,i_t\left[x;\frac{r'}{r}\right]dr' = 2u_t(0,r) \tag{12.15}$$

Thus we have

$$\delta T = \rho Z\int_{R_h}^{R}\delta\Gamma(\omega r - 2u_t(0,r))\,dr \tag{12.16}$$

Applying the same steps to δQ and substituting in (12.5) gives

$$\int_{R_h}^{R} \delta\Gamma\left[\omega r - 2u_t(0,r) - \ell\, r\left[U + 2|u_a(0,r)|\right]\right]dr = 0 \tag{12.17}$$

As the variation $\delta\Gamma$ is arbitrary we have

$$\frac{\omega r - 2u_t(0,r)}{r(U + 2|u_a(0,r)|)} = \ell, \quad \text{a constant} \tag{12.18}$$

This is Betz's (1919) result which states that the ultimate forms of the vortices "far" downstream are true helices, the same result as found in Chapter 9 for the actuator disc, Equation (9.101), p. 192.

We may now manipulate this as follows

$$\frac{\omega r - 2u_t}{r(u + 2|u_a|)} = \frac{\omega r}{rU}\,\frac{1 - \dfrac{2u_t}{\omega r}}{1 + \dfrac{2|u_a|}{U}} = \frac{\omega}{U}\,\frac{\left[1 - \dfrac{u_t}{\omega r}\right]^2 - \left[\dfrac{u_t}{\omega r}\right]^2}{\left[1 + \dfrac{|u_a|}{U}\right]^2 - \left[\dfrac{u_a}{U}\right]^2} = \ell$$

and upon neglecting the squares of the induced velocities this reduces to

$$\frac{\omega}{U}\left[\frac{U}{r\omega}\right]^2\left[\frac{\omega r - u_t(0)}{U + |u_a(0)|}\right]^2 = \ell = \frac{\omega}{U}\,\ell^* \tag{12.19}$$

where ℓ^* is a dimensionless constant (ℓ has the dimensions of (length)$^{-1}$). Comparing with the expressions for β and β_i we see that this criterion can be expressed as

$$\tan\beta_i = \frac{\tan\beta}{\sqrt{\ell^*}} \quad \text{or} \quad \beta_i(r) = \tan^{-1}\left\{\frac{\tan\beta(r)}{\sqrt{\ell^*}}\right\} \tag{12.20}$$

which says that the pitch angles of the vortices *at the propeller* are those of a pure helix, being related to the undisturbed fluid pitch by this relation. As noted before ℓ or ℓ^* is to be found from the specified thrust (or power).

The expression (12.20) is frequently referred to as the Betz "condition" whereas (12.18) is the result of Betz's development. It would be more precise to refer to (12.20) as the linearized Betz condition as we found it necessary to drop terms $(u_t/\omega r)^2$ and $(u_a/U)^2$ to achieve this result. This condition has a simple kinematic interpretation. Defining the axial velocity component u_a^h (Figure 12.1) we have from the geometry

$$\tan\beta_i = \frac{u_a^h + U}{r\omega} = \frac{|u_a| + U}{r\omega - u_t} = \frac{U}{\omega r\sqrt{\ell^*}} \tag{12.21}$$

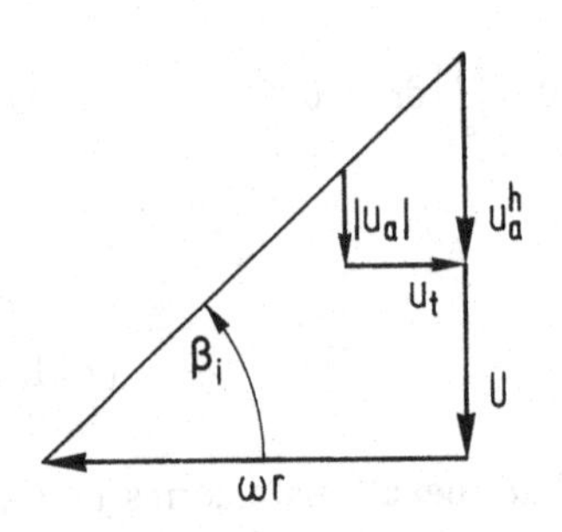

Figure 12.1 Axial velocity component u_a^h.

from (12.19). Multiplying by $r\omega$ gives

$$u_a^h + U = \frac{U}{\sqrt{\ell^*}} = \text{a constant} \tag{12.22}$$

Thus u_a^h is a constant and under the linearized Betz condition the axial velocity convecting the helices is independent of the radius.

Propellers in Hull Wakes

When a propeller operates in a hull wake the thrust must be multiplied by $(1-t)$ (t = thrust deduction) and the speed U replaced by $U_a(r) = U(1 - w(r))$ where $w(r)$ is the local wake fraction. The process of finding a criterion for the optimum circulation distribution is the same as before, hence we modify Equation (12.18) accordingly to obtain

$$\frac{(\omega r - 2u_t)(1-t)}{r(U(1-w) + 2|u_a|)} = \ell = \frac{\omega \ell^*}{U} \tag{12.23}$$

We manipulate in the same fashion

$$\frac{1 - 2\dfrac{u_t}{\omega r}}{1 - w + 2\dfrac{|u_a|}{U}} \approx \frac{\left[1 - \dfrac{u_t}{\omega r}\right]^2}{\dfrac{\left[1 - w + 2\dfrac{|u_a|}{U}\right]^2}{1-w}}$$

to get, analogously to (12.19)

$$\left[\frac{U}{\omega r}\right]^2 \left[\frac{r\omega - u_t}{U(1-w) + |u_a|}\right]^2 (1-t)(1-w) = \ell^* \tag{12.24}$$

We now define

$$\tan\beta_a = \frac{U(1-w)}{\omega r} \quad \text{or} \quad \left[\frac{U}{\omega r}\right]^2 = \frac{\tan^2\beta_a}{(1-w)^2} \tag{12.25}$$

$$\tan\beta_i = \frac{U(1-w) + |u_a|}{\omega r - u_t} \tag{12.26}$$

Use of (12.25) and (12.26) in (12.24) yields

$$\frac{\tan\ \beta_a}{\tan\ \beta_i} = \sqrt{\ell^*}\sqrt{\frac{1\ -\ w(r)}{1\ -\ t(r)}} \tag{12.27}$$

This condition was first given by Lerbs, (1952). It is seen to yield the open-water condition when $w = t = 0$ and the β's return to their open-water definitions.

OPTIMUM DIAMETER AND BLADE-AREA-RATIO DETERMINATIONS

Before turning to the design of a propeller by lifting-line procedures, it is necessary to obtain a series of starting data. These data are typically obtained by a number of calculations with series propellers.

It should be pointed out that the modern calculation methods provide for much greater flexibility in the design of propellers as compared to the use of data from systematic propeller series as produced in the past by MARIN (Wageningen series), Troost (1937–38), (1939–40), (1950–51), Oosterveld & van Oossanen (1975), Admiralty Experiment Works (Gawn series), Gawn (1937, 1953), SSPA (SSPA standard series), Lindgren & Bjärne (1967) and DTRC (Admiral Taylor series), Taylor (1933). The use of these series restricts one to the fixed geometry, section characteristics and hub radius employed and does not allow for distributions in the inflow to the propeller disc as experienced in the wake behind a ship. Moreover, they do not include skew as a parameter as can be incorporated in lifting-line and lifting-surface theories.

The most efficient propeller is generally that having the greatest diameter permissible in the aperture and turning as slowly as possible. To see this, remember from figure 9.16 that the highest efficiency was obtained for $C_{Th} = T/(\frac{1}{2}\rho U_A^2 \pi(D^2/4))$ as small as possible. For a fixed thrust T and mean inflow velocity U_A the smallest possible C_{Th} is obtained by as big a diameter, D, as possible. T and U_A are both given when the speed of the ship is specified (through t and w), but they generally show a weak dependence on the propeller diameter, cf. for example Harvald (1983). There may therefore be cases where the entire propulsion efficiency will not increase even if the propeller diameter is made bigger.

The determination of optimum revolutions for a fixed diameter (as big as possible) from systematic series is made by eliminating the unknown rate of revolutions N between the thrust coefficient K_T and advance ratio $J_A = U_A/ND$

$$K_T = \frac{\pi}{8} C_{Th} J_A{}^2 \tag{12.28}$$

where now C_{Th} will be a constant for fixed T, speed of advance U_A and diameter D.

When this relation is graphed as a function of J_A across the manifold of K_T-, K_Q- and η-curves of a chart for fixed blade-area ratio, as in Figure 12.2, a possible solution is secured at each intersection with the K_T-curves which are parametrically dependent on the pitch-diameter ratio, P/D. The optimum solution is secured at that P/D corresponding to the maximum efficiency given at the point where the locus of efficiencies osculates the envelope of efficiencies. The advance ratio at the optimum determines the rate of revolutions.

A similar procedure yields the optimum diameter for a fixed rate of revolutions by eliminating D between K_T and J_A to obtain

$$K_T = b_T J_A^{\ 4} \quad ; \quad b_T = \frac{TN^2}{\rho U_A^{\ 4}} \tag{12.29}$$

Finally, by a similar procedure, the optimum diameter can be found, if the torque delivered by the engine at a given rate of revolutions is known, by eliminating D between K_Q and J_A

$$K_Q = b_Q J_A^{\ 5} \quad ; \quad b_Q = \frac{QN^3}{\rho U_A^{\ 5}} \tag{12.30}$$

In this case a further check must be made to secure that the thrust delivered by the propeller corresponds to the ship speed and hence to the speed of advance specified at the outset.

Required blade-area ratios can generally be determined through the use of Burrill's empirical criterion shown in Figure 12.3.

The series propellers with characteristics presented in the form of charts, as used in the above processes have been tested in open-water conditions. Experience as well as model experiments, as reported by van Manen & Troost (1952), have shown that the optimum diameter as obtained from the charts is not the diameter for optimum results in the "behind" condition. A propeller of smaller diameter will give the highest efficiency. This can also be found by calculation, as shown in Figure 12.4. The curves indicated lifting-line calculation are the results of a series of calculations of an optimum propeller by the lifting-line theory (which will be explained in detail shortly). Each point of the curve has been obtained by several calculations with systematic variation of the diameter, and the optimum (with respect to efficiency) has been selected.

All of the above calculations including optimization of diameter or rate of revolutions are most efficiently effected by computer with the series propeller characteristics given in polynomial form (MARIN-series: Oosterveld & van Oossanen (1973), (1975), MARIN-ducted propellers: Oosterveld (1970)).

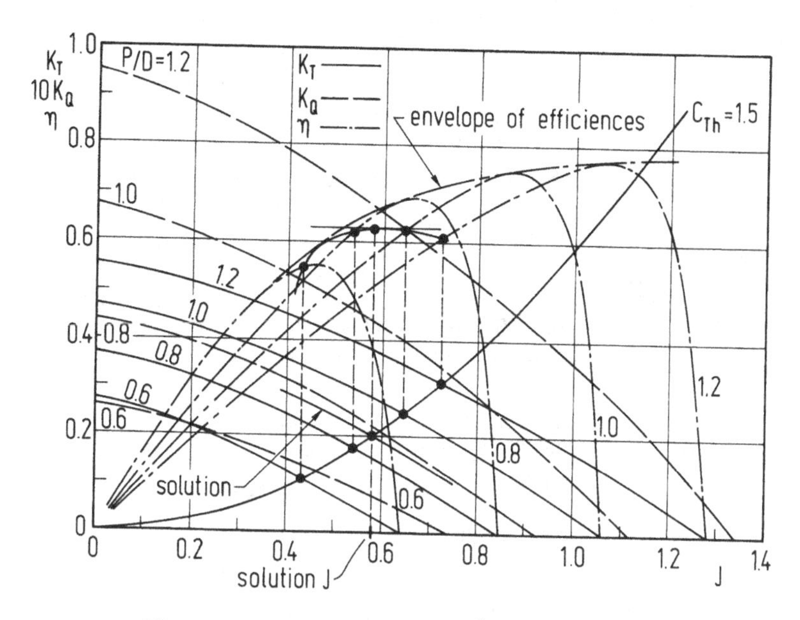

Figure 12.2 Graphical solution for optimum diameter.

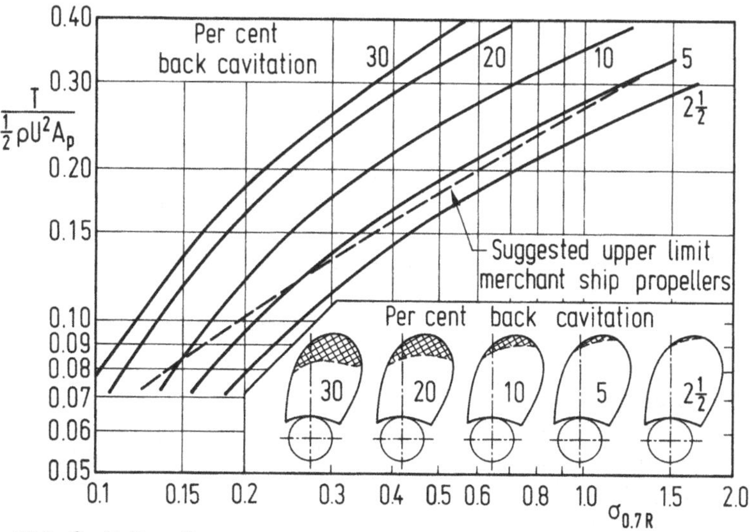

Figure 12.3 Cavitation diagram.

Figure 5 from Burrill, L.C. & Emerson, A. (1962). Propeller cavitation: further tests on 16 in propeller models in the King's College cavitation tunnel. *Trans. North East Coast Institution of Engineers and Shipbuilders*, vol. 79, pp. 295–320. Newcastle upon Tyne: North East Coast Institution of Engineers and Shipbuilders.
By courtesy of North East Coast Institution of Engineers and Shipbuilders, Great Britain.

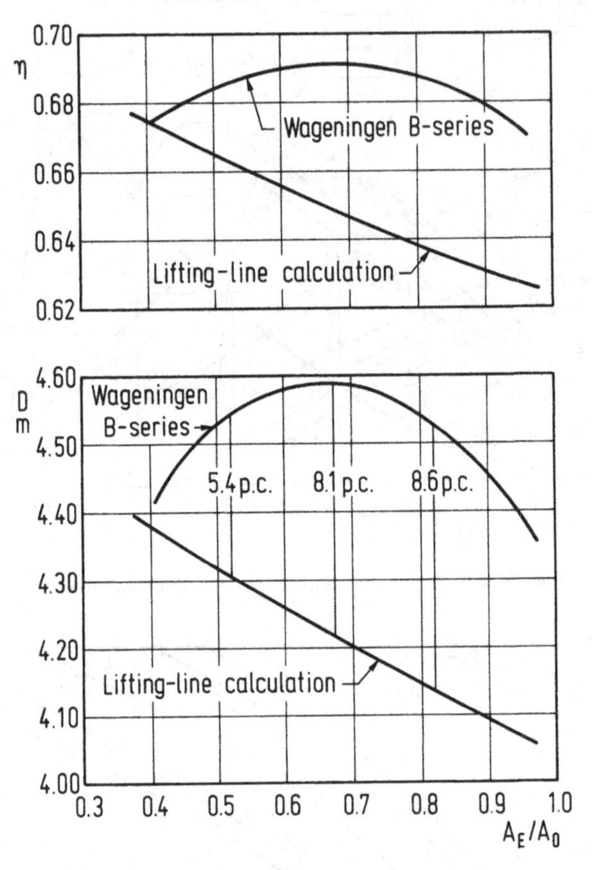

Figure 12.4 Optimum diameter by series propeller calculations and by lifting–line calculation.

On completion of the calculations the data should comprise

- Diameter
- Hub Diameter
- Number of blades
- Distribution of chord lengths
- Rate of revolutions
- Required thrust
- Propeller submergence
- Distribution of inflow velocities
  (radially varying wake, cf. for example Harvald & Hee (1978))

These are the starting data for an optimization by lifting–line method.

CALCULATION PROCEDURES

To effect the calculations required to solve the design problem where we seek to find the propeller which has the highest possible efficiency while producing a specified thrust, it is customary to express the induced velocities and the circulation distribution in the forms evolved in the following. (These expressions are also used in a calculation of the characteristics of a propeller of specified geometry as will also be described).

The induced velocities will be expressed using induction factors, (11.63) and it is important to recall that the induction factors $i_{a,t}$ are parametrically dependent upon tan β_i (cf. (11.64), (11.65), (11.66)) for moderately loaded cases and should be written

$$i_{a,t} = i_{a,t}\left[\frac{r'}{r};\beta_i\right] \tag{12.31}$$

To reduce the calculation of the induced velocities to a finite series, new variables θ and θ' are introduced by the following definitions

$$\frac{r}{R} = \frac{1}{2}\left[1 + \frac{R_h}{R}\right] - \frac{1}{2}\left[1 - \frac{R_h}{R}\right] \cos\theta \tag{12.32}$$

and similarly for r' corresponding to θ'. Here we see that $\theta = 0$ corresponds to the radius at the boss or hub ($r = R_h$) and $\theta = \pi$ to the tip ($r = R$).

The circulation and induction factors may then be conveniently expressed by the following trigonometric series

$$\Gamma(\theta) = \pi DU \sum_{n=1}^{\infty} G_n \sin n\theta \tag{12.33}$$

$$i_{a,t}(\theta,\theta') = \sum_{m=0}^{\infty} I_m^{a,t}(\theta) \cos m\theta' \tag{12.34}$$

The expansion of Γ is limited to sine terms because $\Gamma(R_h) = \Gamma(R) = 0$. The induction factors do not vanish at R_h and R, hence admit of a cosine expansion.

Use of (12.33) and (12.34) in (11.63) produces

$$u_{a,t} = -\frac{U}{1 - \frac{R_h}{R}} \sum_{n=1}^{\infty} \sum_{m=1}^{\infty} nG_n I_m^{a,t}(\theta) \int_0^{\pi} \frac{\cos n\theta' \cos m\theta'}{\cos\theta - \cos\theta'} d\theta' \tag{12.35}$$

The θ'-integral can be written as

$$J^*_{n,m}(\theta) = \frac{1}{2}\int_0^{\pi} \frac{\cos(n+m)\theta' + \cos(n-m)\theta'}{\cos\theta - \cos\theta'}\, d\theta' \tag{12.36}$$

These are so-called Glauert (1948) integrals, of which a direct evaluation by the calculus of residues is given in Appendix A, p. 485–487. Applying (A.11) successively gives

$$J^*_{n,m}(\theta) = -\frac{\pi}{2}\left\{\frac{\sin|n+m|\theta + \sin|n-m|\theta}{\sin\theta}\right\} \tag{12.37}$$

and as n and m are positive $|n + m| = n + m$ for all n, m and $|n - m| = n - m$ for $m \leq n$ but $|n - m| = m - n$ for $n + 1 \leq m < \infty$. Thus from the sum-into-product identities of trigonometry

$$J^*_{n,m} = \begin{cases} -\dfrac{\pi \sin n\theta \cos m\theta}{\sin\theta} \ ; & 0 < m \leq n \\[2ex] -\dfrac{\pi \cos n\theta \sin m\theta}{\sin\theta} \ ; & n + 1 \leq m < \infty \end{cases} \tag{12.38}$$

The m-summation in (12.35) must be segregated in accordance with (12.38) to give the induced velocities in the form

$$u_{a,t} = \frac{U}{1 - \dfrac{R_h}{R}} \sum_{n=1}^{\infty} n\, G_n\, h_n^{a,t}(\theta) \tag{12.39}$$

where

$$h_n^{a,t}(\theta) = \frac{\pi}{\sin\theta}\left\{\sin n\theta \sum_{m=0}^{n} I_m^{a,t}(\theta)\cos m\theta + \cos n\theta \sum_{m=n+1}^{\infty} I_m^{a,t}(\theta)\sin m\theta\right\}$$
$$\text{for } \theta \neq 0, \pi \tag{12.40}$$

in agreement with Lerbs, (1952).

At $\theta = 0$ and π, (12.40) is seen to be indeterminate. Applying l'Hospital's rule (differentiating numerator and denominator with respect to θ and evaluating each at 0 and π) we have

$$h_n^{a,t}(0) = \pi\left\{n \sum_{m=0}^{n} I_m^{a,t}(0) + \sum_{m=n+1}^{\infty} m I_m^{a,t}(0)\right\}$$

and

$$h_n^{a,t}(\pi) = -\pi \cos n\pi \left\{n \sum_{m=0}^{n} I_m^{a,t}(\pi) \cos m\pi + \sum_{m=n+1}^{\infty} m I_m^{a,t}(\pi) \cos m\pi\right\} \tag{12.41}$$

All is not yet in readiness for our calculations as the Fourier cosine coefficients of the induction-factor expansions, $I_m^{a,t}(\theta)$ have not been explicitly given. These coefficients are obtained from (12.34) by multiplying both sides by $\cos\nu\theta'$ and integrating over θ' from 0 to 2π giving

$$\frac{1}{2}\sum_{m=0}^{\infty} I_m^{a,t}(\theta)\int_0^{2\pi}(\cos(m+\nu)\theta' + \cos(m-\nu)\theta')d\theta' = \int_0^{2\pi} i_{a,t}(\theta,\theta')\cos\nu\theta'\, d\theta' \tag{12.42}$$

where the product-to-sum identity has been applied. The left side is zero except for $\nu = -m$ and $\nu = m$ for which the θ' integrals are 2π. For $0 \leq m \leq \infty$ and $0 \leq \nu \leq \infty$, i.e. both m and ν positive then we encounter $\nu = -m$ only for both $\nu = 0$ and $m = 0$. For all other orders only the integral involving $\cos(m - \nu)\theta'$ contributes and then only for $\nu = m$. Thus we have in general

$$I_m^{a,t}(\theta;\beta_i,Z) = \frac{\varepsilon_m}{2\pi}\int_0^{2\pi} i_{a,t}(\theta,\theta';\beta_i,Z)\cos m\theta'\, d\theta' \; ; \begin{cases}\varepsilon_0 = 1 \\ \varepsilon_m = 2,\ m > 0\end{cases} \tag{12.43}$$

where we put in evidence the parametric dependence on blade number Z and on β_i. We recall that β_i is a function of the induced velocities which are in turn functional of the circulation distribution. Thus the coefficients $I_m^{a,t}$ must be determined by iteration for moderately loaded propellers. For lightly loaded cases $\beta_i = \beta$, allowing the coefficients to be determined at once.

The operation posed by (12.43) can be carried out in two ways, one by direct computer evaluation of (12.43) using (11.64) and (11.66). The other way, Lerbs (1952), is to calculate i_a and i_t by evaluating far downstream which as we have indicated gives double the induced velocities at the bound vortices. For sake of completeness, we give the development of this type of approximation by treating only i_a, (the result for i_t being similar) using our expressions for $x = 0$, evaluated on the bound vortex.

Inspection of (11.64) with (11.67) shows that we can write i_a as

$$i_a = -\sum_{\nu=0}^{Z-1} r'(r-r')\frac{\partial}{\partial r'}\int_0^{\infty}\frac{d\psi'}{\sqrt{(r'\psi'\tan\beta_i)^2 + r^2 + r'^2 - 2rr'\cos(\psi'+\gamma_\nu)}} \tag{12.44}$$

where $r'\psi'\tan\beta_i = x'$ is held constant when differentiating with respect to r' (cf. (11.54), (11.55)). This is because the dummy x-wise variable (along the helix) is held fixed as is apparent had we operated on (11.54) in this fashion.

We may now replace the integrand of (12.44) by the form given by (11.31) (with $x = 0$, $x' = r'\psi'\tan\beta_i$ and $\gamma - \gamma_0 - (x-x')\omega/U = \psi' + \gamma_\nu$). Here we must employ alternate forms applicable to the radial regions $R_h < r' \leq r$ and $r \leq r' < R$. With this replacement (12.44) becomes

$$i_a = -\frac{1}{\pi}\sum_{\nu=0}^{Z-1}\sum_{m=-\infty}^{\infty} r'(r-r')\int_0^{\infty} d\psi' \int_{-\infty}^{\infty} |k| \begin{Bmatrix} I'_{|m|}(|k|r')\ K_{|m|}(|k|r) \\ I_{|m|}(|k|r)\ K'_{|m|}(|k|r') \end{Bmatrix} \cdot e^{i[k\psi' r'\tan\beta_i + m(\psi' + \gamma_\nu)]} dk \qquad (12.45)$$

where the upper product of the modified Bessel function applies in $R_h < r' \leq r$ and the lower in $r \leq r' \leq R$ and the r'-derivative has been taken (holding $\psi' r'\tan\beta_i$ fixed as explained above).

The ν-summation is now effected (recalling that $\gamma_\nu = 2\pi\nu/Z$) as demonstrated by (11.18) and (11.20). The resulting q-summation is folded (cf. p. 212) to achieve

$$i_a = -\frac{r'(r-r')}{\pi k_0^3}\left\{\int_0^{\infty} d\psi' \int_{-\infty}^{\infty} |k'| \begin{Bmatrix} I'_0\left[\frac{|k'|r'}{k_0}\right] K_0\left[\frac{|k'|r}{k_0}\right] \\ I_0\left[\frac{|k'|r}{k_0}\right] K'_0\left[\frac{|k'|r'}{k_0}\right] \end{Bmatrix} e^{ik'\psi'} dk' \right.$$

$$+ \sum_{q=1}^{\infty}\int_0^{\infty} d\psi' \int_{-\infty}^{\infty} |k'| \begin{Bmatrix} I'_{qZ}\left[\frac{|k'|r'}{k_0}\right] K_{qZ}\left[\frac{|k'|r}{k_0}\right] \\ I_{qZ}\left[\frac{|k'|r}{k_0}\right] K'_{qZ}\left[\frac{|k'|r'}{k_0}\right] \end{Bmatrix}$$

$$\left. \cdot \left[e^{i[k'+qZ]\psi'} + e^{i[k'-qZ]\psi'}\right] dk' \right\} \qquad (12.46)$$

where

$$k_0 = r' \tan\beta_i \qquad (12.47)$$

and we have substituted

$$k' = kk_0 \qquad (12.48)$$

Now the ψ'-integrals can be carried out using the known results (Equation (11.43) and (11.45)) that

$$\int_0^{\infty} e^{ik'\psi'} d\psi' = \pi\delta(k') + \frac{i}{k'} \qquad (12.49)[21]$$

[21] If the ψ'-integral were to be carried out before separating out the $m = 0$ term the limits given later in (12.52) and (12.53) would be different and in error. It is not frequently found that the order of operations is important.

and

$$\int_0^\infty e^{i[k'\pm qZ]\psi'}d\psi' = \pi\delta(k'\pm qZ) + \frac{i}{k' \pm qZ} \qquad (12.50)^{21}$$

It is important to note that without the substitution $k' = kk_0$ we would not retain the $(\text{length})^{-2}$ dimension of the k-integral.

We can see that the term i/k' will integrate to zero because the integrands of the k'-integrals are even in k' whereas $1/k'$ is odd. The second term in (12.50) also produces zero as has been shown previously (cf. (11.42) and sequel) when the induced velocity is evaluated on the bound vortex as we are doing here. Hence the use of (12.49) in the zero order term of (12.46) and (12.50) (sans the second term) in the q-summation of (12.46) allows us to evaluate the k'-integrals directly through the property of the δ-function. Thus the values of the k'-integrals are the integrands evaluated at $k' = 0$, $k' = -qZ/k_0$ and $k' = qZ/k_0$ respectively. In this way we can reduce (12.46) to

$$i_a = -\frac{Zr'(r-r')}{k_0^3}\left\{\lim_{k\to 0} k'\left\{\begin{matrix} I_0'\left[\frac{k'r'}{k_0}\right] & K_0\left[\frac{k'r}{k_0}\right] \\ I_0\left[\frac{k'r}{k_0}\right] & K_0'\left[\frac{k'r'}{k_0}\right]\end{matrix}\right\}\right.$$

$$\left. + 2\sum_{q=1}^{\infty} qZ\left\{\begin{matrix} I_{qZ}'\left[\frac{qZ}{k_0}r'\right] & K_{qZ}\left[\frac{qZ}{k_0}r\right] \\ I_{qZ}\left[\frac{qZ}{k_0}r\right] & K_{qZ}'\left[\frac{qZ}{k_0}r'\right]\end{matrix}\right\}\right\} \quad ; \quad \begin{matrix} r' < r \\ r' > r \end{matrix} \qquad (12.51)$$

From the properties of the modified Bessel functions (Abramowitz & Stegun, (1972)), the behaviors become algebraic, i.e. taking $y = r/k_0$ and $y' = r'/k_0$

$$I_0(k'y') \to 1 + \frac{(k'y')^2}{4} + \ldots \quad ; \quad K_0(k'y) \to -\left(\gamma + \ln\left[\frac{k'y}{2}\right] + \ldots\right)$$

(γ being the Euler constant) so that

$$\lim_{k'\to 0}\left[k'I_0'(k'y')\, K_0(k'y)\right]$$

$$= \lim_{k'\to 0} -\left[\frac{k'(k'y')}{2} + ..\right]\left[\gamma + \ln\frac{k'y}{2}\right] = 0 \quad \text{for } r' < r \qquad (12.52)$$

(as $\lim_{k'\to 0} k'^2 \ln(k') = 0$). Whereas

$$\lim_{k'\to 0} k'I_0(k'y)\, K_0'(k'y')$$

$$= \lim_{k'\to 0}\left\{k'\left[1+\frac{(k'y)^2}{4}+\ldots\right]\left[-\frac{1}{k'y'}\right]\right\} = -\frac{1}{y'} \tag{12.53}$$

Thus (12.51) with these limits give

$$i_a = -2Zy'(y-y')\sum_{q=1}^{\infty} qZI_{qZ}'(qZy')\, K_{qZ}(qZy) \quad \text{for } 0 < r' < r \tag{12.54}$$

and

$$i_a = -Zy'(y-y')\left\{-\frac{1}{y'} + 2\sum_{q=1}^{\infty} qZI_{qZ}(qZy)\, K_{qZ}'(qZy')\right\} \text{ for } r \le r' \le R \tag{12.55}$$

Lerbs (1952) then approximated these functions by exploiting the fact that the order numbers qZ are large, permitting the use of following asymptotics, Abramowitz & Stegun, (1972) (see also Nicholson (1910)):

$$\left.\begin{aligned} I_{qZ}(qZy) &\approx \frac{1}{\sqrt{2\pi qZ}}\frac{e^{qZY}}{(1+y^2)^{1/4}} \\ K_{qZ}(qZy) &\approx \sqrt{\frac{\pi}{2qZ}}\frac{e^{-qZY}}{(1+y^2)^{1/4}} \end{aligned}\right\} \tag{12.56}$$

$$\left.\begin{aligned} I_{qZ}'(qZy) &\approx \frac{1}{\sqrt{2\pi qZ}}\frac{(1+y^2)^{1/4}}{y}e^{qZY} \\ K_{qZ}'(qZy) &\approx -\sqrt{\frac{\pi}{2qZ}}\frac{(1+y^2)^{1/4}}{y}e^{-qZY} \end{aligned}\right\} \tag{12.57}$$

where

$$Y = \sqrt{1+y^2} + \ln\left[\frac{y}{1+\sqrt{1+y^2}}\right] \quad ; \quad y = \frac{r}{k_0} \tag{12.58}$$

Using these in (12.54) this expression reduces to

$$i_a = -Z(y-y')\left[\frac{1+y'^2}{1+y^2}\right]^{1/4}\sum_{1}^{\infty}\left[e^{Z(Y(y')-Y(y))}\right]^q$$

This infinite sum can be recognized as the sum of a geometric series whose value is given by the first term divided by one minus the common ratio. Thus

$$i_a = -Z(y-y')\left[\frac{1+y'^2}{1+y^2}\right]^{1/4}\frac{e^{-AZ}}{1-e^{-AZ}} \quad ; \quad r' < r \qquad (12.59)$$

where

$$A = Y(y) - Y(y') \qquad (12.60)$$

With a notation very similar to Lerbs' (*ibid.*) (12.59) is succinctly written

$$i_a = -Z(y-y')B_1 \quad ; \quad r' < r \qquad (12.61)$$

with

$$B_1 = \left[\frac{1+y'^2}{1+y^2}\right]^{1/4}\frac{e^{-AZ}}{1-e^{-AZ}} \qquad (12.62)$$

Similarly (12.55) for $r < r'$ reduces to

$$i_a = Z(y-y')(1+B_2) \qquad (12.63)$$

with

$$B_2 = \left[\frac{1+y'^2}{1+y^2}\right]^{1/4}\frac{e^{AZ}}{1-e^{AZ}} \qquad (12.64)$$

Note, that with the present definition of directions i_a will be negative while i_t will be positive. i_t can be derived in a similar fashion

$$i_t = \frac{Z(y-y')}{y}(1+B_1) \quad ; \quad r' < r \qquad (12.65)$$

and

$$i_t = -\frac{Z(y-y')}{y}B_2 \quad ; \quad r' > r \qquad (12.66)$$

Greater precision in the evaluation of the induction factors $i_{a,t}$ has been obtained by Wrench (1957) (also Morgan & Wrench (1965)) who used a different approximation of the Bessel functions than is used in (12.56) and (12.57). The greater precision is paid for by greater complication of the formulas.

It is not an easy task to find the limits of i_a and i_t as $y \to y'$. Returning to (12.59) and writing $y' = y - \varepsilon$ where ε is a small positive quantity and noting that as $\varepsilon \to 0$ $A \to 0$, we can expand the exponentials keeping the first two terms. Noting that the quarter root factor goes to unity linearly with ε we can write

$$\lim_{\varepsilon \to 0} i_a = -Z \lim_{\varepsilon \to 0} \left[\varepsilon \frac{1 - AZ + \dots}{AZ + \dots} \right]$$

$$= -\lim_{\varepsilon \to 0} \varepsilon \left\{ \frac{1}{\sqrt{1+y^2} - \sqrt{1+(y-\varepsilon)^2} + \ln \left[\left[\frac{y}{y-\varepsilon} \right] \left[\frac{1+\sqrt{1+(y-\varepsilon)^2}}{1+\sqrt{1+y^2}} \right] \right]} \right\}$$

$$= -\lim_{\varepsilon \to 0} \varepsilon \left\{ \frac{1}{\dfrac{\varepsilon y}{\sqrt{1+y^2}} + \dots + \dfrac{\varepsilon}{y} + \dots - \dfrac{\varepsilon y}{\sqrt{1+y^2}\left[1-\sqrt{1+y^2}\,\right]} + \dots} \right\}$$

where we retain only terms in ε and have used the binominal expansion and $\ln(1+x) \approx x$ as $x \to 0$. This reduces to

$$\lim_{\substack{y \to y' \\ \text{or } r \to r'}} i_a = -\frac{y}{\sqrt{1+y^2}} = -\frac{r}{\sqrt{k^2+r^2}} = -\frac{r'}{\sqrt{k^2+r'^2}} = -\cos\beta_i \qquad (12.67)$$

Similarly

$$\lim_{\substack{y \to y' \\ \text{or } r \to r'}} i_t = \sin\beta_i \qquad (12.68)$$

We can now see that the coefficients in the expansion of $i_{a,t}$ can be evaluated by using the transformation (12.32) to write (12.59) through (12.66) in terms of $\cos\theta$ and $\cos\theta'$ and the integral operations called for by (11.63) carried out via computer.

Introducing the overall advance ratio $J = U/ND$ and the usual thrust and torque coefficients, and using (12.26) and (12.33) in (11.74) and (11.75) gives after some manipulation

$$K_T = \frac{\pi}{2} Z J^2 \frac{1}{R} \int_{R_h}^{R} \left\{ \left[\sum_{n=1}^{\infty} G_n \sin(n\theta') \right] \left[\frac{\pi r'}{JR} - \frac{u_t}{U} \right] \right.$$
$$\left. - \frac{1}{2\pi} \sqrt{\left[1-w+\frac{|u_a|}{U} \right]^2 + \left[\frac{\pi r'}{JR} - \frac{u_t}{U} \right]^2} \left[1-w+\frac{|u_a|}{U} \right] \frac{c(r')}{D} C_D(r') \right\} dr' \qquad (12.69)$$

$$K_Q = \frac{\pi}{4} Z J^2 \frac{1}{R} \int_{R_h}^{R} \left\{ \left[\sum_{n=1}^{\infty} G_n \sin(n\theta') \right] \left[1-w+\frac{|u_a|}{U} \right] \right.$$
$$\left. + \frac{1}{2\pi} \sqrt{\left[1-w+\frac{|u_a|}{U} \right]^2 + \left[\frac{\pi r'}{JR} - \frac{u_t}{U} \right]^2} \left[\frac{\pi r'}{JR} - \frac{u_t}{U} \right] \frac{c(r')}{D} C_D(r') \right\} \frac{r'}{R} dr' \qquad (12.70)$$

Recall that the induced velocities $u_{a,t}$ depend on the circulation through (12.39). This equation implicitly gives a relation between the circulation G_n and the hydrodynamic pitch angle β_i.

Design Problem

To design a propeller of highest efficiency the criterion of Lerbs (12.27) is used. Several other criteria exist, e.g. that of van Manen & Troost (1952) and other, cf. Minsaas (1967). All criteria give a distribution of the hydrodynamic pitch angle β_i of the form

$$\tan \beta_i = kF(r) \tag{12.71}$$

where k is an unknown factor and F is a function depending on the optimum criterion.

If k were known, and hence $\tan\beta_i$, use of (12.39) into (12.26) would give the following equation in G_n

$$\sum_{n=1}^{\infty} nG_n\left\{-h_n^a(\theta) + \tan\beta_i\, h_n^t(\theta)\right\} = \left[1-\frac{R_h}{R}\right]\left[\frac{\pi r}{JR}\tan\beta_i - 1+w(r)\right] \tag{12.72}$$

This equation is valid for $0 \leq \theta \leq \pi$, (θ corresponding to r).

In the practical solution of this equation, the series in circulation Γ and consequently the number of G_n's must be truncated. If the number of collocation points or θ's, where (12.72) is fulfilled, equals the number of G_n's (12.72) is a linear system of equations which can be solved easily. When G_n is found the thrust and torque coefficients can be calculated by use of (12.69) and (12.70).

The entire procedure outlined above must be carried out iteratively, since the factor k of (12.71) is not known from the beginning. k can then be adjusted during the calculations until the specified thrust or torque coefficient is obtained.

The lift coefficients which the various blade sections are required to give are calculated by

$$C_L = \frac{2\pi D \sum_{n=1}^{\infty} G_n \sin(n\theta)}{c\sqrt{\left[\frac{\pi r}{JR} - \frac{u_t}{U}\right]^2 + \left[1 - w + \frac{|u_a|}{U}\right]^2}} \tag{12.73}$$

This is the basis for selection of the blade-section profiles.

Quite often optimization of propellers with respect to efficiency gives geometries and flows which in other aspects are not feasible. In such cases it is often necessary to reduce the loadings near the blade tip (and hence reduce the efficiency). Although this is not an optimal design in its mathematical sense it may be optimal from an engineering point of view since it is the task of the naval architect to design the propeller to perform as well as possible in the given conditions. This aspect will be discussed in the section on Pragmatic Considerations at the end of this chapter. The calculations in such cases will start with a specified distribution of blade loading over the radius, save for a constant. Again an iterative procedure is used to find this constant and to make sure, the proper thrust or torque is obtained. Details about this can be found in Lerbs (1952).

Analysis Problem

We shall now for a moment depart from the design problem and consider the *characteristics calculation* where we seek to find K_T and K_Q over a range of J's for a specified propeller. Opposed to the design case where we sought the pitch and loading that would give the required (optimum) distribution of hydrodynamic pitch angle β_i, the pitch is now known and we must calculate β_i. We must then establish a relationship between the pitch and the local load which is also unknown. We use the blade-section characteristics and their lift properties of the form

$$C_L = \frac{\partial C_L}{\partial \alpha}(\alpha - \alpha_0)$$
$$= \frac{\partial C_L}{\partial \alpha}\left\{\tan^{-1}\left[\frac{P(r)}{D}\frac{R}{\pi r}\right] - \beta_i - \alpha_0\right\} \tag{12.74}$$

where $\partial C_L/\partial \alpha$ and α_0 are profile properties, α_0 being the angle of zero lift, cf. (5.44) and (5.46). By use of the Kutta-Joukowsky law (11.72) and (12.33) we have

$$\pi D U \sum_{n=1}^{\infty} G_n \sin(n\theta) = C_L\, c\, \tfrac{1}{2} V \tag{12.75}$$

where V is the resulting inflow velocity.

After some manipulations of formulas, this equation can be transformed into

$$\sum_{n=1}^{\infty} G_n \left\{\sin(n\theta) + \frac{c}{D}\frac{1}{2\pi}\frac{\partial C_L}{\partial \alpha}\frac{nR}{R-R_h}\left\{\left[-h_n^a \cos\beta_i + h_n^t \sin\beta_i\right] C_\alpha + \frac{h_n^t}{\cos\beta_i}\left[\tan^{-1}\left[\frac{P}{D}\frac{R}{\pi r}\right] - \beta - \alpha_0\right]\right\}\right\}$$
$$= \frac{c}{D}\frac{\partial C_L}{\partial \alpha}\frac{\pi r}{JR}\left[\tan^{-1}\left[\frac{P}{D}\frac{R}{\pi r}\right] - \beta - \alpha_0\right]\frac{1}{\cos\beta_i} \tag{12.76}$$

where

$$C_\alpha = \frac{(\beta_i - \beta)\left[\dfrac{\pi r}{JR} - \dfrac{u_t}{U}\right]}{\cos\beta_i\left[\dfrac{|u_a|}{U}\cos\beta_i + \dfrac{u_t}{U}\sin\beta_i\right]}$$

Like (12.72) this equation can be transformed into a system of linear equations. The entire procedure is iterative and stops when the largest difference in β_i from two successive loops is sufficiently small. Results of a lifting-line-characteristics calculation are shown in Figure 12.5.

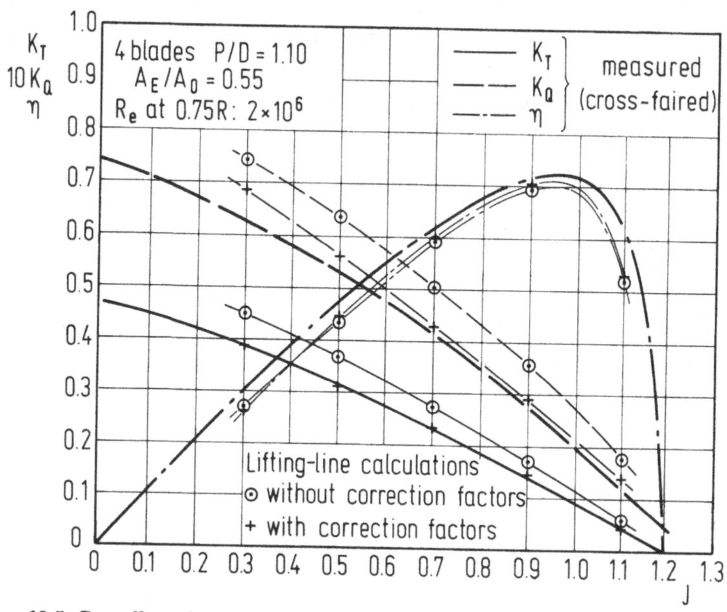

Figure 12.5 Propeller characteristics for MARIN-series propeller measured (cross-faired curves as by Oosterveld & van Oossanen (1975)) and computed by lifting-line procedures with and without lifting-surface correction factors (factors by Walderhaug (1972)). The blade sections have been modelled as NACA a = 0.8 profiles although they are not of this geometry.

The induction-factor method was initially developed for the design of propellers and the analysis problem is then often referred to as *the inverse problem*. This is considered to be more difficult to solve, in particular when the off-design condition of the propeller is treated. Here effects of viscosity and maybe cavitation in combination with lifting-surface effects, when lifting-line procedures as above are used, play important rôles. A comprehensive treatment of these aspects has been given by van Oossanen (1975).

Lifting-Surface Correction Factors

When we calculated the induced velocity by (11.63) we evaluated it in only one point on the lifting line. But the induced velocity and hence the total inflow velocity will vary along the chord length of the blade section. To allow for such three-dimensional effects and to account for the effect of thickness which we also omitted and still be able to use results for two-dimensional wing sections, lifting-surface correction factors should be used. We postpone the development of lifting-surface theory until later because the steady form is subsumed in unsteady lifting-surface theory, thereby avoiding unnecessary redundance.

Various correction factors exist, e.g. due to Morgan, Silovic & Denny, (1968) and Cumming, Morgan & Boswell (1972), and are based on results obtained by lifting-surface theory. Correction factors in polynomial form mainly based on Morgan *et al.* (*ibid.*) have been given by van Oossanen (1975). The factors correct ideal inflow angle (cf. p. 75) and camber by

$$\begin{aligned} \alpha_{i,3} &= k_\alpha \, \alpha_{i,2} + \alpha_t \\ f_3 &= k_f \, f_2 \end{aligned} \qquad (12.77)$$

where k_α, k_f, and α_t are the correction factors. $\alpha_{i,2}$ and f_2 are the values used when calculating lift and drag properties, while $\alpha_{i,3}$ and f_3 are the actual geometrical properties. Application of correction factors has been described by van Oossanen (*ibid.*) who also demonstrates how they should be used in the case of a characteristics calculation. Results of such a calculation with correction factors (although a simpler procedure than van Oossanen's is used) and without are shown in Figure 12.5.

Cavitation Prediction

The blade sections can be selected from a profile series, e.g. the NACA **a** = 0.8 (modified), cf. p. 89. Most often the profiles should be designed to operate at ideal angle of attack. But this may not always be practical, maybe because the profile then must have an exceptionally high camber ratio. The lift should then be produced not by camber alone but by a combination of camber and inflow angle. In such a case the danger of cavitation is greatly increased because of the pressure peak at the leading edge.

To design for no cavitation the requirement must be met, cf. (8.10)

$$\sigma > -C_{pmin} \qquad (12.78)$$

If we consider Figure 12.6 this means that the angle of incidence shall be within the limits indicated. Beyond these limits the profile will suffer from back side cavitation ($\alpha > \alpha_S$) or face side cavitation ($\alpha < \alpha_P$). For design purposes Figure 12.7 is useful by giving the optimum foil geometry with respect to maximum variation in inflow angle combined with minimum pressure coefficient.

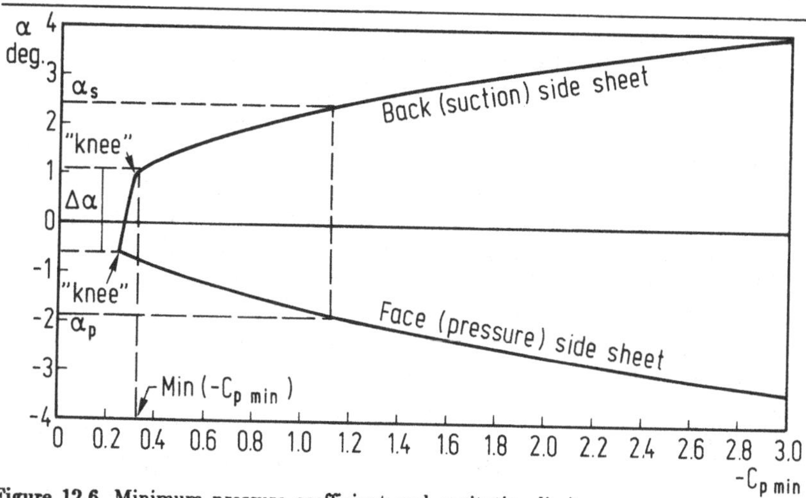

Figure 12.6 Minimum pressure coefficient and cavitation limits.

From Brockett, T. (1966). *Minimum Pressure Envelopes for Modified NACA-66 Sections with NACA a = 0.8 Camber and BuShips Type I and Type II Sections.* Hydromechanics Laboratory, Research and Development Report 1790. Washington, D.C.: David Taylor Model Basin — Department of the Navy.
By courtesy of David Taylor Research Center, USA.

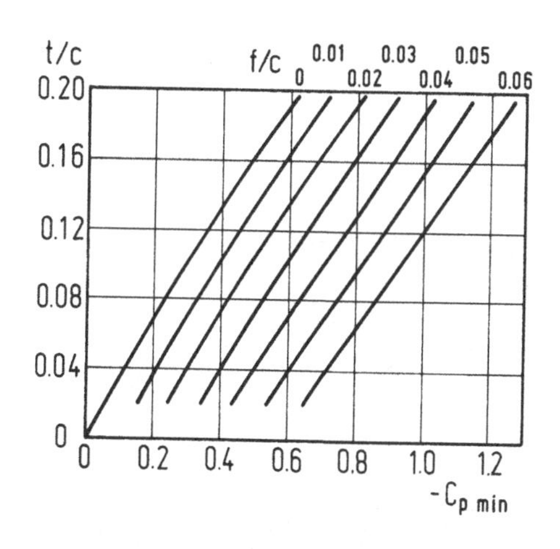

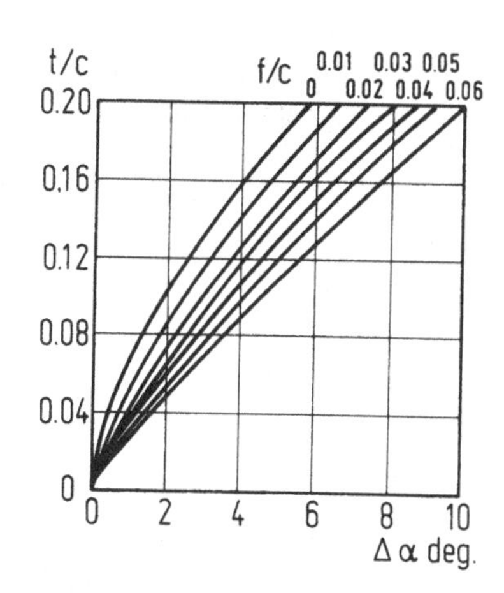

Figure 12.7 Optimum foil geometry with respect to maximum variation in inflow angle α and minimum pressure coefficient. These curves have been made by collecting the data of the "bilges" of the "buckets" or the "knees" of minimum pressure curves, as shown in Figure 12.6. Foil type: NACA 66 (TMB mod.) **a** = 0.8.

Figure 19 from Brockett, T. (1966). *Minimum Pressure Envelopes for Modified NACA-66 Sections with NACA a = 0.8 Camber and BuShips Type I and Type II Sections.* Hydromechanics Laboratory, Research and Development Report 1790. Washington, D.C.: David Taylor Model Basin — Department of the Navy.
By courtesy of David Taylor Research Center, USA.

PRAGMATIC CONSIDERATIONS[22]

In the preceding paragraphs we have dealt with mainly idealized conditions that have made possible exact definitions of what is the optimum, such as the optimum propeller with respect to efficiency. But in real life succesful design means a compromise between many conditions. While some conditions must necessarily be fulfilled entirely, such as the requirements of sufficient strength, others can only be partly met. It will be much too complex to put all these conditions, and how far they must be met, into one single criterion. The designer must therefore rely on a series of repeated calculations with systematically varied parameters under idealized conditions combined with his experience and insight into the problems.

The pragmatic considerations which will be presented in this chapter are in many ways based on cavitation and vibration analyses. They are brought here in the chapter on design of propellers as an overview of the problems and what can be done about them while the reader should obtain deeper insight from the next ten chapters on unsteady propeller theory.

The paramount conditions to be met by successful propeller design are

- High efficiency
- Sufficient strength
- No cavitation erosion
- Acceptably low vibration and noise excitations

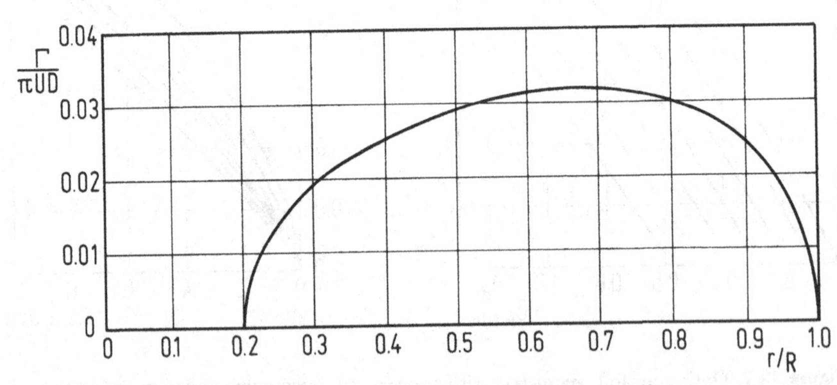

Figure 12.8 Optimum circulation distribution, $Z = 4$, $J = 0.628$, $u_a^h/U = 0.3$.
Figure 19 from Lerbs, H. (1952). Moderately loaded propellers with a finite number of blades and an arbitrary distribution of circulation. *Trans. SNAME*, vol. 60, pp. 73–123. New York, N.Y.: SNAME

[22] The authors are indebted to Senior Scientist Carl-Anders Johnsson, SSPA Maritime Consulting AB, Göteborg, for much of the material in this section provided by him in Johnsson, (1983). Mr Johnsson has had extensive experience in design and research of propellers.

Tip-Region Unloading

The derived criteria for optimum efficiency produce designs having relatively large loading on the outer sections. These loadings are typically like that shown in Figure 12.8 as calculated by Lerbs, (1952). The associated pitch distribution is nearly constant and yields larger pitch for the outer sections. Such pitch distributions are generally characteristic of propeller designs comprising the several existing systematic series referred to earlier. Such propellers require favorable wake distributions (small spatial wake variations) in order to avoid intermittent cavitation.

Prior to the advent of heavier loadings such optimum propellers were successful. However, nowadays adherence to designs for optimum efficiency because of the stress on low fuel consumption frequently leads to unacceptable unsteady cavitation with concurrent erosion, high vibratory forces and noise. Thus it is necessary to compromise efficiency to secure compatible performance with respect to vibratory levels of excitation. To achieve such compromises in a rational, predictable way one must exploit lifting-line and lifting-surface theory methods.

One obvious way of reducing the extent of cavitation or possibly to postpone inception to speeds above the design speed is to unload the outer blade sections by reducing the radial pitch distribution and camber in that region. However, a numerical design method has then to be used in order to secure a compatible relation between camber and pitch as stressed by Johnsson, (1983) who we quote (nearly exactly) in the following:

"What can happen, if such a procedure is not used, is illustrated in Figure 12.9. Here the original propeller (picked out of a systematic series) was rejected by the shipowner mainly because unacceptable vibration and noise were experienced on the ship trials. To improve the situation in these respects, the pitch distribution and blade form were changed, but the flat face of the blade sections was maintained, i.e. the radial camber distribution was not changed, see Figure 12.9. The results of the full-scale tests with the new propeller (P1868), which were confirmed at model tests in the SSPA cavitation tunnel, showed that a certain reduction of the vibration excitation was obtained at the blade frequency. The tendency to cavitation erosion (present already on the original propeller, according to the model tests), was however seriously aggravated, the result being that the new propeller was regarded as unacceptable in this respect. An analysis of the design showed that the camber of the outer sections, (having been determined from the hydrodynamically irrevelant condition that the face of the blade sections should be flat), was too large compared to the camber of the inner sections. As a result of this, the extension of cavitation became large in the middle part of the blade, which in turn gave a cavity having a convex trailing edge. This is known to promote cavitation damage. This effect was present already at the original propel-

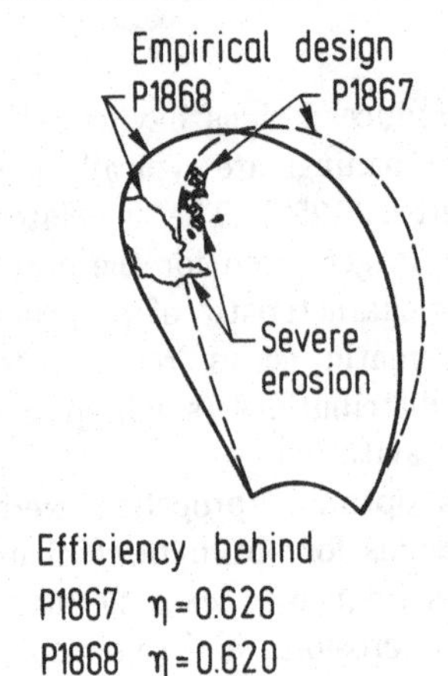

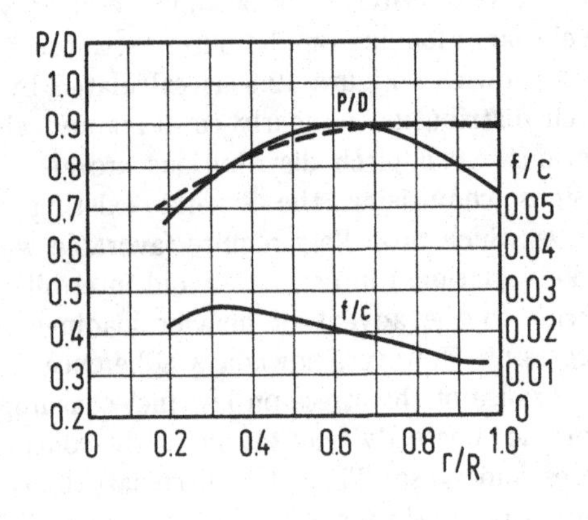

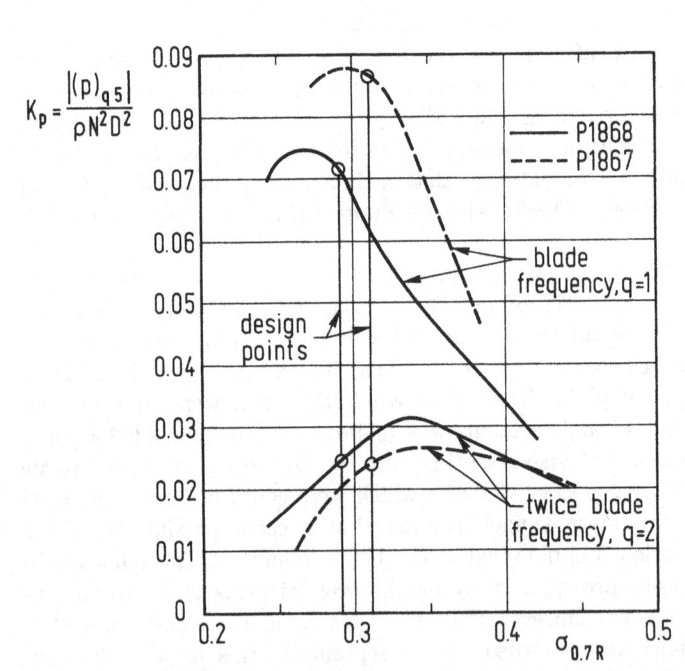

Figure 12.9 Comparison of two empirical propeller designs. Results of measurements in cavitation tunnel. Ship stern shape and pressure-gage positions, cf. Figure 22.6, p. 437.

Figure 5 from Johnsson, C.-A. (1983). Propeller with reduced tip loading and unconventional blade form and blade sections. In *Eighth School Lecture Series, Ship Design for Fuel Economy. West European School Graduate Education in Marine Technology*. Swedish Maritime Research Centre. Göteborg: SSPA.
By courtesy of SSPA Maritime Consulting, Sweden.

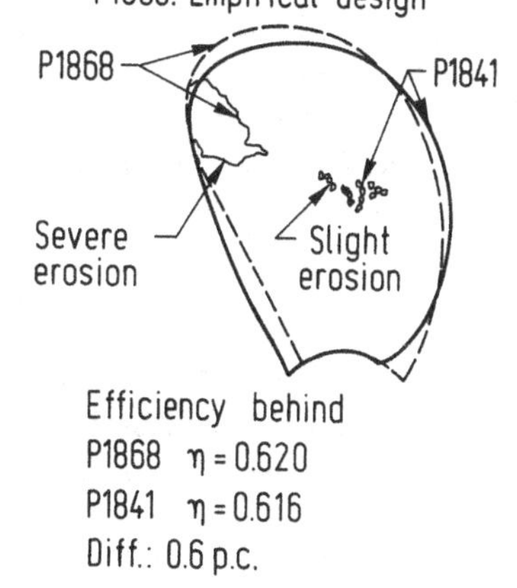

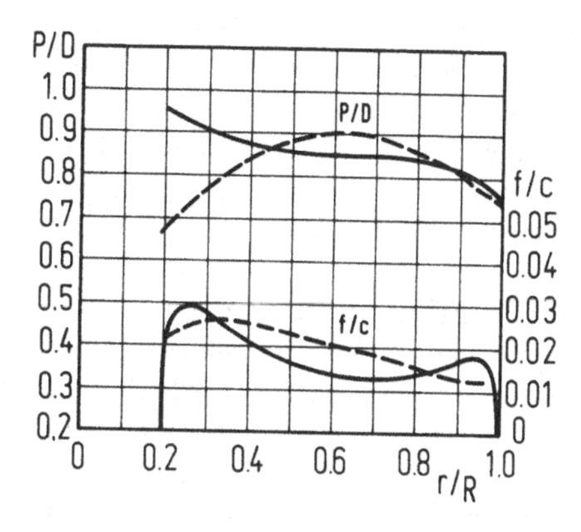

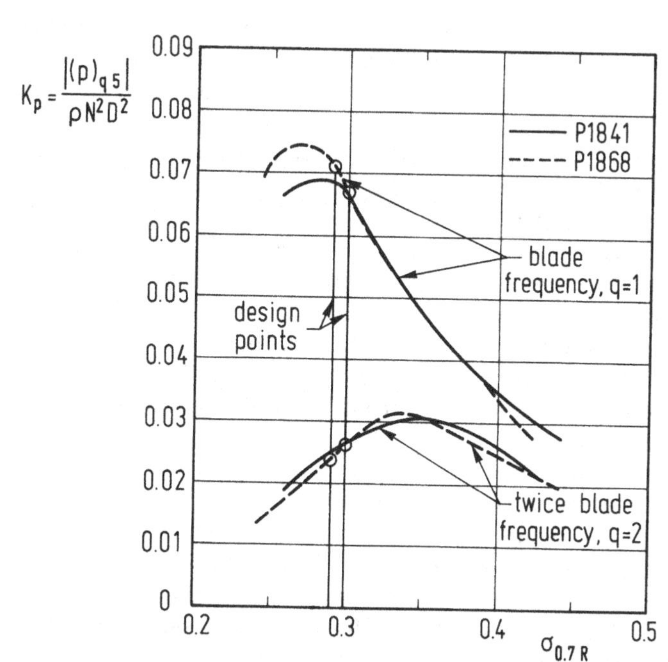

Figure 12.10 Comparison of empirical and lifting–surface propeller designs. Ship stern shape and pressure–gage positions, cf. Figure 22.6, p. 437.

Figure 6 from Johnsson, C.–A. (1983). Propeller with reduced tip loading and unconventional blade form and blade sections. In *Eighth School Lecture Series, Ship Design for Fuel Economy. West European School Graduate Education in Marine Technology.* Swedish Maritime Research Centre. Göteborg: SSPA.
By courtesy of SSPA Maritime Consulting, Sweden.

ler but was amplified when reducing the pitch of the outer sections, without making the corresponding reduction of the camber.

A more successful design for the same case is shown in Figure 12.10. For this propeller design calculations were made, using lifting-surface theory, the result being that the pitch and camber fit each other. It is evident from Figure 12.10 that this propeller has much less cavitation damage than the empirically designed propeller, the vibration excitation level and efficiency being the same."

Effects of Blade Form

An effective measure to reduce vibratory pressures and shaft forces is to employ extreme skew. (A prediction that a relative reduction in vibratory thrust of 100 per cent could be achieved by 100 per cent skew (blade tips of each blade at same angular location as the root of the following blade) was made by Ritger & Breslin (1958) via an approximate theory employing unsteady-section theory combined with the steady-state procedure of Burrill (1944)). It is obvious that when the blades are sufficiently skewed the sections pass through the wake "spike" or valley in a staggered fashion thus reducing the forces remarkably as compared with a blade whose locus of mid-chords is radially straight. When combining skew and tip-region unloading Johnsson's experience indicates the necessity of employing a "rigorous calculation" (procedure) as a drastic change in blade form affects the relation between radial distribution of circulation, pitch and camber. His point is made in Figure 12.11 where three widely different blade forms are shown which all have the same radial distribution of circulation. The pitch and camber distributions obtained by two different design methods are also shown. The approximate method is based on lifting-line calculations with camber-correction factors from the literature being used for determining the final values of pitch and camber at each section. Using this process, *the same radial distributions of pitch and camber are obtained for all three designs.*

The other design method, based on a rigorous lifting-surface representation, determines different radial distributions of camber, and particularly pitch, for the three blade forms. It is evident from Johnsson's calculations that the lifting-line method with so-called lifting-surface corrections yields much too large pitches at the outer sections of the skewed blades. In this way the beneficial influence of skew on vibration excitation ".... is reduced and sometimes eliminated".

This experience should not be surprising as the influences of the chordwise geometry and especially the radially induced velocity are not accounted for in lifting-line theory with approximate corrections.

We may encapsulate findings from recent extensive experimental studies at SSPA Maritime Consulting, Sweden of the influence of radial load reduction and skew in the following:

Efficiency (Refer to Figure 12.13)

i. The radial pitch distribution is the most important parameter in regard to efficiency of propellers with conventional blade form and equal blade area.

ii. Loss of efficiency with tip unloading is smaller in "behind" condition than in open water.

iii. Efficiency loss from tip unloading is reasonably predicted by calculation.

iv. A propeller having extreme skew has efficiency equal to the corresponding propeller with conventional blade form (i.e., having the same radial circulation distribution).

Pressure Fluctuations (Refer to Figure 12.14)

i. Radial pitch variation is the most important parameter for reduction of vibratory pressures at blade frequencies for conventional blade forms; design principles and blade area being of less importance.

ii. At twice and higher-order blade frequencies, unloading of the tips is not effective in reducing these amplitudes.

iii. Designs with extreme skew show significant reductions in vibratory pressure amplitudes over the entire frequency range.

Cavitation Damage

i. An increasing extent of cavitation damage attends reduction of pitch without a corresponding reduction of camber in the tip region. (Refer to Figure 12.9)

ii. In the designs tested the extent (area) of cavitation damage is only slightly reduced with increasing tip unloading and skew.

It is necessary to note, as has Johnsson, (1983), that structural analysis via finite elements of extremely skewed blades showed that high stresses can be expected in some parts of the blades, particularly in the reversing and backing modes.

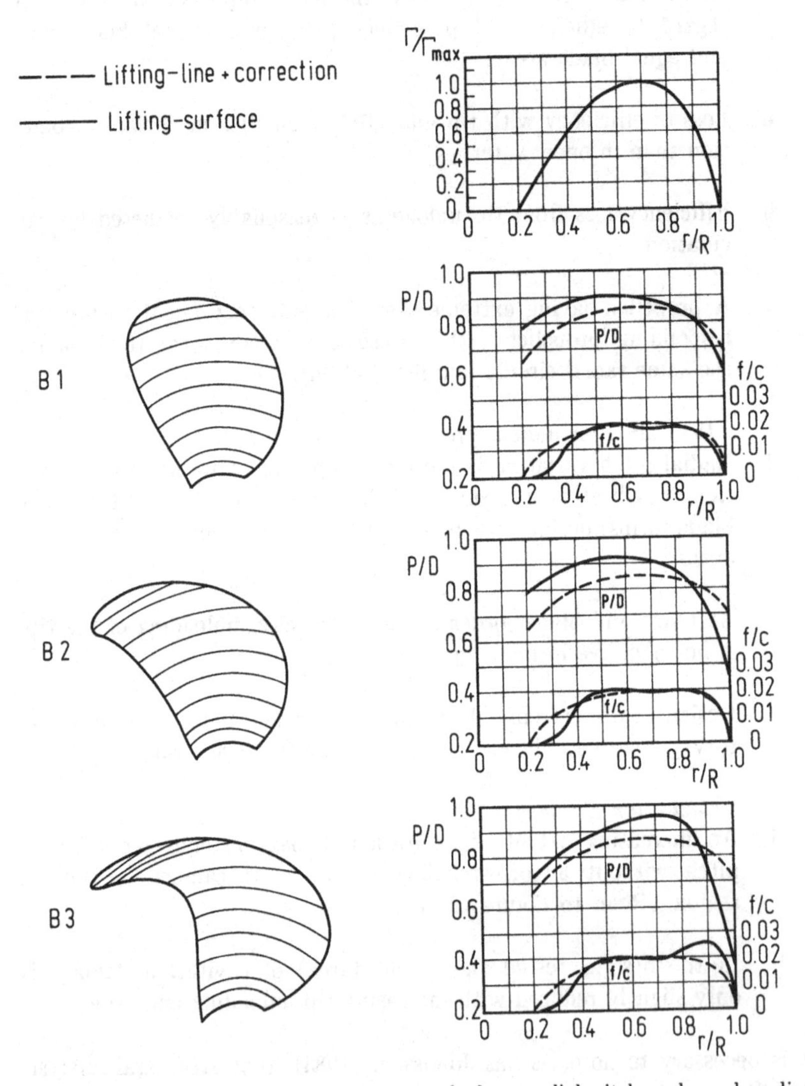

Figure 12.11 Influence of propeller design method on radial pitch and camber distributions for different blade forms.

Figure 7 from Johnsson, C.-A. (1983). Propeller with reduced tip loading and unconventional blade form and blade sections. In *Eighth School Lecture Series, Ship Design for Fuel Economy. West European School Graduate Education in Marine Technology.* Swedish Maritime Research Centre. Göteborg: SSPA.
By courtesy of SSPA Maritime Consulting, Sweden.

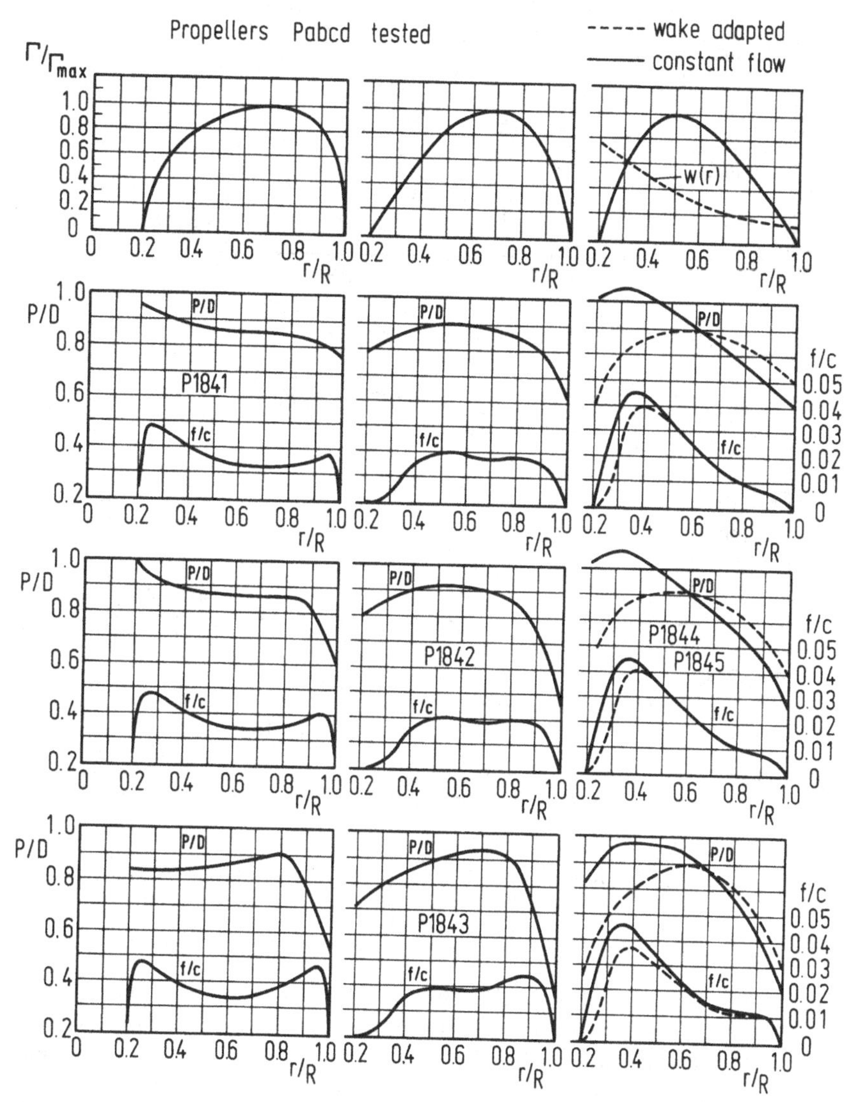

Figure 12.12 Different propeller designs for the same loading case (5–bladed container ship propellers). Variation of radial circulation distribution and blade form. For blade forms see Figure 12.11. More details of geometry of propeller model P1841 and model test results are given in Chapter 22.

Figure 8 from Johnsson, C.–A. (1983). Propeller with reduced tip loading and unconventional blade form and blade sections. In *Eighth School Lecture Series, Ship Design for Fuel Economy. West European School Graduate Education in Marine Technology.* Swedish Maritime Research Centre. Göteborg: SSPA.
By courtesy of SSPA Maritime Consulting, Sweden.

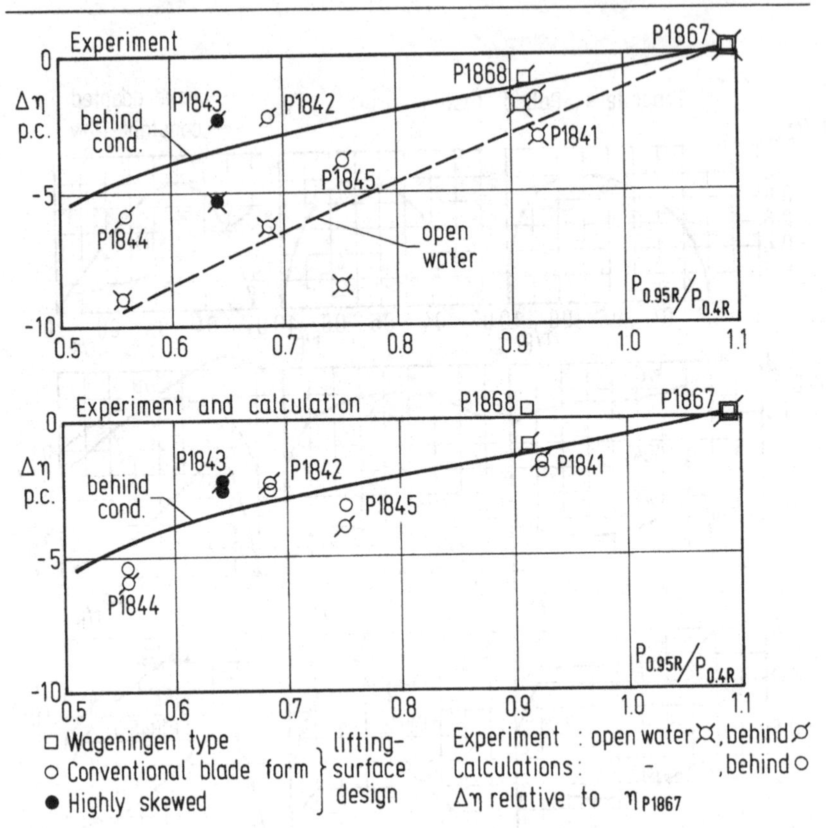

Figure 12.13 Influence of radial load distribution (tip unloading) and skew on efficiency. For details of propeller models see Figure 12.11 and 12.12.

Figure 13 from Johnsson, C.-A. (1983). Propeller with reduced tip loading and unconventional blade form and blade sections. In *Eighth School Lecture Series, Ship Design for Fuel Economy. West European School Graduate Education in Marine Technology.* Swedish Maritime Research Centre. Göteborg: SSPA.
By courtesy of SSPA Maritime Consulting, Sweden.

By choosing to treat the fundamentals in great detail the authors have so far omitted many topics of importance to steady-state propeller theory. Instead of treating these topics in equally great detail we shall give an overview of some of them in Chapter 23 on special propulsion devices. Meanwhile we shall leave the steady cases and turn to aspects of unsteady hydrodynamics, describing the flow around a propeller operating abaft a ship.

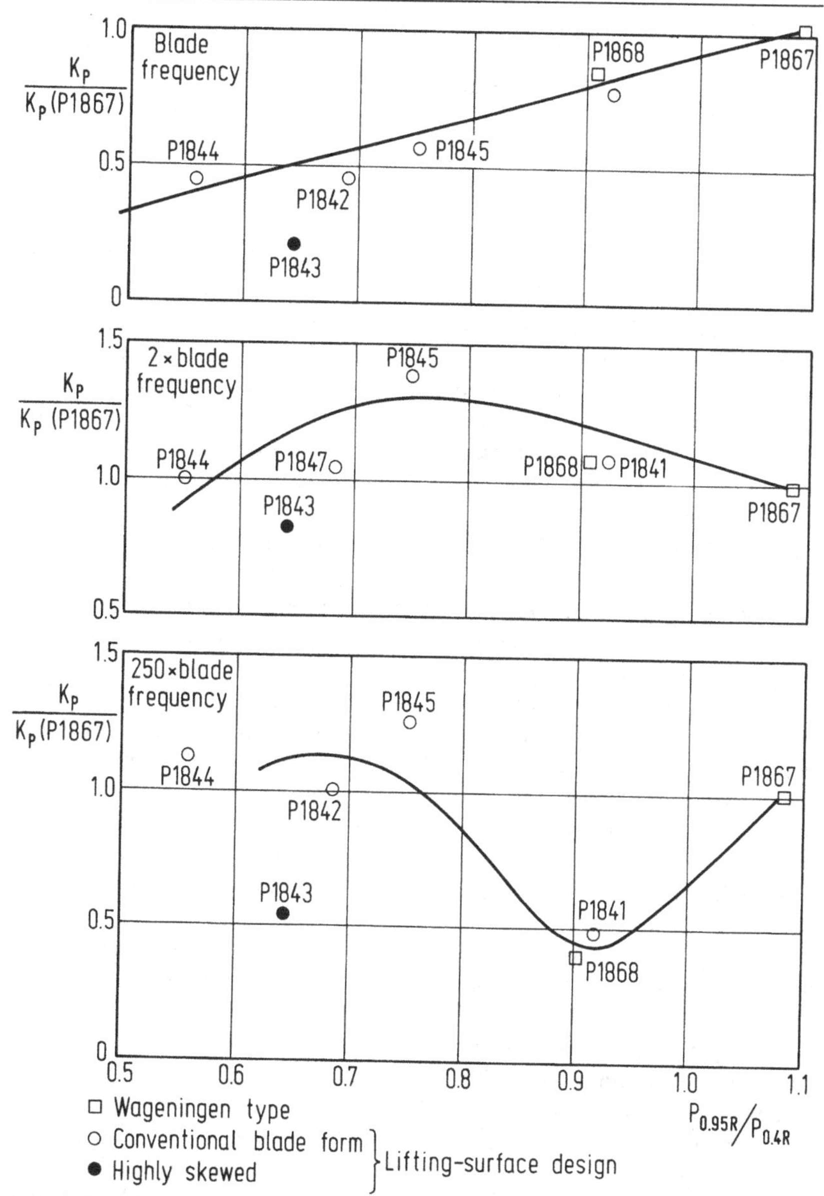

Figure 12.14 Influence of radial load distribution (tip unloading) and skew on pressure fluctuations. For details of propeller models, cf. Figure 12.11 and 12.12.

Figure 14 and 15 from Johnsson, C.-A. (1983). Propeller with reduced tip loading and unconventional blade form and blade sections. In *Eighth School Lecture Series, Ship Design for Fuel Economy. West European School Graduate Education in Marine Technology.* Swedish Maritime Research Centre. Göteborg: SSPA.
By courtesy of SSPA Maritime Consulting, Sweden.

13 Hull-Wake Characteristics

It is well known that the flow abaft of ships is both spatially and temporally varying. This variability arises from the "prior" or upstream history of the flow produced by the action of viscous stresses and hull-pressure distribution acting on the fluid particles as they pass around the ship from the bow to the stern. Thus the blade sections "see" gust patterns which over long term have mean amplitudes but from instant-to-instant change rapidly with time because of the inherent unsteadiness of the turbulent boundary layer.

Our knowledge of the distribution of flow in the propeller disc is almost entirely based on pitot-tube surveys conducted on big ($\approx$ 6 m) models in large towing tanks and in the absence of the propeller. These are termed nominal wake flows. As is well appreciated, pitot-tube measurements provide only long-term averages of the velocity components at various angular and radial locations in the midplane of the propeller. These measurements depend upon the calibration of the pitot-tube in uniform flow whereas the wake flow radially and tangentially has the effect of shifting the stagnation point on the pitot-tube head, a mechanism not operating in the calibration mode. Thus there is a systematic error which is, to the authors' knowledge, not generally corrected. Moreover, wake-fraction (and thrust-deduction) calculations based upon tests with the same model in several large model basins and upon repetitive tests with the same model in the same large model basin, have shown remarkably different results. A similar scatter was also found in results of wake surveys. These findings were reported by Harvald & Hee (1984, 1988 and 1978).

To compound our ignorance of ship wakes we know from the general theory of viscous flows that the hull boundary-layer characteristics are dependent upon Reynolds number UL/ν (U the speed, L the length, and ν, the kinematic viscosity) which is of order 10^6 in model scale and 10^9 in full scale. Both theory and experiment lead us to expect (for both model and ship with hydraulically smooth surfaces) that the model wake will be thicker. This implies that minimum velocities relative to reference speed will be less on model than on ship and that the dimensionless wake-harmonic amplitudes (to be defined later) will be smaller in model scale because the model flow is more viscous hence less rapidly changing or less "sharp" than the ship. Still another bleak aspect is provided by our nearly complete ignorance of the effective wake, i.e., the wake "seen" by the propeller when driving the ship. Some knowledge of the mean effective wake is secured by using the open-water propeller curves together with thrust and torque measurements on propeller models operating abaft hull models. This tells us nothing in regard to those aspects or details of the distribution of effective wake which are the root cause of propeller-generated vibratory forces and pressure.

We do know that gross observations of the flow patterns on full-form models having separated-boundary-layer flows, generally forward and somewhat above the propeller, without the propeller operating have revealed that such "pockets" of separation are removed when power is applied to the propeller and often a new region of separation appears directly abaft and above the operating propeller. An increasing use of laser-doppler velocity meters will, in time, lessen our ignorance of model and ship effective wake.

The development of computational methods directed at computing hull boundary layers is impeded by the complexity of the problem. An efficient (and necessary, with respect to use of computer resources) method is to divide the flow zone around the hull into three zones, with necessary interactions, cf. Larsson, Broberg, Kim & Zhang (1990). Outside the boundary-layer and wake regions a potential-flow description is adequate and can more efficiently include the effects of the free surface. A special boundary-layer method can then be used in the region from the bow and appr. 70 per cent of the ship length downstream. In the stern and near-wake region the thickening of the viscous flow cannot be predicted by boundary-layer theory and the flow is calculated by solution of the Navier-Stokes equations. More details about this and similar procedures and references to the literature can be found in Larsson *et al.* (*ibid.*), Patel (1989) and ITTC (1990a). We may conclusively quote Larsson *et al.* (*ibid.*) that "the usefulness of the new numerical methods does not mean that they are perfect, or that computations can replace model testing in general". But so far they have "...been developed to a stage, where they can be used side by side with towing tanks to aid in the design of ships". Further developments in computer technology and computational procedures will no doubt remove the limits of these statements.

In the face of all the foregoing "bad press" about our abject knowledge of ship-scale wakes and even model-scale effective wakes, what are we to do in regard to predicting vibratory blade and hull forces which depend on certain detailed features of the wake (soon to be defined)? This ignorance has been used by many pragmatic naval architects and engineers as a basis for disparaging efforts to develop totally rational mathematical models since the accuracy of the required input is not comparable to that of the complex representations. The answer to them is that one should not deal with simple *ad-hoc* theories in which systematic errors or lack of realisms are imbedded but do the very best we can, so that the level of comparative results (using the same wake as input) can be relied upon. Then, when accuracy of the wake input data is achieved we shall have at hand tools which are as rational, consistent and effective as we can make them. As matters now stand comparisons of results from the models to be described with relevant data when the wake is well established (or can be reasonably modified) are good to excellent. However, the number of correlations is far from legion. Meanwhile, it is fair to say that the sophistication of the mathematical models is in advance of that of the input data particularly for full form ships.

ANALYSIS OF THE SPATIAL VARIATION OF HULL WAKES

We put aside the temporal variability of hull wakes for the present and deal only with the steady or long-time averages of the velocity field as a function of radius r and angular position γ in the plane of the propeller. Such a velocity field for a relatively modern 125 m cargo ship is shown in Figure 13.1 and 13.2. From Figure 13.1 one observes a region of relatively low inflow velocity ($\approx$ large wake) at the 12 o'clock position. An interesting observation can be made from Figure 13.2 where it is apparent that the water to the propeller flows over the bottom of the ship aft and upwards into the disc (at least for a ship with this breadth-draft ratio of 3.8 (ballast condition)). (One can perceive anomalies in the neighborhood of 6 and 12 o'clock ($\gamma = 180^\circ$ and 0°) which are believed to be due to the effects of the shear referred to previously, since the tangential flow velocity should be zero here due to symmetry).

By a different presentation of the wake-field data (Figure 13.3) we can envision that we have a "standing wave" through which the blade sections cyclically "orbit". We designate the varying part of the axial wake component by $u_a^{w'}(\gamma)$ and the varying tangential component by $u_t^{w'}(\gamma)$ (both relative to the mean velocity). Both components vary about mean values. We then see from Figure 13.4 that the geometric wake flow angle α_g will oscillate about the mean geometric angle β_a as we follow the section around the "clock" or across the total wake variation in Figure 13.3. It is of interest to calculate the variation in the angle of attack defined by the component normal to the pitch plane and the nominal resultant velocity which from the figure is

$$\alpha_g \approx \tan^{-1} \frac{1}{V} \left[u_a^{w'} \cos \beta_P - u_t^{w'} \sin \beta_P \right]$$

$$= \tan^{-1} \frac{U}{V} \left[\frac{u_a^{w'}}{U} \cos \beta_P - \frac{u_t^{w'}}{U} \sin \beta_P \right] \tag{13.1}$$

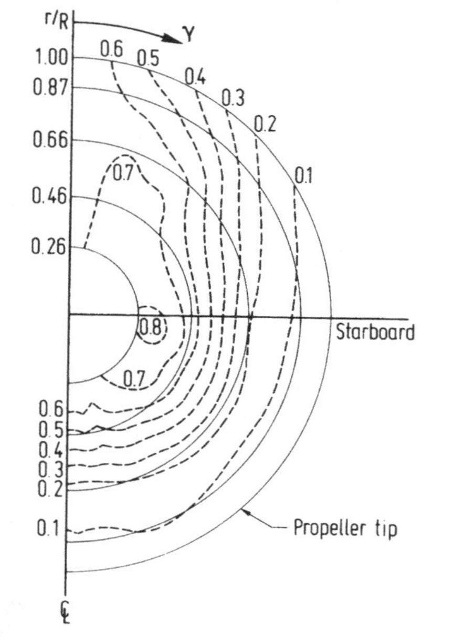

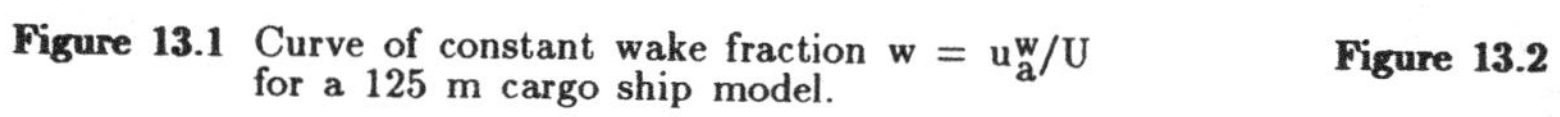
Figure 13.1 Curve of constant wake fraction $w = u_a^w/U$ for a 125 m cargo ship model.

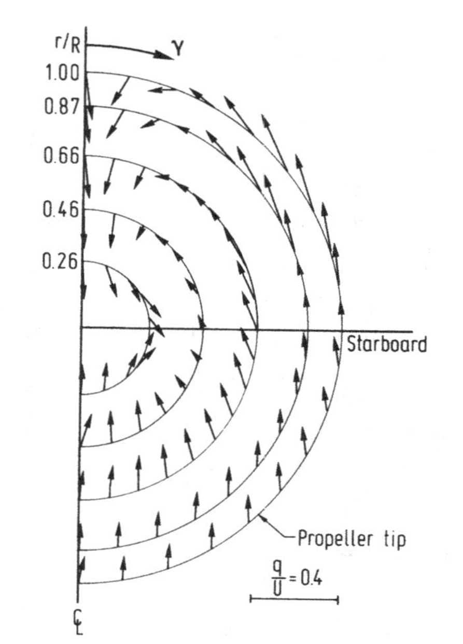

Figure 13.2 Transverse velocity q/U in the propeller disc of 125 m cargo ship model.

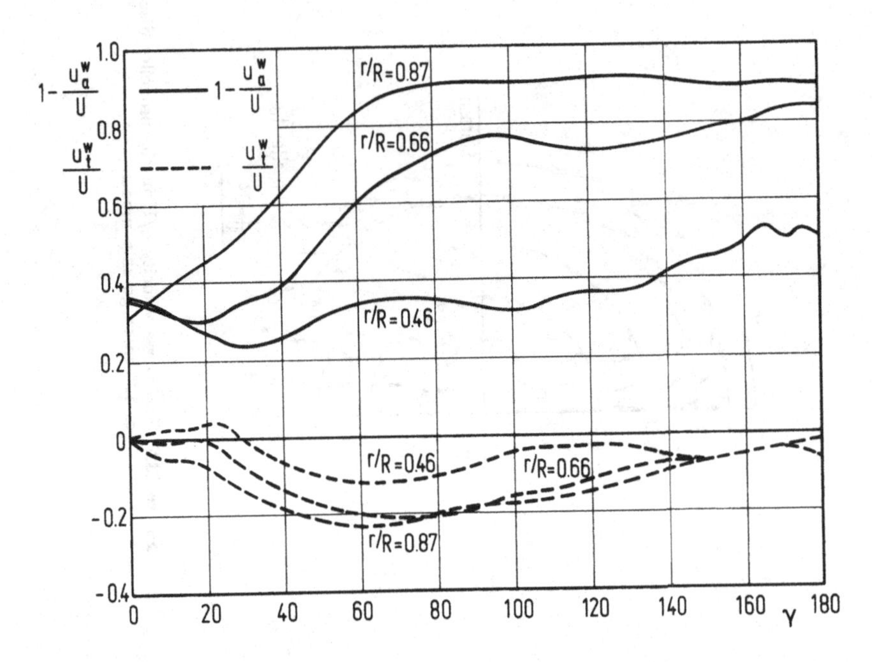

Figure 13.3 Axial and tangential velocity distribution of 125 m cargo ship model.

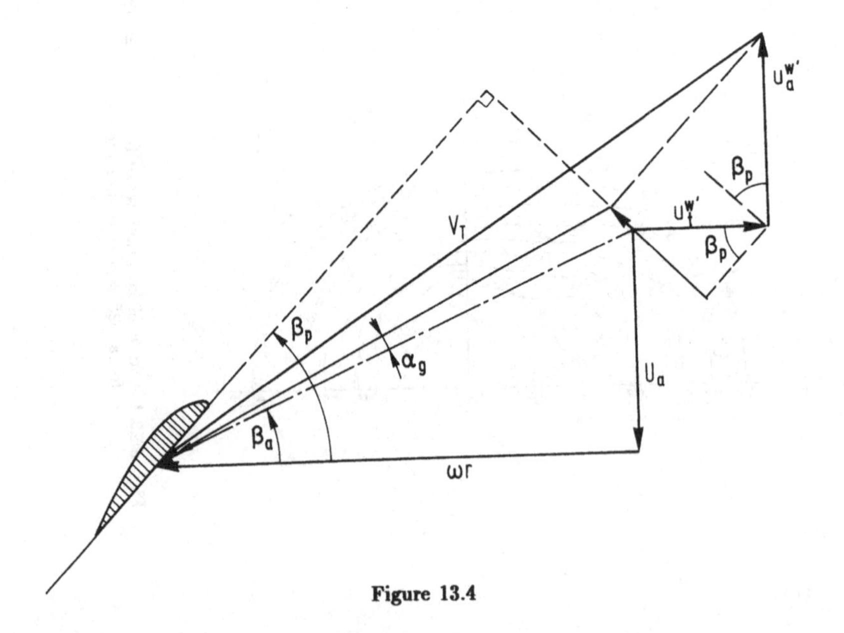

Figure 13.4

where $V = \sqrt{(\omega r)^2 + U_a^2}$; U = ship speed; U_a = the mean axial wake velocity at the radius r. The ship speed, U, is introduced because the wake velocities are given in fraction of U as in Figure 13.3. From this figure we estimate $U_a = 0.634$ U at r/R = 0.664 and by (13.1) can then obtain the data shown in Table 13.1.

γ deg.	$1 - \frac{u_a^w}{U}$	$\frac{u_a^{w'}}{U}$	$\frac{u_t^{w'}}{U}$	α_g deg.
0	0.32	−0.284	0	−5.3
90	0.77	0.136	−0.190	3.8
180	0.84	0.206	0	3.9
270	0.77	0.136	0.190	1.3

Ship speed: 16 knots
Rate of revolutions: 140 RPM
Propeller diameter: 4.72 m
Calculation radius r: 0.664 R
Pitch angle β_p: 20 deg.

Table 13.1 Estimated geometry angle α_g based on variation of wake angle from Figure 13.3.

Thus we see that the change in geometric angle

$$\alpha_g(180^\circ) - \alpha_g(0^\circ) = 3.9^\circ + 5.3^\circ = 9.2^\circ \tag{13.2}$$

If induced angles were to be included the change in hydrodynamic angle would be about 5° which is a large deviation as viewed from our considerations of the cavitation sensitivity of thin sections (cf. Figure 6.8).

We also see from the details exhibited above that the tangential component produces a weak contribution relative to that from the axial component because $u_t^{w'}$ is generally less than $u_a^{w'}$ and *moreover because the tangential component is weighted by* $\sin\beta_p$ *which is generally about 1/3 of* $\cos\beta_p$. Thus the *major* contributor to the generation of unsteady propeller blade loading is the varying *axial component* of the wake.

Under the assumption that the variation in the wake is only spatial we can exploit the cyclic behavior of such patterns and write for the axial and tangential wake velocity components

$$\frac{u_a^w(r,\gamma)}{U} = \sum_{n=0}^{\infty} \frac{u_{an}^c(r)}{U} \cos n\gamma + \sum_{n=1}^{\infty} \frac{u_{an}^s(r)}{U} \sin n\gamma \tag{13.3}$$

$$\frac{u_t^w(r,\gamma)}{U} = \sum_{n=0}^{\infty} \frac{u_{tn}^c(r)}{U} \cos n\gamma + \sum_{n=1}^{\infty} \frac{u_{tn}^s(r)}{U} \sin n\gamma \tag{13.4}$$

where the coefficients of the cosine and sine series are functions only of r and are calculated from given wake data by the usual operations

$$\left.\begin{matrix} u^{c}_{an}(r) \\ u^{s}_{an}(r)\end{matrix}\right\} \frac{1}{U} = \frac{\varepsilon_n}{2\pi}\int_0^{2\pi} \frac{u^{w}_{a}(r,\gamma)}{U}\left\{\begin{matrix}\cos n\gamma\\ \sin n\gamma\end{matrix}\right\} d\gamma \;\; ; \qquad \begin{matrix}\varepsilon_0 = 1\\ \varepsilon_n = 2 \quad n > 0\end{matrix}$$

and similarly for u^{c}_{tn} and u^{s}_{tn}. Note, that the wake velocity (u^{w}_{a}, u^{w}_{r}, u^{w}_{t}) is a perturbation velocity. The velocity, measured for example by a gage moving with the ship (at speed U) is then ($-U + u^{w}_{a}$, u^{w}_{r}, u^{w}_{t}).

For single screw ships u^{s}_{an} and u^{c}_{tn} should be zero although measurements made from 0 to 359° often show some asymmetry in u^{w}_{a} and lack of pure asymmetry in u^{w}_{t} (An example of this can be seen in the wake measurements, shown in Chapter 22 (e.g. Figure 22.8)). These aberrations may be considered to be anomalies in the measurements due to lack of model symmetry and from imperfect alignment of the model. They may also be real, arising from mean unsymmetrical vortex shedding which may be particularly the case in large beam-to-draft hulls which are known to shed longitudinal "bilge" vortices which meander in an aperiodic fashion. For simplicity we shall consider the axial flow for single screws (on symmetrical hulls) to be even functions of γ and the tangential components to be odd functions for which $u^{s}_{an} \equiv 0$ and $u^{c}_{tn} \equiv 0$. Most often data for center-line propellers are only taken over half the disc and then we have no way of knowing u^{s}_{an} and u^{c}_{tn}. For off-center-line propellers these components are neither even nor odd, thus all terms must be retained.

A large collection of model wake data has been analyzed by Hadler & Cheng (1965). They present both the three-dimensional velocity fields and the harmonic amplitudes of these velocities and they also compare the influence of various types of sterns upon the wake field.

An example of harmonically analyzed model wake data is shown in Figure 13.5 and 13.6 showing the variation of the amplitudes with radius and harmonic order. From these typical results we can secure an idea of the numerical size of the amplitudes and also note that the odd orders often show a 180° phase change from the inner to the outer radii. The total angle-of-attack variation on any one blade is about 85 per cent accounted for by the sum of the first three harmonics and hence these are principally responsible for the occurrence of intermittent cavitation.

In contrast the shaft forces and moments generated at frequencies qZ by a Z-bladed propeller are produced by only the (qZ − 1)-th, the qZ-th and the (qZ + 1)-th wake harmonic amplitudes as will be seen later, p. 365 and sequel. It is also to be appreciated that the accuracy of the harmonics of higher orders is not high say for order numbers greater than 6.

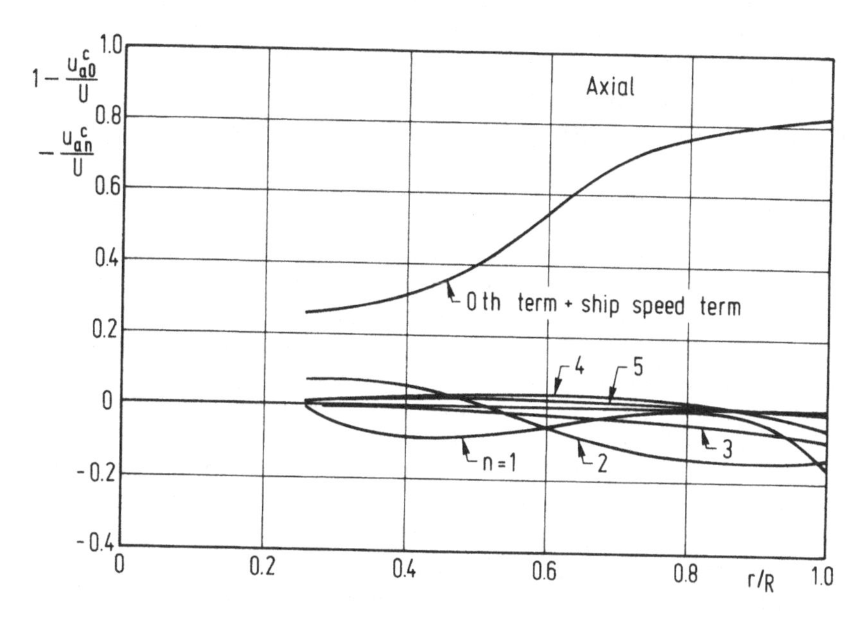

Figure 13.5 Amplitudes of various harmonics of axial (u_a^w/U) velocity, single-screw 125 m cargo ship model.

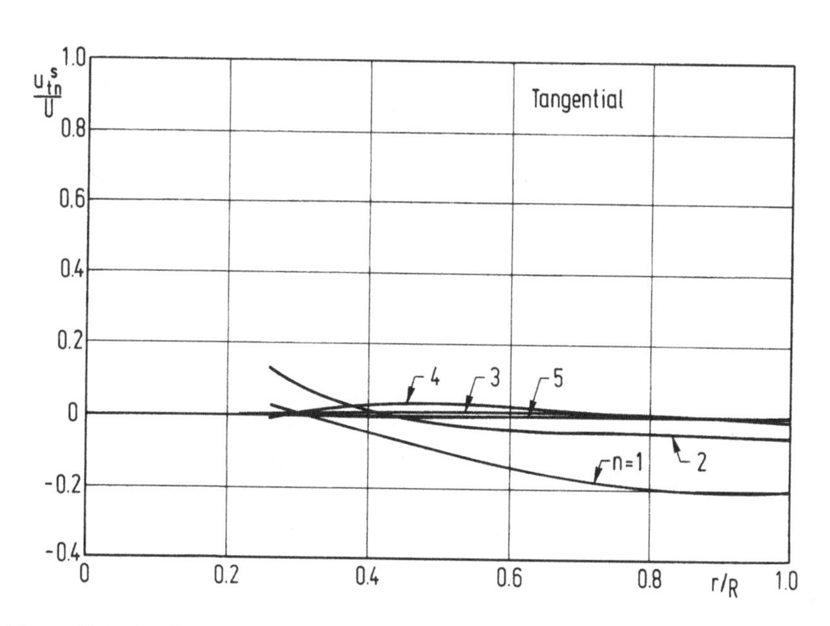

Figure 13.6 Amplitudes of various harmonics of tangential (u_t^w/U) velocity, single-screw 125 m cargo ship model.

TEMPORAL WAKE VARIATIONS

As noted earlier the wake field is not steady in time. No systematic data exist to quantify this phenomenon, nor do any existing theories take into account this temporal variability. We do know from high-frequency measurements of shaft forces and near-field pressures that the amplitudes of these quantities are not constant but (in an aperiodic fashion) modulate considerably; the ratio of maximum to minimum amplitudes being of the order of 3.

It is the authors' conjecture that temporal variations arise from turbulent intermittency in the outer regions of the boundary layer which gives rise to coherent vortical flows. This behavior is suggested in Figure 13.7 where we can see vortical flows between regions of laminar flow with non-uniform size and spacing. These non-uniform patterns must give rise to varying spatial harmonics in the propeller disc which produce temporal variations in the blade and shaft forces.

In possible support of this general conception, the experimental work in thin shear layers by Brown & Roshko (1974) identified large coherent structures in the turbulent mixing of thin parallel jets. Any large scale disturbance such as U-shaped vortices shed around the periphery of the hull would wash the propeller with a coherent distribution which has blade-frequency (and higher) spatial harmonics which the propeller would filter out and give a response.

Japanese observations of the flow about ship models showed the existence of discrete longitudinal vortices generated by hulls of large beam-draft ratios extending past the propeller plane into the wake. Through flow visualization these discrete vortices were found to meander up and down on either side which thereby explained the observed large modulations in unsteady propeller forces on such models.

The only practical conclusion that can be drawn from these foregoing observations is that the presence of temporal variability in model and ship wakes determines a lower bound to the level to which we can reduce propeller excitations by use of passive design measures such as skewing the blades and the use of ducts with peripheral camber variations. The mitigation of temporal-flow effects would require active controls coupled with flow sensors.

Having outlined the principal characteristics of hull wakes we next ignore all spatial and temporal variations by considering (for simplicity) the pressure field about a propeller in uniform flow or open water sans boundaries.

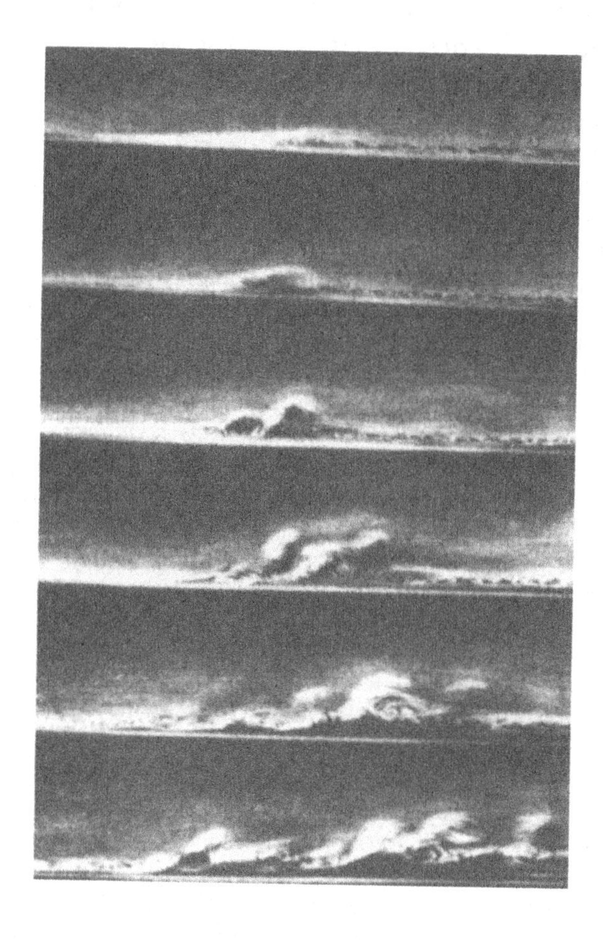

Figure 13.7 Creation of turbulence in the flow along a flat plate. The photographs show the same group of vortices as they are originated (top photograph) and develop downstream (from top to bottom). The flow which enters from the left is made visible by aluminum dust.

Figure 2 from Prandtl, L. (1933). Neuere Ergebnisse der Turbulenzforschung. *Zeitschrift des Vereines Deutscher Ingenieure*, Band 77, No. 5, 105–14.
By courtesy of VDI–Verlag, Germany.

14 Pressure Fields Generated by Blade Loading and Thickness in Uniform Flows; Comparisons with Measurements

In this chapter the pressure fields induced by the loading and thickness distributions on a single blade will be derived. They are found to be composed of pressure dipoles (for the loading) distributed with axes normal to the fluid helical reference surface in way of the blade and tangentially directed dipoles along this surface for the thickness. The expressions are then expanded in exponential Fourier series to facilitate determination of the total contributions from Z blades. This reveals that the pressure "signature" contains only components at integer multiples of blade number. The behaviour at large axial distances at blade frequency is examined analytically and the variation of the weighting functions in the integrals for small axial distance is displayed graphically via computer evaluation. The chapter concludes with comparisons with measurements and with various approximate evaluations of the integrals involved made in the past.

PRESSURE RELATIVE TO FIXED AXES

We shall derive the induced pressures in a fixed coordinate system. In this system the propeller sees an axial inflow (in the negative direction) while it rotates about the x-axis. The pressure is derived at a fixed point so this situation corresponds to finding the pressure induced by the propeller at a point which travels along with the ship and is fixed, for example on the ship surface. Note however that neither the varying wake, mirror effects of the ship surface nor the influence of the free surface will be considered at present as they will be dealt with later (in Chapters 15, 21 and 22).

Pressure Due To Loading

Description of flows which are non-stationary when viewed relative to a fixed frame of reference requires retention of the temporal derivatives in the equations of motion. We may return to Equation (9.8)[23], and linearize. In this process we assume that the perturbation velocities u_a', u_r' and u_t' are small relative to the axial inflow velocity $-U$. We then insert $u_a =$

[23] In (9.8) u_a was the total axial velocity but, in general, u_a designates the axial velocity induced by the propeller.

$-U + u'_a$ [23], $u_r = u'_r$ and $u_t = u'_t$ into (9.8) and drop the products of perturbation velocities and their derivatives. The primes are used *to indicate differential quantities* induced by the loading on the single element of the blade. The primes will be removed later by integration over the blade surface. Then for an element of a propeller blade rotating about the x-axes in a stream of speed $-U$ we have the following set of equations for the velocity and pressure perturbations arising from *blade loading*

$$\begin{aligned} \frac{\partial}{\partial t} u'_a - U \frac{\partial}{\partial x} u'_a + \frac{1}{\rho}\frac{\partial p'_\ell}{\partial x} &= \frac{F'_a}{\rho} \\ \frac{\partial}{\partial t} u'_r - U \frac{\partial}{\partial x} u'_r + \frac{1}{\rho}\frac{\partial p'_\ell}{\partial r} &= 0 \\ \frac{\partial}{\partial t} u'_t - U \frac{\partial}{\partial x} u'_t + \frac{1}{\rho}\frac{1}{r}\frac{\partial p'_\ell}{\partial \gamma} &= \frac{F'_t}{\rho} \end{aligned} \tag{14.1}$$

We have here neglected effects of rake by setting the radial component of the blade force $F'_r = 0$.

When integrating over the blade it is useful to operate in a helicoidal-normal coordinate system, cf. Figure 14.1, lying on a cylindrical surface with the axis coinciding with the x-axis. The system is inclined with the fluid pitch angle β and rotates about the x-axis with the blade reference line (key blade). We then have (Figure 14.2)

$$\begin{aligned} h &= x \sin\beta + r\, \psi \cos\beta \\ n &= x \cos\beta - r\, \psi \sin\beta \end{aligned} \tag{14.2}$$

On the blade, where $n' = 0$, (these primes indicate dummy points or points on the blade), we have

$$x' = r'\, \psi' \tan\beta' = h' \sin\beta' \tag{14.3}$$

$$h' = \frac{r'\psi'}{\cos\beta'} \tag{14.4}$$

From this we have the blade normals in the axial, radial and tangential directions

$$(n_a, n_r, n_t) \approx (\cos\beta', 0, -\sin\beta') \tag{14.5}$$

where we for the moment neglect the blade camber and local angle of incidence, and moreover assume that the blade is non-raked, as in (14.1). From Equation (9.13) we then have the forces from the blade element on the fluid, represented by

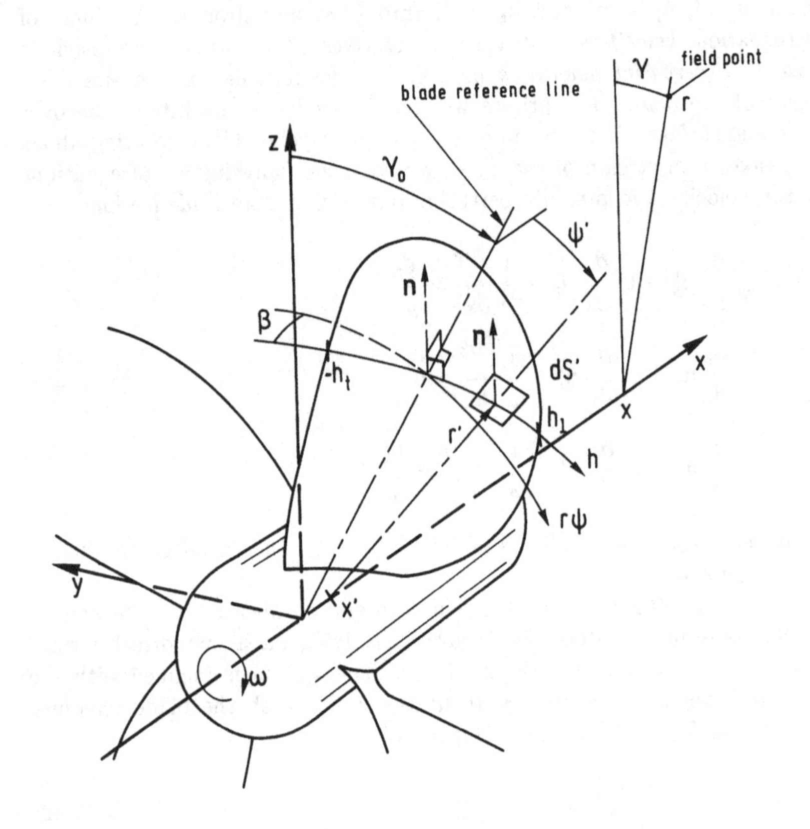

Figure 14.1 Definition of blade coordinates.

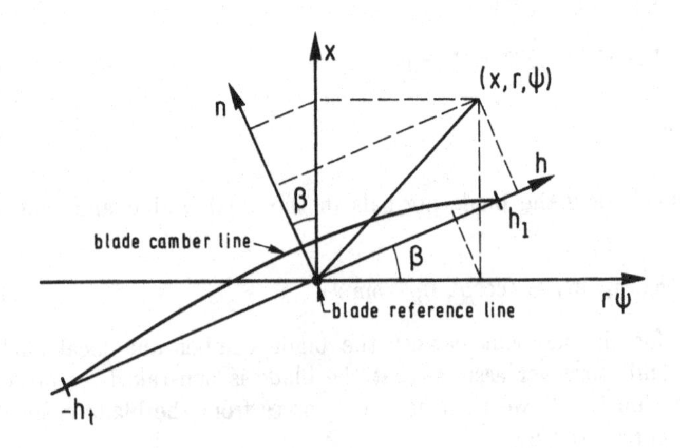

Figure 14.2 Helicoidal coordinate system.

$$F_a' = -\frac{\Delta p(h',r')}{r} \cos\beta' \; \delta(x-x') \; \delta(r-r') \; \delta(\gamma-\gamma_0-\psi') \; dS'$$

$$F_t' = \frac{\Delta p(h',r')}{r} \sin\beta' \; \delta(x-x') \; \delta(r-r') \; \delta(\gamma-\gamma_0-\psi') \; dS' \tag{14.6}$$

where

$\Delta p(h',r')$	is the pressure jump across the element or reference surface area $dS' = dh'dr'$;
h'	is the dummy helical coordinate;
$\beta' =$	$\tan^{-1} U/r'\omega$, the pitch angle of the fluid reference surface;
$\gamma_0 =$	ωt, the angle of the blade reference line from the vertical;
x,r,γ	are the axial, radial and tangential coordinates of any field point;
x',r',γ'	are the axial, radial and tangential coordinates of any point on the blade (dummy point);
ψ'	is the angle of the point on the blade measured from the blade reference line $\psi' = \gamma' - \gamma_0$.

Note that in (14.1), the forces acting on the fluid are opposite and equal to those on the blade element

As our prime interest here is in the pressure we proceed to determine a field equation for the pressure by elimination of the velocity components between (14.1) and the equation of continuity (9.7). This is accomplished by applying the operations indicated in (9.7) to the components in (14.1) and summing after interchange of the order of differentiations. This gives

$$\left[\frac{\partial}{\partial t} - U\frac{\partial}{\partial x}\right]\left[\frac{\partial}{\partial x} u_a' + \frac{1}{r}\frac{\partial}{\partial \gamma} u_t' + \frac{\partial}{\partial r} u_r' + \frac{u_r'}{r}\right] + \frac{1}{\rho}\nabla^2 p_\ell'$$
$$= \frac{1}{\rho}\left[\frac{\partial}{\partial x} F_a' + \frac{1}{r}\frac{\partial}{\partial \gamma} F_t'\right] \tag{14.7}$$

As the first term vanishes by (9.7) we are left with the inhomogeneous Laplace or Poisson equation, cf. p. 19

$$\nabla^2 p'_\ell = -\frac{\Delta p(h',r')}{r}\left\{\frac{d}{dx}\,\delta(x-x')\delta(\gamma-\gamma')\cos\beta' - \frac{\delta(x-x')}{r}\frac{d}{d\gamma}\,\delta(\gamma-\gamma')\sin\beta'\right\}\delta(r-r')dS' \tag{14.8}$$

where ∇^2 is the Laplacian operator, viz., in these coordinates

$$\nabla^2 = \frac{\partial^2}{\partial x^2} + \frac{\partial^2}{\partial r^2} + \frac{1}{r}\frac{\partial}{\partial r} + \frac{1}{r^2}\frac{\partial^2}{\partial \gamma^2} \tag{14.9}$$

In a boundless domain the solution of the Poisson equation of the form

$$\nabla^2 p'_\ell = G(x,r,\gamma) \tag{14.10}$$

is, in cylindrical coordinates (analogous to Equation (1.78)),

$$p'_\ell(x,r,\gamma) = -\frac{1}{4\pi}\int_{-\infty}^{\infty} dx''\int_0^{\infty} dr'' \cdot r''\int_0^{2\pi}\frac{G(x'',r'',\gamma'')\,d\gamma''}{\sqrt{(x-x'')^2 + r^2 + r''^2 - 2rr''\cos(\gamma-\gamma'')}} \tag{14.11}$$

i.e., a volume integral of sources of strength G.

With G given by the right-hand side of (14.8) we must integrate the δ-functions, cf. Mathematical Compendium, Section 2. The integration over r'' is straightforward yielding the only contribution when $r'' = r'$. For the other integrations where we have derivatives of δ-functions we integrate by parts, for example over x''

$$\int_{-\infty}^{\infty}\frac{1}{R(x,r,\gamma;x'',r'',\gamma'')}\frac{d}{dx''}\,\delta(x''-x')dx''$$

$$= \left[\frac{1}{R(x,r,\gamma;x'',r'',\gamma'')}\,\delta(x''-x')\right]_{x''=-\infty}^{\infty} - \int_{-\infty}^{\infty}\frac{\partial}{\partial x''}\frac{1}{R}\,\delta(x''-x')dx''$$

$$= 0 - \frac{\partial}{\partial x''}\frac{1}{R}\bigg|_{x''=x'} = \frac{\partial}{\partial x'}\frac{1}{R}\bigg|_{x''=x'} \tag{14.12}$$

Because the pressure difference is over the blade, where $\psi' = x'\omega/U$ by (14.3) and $r'\tan\beta' = U/\omega$, we now have

$$p'_\ell = -\frac{1}{4\pi}\,\Delta p(h',r')dS'\left\{\cos\beta'\,\frac{\partial}{\partial x'} - \frac{\sin\beta'}{r'}\,\frac{\partial}{\partial\psi'}\right\}\frac{1}{R}\Bigg|_{\psi'=\frac{\omega}{U}x'} \tag{14.13}$$

where $R = \sqrt{(x-x')^2 + r^2 + r'^2 - 2rr'\cos(\gamma-\gamma_0-\psi')}$.

Here we recognize axially- and tangentially-directed pressure dipoles which together comprise a dipole of strength $\Delta p dS'$ with axis directed along the normal to the fluid reference helical surface. Observe the analogy between pressure dipoles and "velocity" dipoles, p. 36.

We have here followed the procedure used in airfoil theory (Chapters 4–6) by placing the pressure dipoles on the fluid reference line. In the two-dimensional section theory wherein the singularities are colinear with the reference velocity, the extent and orientation of the singularities remain unchanged when the reference velocity ($-U$) in that case is varied. However, in the case of the propeller the reference velocity is a function of two parameters ($-U,\omega$) and by adherence to the tenets we have for consistency a different geometry for each $J = J(U,\omega)$. Ideally the singularities should be located along the camber line of each radius. However, the consistency requirements of small perturbations theory impose location on the fluid reference surface.

Integration of (14.13) over the area of the blade ($dS' = dh'dr'$) gives the pressure due to loading from a single blade

$$p_\ell(x,r,\gamma) = -\frac{1}{4\pi}\int_{R_h}^{R} dr' \int_{-h_t(r')}^{h_l(r')} \Delta p(h',r')\,\frac{\partial}{\partial n'}\,\frac{1}{R}\,dh' \tag{14.14}$$

where

$$\frac{\partial}{\partial n'} = \cos\beta'\,\frac{\partial}{\partial x'} - \frac{\sin\beta'}{r'}\,\frac{\partial}{\partial\psi'} \tag{14.15}$$

The connection between x' and h' is $x' = h'\sin\beta'$, (14.3) and R_h is the hub radius; R the propeller radius; h_l and h_t are the lengths to the leading and trailing edges from the origin on the blade.

Pressure Due To Blade Thickness

To obtain the linearized pressure due to thickness we can write the same equations as (14.1) with $\Delta p = 0$ (no forces normal to blade due to thickness) but the continuity equation must now reflect the presence of distributed sources.

The strengths of these sources are used in the right side of the continuity equation, as in Equation (1.57), where we express the left side by velocities (cf. (9.7)) instead of the velocity potential to obtain

$$\frac{\partial}{\partial x}u'_a + \frac{1}{r}\frac{\partial}{\partial\gamma}u'_t + \frac{\partial}{\partial r}u'_r + \frac{u'_r}{r}$$

$$= m(x',r')dS'\,\frac{\delta(x-x')}{r}\,\delta(r-r')\,\delta(\gamma-\gamma') \tag{14.16}$$

Inserting this into (14.7) where now $F'_a = F'_t = 0$ since there are no forces normal to the blade gives

$$\nabla^2 p'_\tau = -\rho m(x',r')dS'\left[\frac{\partial}{\partial t} - U\frac{\partial}{\partial x}\right]\frac{\delta(x-x')}{r}\,\delta(r-r')\,\delta(\gamma-\gamma_0-\psi') \tag{14.17}$$

Here we note that $\gamma_0 = \gamma_0(t) = \omega t$ (assuming $\gamma_0 = 0$ at $t = 0$) so that $\partial\delta(\gamma-\gamma_0-\psi')/\partial t = -\omega\partial\delta(\gamma-\gamma_0-\psi')/\partial\gamma$. Then solving (14.17) as previously and making use of the sifting properties of the δ-functions and their derivatives we obtain

$$p'_\tau(x,r,\gamma) = -\frac{2\rho}{4\pi}\frac{d\tau(h',r')}{dh'}V^2\left[\sin\beta'\frac{\partial}{\partial x'} + \frac{\cos\beta'}{r'}\frac{\partial}{\partial\psi'}\right]\frac{1}{R}\,dS' \tag{14.18}$$

where $V = \sqrt{U^2 + (r'\omega)^2}$, the resultant relative velocity, and $m = -2V \cdot d\tau/dh'$ has been used in analogy to Equation (4.14) (τ being the semi-thickness of the blade section). We observe that the pressure generated by thickness is also a distribution of dipoles with axes along the tangent to the fluid reference surface in way of the blade, the strength of the dipoles being $2\rho V^2(r')d\tau/dh'$.

Pressure Due to Loading and Thickness

The sum of the loading- and thickness-generated pressures by a single blade can now be found by adding (14.14) and (14.18) where the integration must be carried out

$$p = p_\ell + p_\tau$$

$$p = -\frac{1}{4\pi}\int_{R_h}^{R}dr'\int_{-h_t}^{h_l}\frac{1}{\sqrt{P_f^{*2} + r'^2}}$$

$$\cdot\left\{\left[2\rho\tau'(h',r')V^2P_f^* + \Delta p(h',r')r'\right]\frac{\partial}{\partial x'}\frac{1}{R}\right.$$

$$\left. + \left[2\rho\tau'(h',r')V^2r' - \Delta p(h',r')P_f^*\right]\frac{1}{r'}\frac{\partial}{\partial\psi'}\frac{1}{R}\right\}dh' \tag{14.19}$$

where $\tau' = d\tau/dh'$ and we have introduced the *fluid pitch* P_f and $P_f^* = P_f/2\pi = U/\omega$ which gives

$$\cos\beta' = \frac{r'}{\sqrt{P_f^{*2} + r'^2}} \quad ; \quad \sin\beta' = \frac{P_f^*}{\sqrt{P_f^{*2} + r'^2}} \tag{14.20}$$

To obtain the total pressure from a Z-bladed propeller we expand $1/R$ in the exponential Fourier series (cf. Mathematical Compendium, Section 5)

$$\frac{1}{R} = \sum_{m=-\infty}^{\infty} A_{|m|}(|x-x'|,r,r')e^{im(\gamma-\psi')}\ e^{-im\gamma_0} \tag{14.21}$$

where $A_{|m|}$ is the propagation-amplitude function having the three well-known possible forms, given by Equations (M5.22)-(M5.24).

The angular location of the ν-th blade is at $\gamma_\nu = \gamma_0 + 2\pi\nu/Z$, γ_0 being the angle of the key blade. The sum of the contributions to the pressure is seen to involve

$$S = \sum_{\nu=0}^{Z-1} e^{-im(\gamma_0+2\pi\nu/Z)} = e^{-im\gamma_0}\sum_{\nu=0}^{Z-1} e^{-i2\pi\nu m/Z} \tag{14.22}$$

From Equation (11.20) we see that only those m contribute such that m = qZ, q = 0, ±1, ±2,.... Hence the total pressure *from a Z-bladed propeller* is of the form

$$p = -\frac{Z}{4\pi}\sum_{q=-\infty}^{\infty}\int_{R_h}^{R} dr'\int_{-h_t}^{h_l}\frac{1}{\sqrt{P_f^{*2}+r'^2}}$$

$$\cdot\left\{\left[2\rho\tau'(h',r')V^2(r')P_f^*+\Delta p(h',r')r'\right]\frac{\partial}{\partial x'}A_{|qZ|}\right.$$

$$\left.-\frac{iqZ}{r'}\left[2\rho\tau'(h',r')V^2(r')r'-\Delta p(h',r')P_f^*\right]A_{|qZ|}\right\}e^{iqZ(\gamma-\gamma_0-\psi')}dh' \tag{14.23}$$

where $\psi' = h'\cos\beta'/r'$ and $x' = h'\sin\beta'$. Note that after differentiation we must replace ψ' by $h'\cos\beta'/r'$ or x'/P_f^* and x' by $h'\sin\beta'$ to place these dummy variables upon the fluid reference surface, cf. (14.3) and (14.4).

The principal component of interest is the pressure at blade frequency given by the sum of the terms q = 1 and q = −1. Then the dimensionless pressure ($K_P = p/\rho N^2D^2$) arising from blade thickness at blade frequency can be expressed as

$$\left[K_{p_\tau}\right]_Z = -\frac{\pi Z}{R^2}\int_{R_h}^{R} dr'\int_{-h_t}^{h_l}\tau'(h',r')\left[W_\tau^c\cos Z(\gamma-\gamma_0) + W_\tau^s\sin Z(\gamma-\gamma_0)\right]dh' \tag{14.24}$$

where the weighting functions are

$$W_\tau^c = \sqrt{P_f^{*2} + r'^2}\left[P_f^* \cos Z\psi' \frac{\partial}{\partial x'} A_Z - Z \sin Z\psi' \, A_Z\right]$$
$$W_\tau^s = \sqrt{P_f^{*2} + r'^2}\left[P_f^* \sin Z\psi' \frac{\partial}{\partial x'} A_Z + Z \cos Z\psi' \, A_Z\right] \qquad (14.25)$$

Similarly the dimensionless pressure from loading at blade frequency is

$$\left[K_{p\ell}\right]_Z = -\frac{Z}{2\pi}\frac{1}{R^2}\int_{R_h}^{R} dr' \int_{-h_t}^{h_l} \frac{\Delta p(h',r')}{\rho N^2 D^2}$$
$$\cdot\left[W_\ell^c \cos Z(\gamma-\gamma_0) + W_\ell^s \sin Z(\gamma-\gamma_0)\right] dh' \qquad (14.26)$$

where

$$W_\ell^c = \frac{R^2}{\sqrt{P_f^{*2} + r'^2}}\left[r' \cos Z\psi' \frac{\partial}{\partial x'} A_Z - Z\, P_f^* \frac{1}{r'} \sin Z\psi' \, A_Z\right]$$
$$W_\ell^s = \frac{R^2}{\sqrt{P_f^{*2} + r'^2}}\left[r' \sin Z\psi' \frac{\partial}{\partial x'} A_Z + Z\, P_f^* \frac{1}{r'} \cos Z\psi' \, A_Z\right] \qquad (14.27)$$

$$\sin Z\psi' = \sin\left[\frac{Zh'}{r'}\cos\beta'\right] \quad ; \quad \cos Z\psi' = \cos\left[\frac{Zh'}{r'}\cos\beta'\right]$$

Note that W_τ^c, W_τ^s, W_ℓ^c and W_ℓ^s are all non-dimensional.

The only form of the amplitude function A_Z which does not introduce an additional integration is represented by Equation (M5.23) of the Mathematical Compendium, p. 518, or

$$A_Z = \frac{1}{\pi\sqrt{rr'}} Q_{Z-1/2}(\mathcal{Z}) \quad ; \quad \mathcal{Z} = \frac{(x-x')^2 + r^2 + r'^2}{2rr'} \qquad (14.28)$$

The function $Q_{Z-1/2}(\mathcal{Z})$ is the associated Legendre function of the second kind of zero order and half-integer degree referred to as a toroidal function since surfaces of $\mathcal{Z}$ = constants are tori (cf. Mathematical Compendium, Section 6). We see that the weighting functions also involve $\partial A_Z/\partial x'$ which is,

$$\frac{\partial}{\partial x'} A_Z = -\frac{x-x'}{\pi(rr')^{3/2}} \frac{Z-\frac{1}{2}}{\mathcal{Z}^2-1}\left[\mathcal{Z} Q_{Z-1/2} - Q_{Z-3/2}\right] \qquad (14.29)$$

It is of interest to examine how these pressures attenuate with axial distance from the propeller for fixed r. Taking $x >> \sqrt{r^2 + r'^2}$ so that $Z \approx x^2/2rr' >> 1$ we use the approximation (M6.5) to obtain

$$\frac{1}{\sqrt{rr'}} Q_{Z-1/2} \approx \text{const.} \frac{(rr')^Z}{x^{2Z+1}}$$

which indicates that the loading and the derivatives of the thickness in the outer radii are heavily weighted and the descent to zero is extremely rapid. The terms involving $\partial Q/\partial x'$ decay even faster as x^{-2Z-2}.

In order to determine the character of the weighting functions for points of interest near the propeller these functions given by (14.25) and (14.27) (with (14.28) and (14.29)) are graphed in Figure 14.3 (for $x = 0$ and $x = \pm R$ on $r = 1.5 R$ for $h' = 0$) as functions of the dummy radius r'. Here again we see that the weighting functions for the field point $x = 0$, $r = 1.5 R$ and blade point $h' = 0$ rise very rapidly emphasizing the thickness and loading at the outer radii. It is also clear that the magnitudes of these functions are greatly reduced at $x = \pm R$, $r = 1.5 R$ from $h' = 0$, reflecting the behaviour of the first term in the asymptotic expansion. For this field point $1.42 \leq Z \leq 5.48$, as r' varies from 0.20 R to R.

Calculation of blade-frequency pressures can be made after selection of the blade-thickness distribution $2\tau(h',r')$ and the pressure jump $\Delta p(h',r')$, which can be specified from our general knowledge of the distribution of loading at, say design J. In general, for a specified propeller and operating advance ratio J, Δp must be found from the inversion of an integral equation arising from the kinematic condition on the key-blade surface. (This will be dealt with in Chapter 18). This task and evaluation of the double integrals in (14.24) and (14.26) can be carried out via computer. But in the era when these formulas were developed, the loading was represented by a lifting line, and the blade outline by sectors and the thickness specified by circular arcs (fitted by parabolas). Surprisingly, these crude representations gave quite good correlation with measurements as displayed together with more recent evaluations shown in the following section.

COMPARISONS WITH MEASUREMENTS

We compare the total blade-frequency calculated pressures with those measured by Denny (1967) at DTRC using a three-bladed propeller in their 24-inch water tunnel. As the pressures were measured on a flat steel plate above the propeller and extending fore and aft of the propeller it is assumed that the effect of this boundary is to double the free-space or boundless-fluid pressures. Hence for this comparison, all calculated amplitudes are multiplied by 2.

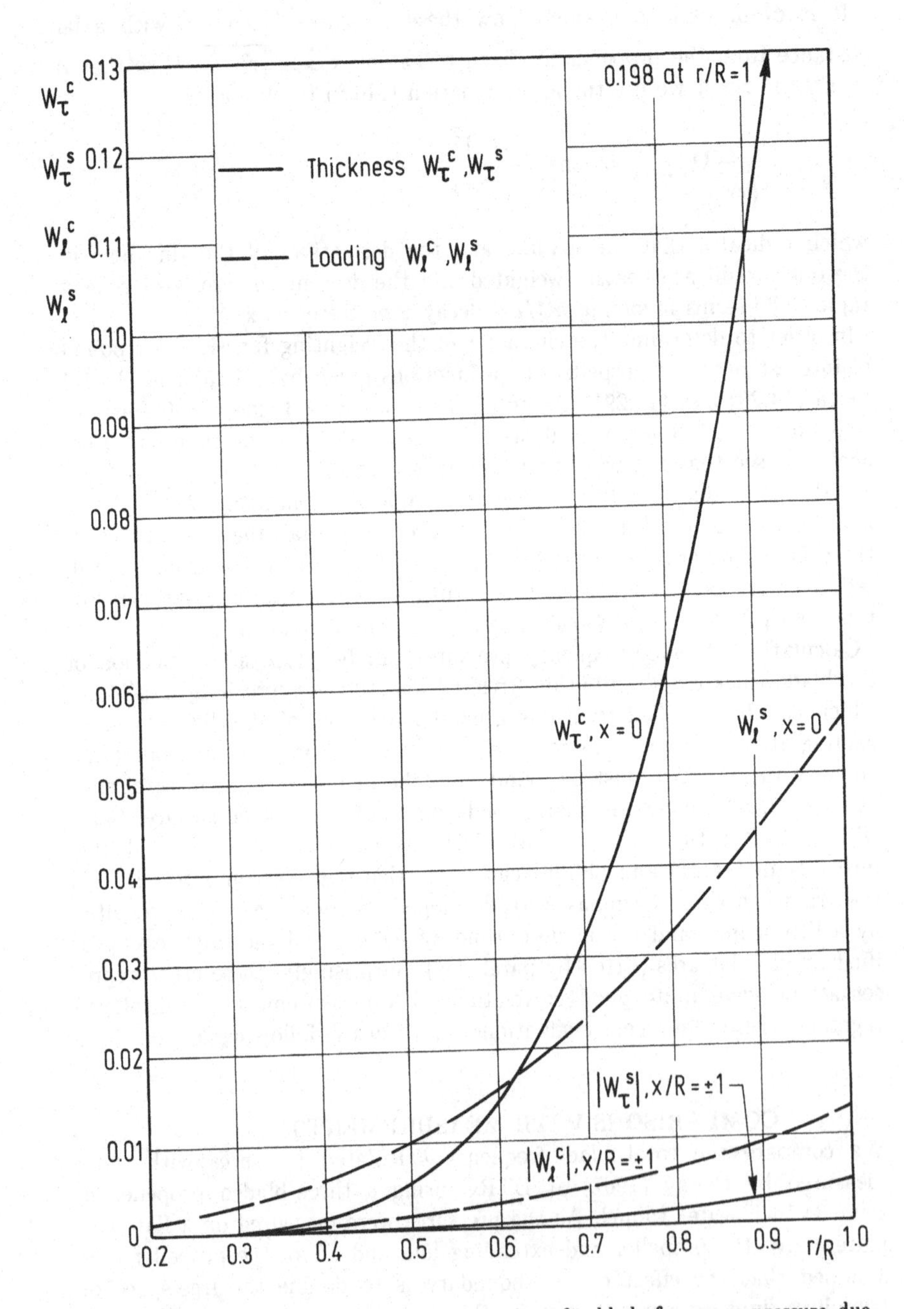

Figure 14.3 Radial variation of weighting functions for blade-frequency pressure due to thickness W_T^c, W_T^s and to loading W_ℓ^c, W_ℓ^s. The blade loading is distributed at h' = 0. Advance ratio J = 1.0, 4 blades.

To comprehend the reflecting effect (ignoring acoustical phenomena) of a rigid, non-porous doubly infinite plane inserted above the propeller at a distance d from the axis we can employ the method of images. Thus if $\phi(x,y,z;\gamma)$ is the potential of the propeller in a boundless fluid, then the non-porous, rigid boundary in $z \equiv d$ can be accounted for by

$$\phi_T = \phi(x,y,z;\gamma) + \phi(x,y,2d-z;\gamma_i) \tag{14.30}$$

We see that the boundary condition

$$\frac{\partial \phi_T}{\partial z} \equiv 0$$

for all (x,y,γ) is satisfied by (14.30) since

$$\frac{\partial \phi_T}{\partial z} = \frac{\partial \phi(x,y,d;\gamma)}{\partial z} - \frac{\partial \phi(x,y,2d-d,\gamma_i)}{\partial z} = 0 \tag{14.31}$$

provided that $\gamma_i = \gamma$ as defined by Figure 14.4 as a mirror image in the plane $z \equiv d$.

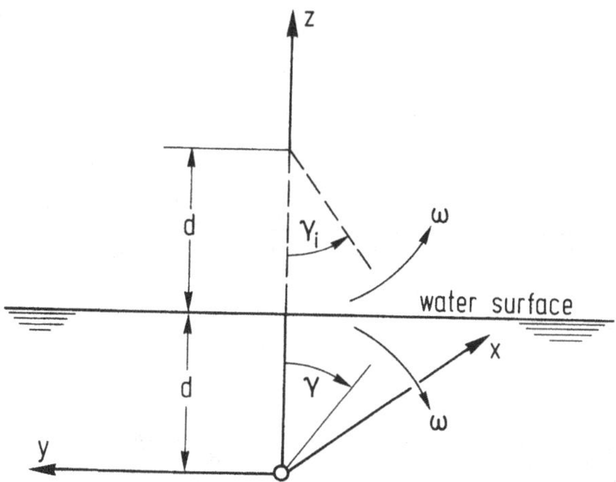

Figure 14.4 Schematic of propeller and image.

From (14.30) we see that on the plane

$$\phi_T(x,y,d;\gamma) = 2\phi(x,y,d;\gamma) \tag{14.32}$$

and that

$$\left.\frac{\partial \phi_T}{\partial x}\right|_{z=d} = 2\,\frac{\partial \phi(x,y,d;\gamma)}{\partial x} \tag{14.33}$$

Hence the linearized pressures on the plane are double the values obtained on the same locus in a boundless fluid.

We may note, in passing, that the transverse induced velocity is also doubled so that the exact pressure including quadratic terms is not double the exact pressure in a boundless fluid on $z \equiv d$. However, the quadratic terms can give rise only to mean pressures (of no interest) and pressures of twice blade frequency which are negligible.

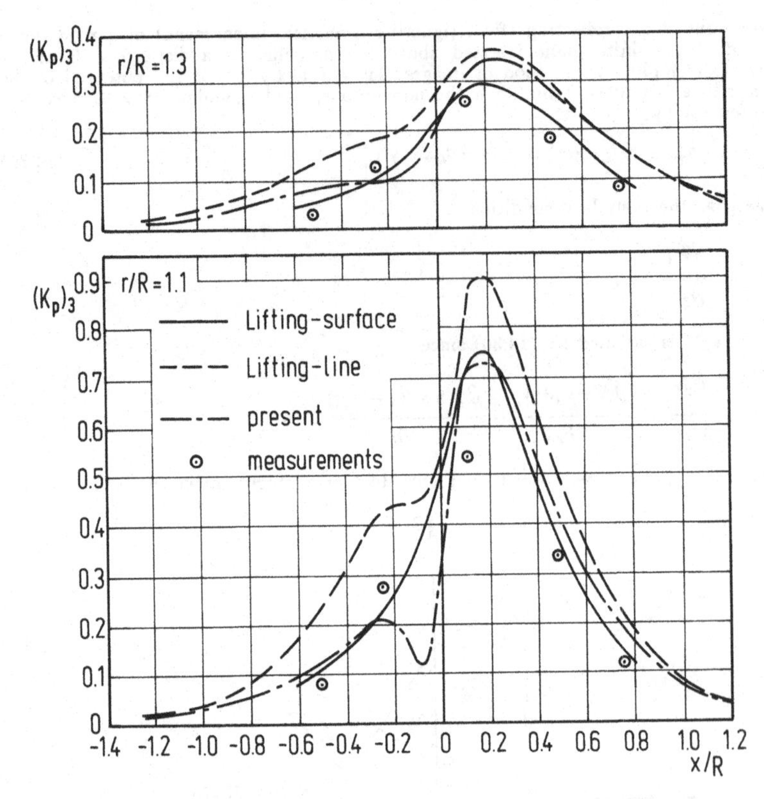

Figure 14.5 Total blade-frequency pressure-amplitude variation with axial distance as determined from experiment and theories for points on a flat plate above propeller.

Figure based on Figure 14 from Breslin, J.P. (1970). Theoretical and experimental techniques for practical estimation of propeller-induced vibratory forces. *Trans. SNAME*, vol. 78, pp. 23–40. New York, N.Y.: SNAME.

Computer evaluations for the case of a three-bladed propeller are compared with pressures measured by Denny (1967) in the centerplane $\gamma = 0$ at various x/R for r/R = 1.10 and 1.30 as shown in Figure 14.5 together with results of earlier evaluations of theory by Breslin (1959), (lifting-line theory plus thin-body thickness) and by Jacobs, Mercier & Tsakonas (1972), (lifting-surface theory plus thickness). The early calculations made by Breslin used very approximate evaluations of the integrals involved in each component by assuming an effective dummy radius for evaluating the Legendre function and by assuming the circulation density to be distributed elliptically. The thickness was taken as distributed in pie-shaped sections in the plane x = 0. In the present evaluations the circula-

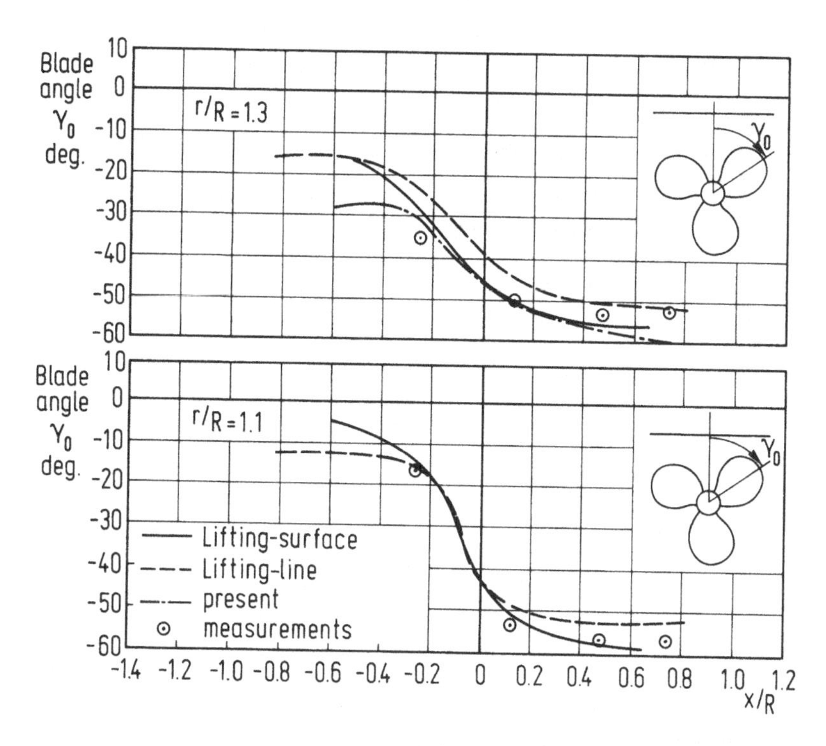

Figure 14.6 Blade position angles at maximum blade–frequency total pressures as determined from experiment and theories for points on flat plate above propeller.

Figure based on Figure 15 from Breslin, J.P. (1970). Theoretical and experimental techniques for practical estimation of propeller–induced vibratory forces. *Trans. SNAME*, vol. 78, pp. 23–40. New York, N.Y.: SNAME.

tion density was calculated using the optimum criterion of Betz and the thickness distributed over the helical skeleton of the blades. Thus these new calculations show better agreement with both data and the lifting-surface results of Jacobs *et al.* (*ibid.*) as should be expected.

It is important to note that the maximum pressures occur just forward of the propeller because of the constructive combination of loading- and thickness-generated components whereas abaft of the propeller these constituents destructively interfere with each other to produce a rapid diminution of the amplitudes.

The variation of phase angle is depicted in Figure 14.6. The present results are in agreement with those ascertained at DTRC by Denny (1967).

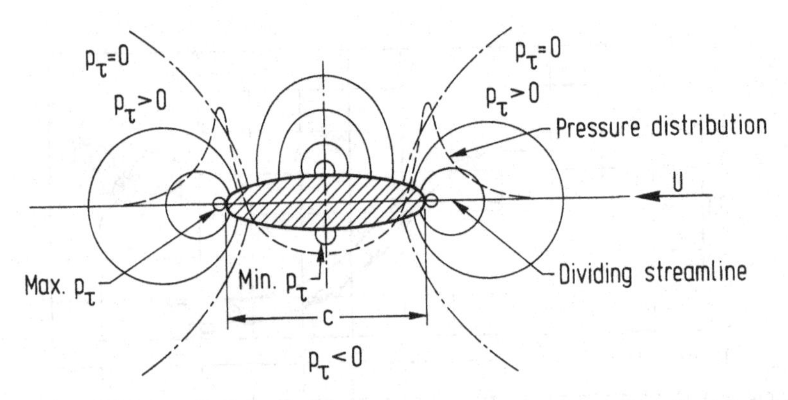

Theoretical pressure pattern about a typical symmetrical section placed in a uniform stream.

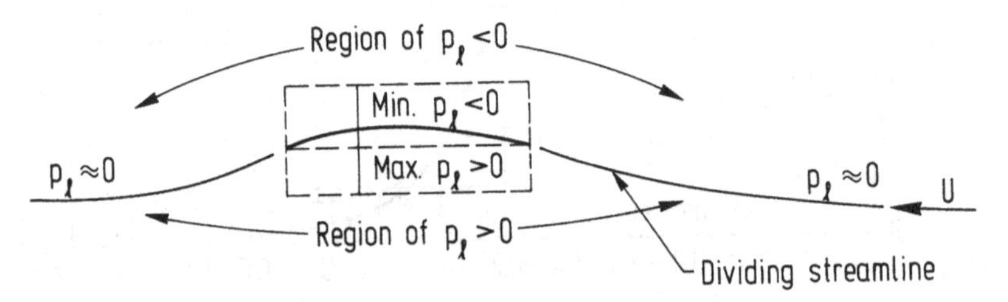

Broad aspects of the idealized theoretical pressure field about a very thin cambered section at optimum (designed) attitude.

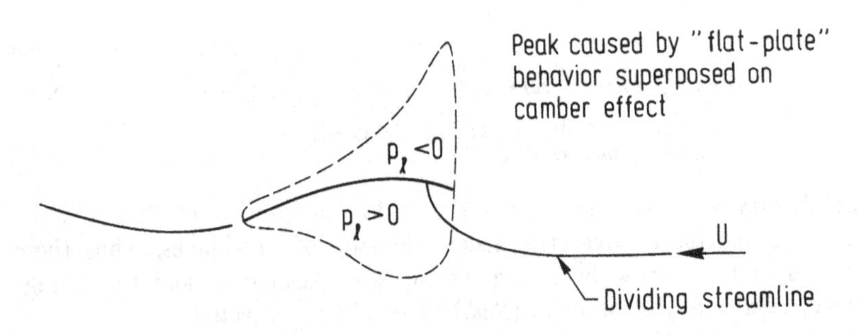

Typical pressure distribution on a very thin cambered section at an angle greater than the design or ideal angle of attack.

Figure 14.7 Regions of positive and negative pressure change due to thickness, camber and angle of attack.

Figures 1 and 2 from Breslin, J.P. & Tsakonas, S. (1959). Marine propeller pressure field due to loading and thickness effects. *Trans. SNAME*, vol. 67, pp. 386–422. New York, N.Y.: SNAME.

The region of blade angles at which the pressure is a positive maximum can be envisioned by applying our general knowledge of the regions of positive and negative pressure change around sections as shown schematically in Figure 14.7.

Using these sketches we can envision that as a right-handed propeller blade approaches 12 o'clock (as viewed from aft, say) a pressure gage at $\gamma = 0$ (center plane) will sense the positive pressures "preceding" the blade. Hence the angle of the key blade at which the pressure is a positive maximum is negative in our notation, i.e., the blade angle is to the left of 12 o'clock (as viewed from aft). The most negative signature ensues when the blade has passed 12 o'clock. This is also clear from Figure 14.6.

The determination of phase angle requires care and one should not take most computer-generated angles which evaluate the inverse tangent only in the interval -90° to 90°. The proper definition of phase angle α is determined as follows.

In general

$$(p)_Z = a_z \cos Z\gamma + b_z \sin Z\gamma \tag{14.34}$$

$$= \sqrt{a_z^2 + b_z^2} \cos (Z\gamma - \alpha) \tag{14.35}$$

where

$$\alpha = \tan^{-1}\left[\frac{b_z}{a_z}\right] \tag{14.36}$$

In determining the angle α the signs of both b_z and a_z must be accounted for as they give the location of the vector in the complete Argand plane. Thus for example, when b_z and a_z are both negative the vector is in the third quadrant and α is the angle as shown in Figure 14.8 so that

$$\alpha = 180^\circ + \tan^{-1}\left|\frac{b_z}{a_z}\right|$$

where the last term is that given by a determination of the arc tangent limited to the branch in -90° to 90°.

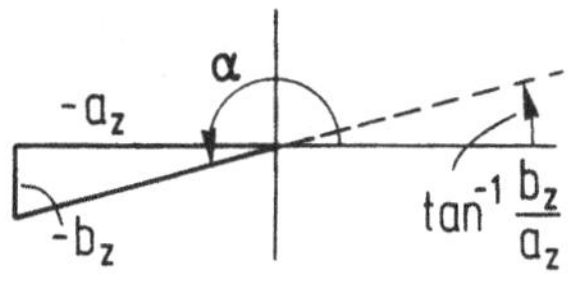

The blade position angle at which the cosine function in (14.35) first achieves plus unity is given by

Figure 14.8 Phase angle when $a_z < 0$ and $b_z < 0$.

$$Z\gamma - \alpha = 0 \quad \text{or} \quad \gamma = \frac{\alpha}{Z} \tag{14.37}$$

Thus for the foregoing example, the blade position angle (say for Z = 3 blades) is

$$\gamma = \frac{1}{3}\left[180^\circ + \tan^{-1}\left[\frac{b_Z}{a_Z}\right]\right]$$

For a propeller in uniform flow we have seen that the dependence on angle is always of the form $\gamma - \gamma_0$ so that

$$(p)_Z = \sqrt{a_Z^2 + b_Z^2}\cos\left(Z(\gamma-\gamma_0)-\alpha\right) \tag{14.38}$$

So the blade position for maximum pressure is given by

$$\gamma_0 = \gamma - \frac{\alpha}{Z} \tag{14.39}$$

In our correlation $\gamma = 0$, and α is determined using the entire Argand diagram.

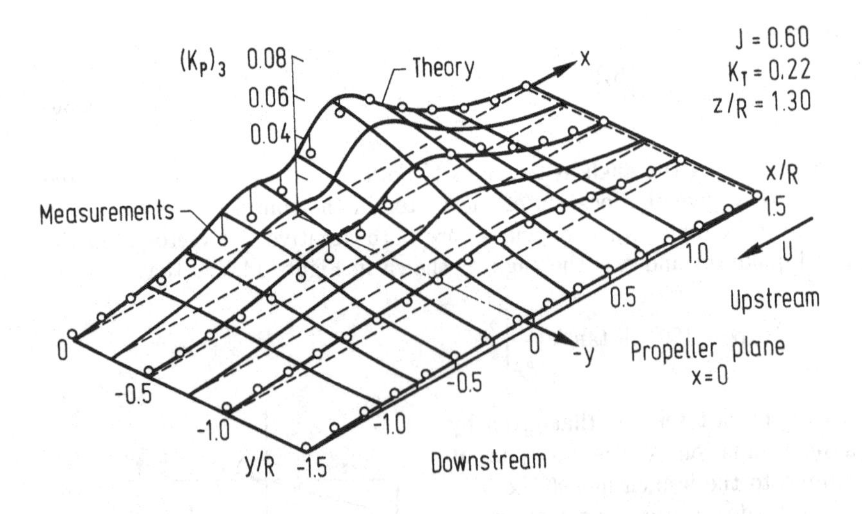

Figure 14.9 Comparison of blade frequency pressure amplitudes on a flat plate with lifting-line plus thin-blade theory.

Figure based on Figure 16 from Breslin, J.P. (1970). Theoretical and experimental techniques for practical estimation of propeller-induced vibratory forces. *Trans. SNAME*, vol. 78, pp. 23–40. New York, N.Y.: SNAME.

We conclude our look at correlation with measurement by reference to Figure 14.9 which shows the "half-carpet" of pressure amplitudes measured in Tank III of Davidson Laboratory by running a three-bladed propeller past a submerged table with flush mounted pressure gages. The theoretical values were computed by Breslin (Breslin & Kowalski (1964)) using the approximations alluded to above. We see that the correlation is very good except right abaft the propeller in the center plane and that the "signatures" become quite negligible beyond $|x|/R > 1.5$ and $|y|/R$ somewhat larger.

15 Pressure Fields Generated by Blade Loadings in Hull Wakes

The pressure fluctuations generated by propellers in the wake of hulls are markedly different from those produced in uniform inflow. The flow in the propeller plane abaft a hull varies spatially as well as temporally. Here we deal only with the effects attending spatial variations peripherally and radially as provided by wake surveys which give the averaged-over-time velocity components as a function of r and γ for a fixed axial location. Temporal variations in the components are aperiodic and cannot be addressed until sufficient measurements have been made to determine their frequency spectra. Ultimately, numerical solutions of the Navier-Stokes equations may provide both spatial and temporal aspects of hull wakes.

Here the spatial variations in the axial and tangential components are reflected in the pressure jump Δp which is taken to vary harmonically with blade position angle γ_0. Then we discover a coupling between the harmonics of $\Delta p(\gamma_0)$ and the harmonics of the propagation function yielding a plethora of terms all at integer multiples of blade frequency. Graphical results are given for pressure and velocity fields showing the effect of spatial non-uniformity of the inflow.

We have seen in the previous chapter that the pressure field arising from a lifting-surface model of a propeller in a uniform flow is that due to pressure and velocity dipoles distributed over the blade. Both dipole strengths were constant in time since we considered uniform and stationary inflow. To account for varying blade loading when the propeller works in the wake abaft a hull we must assume a temporal variation in the strength of the pressure dipoles. (Had we wished to include effects of cavitation it would have been necessary to use sources to describe thickness and the varying cavity volumes as described in Chapter 8). Now that the inflow is stationary but non-uniform we must let the pressure vary in terms of the blade position angle γ_0, or

$$\Delta p(h',r';\gamma_0) = \sum_{n=0}^{\infty} (a_n \cos n\gamma_0 + b_n \sin n\gamma_0)$$

This can be written in the complex exponential form (by expressing the trigonometric functions in terms of $e^{\pm in\gamma_0}$)

$$\Delta p(h',r';\gamma_0) = \sum_{n=-\infty}^{\infty} \Delta p_n(h',r')e^{in\gamma_0} \tag{15.1}$$

where

$$\Delta p_n = \frac{a_n - ib_n}{2} \quad ; \quad \Delta p_{-n} = \frac{a_n + ib_n}{2} \quad ; \quad \Delta p_0 = \frac{a_0}{2}$$

It is important to realize that (15.1) is purely real and that $2\Delta p_n$ is the complex conjugate of the pressure jump arising from the n-th harmonic of the spatially distributed wake components resolved normal to the key blade.

The reason why we consider the pressure jumps to be complex is that when the loading is unsteady, components arise which are in-phase and also those which are in quadrature with the incident flow angle of attack. Thus when a section encounters an angle of attack varying as $\cos n\gamma_0$ a loading is generated which in part varies as $\cos n\gamma_0$ and also in part varies as $\sin n\gamma_0$ as will become evident later when we examine the unsteady lift of sections beset by transverse, traveling gusts.

The total field pressures from loading are then, as a generalization of (14.14), given by the operation

$$p_\ell(x,r,\gamma,\gamma_0) = -\frac{1}{4\pi}\int_{R_h}^{R} dr' \int_{-h_t}^{h_l} \sum_{n=-\infty}^{\infty} \Delta p_n \, e^{in\gamma_0}$$

$$\cdot \frac{\partial}{\partial n'} \sum_{m=-\infty}^{\infty} \frac{1}{\pi\sqrt{rr'}} Q_{m-\frac{1}{2}}(Z)\, e^{im(\gamma-\gamma_0-\psi')} dh' \tag{15.2}$$

where the m-summation, as before, is a Fourier expansion of $1/R$, cf. (14.21) and (14.28) and the operator $\partial/\partial n'$ is given by (14.15).

We now may note a very significant coupling in (15.2) (first observed by the senior author, circa 1961[24]) when we seek to sum the effect of all such blades. We see that the dependence on the blade position angle γ_0 is

$$e^{i(n-m)\gamma_0}$$

and replacing γ_0 by $\gamma_0 + 2\pi\nu/Z$ and summing over ν from zero to $Z - 1$ we have a sum similar to that of Equation (11.20). Hence only those n and m contribute such that $n - m = qZ$; $q = 0, \pm 1, \pm 2, \ldots$. Thus we can eliminate m in favor of n and qZ and express the total pressure as

[24] This coupling was also exhibited by Bavin, Vashkevich & Miniovich (1968) who also showed comparisons with experimental data.

$$p_\ell = -\frac{Z}{4\pi^2}\sum_{q=-\infty}^{\infty}\sum_{n=-\infty}^{\infty}\int_{R_h}^{R} dr' \int_{-h_t}^{h_l} \Delta p_n$$

$$\cdot \frac{\partial}{\partial n'}\left\{\frac{1}{\sqrt{rr'}} Q_{|n-qZ|-\frac{1}{2}}(\mathcal{Z})\, e^{i(n-qZ)(\gamma-\psi')}\right\} e^{iqZ\gamma_0} dh' \quad (15.3)$$

Here we see that wake-generated loadings of harmonic order n are radiated to field points by propagation-function amplitudes of harmonic orders $n - qZ$. Thus, as for the propeller in uniform inflow corresponding to $n = 0$, these loadings are propagated by the Legendre function of degree $|qZ|-\frac{1}{2}$ and because the degree number is large they decay rapidly with distance from the propeller. However, the loadings arising from wake frequencies $n = \pm qZ$ will give rise to pressures which fall away slowly as well as terms which fall very rapidly.

To see this we must first carry out the indicated differentiation (using (14.15) and (14.20)). Then

$$p_\ell = -\frac{Z}{4\pi^2}\sum_{q=-\infty}^{\infty}\sum_{n=-\infty}^{\infty}\int_{R_h}^{R} dr' \int_{-h_t}^{h_l} \frac{\Delta p_n}{\sqrt{P_f^{*2}+r'^2}}$$

$$\cdot \frac{1}{\sqrt{rr'}}\left[r' \frac{\partial Q_{|n-qZ|-\frac{1}{2}}(\mathcal{Z})}{\partial x'} + i\,\frac{P_f^*(n-qZ)}{r'}\, Q_{|n-qZ|-\frac{1}{2}}(\mathcal{Z})\right]$$

$$\cdot e^{i(n-qZ)(\gamma-\psi')}\, e^{iqZ\gamma_0} dh' \quad (15.4)$$

For $n = qZ > 0$ we see that the propagation function is $\partial Q_{-\frac{1}{2}}(\mathcal{Z})/\partial x'$ which is like a simple dipole. For $n = -qZ$ the degree of the Legendre function and its derivative is $|2qZ|-\frac{1}{2}$ so they vanish rapidly for $q = 1, 2,$ .... In practice the loadings corresponding to wake orders in excess of $|n| = 7$ or 8 are negligible. We may therefore limit our attention to the terms given by $q = 1, -1$.

It is important to observe that this coupling between the harmonics of the loading and the propagation functions has a distinct bearing on the forces generated on the hull. Thus the very small blade loadings arising from the harmonics of orders $n = Z-1, Z, Z+1$ of the wake will contribute greatly to the vibratory force since they decay slowly with distance and because the area of the hull increases rapidly with distance forward of the propeller.

Thus the blade frequency pressures due to loadings in a wake are (taking only the terms $q = -1$ and 1)

$$(p_\ell)_Z = -\frac{Z}{4\pi^2}\sum_{n=-\infty}^{\infty}\int_{R_h}^{R} dr' \int_{-h_t}^{h_l} \frac{\Delta p_n}{\sqrt{P_f^{*2} + r'^2}} \frac{1}{\sqrt{rr'}}$$

$$\cdot \left\{ r' \left[\frac{\partial}{\partial x'} Q_{|n-Z|-\frac{1}{2}} \; e^{i[(n-Z)(\gamma-\psi')+Z\gamma_0]} \right.\right.$$

$$\left. + \frac{\partial}{\partial x'} Q_{|n+Z|-\frac{1}{2}} \; e^{i[(n+Z)(\gamma-\psi')-Z\gamma_0]} \right]$$

$$+ \frac{iP_f^*}{r'} \left[(n-Z) Q_{|n-Z|-\frac{1}{2}} \; e^{i[(n-Z)(\gamma-\psi')+Z\gamma_0]} \right.$$

$$\left.\left. + (n+Z) Q_{|n+Z|-\frac{1}{2}} \; e^{i[(n+Z)(\gamma-\psi')-Z\gamma_0]} \right] \right\} dh' \tag{15.5}$$

To simplify, let us fold the n-summation, taking the term $n = 0$ separately. We obtain

$$(p_\ell)_Z = -\frac{Z}{4\pi^2}\int_{R_h}^{R} dr' \frac{1}{\sqrt{P_f^{*2}+r'^2}} \frac{1}{\sqrt{rr'}}$$

$$\cdot \int_{-h_t}^{h_l} \left\{ 2\Delta p_0 \left[\frac{\partial}{\partial x'} Q_{Z-\frac{1}{2}} \cos Z(\gamma-\gamma_0-\psi') \right.\right.$$

$$\left.\left. - \frac{zP_f^*}{r'} Q_{Z-\frac{1}{2}} \sin Z(\gamma-\gamma_0-\psi') \right] + \sum_{n=1}^{\infty} [\ldots] \right\} dh' \tag{15.6}$$

where the terms in the summation are

$$\Delta p_n \left[r' \frac{\partial}{\partial x'} Q_{|n-Z|-\frac{1}{2}} + \frac{iP_f^*}{r'} (n-Z) Q_{|n-Z|-\frac{1}{2}} \right] e^{i(n-Z)(\gamma-\psi')} \; e^{iZ\gamma_0} \tag{a}$$

$$\Delta p_n \left[r' \frac{\partial}{\partial x'} Q_{|n+Z|-\frac{1}{2}} + \frac{iP_f^*}{r'} (n+Z) Q_{|n+Z|-\frac{1}{2}} \right] e^{i(n+Z)(\gamma-\psi')} \; e^{-iZ\gamma_0} \tag{b}$$

$$\Delta p_{-n} \left[r' \frac{\partial}{\partial x'} Q_{|n+Z|-\frac{1}{2}} - \frac{iP_f^*}{r'} (n+Z) Q_{|n+Z|-\frac{1}{2}} \right] e^{-i(n+Z)(\gamma-\psi')} \; e^{iZ\gamma_0} \tag{c}$$

$$\Delta p_{-n} \left[r' \frac{\partial}{\partial x'} Q_{|n-Z|-\frac{1}{2}} - \frac{iP_f^*}{r'} (n-Z) Q_{|n-Z|-\frac{1}{2}} \right] e^{-i(n-Z)(\gamma-\psi')} \; e^{-iZ\gamma_0} \tag{d}$$

The lines (c) and (d) come from the negative n. We see by inspection that line (d) is the complex conjugate of line (a) and line (c) is the complex conjugate of line (b) (i.e., wherever there is a –i in (d) and (c) there is a +i in (a) and (b)). As the sum of complex conjugates is twice the real part of either we can compactly write the result as

$$
\begin{aligned}
(p_\ell)_Z = -\frac{Z}{2\pi^2}\int_{R_h}^{R} dr' \frac{1}{\sqrt{P_f^{*2}+r'^2}}\frac{1}{\sqrt{rr'}} \\
\cdot \int_{-h_t}^{h_l}\Bigg\{\Delta p_0\left[r'\frac{\partial}{\partial x'}Q_{Z-\frac{1}{2}}\cos Z(\gamma-\gamma_0-\psi') - Z\,Q_{Z-\frac{1}{2}}\sin Z(\gamma-\gamma_0-\psi')\right] \\
+ Re\sum_{n=1}^{\infty}\Delta p_n\Bigg[\left[r'\frac{\partial}{\partial x'}Q_{|n-Z|-\frac{1}{2}} + \frac{iP_f^*}{r'}(n-Z)Q_{|n-Z|-\frac{1}{2}}\right] \\
\cdot e^{i[(n-Z)(\gamma-\psi')+Z\gamma_0]} \\
+\left[r'\frac{\partial}{\partial x'}Q_{|n+Z|-\frac{1}{2}} + \frac{iP_f^*}{r'}(n+Z)Q_{|n+Z|-\frac{1}{2}}\right] \\
\cdot e^{i[(n+Z)(\gamma-\psi')-Z\gamma_0]}\Bigg]\Bigg\}dh'
\end{aligned}
\qquad (15.7)
$$

where *Re* indicates that only the real part is to be retained.

The contributions from the terms involving the orders n + Z will vanish rapidly with distance from the propeller. The dominant terms at distance are those for which n = Z–1, Z and Z+1 stemming from terms involving only $|n-Z|$. Designating the contribution from the n-th harmonic of the wake by $(p_\ell)_Z^{(n)}$ we have for n = Z

$$
\begin{aligned}
(p_\ell)_Z^{(Z)} = -\frac{Z}{2\pi^2}Re\Bigg\{\int_{R_h}^{R} dr' \frac{1}{\sqrt{P_f^{*2}+r'^2}}\frac{1}{\sqrt{rr'}}\int_{-h_t}^{h_l}\Delta p_Z\Bigg[\left[r'\frac{\partial}{\partial x'}Q_{-\frac{1}{2}}\right]e^{-iZ\gamma_0} \\
+ \text{ terms involving orders } 2Z\Bigg]dh'\Bigg\}
\end{aligned}
\qquad (15.8)
$$

We see that for n = Z there is no dependence on the space angle γ and the dummy angle ψ'. Thus the propagation is axially symmetric (ignoring the higher order (2Z) terms). We can now show that this term behaves like an x-directed dipole.

For the argument Ƶ large $Q_{-\frac{1}{2}}$ has the asymptotic approximation given by (M6.6) of the Mathematical Compendium, p. 520 so for $x^2 + r^2 >> R^2$ (the maximum value of r') and as then x >> maximum of x', the asymptotic behavior of (15.8) is

$$(p_\ell)_Z^{(Z)} \to -\frac{Z}{2\pi}\, Re\left\{\left[\int_{R_h}^{R} dr'\frac{r'}{\sqrt{P_f^{*2}+r'^2}}\int_{-h_t}^{h_l}\Delta p_Z(h',r')dh' e^{iZ\gamma_0}\right\}\frac{x}{[x^2+r^2]^{3/2}} \tag{15.9}$$

This is seen to be an x-directed pressure dipole located at $x = 0$ and $r = 0$ (compare with the expression for a normal dipole (1.67)). Its strength is weak because the pressure jumps at blade frequency Δp_Z are small, being the order $\Delta p_0/50$, Δp_0 being the mean pressure jump. *However, because of its independence of the space angle γ and its relatively small rate of attenuation, i.e., like $1/x^2$, Δp_Z can contribute to the vertical force on the ship an amount comparable to the mean loading!* The pressures $(p_\ell)_Z^{(0)}$ decay like $1/x^{2Z+2}$ and $1/x^{2Z+1}$ and in addition have the angular variations as $\cos Z\gamma$ and $\sin Z\gamma$. Because of lateral symmetry (for single screws) $\sin Z\gamma$ cancel and $\cos Z\gamma$ partially annul when integrated over the hull.

In addition, the weighting functions $\cos Z\psi'$ and $\sin Z\psi'$ cause partial annulling the integral of Δp_0 (in the first term, $n = 0$, of (15.7)) for large Z. Thus the components arising from the zero harmonic of the wake, $(p_\ell)_Z^{(0)}$ are "poor propagators" of the Δp_0, affecting only the local regions about the propeller.

The foregoing behavior cannot be "pushed" too far because the presence of the water surface "quenches" these pressures at distant points. To satisfy the free-surface condition for an incompressible fluid at high frequencies, perturbation pressures must vanish there. A proof of this is given in Appendix F.

To secure this condition on the water surface a "negative" image of the propeller is employed. To illustrate this only for our asymptotic expression we replace the axial dipole function in (15.9) by

$$x\left[\frac{1}{[x^2+y^2+z^2]^{3/2}}-\frac{1}{[x^2+y^2+(2d-z)^2]^{3/2}}\right]$$

where d is the depth of the propeller axis below the free water surface. We see that this function vanishes identically for all x, y along $z = d$.

Now for y and z bounded (as they are along the hull) expanding by the binomial theorem we see that the dipole plus its negative image for large x behave like

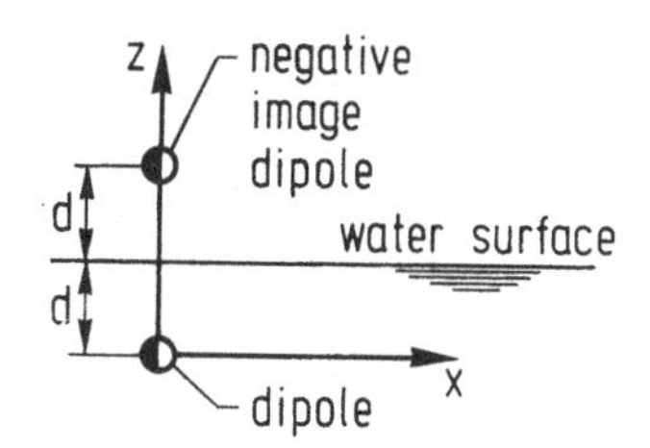

Figure 15.1 Dipole and negative image.

$$-\frac{3}{2}\frac{4d(d-z)x}{(x^2 + r^2)^{5/2}}$$

and vanish as

$$-\frac{6d(d-z)}{x^4}$$

Thus the effect of the water surface is to quench the pressures as $1/x^4$ as compared with $1/x^2$ in a boundless fluid.

It is interesting to find the value of x relative to r, d and z at which this behavior sets in. If we take as a criterion that the second term in the binomial expansion must be less than 1/10, we have upon considering all lengths to be in propeller radii, (indicated by ˇ)

$$\frac{6\check{d}(\check{d}-\check{z})\check{x}}{(\check{x}^2 + \check{r}^2)^{5/2}} < \frac{1}{10}$$

or

$$\frac{(\check{x}^2 + \check{r}^2)^{5/2}}{\check{x}} = \check{x}^4\left[1 + \frac{5}{2}\left[\frac{\check{r}}{\check{x}}\right]^2 + ...\right] < 60\check{d}(\check{d}-\check{z})$$

Then for $\check{x} > 5\check{r}$ (same criterion)

$$\check{x} > (60\check{d}(\check{d}-\check{z}))^{1/4} \tag{15.10}$$

For large ships $\check{d} \approx 3$ and for $\check{z} \approx 1$

$$x > 4.36 \text{ (radii)} \tag{15.11}$$

It is also interesting to note that propellers in wakes produce pressures and also induced velocities that, at frequency order $q = 0$, vary spatially as may be seen from (15.4) by taking only the term $q = 0$. These terms have the appearance

$$(p_\ell)_0 = -\frac{Z}{4\pi^2} Re\left\{\sum_{n=-\infty}^{\infty}\int_{R_h}^{R} dr'\int_{-h_t}^{h_l}\frac{\Delta p_n}{\sqrt{P_f^{*2}+r'^2}}\left[r'\frac{\partial}{\partial x'} + in\frac{P_f^*}{r'}\right]\right.$$
$$\left.\cdot\frac{1}{\sqrt{rr'}}Q_{|n|-\frac{1}{2}}e^{in(\gamma-\psi')}dh'\right\} \tag{15.12}$$

Hence, the propeller impresses, as it were, a set of steady imprints at field points, invariant with blade position angle or time which vary with the n-th harmonics of the field point angle γ.

Thus if one measures the flow components in the neighborhood of a propeller operating in the ship wake each spatial harmonic will contain contributions from the unsteady loading of the propeller, one from each wake harmonic, which should be subtracted from the measured effective wake to secure the modification of the nominal wake sans the propeller inductions. Of course, the Δp_n from the wake are themselves "creatures" of the effective wake which differs from the nominal wake because of the alteration of the local boundary layer caused by the propeller mean pressure gradients and the coupling of the induced velocities with radial and tangential shears. Our ignorance of the harmonics of the effective wake hampers our ability to calculate blade- and shaft-loading and near-field pressures since all we have available are the harmonics of the model nominal wake. Hopefully laser measurements with operating propellers will improve our knowledge.

A few correlations with measured blade-frequency pressures about propellers in screen-generated wakes have been made. The comparisons with computed value show good to poor agreement. Comparative calculations of pressures reported to the ITTC (1990b) show lack of agreement amongst several towing tanks presumably due to difference in computing Δp. No one, including the authors, has included the possible contributions from blade thickness arising from the variations in the resultant velocity with blade angular position because this mechanism has been thought to be weak. Alteration of the harmonics of the nominal wake due to shear has not as yet been undertaken.

A global property of the pressure field about a propeller in a wake in contrast to that for uniform flow can be observed by considering the amplitude or modulus of the sum of the thickness and loading pressures in uniform flow and in a wake. For uniform flow we can see from summing (14.24) (from thickness) with (14.26) (from loading) that the amplitude of this sum (adding coefficients of cos $Z(\gamma-\gamma_0)$ and sin $Z(\gamma-\gamma_0)$ in-phase) can be expressed in the form

$$\left[K_{p_T}\right]_{Z\ \text{uniform}} = \sqrt{a_Z^2(x,r) + b_Z^2(x,r)} \tag{15.13}$$

which says that for fixed x *the loci of constant pressure amplitudes in uniform flow are circles with centers on the axis of rotation* $(r = 0)$.

In contrast to this, the sum of (14.24) and (15.7) (from loading in wakes) when both are expanded to give coefficients of cos $Z\gamma_0$ and sin $Z\gamma_0$ and in-phase components summed, yields an amplitude variation of the form

$$\left[K_{p_T}\right]_{Z\ \mathrm{wake}} = \sqrt{\left[\sum_{n=0}^{\infty} A_n(x,r,\gamma)\right]^2 + \left[\sum_{n=0}^{\infty} B_n(x,r,\gamma)\right]^2} \qquad (15.14)$$

Hence, for fixed x and r the pressure amplitudes vary with angular position γ, i.e.,

$$\left[K_{p_T}\right]_{Z\ \mathrm{wake}} = f_Z(\gamma) \quad ; \quad x, r \ \text{fixed}$$

Examples of the variation in velocity component amplitudes which behave in a similar way to the pressure are shown in Figure 15.2 for a three-bladed propeller in a three-cycle screen wake. Here we see the effect of linear combinations of blade-frequency loadings and mean loadings and thickness giving a three-cycle pattern. However, in a ship wake the combination of many frequencies yields the pattern shown in Figure 15.3. These curves are the result of evaluations of a lifting-surface theory by Jacobs & Tsakonas (1975).

Observations of the effect on pressures due to non-uniformity of the inflow as measured and calculated in the USSR are presented from a paper given by Bavin, Vashkevich & Miniovich (1968) in Figure 15.4 and 15.5. These graphs reveal that the effect of non-uniformity of inflow in the near-field pressure is large and also that lifting-line theory shows poor agreement in comparison to lifting-surface theory relative to measurements.

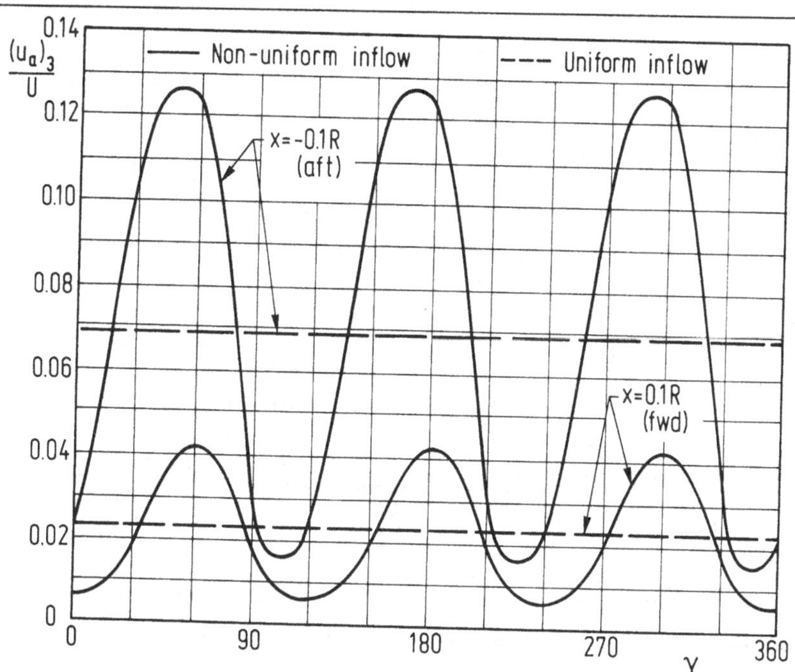

Figure 15.2 Blade–frequency axial–velocity–component variation with angular location γ, due to loading and thickness of 3–bladed DTRC propeller model 4118 in uniform inflow and non–uniform inflow of a 3–cycle screen wake (r = 1.1 radius).

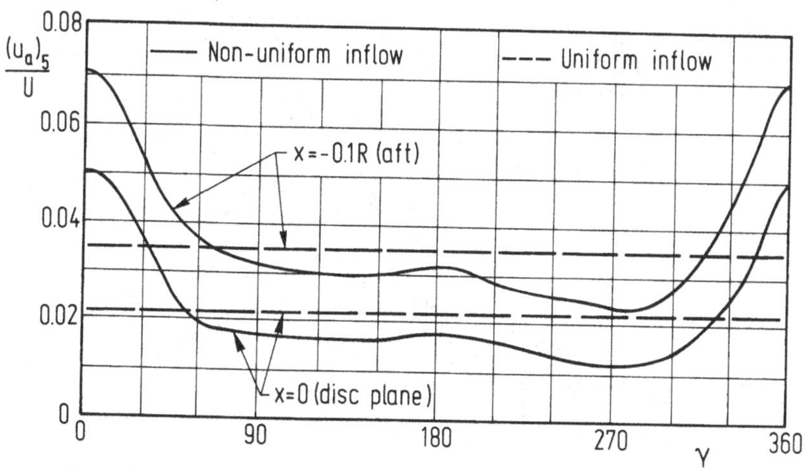

Figure 15.3 Blade–frequency axial–velocity–component variation with angular location γ, due to loading and thickness of 5–bladed propeller in uniform and non–uniform inflows (ship wake) (r = 1.1 radius).

Figure 15.2 and 15.3 from Jacobs, W.R. & Tsakonas, S. (1975). Propeller–induced velocity field due to thickness and loading effects. *Journal of Ship Research*, vol. 19, no. 1, 44–56.

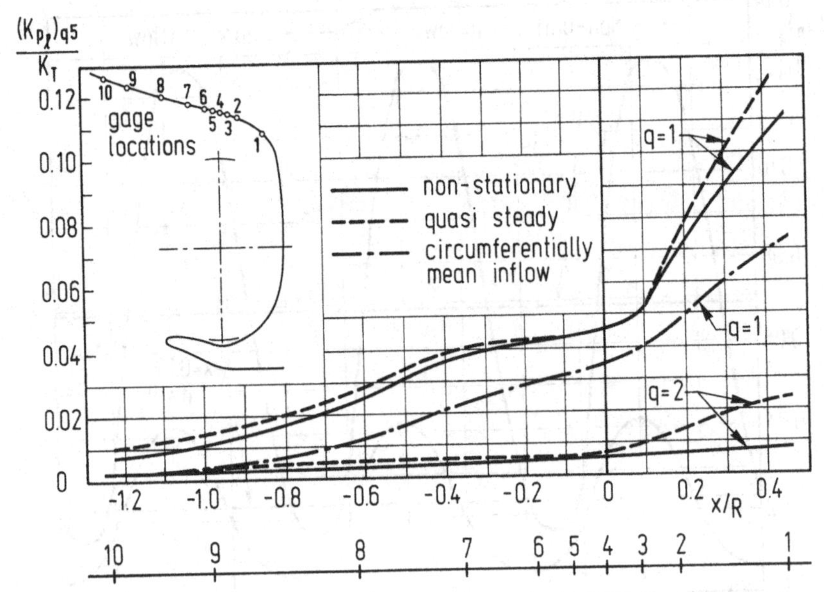

Figure 15.4 Effect of flow non-uniformity calculated on pressures due to loading.

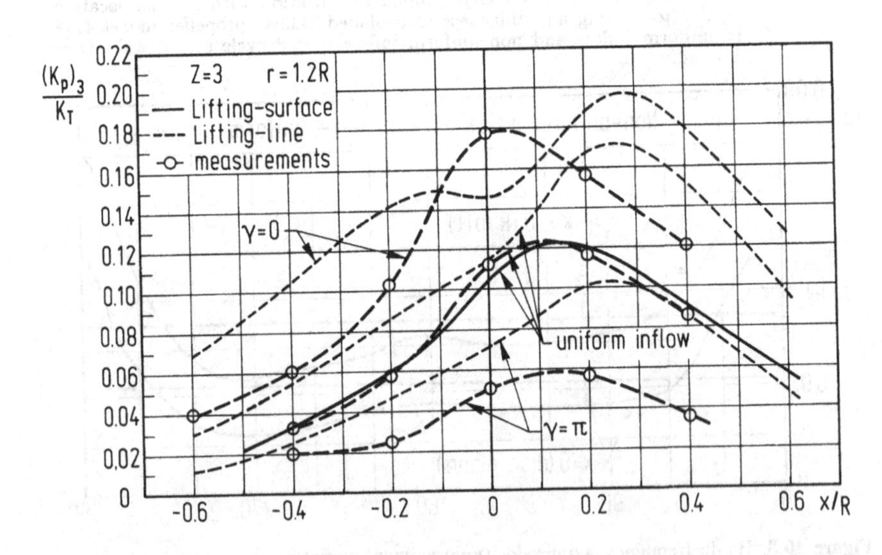

Figure 15.5 Effect of flow non-uniformity on pressures generated by a 3-bladed propeller operating in free stream and behind a screen.

Figure 15.4 and 15.5 from Bavin, V.F., Vashkevich, M.A. & Miniovich, I.Y. (1968). Pressure field around a propeller operating in spatially non-uniform flow. In *Proc. Seventh Symp. on Naval Hydrodynamics*, ed. R.D. Cooper & S.W. Doroff, pp. 3-15. Arlington, Va.: Office of Naval Research – Department of the Navy. By courtesy of Office of Naval Research, USA.

16 Vibratory Forces on Simple Surfaces

Armed with our knowledge of the structure of the pressure fields arising from propeller loading and thickness effects (in the absence of blade cavitation) we can seek to determine blade-frequency forces on simple "hulls". There are pitfalls in so over-simplifying the hull geometry to enable answers to be obtained by "hand-turned" mathematics, giving results which may not be meaningful. Yet the problem which can be "solved" in simple terms has a great seduction, difficult to resist even though the required simplifications are suspected beforehand to be too drastic. One then has to view the results critically and be wary of carrying the implications too far.

From our knowledge that most of the terms in the pressure attenuate rapidly with axial distance fore and aft of the propeller we are tempted to assume that a ship with locally flat, relatively broad stern in way of the propeller may be replaced by a rigid flat plate. The width of the plate is taken equal to that of the local hull and the length extended to infinity fore and aft on the assumption that beyond about two diameters the load density will virtually vanish. As we shall note later, this assumption of fore-aft symmetry of the area is unrealistic as hulls do not extend very much aft of the propeller. We shall also assume at the outset that the submergence of the flat surface above the propeller is large so we might ignore the effect of the water surface. We shall assume that the propeller is immersed in a wake and that the pressure loadings arising from all harmonic orders of the wake are known. The case of uniform flow is encompassed, being given by the term for wake order n = 0. Finally, we shall further assume that the effect of this flat boundary is to double the induced pressures thereon (as explained on p. 283) although we know this cannot be correct near the lateral edges of the plate.

All of our foregoing expansions in terms of cylindrical coordinates are not of use here because the plate is not a portion of a cylinder. We return to (15.2) and replace the m-sum by $1/R$ in cylindrical coordinates and write the pressure on the plate induced by the loading on a single blade as

$$p_\ell(x,y,d;\gamma) = \frac{2}{4\pi}\int_{R_h}^{R} dr' \int_{-h_t}^{h_l} \frac{\sum_{n=-\infty}^{\infty} \Delta p_n e^{in\gamma_0}}{\sqrt{P_f^{*2} + r'^2}} \left[r'\frac{\partial}{\partial x'} - \frac{P_f^*}{r'}\frac{\partial}{\partial \psi'}\right]\frac{1}{R}\, dh' \tag{16.1}$$

where now $R = \sqrt{(x-x')^2 + r^2 + r'^2 - 2rr' \cos(\gamma-\gamma_0-\psi')}$ in which $x' = P_f^*\psi'$; $r^2 = y^2 + d^2$; and $z = d$ is the vertical height of the plate (ship counter) above the propeller axis. (The assumed doubling effect has been put in evidence by the factor of 2 in (16.1)).

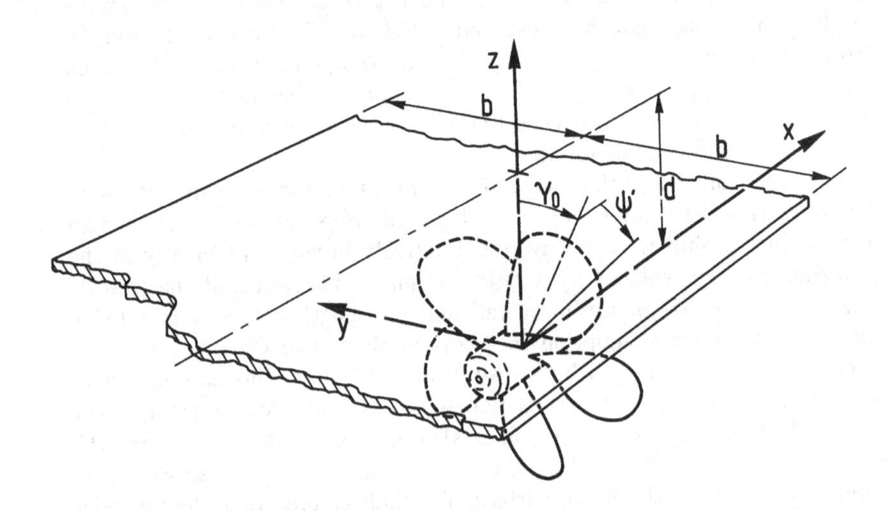

Figure 16.1 Definition sketch for a propeller below a flat plate of width 2b.

Now consider the induced vertical force, $F_{z\ell}$, due to loading on this single blade. It is simply

$$F_{z\ell} = \int_{-\infty}^{\infty} dx \int_{-b}^{b} p_\ell(x,y,d;\gamma)dy \tag{16.2}$$

with the plate of beam 2b symmetrically disposed port and starboard with respect to the propeller axis as shown in Figure 16.1.

Inspection of (16.1) and R above reveals that $\partial/\partial x'$ can be replaced by $-\partial/\partial x$ and then inserted to (16.2). This term will integrate to

$$-\frac{1}{R(\infty,....)} + \frac{1}{R(|-\infty|,....)} = 0$$

Observe! we have just lost the effect of the largest pressure terms because of the assumption that the "ship" extends to infinity aft as well as forward of the "wheel". We shall return to this aspect later. To proceed with the calculation we must carry out $\partial(1/R)/\partial\psi'$ in (16.1) and then see that we are confronted with the x-integral

$$rr' \sin(\gamma-\gamma_0-\psi')\int_{-\infty}^{\infty} \frac{1}{[(x-x')^2+r^2+r'^2-2rr'\cos(\gamma-\gamma_0-\psi')]^{3/2}}\,dx$$
$$= \frac{2rr' \sin(\gamma-\gamma_0-\psi')}{r^2 + r'^2 - 2rr' \cos(\gamma-\gamma_0-\psi')}$$

which we can write as

$$-\frac{\partial}{\partial\psi'}\left\{\ln[r^2 + r'^2 - 2rr' \cos(\gamma-\gamma_0-\psi')]\right\}$$

Hence the calculation of the vertical force on the plate is reduced to

$$F_{z\ell} = \frac{P_f^*}{2\pi}\int_{R_h}^{R} \frac{dr'}{r'}\int_{-h_t}^{h_l} dh' \frac{\sum_{-\infty}^{\infty} \Delta p_n e^{in\gamma_0}}{\sqrt{P_f^{*2} + r'^2}}$$
$$\cdot \frac{\partial}{\partial\psi'}\int_{-b}^{b} \ln[r^2 + r'^2 - 2rr' \cos(\gamma-\gamma_0-\psi')]dy \quad (16.3)$$

Now as $r^2 = y^2 + d^2$ for points on the plate then $r > r'$ and we may factor out r^2 from the logarithm and make use of the expansion of Gegenbauer,

$$\ln\left[1 + \left[\frac{r'}{r}\right]^2 - 2\left[\frac{r}{r'}\right]\cos(\gamma-\gamma_0-\psi')\right]$$
$$= -2\sum_{m=1}^{\infty} \frac{\cos m(\gamma-\gamma_0-\psi')}{m}\left[\frac{r'}{r}\right]^m \quad (16.4)$$

which may be expressed as

$$\ln[...] = -2\, Re\left\{\sum_{m=1}^{\infty} \frac{e^{im\gamma}}{m}\left[\frac{r'}{r}\right]^m e^{-im(\gamma_0+\psi')}\right\} \quad (16.5)$$

To put the y-dependence in evidence we note that

$$e^{i\gamma} = \frac{1}{r}(d-iy) \;\; ; \;\; \frac{d}{r} = \cos\gamma \;\; ; \;\; \frac{y}{r} = -\sin\gamma$$

as may be observed in Figure 16.2 below.

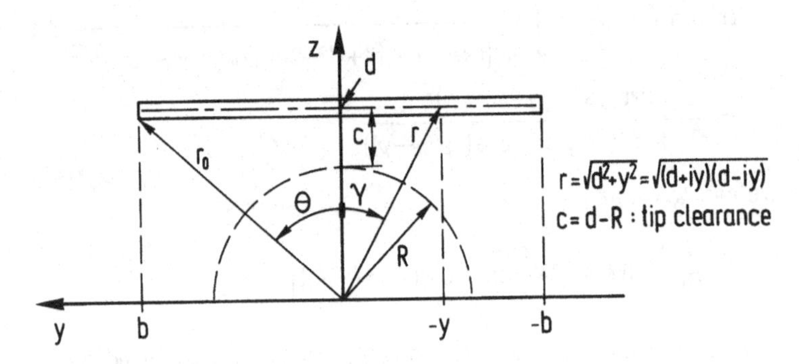

Figure 16.2 Definition of angle γ and radius r for any point on the flat boundary above the propeller.

Hence

$$\frac{e^{im\gamma}}{r^m} = \frac{(d-iy)^m}{r^{2m}} = \frac{(d-iy)^m}{(d^2+y^2)^m} = \frac{(d-iy)^m}{(d-iy)^m}\frac{1}{(d+iy)^m} = \frac{1}{(d+iy)^m} \tag{16.6}$$

Consequently we may write the y-integrand in (16.3) with (16.6) as

$$\ln[r^2+r'^2-2rr'\cos(\gamma-\gamma_0-\psi')] = \ln r^2 - 2Re\left\{\sum_{m=1}^{\infty}\frac{r'^m \; e^{-im(\gamma_0+\psi')}}{m(d+iy)^m}\right\} \tag{16.7}$$

Taking the derivative $\partial/\partial\psi'$ as required in (16.3) removes the first term in (16.7) and brings down $-im$ in the second term so the y-integral in (16.3) becomes

$$I^* = 2Re\left\{\left[\int_{-b}^{b} i\sum_{m=1}^{\infty}\frac{r'^m \; e^{-im(\gamma_0+\psi')}}{m(d+iy)^m}\,dy\right]\right\}$$

where we can integrate directly taking m = 1 separately to achieve (after some elementary manipulations)

$$I^* = 4\theta\, r' \sin(\gamma_0+\psi') + 4 \sum_{m=2}^{\infty} \frac{\sin[(m-1)\,\theta]\, r'^m \sin[m(\gamma_0+\psi')]}{(m-1) r_0^{m-1}} \tag{16.8}$$

where

$$\theta = \tan^{-1}\left[\frac{b}{d}\right] \quad ; \quad r_0 = \sqrt{d^2 + b^2} \tag{16.9}$$

Use of (16.8) in (16.3) yields

$$F_{z\ell} = \frac{2P_f^*}{\pi} \sum_{n=-\infty}^{\infty} \int_{R_h}^{R} dr' \int_{-h_t}^{h_l} \frac{\Delta p_n}{\sqrt{P_f^{*2} + r'^2}}\, Im \left\{ \theta\, e^{i(n+1)\gamma_0}\, e^{i\psi'} - \frac{1}{2} \sum_{\substack{m=-\infty \\ m\neq 0,\pm 1}}^{\infty} \frac{m}{|m|} \frac{\sin[(|m|-1)\theta]}{|m|-1} \left[\frac{r'}{r_0}\right]^{|m|-1} e^{i(n-m)\gamma_0}\, e^{-i\psi'} \right\} dh' \tag{16.10}$$

where the m-sum has been unfolded to facilitate the summation over Z blades. As usual, p. 279, we replace γ_0 by $\gamma_0 + 2\pi\nu/Z$ and sum from $\nu = 0$ to $Z-1$. Thus the factor $e^{i(n+1)\gamma_0}$ yields Z for only $n + 1 = qZ$, $q = 0, \pm 1, \ldots$ and zero otherwise, and the factor $e^{i(n-m)\gamma_0}$ yields Z for only $n-m = qZ$ or $m = n-qZ$. We shall be content to limit our attention to blade frequencies, i.e. $q = \pm 1$

$$(F_{z\ell})_Z = \frac{2P_f^* Z}{\pi} \int_{R_h}^{R} dr' \int_{-h_t}^{h_l} \frac{1}{\sqrt{P_f^{*2} + r'^2}}$$

$$\cdot\, Im \left\{ \theta \left[\Delta p_{Z-1}\, e^{i\psi'}\, e^{iZ\gamma_0} + \Delta p_{-(Z+1)}\, e^{i\psi'}\, e^{-iZ\gamma_0} \right] \right.$$

$$- \frac{1}{2} \sum_{n=-\infty}^{\infty} \Delta p_n(h',r') \left[\frac{n-Z}{|n-Z|} \frac{\sin[(|n-Z|-1)\theta]}{|n-Z|-1} \left[\frac{r'}{r_0}\right]^{|n-Z|-1} e^{-i(n-Z)\psi'} \cdot e^{iZ\gamma_0} \right.$$

$$\left.\left. + \frac{n+Z}{|n+Z|} \frac{\sin[(|n+Z|-1)\theta]}{|n+Z|-1} \left[\frac{r'}{r_0}\right]^{|n+Z|-1} e^{-i(n+Z)\psi'}\, e^{-iZ\gamma_0} \right] \right\} dh' \tag{16.11}$$

where $(n \pm Z) \neq 0, \pm 1$.

It is clear that as the plate width $2b \rightarrow \infty$, $r_0 \rightarrow \infty$ (cf. 16.9) the terms in the n-series all vanish leaving only the first line which reduces to, since $\lim_{b \rightarrow \infty} \theta = \pi/2$

$$\lim_{b \rightarrow \infty} (F_{z\ell})_Z = P_f^* Z \; Im \left\{ \int_{R_h}^{R} dr' \int_{-h_t}^{h_l} \frac{e^{i\psi'}}{\sqrt{P_f^{*2}+r'^2}} \cdot \left[\Delta p_{Z-1} \, e^{iZ\gamma_0} + \Delta p_{-(Z+1)} \, e^{-iZ\gamma_0} \right] dh' \right\} \tag{16.12}$$

This residual or force on the doubly infinite plate can be interpreted by considering the vertical force on the entire propeller or shaft arising from the blade loadings in the wake.

Vertical Blade-Frequency Force on the Shaft

Consider, at first, a single blade with differential forces arising from any wake order n as shown in Figure 16.3.

Thus the vertical force contribution from a single blade is

$$F_z = - P_f^* \; Im \left\{ \sum_{n=-\infty}^{\infty} \int_{R_h}^{R} \frac{dr'}{\sqrt{P_f^{*2}+r'^2}} \int_{-h_t}^{h_l} \Delta p_n \, e^{i(n+1)\gamma_0} \, e^{i\psi'} dh' \right\} \tag{16.13}$$

For Z blades, $n = qZ-1$ and taking only $q = 1, -1$ we have for the total

$$(F_z)_Z = - P_f^* Z \; Im \left\{ \int_{R_h}^{R} \frac{dr'}{\sqrt{P_f^{*2}+r'^2}} \cdot \int_{-h_t}^{h_l} e^{i\psi'} \left[\Delta p_{Z-1} \, e^{iZ\gamma_0} + \Delta p_{-(Z+1)} \, e^{-iZ\gamma_0} \right] dh' \right\} \tag{16.14}$$

Comparison of (16.12) and (16.14) reveals that the blade-frequency force on the doubly infinite plate is opposite and equal to that acting on the shaft[25]. This appears consonant with the result one should achieve from a balance of momentum flux. Early independent calculations by Pohl (1959) and Breslin (1959) gave zero force on the doubly infinite plate from a propeller in uniform flow. The absence of a wake corresponds to setting n

[25] This result was dubbed the "Breslin Condition" by the late Professor F.M. Lewis (1974).

= 0 in (16.11) and there would be no first line in that equation. We see that as b, $r_0 \to \infty$ the force is zero as must be expected as there is no vertical force acting on the propeller.

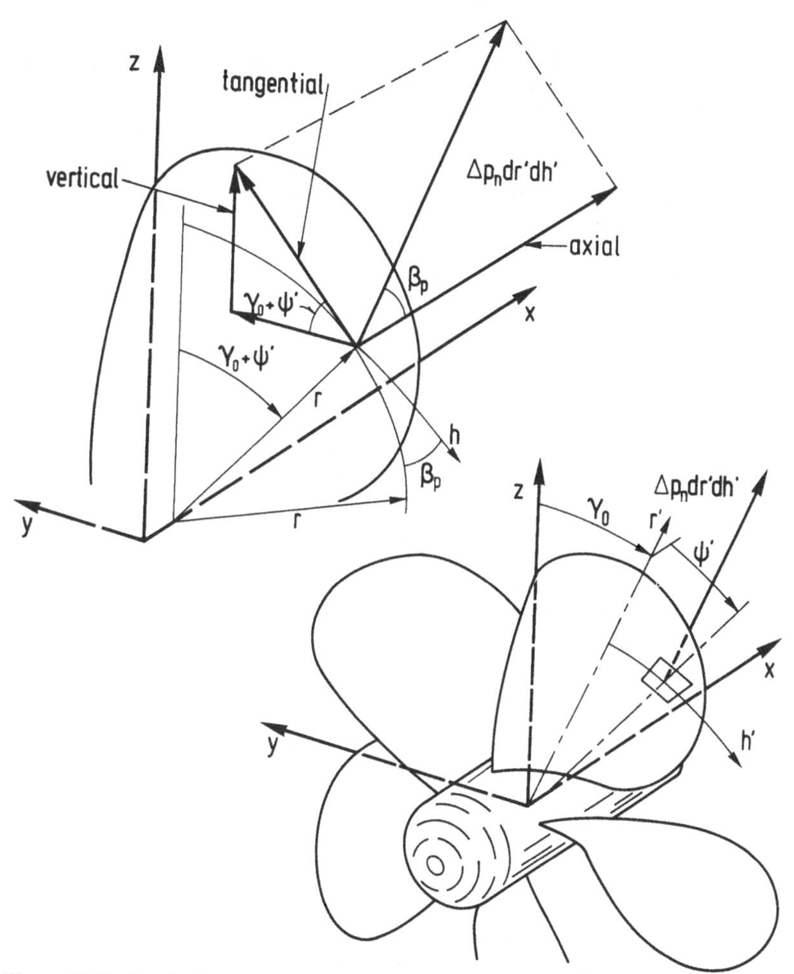

Figure 16.3 Resolution of force normal to blade pitch line to vertical component.

Force on Plate Due to Blade Thickness

It has been found that the blade-frequency force due to thickness on this flat boundary is much larger than that from the mean loading of the blades. We deal with this briefly, by recalling from Equation (14.18) that the pressure from thickness is composed of a term involving $\partial(1/R)/\partial x'$

and one involving $\partial(1/R)/\partial\psi'$. As the first of these integrates to zero and the second is analogous to the term involving $\partial/\partial\psi'$ in (14.13) for the pressure due to loading the calculation of the force due to thickness is completely analogous to that arising from Δp_0. The force from blade thickness (single blade) is then from (14.18)

$$F_{z\tau} = -\frac{\rho\omega}{\pi}\int_{-\infty}^{\infty} dx \int_{-\infty}^{\infty} dy \int_{R_h}^{R} dr' V(r') \int_{-h_t}^{h_l} \tau' \frac{\partial}{\partial\psi'}\left[\frac{1}{R}\right] dh' \tag{16.15}$$

where $\tau' = \partial\tau/\partial h'$. Again a factor of 2 has been applied. Comparing this with the contributing term from the loading as in (16.1) we see that the factor

$$\frac{1}{2}\frac{P_f^*}{r'}\frac{\Delta p_0}{\sqrt{P_f^{*2} + r'^2}}$$

is to be replaced by $\rho\omega\tau' V(r')$. Hence the blade-frequency result for thickness is

$$(F_{z\tau})_Z = \frac{4\rho\omega Z}{\pi}\frac{\sin[(Z-1)\theta]}{(Z-1)r_0^{Z-1}}\int_{R_h}^{R} dr' r'^Z\, V(r') \int_{-h_t}^{h_l} \tau' \sin Z(\gamma_0+\psi')dh' \tag{16.16}$$

This may be compared to that from mean blade loading (n = 0) from (16.11)

$$(F_{zl})_Z^{(0)} = \frac{2P_f^* Z}{\pi}\frac{\sin[(Z-1)\beta]}{(Z-1)r_0^{Z-1}}\int_{R_h}^{R}\frac{dr' r'^{Z-1}}{\sqrt{P_f^{*2}+r'^2}}$$

$$\cdot \int_{-h_t}^{h_l} \Delta p_0(h',r') \sin Z(\gamma_0+\psi')dh' \tag{16.17}$$

These forces are seen to be in phase and have the same dependence on the number of blades except for the dependence on the weighting factors r'^Z and r'^{Z-1}. As $V(r') \approx \omega r'$ for $r' \to R$ the thickness contribution is seen to vary essentially as ω^2 which may account for the relatively large contribution due to thickness found in early evaluations (Tsakonas, Breslin & Jacobs (1962)) made in the absence of hull wake.

It is of interest to examine the increase in force due to blade thickness with decrease of tip clearance. Tip clearance is imbedded in θ and r_0, cf. Figure 16.2 and (16.9). Thus for two tip clearances c_1, c_2 the ratio of the force from thickness is

$$\frac{\left[F_{zT}(c_1)\right]_Z}{\left[F_{zT}(c_2)\right]_Z} = \frac{\sin\left[(Z-1)\tan^{-1}\left[\dfrac{b}{R+c_1}\right]\right]}{\sin\left[(Z-1)\tan^{-1}\left[\dfrac{b}{R+c_2}\right]\right]} \left[\frac{((R+c_2)^2 + b^2}{((R+c_1)^2 + b^2)}\right]^{Z/2} \tag{16.18}$$

For propeller radius, $R = 1$, $b = 3$, $Z = 5$, $c_1 = 0.30$, $c_2 = 0.60$ (i.e., clearances of 0.15 and 0.30·diameter), this ratio evaluates to 1.26. Hence decreasing the tip clearance by a factor of 2 increases this force by 26 per cent for a five-bladed propeller. The corresponding ratio of the force due to blade loading cannot be easily made because the Δp_n depends upon the wake which becomes more severe as clearance is decreased.

It is curious to note that the forces due to thickness and mean loading will be zero for certain values of the plate semi-beam b. They vanish for the argument of $\sin[(Z-1)\theta]$ equal to π (or integer multiplies thereof) or those b for a fixed clearance c which satisfy

$$(Z-1)\tan^{-1}\left[\frac{b}{R+c}\right] = k\pi \quad , \quad k = 1, 2, 3, \ldots$$

The plate beams for which these forces are a maximum occur when the above arguments are multiples of $\pi/2$. Results for the only practical value $k = 1$ are shown in Table 16.1.

No. of blades Z	Vanishing 2b/R	Maximum 2b/R
4	5.20	1.73
5	3.00	1.24
6	2.18	0.974

Table 16.1 Theoretically predicted plate beams for vanishing and maximum forces on plate, for $c = 0.5$ R.

These small values are not to be relied upon because of our neglect of the fall-off of pressures in the vicinity of the lateral edges. We must be skeptical of all of these observations recalling that the extension of the boundary to infinity abaft the propeller is not very realistic. These deductions can be cautiously viewed as giving gross trends. The effect of terminating the aft extent "shortly" downstream of the propeller as evaluated by computer will be seen later.

Results of Experiment and Calculations

In spite of the above cautions, good correlations were found with measurements made by Lehman (1964) in a water tunnel at Oceanics Inc..Vertical forces on flat plates above a propeller model (extending forward and aft parallel to the shaft) were measured over a range of plate-axis clearances for various propeller loadings and two different blade thicknesses. The correlations shown in Figure 16.4 are surprisingly good considering the narrow width of the plate.

It is seen that the agreements of theory with the data are better at the lower thrust coefficient (K_T). The experimental results confirmed several of the basic observations from the theory; one most significant of these is the dominating rôle played by blade thickness. Note from the double-thickness, standard-thickness results in Figure 16.4 that doubling the blade thickness nearly doubles the force amplitude. Curiously, the correlation of theory with measurements made on wider plates was not as good in spite of the expectation that it should be better.

The blade-frequency forces induced by a propeller in uniform flow parallel to an infinite circular cylinder (Breslin (1962)) were extended to non-uniform flow (Breslin (1968)). Here the boundary-value problem was solved exactly, i.e. the artifice of doubling the incident pressures was not employed. The geometry and the results for uniform flow are shown in Figure 16.5. Here we see, again, the dominating contribution due to thickness as the large pressures from thrust loading have integrated to zero because of the fore-aft symmetry of the boundary relative to the propeller.

The transverse force on cylinders of various radii due to a propeller in a typical hull wake are shown in Figure 16.6 for a range of advance ratios. Here the abscissa is the net hull-shaft force so as the cylinder-diameter-propeller-diameter ratio becomes infinite the net force vanishes. It is seen that this limit is approached very slowly.

A curious effect was noted in the case of the circular cylinder in that the force produced by the propeller pressure field was found to be equal to that of the "image" of the propeller in the cylinder. Put another way, the potential required to generate the cylinder in the presence of the propeller produces a force equal to that obtained by integrating the resolved, direct, incident pressures from the propeller over the cylinder. This is not to say that the local incident pressures are doubled, only the force is doubled.

It is recalled that extending the boundary provided by the "hull" (i.e., the flat strip or cylinder) to infinity aft as well as forward had the effect of negating the influence of the larger of the two components of pressure emanating from the propeller loading, i.e., the part associated with the mean and unsteady thrust-producing pressure dipoles. This because the axial dependence of this constituent is essentially an odd function of x with respect to the propeller. To examine this a calculation was made by Breslin (presented as part of a paper, Cox, Vorus, Breslin & Rood (1978)) in which the hull was terminated at one propeller radius abaft the propeller. The results are summarized in Figure 16.7. Here we see that the force arising from the exceedingly small blade-frequency loadings Δp_5 is some

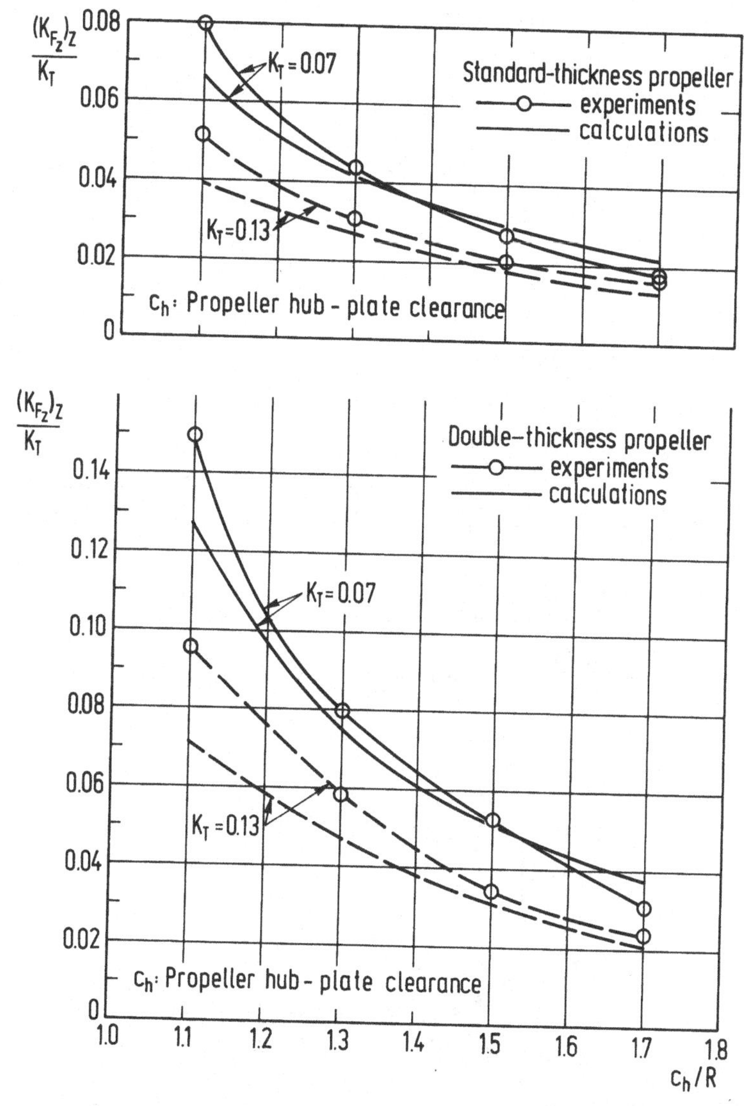

Figure 16.4 Comparison of measured and calculated blade–frequency force amplitudes on a flat plate above and parallel to axis of a model propeller in a water tunnel.

Data from Lehman (1964); figures 26, 27 from Breslin, J.P. (1969). Report on vibratory propeller, appendage and hull forces and moments. In *Proc. Twelfth International Towing Tank Conference (ITTC)*, ed. Istituto Nazionale per Studi ed Esperienze di Architettura Navale, pp. 280–90. Rome: Consiglio Nazionale delle Ricerche.
By courtesy of ITTC.

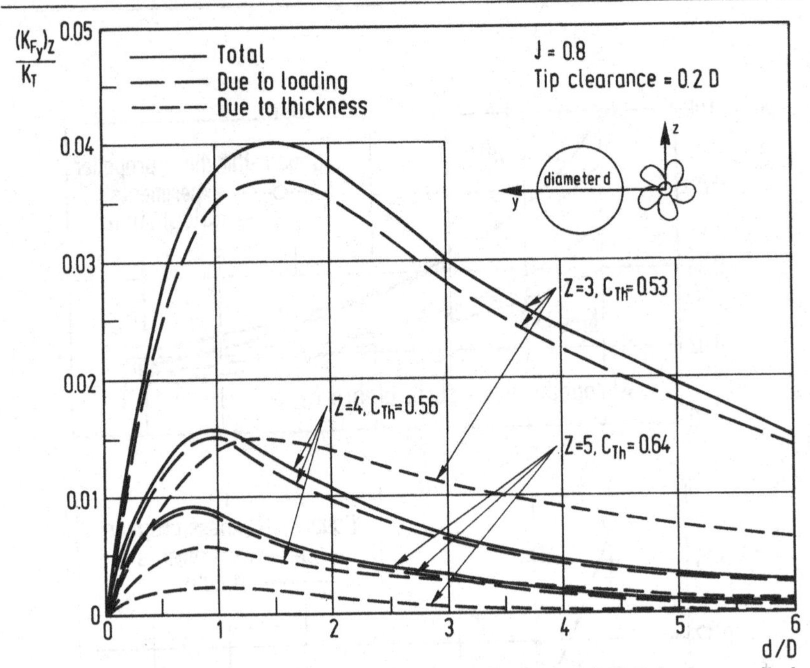

Figure 16.5 Transverse blade–frequency force amplitudes on infinitely long cylinder induced by 3–, 4– and 5–bladed propellers in uniform inflow.

Figure 11, 12 from Breslin, J.P. (1962). Review and extension of theory for near field propeller–induced vibratory effects. In *Proc. Fourth Symp. on Naval Hydrodynamics*, ed. B.L. Silverstein, pp. 603–640. Washington, D.C.: Office of Naval Research – Department of the Navy.
By courtesy of Office of Naval Research, USA.

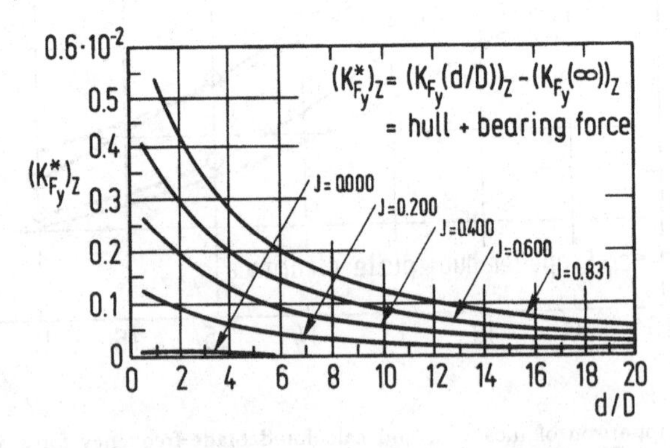

Figure 16.6 Transverse blade–frequency force amplitude on propeller–cylinder system at various advance ratios and non–uniform inflow.

Figure 28 from from Breslin, J.P. (1969). Report on vibratory propeller, appendage and hull forces and moments. In *Proc. Twelfth International Towing Tank Conference (ITTC)*, ed. Istituto Nazionale per Studi ed Esperienze de Architettura Navale, pp. 280–90. Rome: Consiglio Nazionale delle Ricerche.
By courtesy of ITTC.

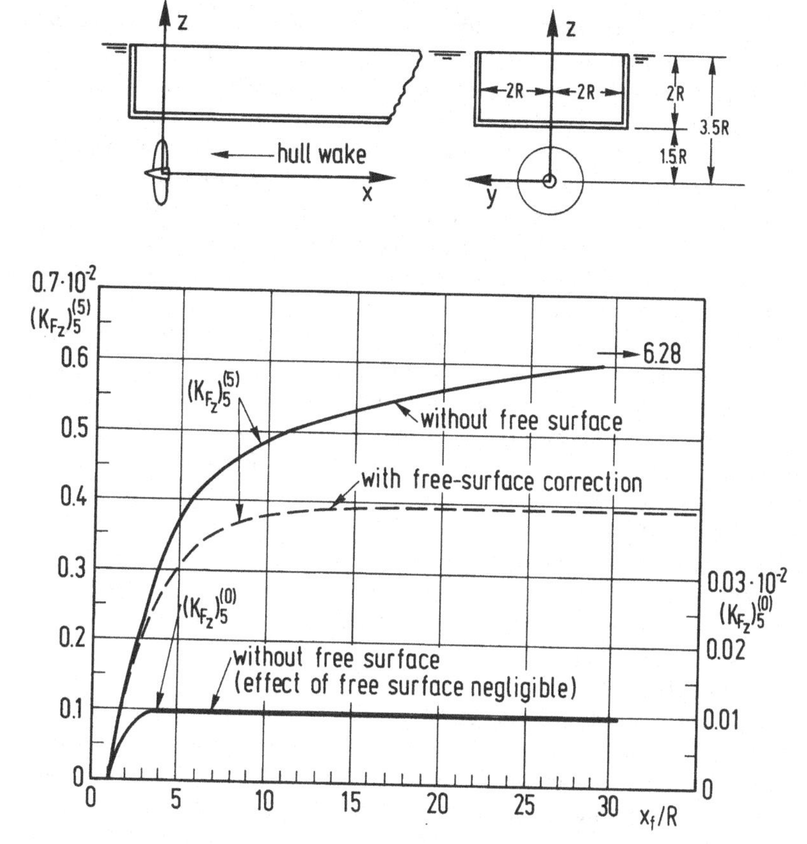

Figure 16.7 Approximate moduli of blade-frequency forces on barge-like ship from pressures emanating from mean and blade-frequency loadings on a 5-bladed propeller (in a single-screw ship wake) as a function of integration length forward of propeller.

Figure 4 from Cox, B.D., Vorus, W.S., Breslin, J.P. & Rood, E.P. (1978). Recent theoretical and experimental developments in the prediction of propeller induced vibration forces on nearby boundaries. In *Proc. Twelfth Symp. on Naval Hydrodynamics*, pp. 278–99. Washington, D.C.: National Academy of Science.
By courtesy of Office of Naval Research, USA.

65 times larger than that from the mean loading (in spite of the fact that the mean loading $\Delta p_0 = 40\Delta p_5$!) when the effect of the free surface is neglected. When the quenching effect of the water surface is approximately accounted for, the force generated by the Δp_5 can be seen to be 38 times larger than that due to Δp_0. These results also show that the integration length must be extended to some 15 diameters even when the water surface effect is included. For ships, this length may be considerably larger because of the growth of the available area with distance forward of the propeller.

These results indicate that the current practice in European model basins to measure the pressures in the vicinity of the propeller over a very limited area and then to infer the exciting force by integration is highly suspect. The relatively slow decay of the pressures, emanating from the blade-frequency loading (particularly those associated with axial pressure dipoles) and moreover from intermittent cavitation the "signature" of which decays as $1/x$ without water surface quenching and as $1/x^2$ with water surface effect, coupled with the "growing" cross-sectional shape as one integrates forward, contributes large sectional force densities. These are not included in the above-cited practice in European model basins. We see that great contribution of the term in the loading coupled with $\partial(1/R)/\partial x'$ whose force contribution is only negated in the region $-$Radius $<$ x $<$ Radius because of its effectively odd-function character. As indicated earlier, oversimplification of the hull boundary can and does lead to non-representative results for practical hull forms. Hence the actual hull geometry must be represented. Methods for accomplishing this will be given in Chapter 22.

Although the forces induced by non-cavitating propellers are sufficient to cause objectionable hull vibration when the frequencies coincide with hull and/or local structural natural frequencies, they are considerably smaller than those arising from intermittent blade cavitation. These will also be dealt with in the later chapters.

As the evaluation of propeller and hull forces is dependent on the blade pressure jumps Δp_n, arising from each of the wake harmonics we must now turn to their determination by first considering the unsteady loads generated by two-dimensional sections and wings beset by travelling, transverse gusts of constant amplitude.

17 Unsteady Forces on Two-Dimensional Sections and Hydrofoils of Finite Span in Gusts

As a preparation for determining unsteady forces on propellers in ship wakes, we first consider two-dimensional sections beset by travelling gusts. Our development of the unsteady force on such sections differs from that given in the seminal work of von Kármán & Sears (1938) by adopting a procedure which is easily extended to wings and propellers. Their formula for unsteady sectional lift is recovered, being that of lift at an effective angle of attack which varies with the parameter $k = \omega c/2U$, the "reduced" frequency. Turning to hydrofoils of finite span, we derive results for low aspect ratio in steady flow. For wings in gusts there is no analytical inversion of the integral equation which involves a highly singular kernel function. Graphical results are given from numerical solutions for a range of aspect ratios which reveal diminishing unsteady effects with decreasing aspect ratio.

Corresponding reduced frequencies for propeller blades in terms of expanded-blade-area ratio are shown to be high relative to aerodynamic experience. This indicates that two-dimensional, unsteady section theory cannot be applied to wide-bladed (low-aspect-ratio) propellers.

TWO-DIMENSIONAL SECTIONS

The blades of a propeller orbit through the spatially non-uniform flow of the hull wake and consequently experience cyclic variations in the flow normal to their sections. For blades of small chord-to-radius, this is analogous to the case of a two-dimensional section moving at constant speed through a stationary, cyclic variation in cross flow distributed as a standing wave along the course of the moving section. Equivalently, the section may be taken as fixed and a travelling wave convected longitudinally at speed $-U$, presenting a vertical or cross flow of the form

$$v_g = v_g\left[\frac{2\pi}{\lambda}(x + Ut)\right] = Re\left\{\tilde{v}\ e^{(i2\pi/\lambda)(x+Ut)}\right\} \qquad (17.1)$$

where $\tilde{v}$ is the complex amplitude and λ is the wavelength of the gust which travels parallel to the negative x-axis. This may be seen by looking at any fixed phase of the wave, say $x + Ut$ = a constant. Then as time goes by $x \approx -Ut$, hence the wave travels from $x = +\infty$ toward $x = -\infty$ at the velocity $-U$.

As the temporal frequency f (Hertz) is

$$f = \frac{U}{\lambda} = \frac{\omega}{2\pi} \quad \text{or} \quad \lambda = \frac{2\pi U}{\omega}, \tag{17.2}$$

we see that the vertical velocity (17.1) can be written

$$v_g(x,t) = \tilde{v}\ e^{i(k\check{x}+\omega t)} \tag{17.3}$$

where we drop the real-part indicator, *Re*, with the understanding that we will retain only the real part and

$$k = \frac{\omega a}{U} \tag{17.4}$$

commonly referred to as the *reduced frequency* and $\check{x} = x/a$, a being the semi-chord of the section.

The fluid mechanics operable when a foil is beset by a varying angle of attack is complicated by the necessity to include the time history of the flow. To illustrate this we can model the flow by vortices and use ***Kelvin's theorem*** of the constancy of circulation which states that when external forces are conservative (e.g. due to gravity) the circulation in any circuit which moves with the fluid is constant (cf. Appendix D for a proof). As the angle of attack of our section varies, the lift, and hence the circulation, changes. To preserve zero net change in vorticity, the foil sheds vorticity in the opposite sense so that the total additional circulation taken around the entire fluid domain vanishes. The physical manifestation of this principle and its implications are depicted in Figure 17.1.

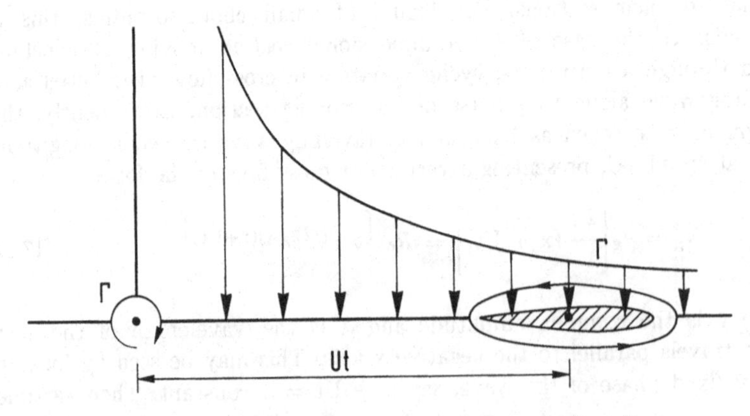

Figure 17.1 Downwash due to vortex shed t seconds earlier.

Here we see that t seconds after a foil was given an impulsive positive change in angle of attack, the shed vortex of opposite and equal circulation is located downstream at a distance Ut, and it induces a down flow back at the foil in a direction to produce a reduced angle of attack. Thus the lift associated with change in circulation is always less in unsteady flow than the quasi-steady lift. In steady flow the conservation of circulation was secured via the starting vortex shed by the foil when the motion was started from its initial state of rest. This starting vortex is indicated in Figure 2.6 (p. 37) where it is assumed to lie at the position $x = -\infty$. Its influence on the foil is then negligible and the flow is stationary.

In addition to the lift associated with circulation there is another lift component (in temporal quadrature with the modified circulatory lift) which is produced from the acceleration imposed by the unsteady-onset flow or time-varying motions imposed on the section (such as heaving or pitching). This force is proportional to the added mass of the section in transverse modes of motion. Finally, there is a net axial force of second order arising from unsteady leading-edge suction which we shall ignore.

The case of a foil in a travelling gust was solved by von Kármán & Sears (1938) and further work was done by Sears (1941). Here we take a different approach which does not require us to deal explicitly with vortices shed downstream although all their effects are encompassed.

The linearized Euler equation for the y-momentum is, by linearizing Equation (1.8)

$$\frac{\partial v(x,y,t)}{\partial t} - U\frac{\partial v(x,y,t)}{\partial x} = -\frac{1}{\rho}\frac{\partial p(x,y,t)}{\partial y}$$

$$-\frac{1}{\rho}\int_{-a}^{a} \Delta p(x',t)\ \delta(x-x')\ \delta(y)\ dx' \qquad (17.5)$$

where we have included the pressure difference over the foil $\Delta p(x',t)$ as a vertically directed external force, acting on the fluid, opposite to that on the foil, Figure 17.2.

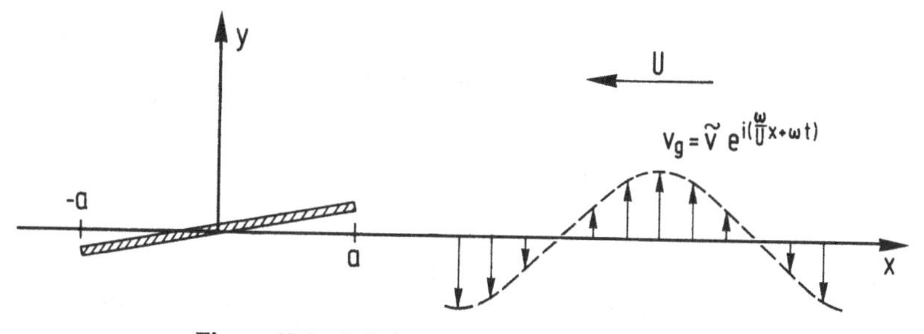

Figure 17.2 Foil (flat plate) in a travelling gust.

Since we can express the pressure $p(x,y,t)$ at any field point as a function of the pressure difference over the foil, (17.5) is a linear, partial differential equation for v in terms of Δp. One way of solving this equation is to use the method of characteristics, see Appendix E, p. 498. We have immediately from (E.16)

$$v(x,y,t) = -\frac{1}{\rho U}\left[\int_x^\infty \left[\frac{\partial}{\partial y} p\left[(x'',y,t + \frac{x-x''}{U}\right]\right.\right.$$
$$\left.\left. + \int_{-a}^{a} \Delta p\left[x',t + \frac{x-x''}{U}\right] \delta(x''-x')\, \delta(y)\, dx'\right] dx''\right] \quad (17.6)$$

In the double integral, interchange order of integration to obtain

$$\int_{-a}^{a} dx' \int_x^\infty \left[\Delta p(x',t + \frac{x-x''}{U}\right] \delta(x''-x')\, \delta(y)\, dx''$$
$$= \int_{-a}^{x} 0\, dx' + \int_x^a \Delta p\left[x',t + \frac{x-x'}{U}\right] \delta(y)\, dx'$$

since there will only be a contribution from the δ-function for $x'' < x'$. For $x < -a$ the integral should be interpreted as $\int_{-a}^{a}$; for $x > a$ it is zero. Using this result (17.6) can be written

$$v(x,y,t) = -\frac{1}{\rho U}\left[\int_x^\infty \frac{\partial}{\partial y} p\left[x',y,t + \frac{x-x'}{U}\right] dx'\right.$$
$$\left. + \int_x^a \Delta p\left[x',t + \frac{x-x'}{U}\right] \delta(y)\, dx'\right] \quad (17.7)$$

We are interested in evaluating the velocity on the surface of the foil where $y = 0$ and where we will encounter the δ-function of the last integral of (17.7). But from (2.24) we see that there is a jump in the pressure over the foil surface and we can then for a point close to the surface (or y sufficiently small) write the pressure as

$$p(x,y,t) = \left[\frac{1}{2} - H(y)\right] \Delta p(x,t) + p_1(x,y,t)$$
$$\text{for } -a < x < a \text{ and } |y| \to 0 \quad (17.8)$$

where $H(y)$ is the Heaviside step function, cf. Section 2 of the Mathematical Compendium (p. 504), and $p_1(x,y,t)$ is an additional pressure which is a continuous function of y and zero for $y = 0$.

When we differentiate the pressure with respect to y, we use (17.8) and we see that the first term annuls the δ-function term of (17.7) leaving

$$v(x,0,t) = -\frac{1}{\rho U}\int_x^\infty \frac{\partial}{\partial y} p_1\left[x',y,t+\frac{x-x'}{U}\right]\Bigg|_{y=0} dx' \tag{17.9}$$

and we observe that the vertical velocity is continuous over the blade surface, as we also expect physically.

That the force term is annulled can be seen in another way by returning to (17.7) and inserting (2.23)

$$v = -\frac{1}{\rho U}\left[\int_x^\infty dx''\int_{-a}^{a}\frac{\partial}{\partial y}\left[-\frac{y}{2\pi}\frac{\Delta p(x',t'')}{(x''-x')^2+y^2}\right]dx' + \int_x^a \Delta p(x',t')\,dx'\,\delta(y)\right] \tag{17.10}$$

where $t'' = t + (x-x'')/U$.

The integrand of the first term can be written as

$$-\Delta p(x',t'')\frac{1}{4\pi}\frac{\partial^2}{\partial y^2}\ln((x''-x')^2+y^2)$$

The ln-term is the two-dimensional source potential (cf. (1.34)) and is, as such, the solution to the equation

$$\left[\frac{\partial^2}{\partial x^2}+\frac{\partial^2}{\partial y^2}\right]\phi(x,y;x',y') = 4\pi\,\delta(x-x')\,\delta(y-y')$$

Using this result in (17.10) gives

$$v = \frac{1}{\rho U}\int_x^\infty dx''\int_{-a}^{a}\Delta p(x',t'') \cdot\left[-\frac{1}{4\pi}\frac{\partial^2}{\partial x''^2}\ln\left((x''-x')^2+y^2\right)+\delta(x''-x')\,\delta(y)\right]dx' - \frac{1}{\rho U}\int_x^a \Delta p(x',t')dx'\,\delta(y) \tag{17.11}$$

By operating on the δ-function term of the double integral in a similar way as from equation (17.6) to (17.7) it is seen that the two δ-function terms annul. Carrying out the differentiation in (17.11) and letting y = 0 for the velocity on the foil yields

$$v(x,0,t) = \frac{1}{2\pi\rho U}\int_x^{\infty} dx'' ⨎_{-a}^{a} \frac{\Delta p(x',t'')}{(x''-x')^2}\, dx' \tag{17.12}$$

where the notation ⨎ indicates that only the finite part of this highly singular integral is taken and the singular part is ignored. The interpretation is based on argumentation given above that the vertical velocity must remain finite. The concept of the finite part is due to Hadamard (1923). Procedures for evaluation of integrals of this type, following the clear exposition by Mangler (1951) are given in Section 4 of the Mathematical Compendium, p. 507 and sequel.

The integral in (17.12) is seen to be a history integral in which time is retarded. This can be seen by letting

$$t'' = t + \frac{x-x''}{U} \quad ; \quad x'' = x - U(t''-t) \quad ; \quad dx'' = -U dt''$$

then

$$v(x,0,t) = \frac{1}{2\pi\rho}\int_{-\infty}^{t} dt'' ⨎_{-a}^{a} \frac{\Delta p(x',t'')}{(x-U(t''-t)-x')^2}\, dx' \tag{17.13}$$

Now we see that the velocity at present time t is an integral of the pressure derivative over all prior time, thus being the summation of the entire history of the flow. This integration over the "past" arises because of the vorticity shed downstream.

We now simplify the foil geometry to that of a flat plate of incidence α. As v(x,y,t) must, on the plate, be such as to counter the flow imposed by the component of –U normal to the plate and the gust we have, for small foil angle, α, the kinematic condition

$$v(x,0,t) \equiv -U\alpha - v_g\left[\frac{x}{U} + t\right] \quad ; \text{ all } t \ ; \ |x| \le a \tag{17.14}$$

or, by use of (17.3)

$$\frac{\partial v(x,0,t)}{\partial t} - U\frac{\partial v(x,0,t)}{\partial x} = -\left[\frac{\partial}{\partial t} - U\frac{\partial}{\partial x}\right] v_g\left[\frac{x}{U} + t\right]$$

$$= 0 \text{ all } t, \text{ and } |x| \le a \tag{17.15}$$

Operating on (17.12) by $\partial/\partial t - U\partial/\partial x$ gives only a contribution from the lower limit of the x''-integral since $(\partial/\partial t - U\partial/\partial x)t'' = 0$, because here we have $t'' = t+(x-x'')/U$. Equation (17.15) then takes the simple form

$$\frac{1}{2\pi} \oint_{-a}^{a} \frac{\Delta p(x',t)}{(x-x')^2} dx' = 0 \qquad |x| \leq a \tag{17.16}$$

and can be recast into

$$\frac{\partial}{\partial x} \int_{-a}^{a} \frac{\Delta p(x',t)}{x-x'} dx' = 0$$

Integrating we obtain

$$\int_{-a}^{a} \frac{\Delta p(x',t)}{x-x'} dx' = C(t), \text{ a function of time}$$

This is the Munk integral equation, the integral being taken as its principal value. The inversion of this equation which satisfies the Kutta condition $\Delta p(-a,t) = 0$ (at the trailing edge, cf. p. 71) is, (cf. Equations (5.13) and (5.24), p. 69 and 71)

$$\Delta p(x,t) = -\frac{C(t)}{\pi^2} \sqrt{\frac{a+x}{a-x}} \int_{-a}^{a} \sqrt{\frac{a-x'}{a+x'}} \frac{1}{x-x'} dx' \tag{17.17}$$

The integrand can be written as $(a-x)/\sqrt{a^2-x'^2}/(x-x')$ and from (M8.6) and (M8.7) of the Table of Airfoil Integrals, p. 524, the first term is zero and the second yields π. Hence

$$\Delta p(x,t) = -\frac{C(t)}{\pi} \sqrt{\frac{a+x}{a-x}} \tag{17.18}$$

Put this to (2.23) (p. 31) to give the pressure at any field point as

$$p(x,y,t) = \frac{C(t)y}{2\pi^2} \int_{-a}^{a} \sqrt{\frac{a+x'}{a-x'}} \frac{1}{(x-x')^2 + y^2} dx' \tag{17.19}$$

The kinematic condition on the plate as specified by (17.14) is

$$v(x,0,t) = -U\alpha - \tilde{v}\, e^{i(\omega x/U+\omega t)} \tag{17.20}$$

where v_g is taken from (17.3), reverting to dimensional x, to avoid notational confusion.

We observe that the boundary condition (17.20) contains a term independent of t, $(-U\alpha)$, and one which varies sinusoidally or harmonically in t, i.e., as $e^{i\omega t}$. The steady part is well known to us and we shall drop consideration of $-U\alpha$ as we can see that the steady and unsteady parts are additive but independent, within the framework of linearized theory. Now we can separate out the time dependence by writing

$$\begin{aligned} p(x,y,t) &= \breve{p}(x,y)\, e^{i\omega t} \\ C(t) &= \breve{C}\, e^{i\omega t} \end{aligned} \tag{17.21}$$

Inserting this to (17.18), remembering that $t'' = t+(x-x'')/U$ when used in (17.12) gives,

$$v(x,0,t) = -\frac{\breve{C}}{2\pi^2\rho U}\int_x^{\infty} dx'' \times\!\!\!\!\!\!\int_{-a}^{a} \sqrt{\frac{a+x'}{a-x'}}\,\frac{e^{ik(x-x'')/a}}{(x''-x')^2}\,dx'\, e^{i\omega t} \tag{17.22}$$

where k is the reduced frequency $(= \omega a/U)$.

When enforcing the kinematic condition (17.20) (sans $-U\alpha$) we first observe that $e^{ikx/a}$ factors out. Moreover, the x''-integral with $y = 0$ is actually (cf. (17.9))

$$\int_x^{\infty} \frac{\partial p}{\partial y}\bigg|_{y=0} dx'' = \int_x^{a} \frac{\partial p}{\partial y}\bigg|_{y=0} dx'' + \int_a^{\infty} \frac{\partial p}{\partial y}\bigg|_{y=0} dx''$$

But since (17.15) expresses that $\partial p/\partial y|_{y=0} = 0$ on the foil, the first integral is zero for $|x| \le a$. The kinematic condition then takes the form

$$\frac{\breve{C}}{2\pi^2\rho U}\int_a^{\infty} dx'' e^{-ikx''/a} \times\!\!\!\!\!\!\int_{-a}^{a} \sqrt{\frac{a+x'}{a-x'}}\,\frac{dx'}{(x''-x')^2} = \breve{v} \;\; ; \;\; |x| \le a \tag{17.23}$$

The partial kernel in (17.23), $1/(x''-x')^2$, is not singular generally as $a \le x'' < \infty$, except for $x'' = a$. We then postpone the interpretation of the finite part to the outer integral of (17.23) which we write with all variables in fraction of the semi-chord a as

$$I^* = -\int_1^{\infty} dx'' e^{-ikx''}\frac{\partial}{\partial x''}\int_{-1}^{1} \frac{1+x'}{\sqrt{1-x'^2}\,(x''-x')}\,dx' \tag{17.24}$$

Let the inner integral of (17.24) be designated I_1^*

$$I_1^* = \int_{-1}^{1} \frac{1+x'}{\sqrt{1-x'^2}\ (x''-x')} dx' = \int_{-1}^{1} \frac{1}{\sqrt{1-x'^2}} \left\{ \frac{1}{x''-x'} - 1 + \frac{x''}{x''-x'} \right\} dx'$$

$$= -\int_{-1}^{1} \frac{1}{\sqrt{1-x'^2}} dx' + (1+x'')\int_{-1}^{1} \frac{1}{\sqrt{1-x'^2}\ (x''-x')} dx' \tag{17.25}$$

Let $x' = -\cos\theta$, then

$$I_1^* = -\pi + (1+x'')\int_0^{\pi} \frac{1}{x'' + \cos\theta} d\theta \tag{17.26}$$

The integral can be found in most tables of integrals, e.g. Peirce & Foster (1956)

$$I_1^* = -\pi + \frac{2(1+x'')}{\sqrt{x''^2-1}} \tan^{-1} \left\{ \frac{\sqrt{x''^2-1}\ \tan\frac{1}{2}\theta}{x''+1} \right\}_0^{\pi}$$

$$= -\pi + \pi\sqrt{\frac{x''+1}{x''-1}} \tag{17.27}$$

Substituting (17.27) into (17.24) yields

$$I^* = -\pi \int_1^{\infty} e^{-ikx''} \frac{\partial}{\partial x''} \sqrt{\frac{x''+1}{x''-1}}\, dx'' \tag{17.28}$$

Upon integration by parts there is clearly a square-root infinity from the lower limit and within the concept of the finite part the singular contribution is simply ignored, as demonstrated in Section 4 of the Mathematical Compendium, p. 510.

We see also that the integral is improper with respect to the upper limit. To extract the value we write (17.28)

$$I^* = -\pi \lim_{\mu \to 0} \int_1^{\infty} e^{-(ik+\mu)x''} \frac{\partial}{\partial x''} \sqrt{\frac{x''+1}{x''-1}}\, dx''$$

$$= -\pi \left\{ e^{-ikx''} \sqrt{\frac{x''+1}{x''-1}} \bigg|_{x''=1} + ik \int_1^{\infty} \sqrt{\frac{x''+1}{x''-1}}\, e^{-ikx''} dx'' \right\}$$

and ignoring the infinity à la Mangler,

$$I^* = -ik\pi \left\{ \int_1^{\infty} \frac{x'' e^{-ikx''}}{\sqrt{x''^2-1}} dx'' + \int_1^{\infty} \frac{e^{-ikx''}}{\sqrt{x''^2-1}} dx'' \right\} \tag{17.29}$$

From Abramowitz & Stegun (1972) the modified Bessel function of the second kind has the following integral representation

$$K_\nu(z) = \frac{\sqrt{\pi}}{\Gamma(\nu+\frac{1}{2})} (\tfrac{1}{2}z)^\nu \int_1^\infty e^{-zx''}(x''^2-1)^{\nu-\frac{1}{2}} dx''$$

and for $\nu = 0$ and $z = ik$,

$$K_0(ik) = \int_1^\infty \frac{e^{-ikx''}}{\sqrt{x''^2-1}} dx'' \quad ; \quad \Gamma(\tfrac{1}{2}) = \sqrt{\pi} \tag{17.30}$$

and

$$\frac{d}{d(ik)} K_0(ik) = -K_1(ik) = -\int_1^\infty \frac{x''e^{-ikx''}}{\sqrt{x''^2-1}} dx'' \tag{17.31}$$

We see by comparison with (17.29) that the first term involves K_1 and the second K_0 so our integral can now be written as

$$I^* = -ik\pi \left\{K_0(ik) + K_1(ik)\right\}$$

and inserting this into (17.23) gives

$$-\frac{\tilde{C}ik}{2\pi\rho U} \left\{K_0(ik) + K_1(ik)\right\} = \tilde{v} \tag{17.32}$$

which can be solved for $\tilde{C}$ to yield

$$\tilde{C} = -\frac{2\pi\rho U \tilde{v}}{ik\{K_0(ik) + K_1(ik)\}} \tag{17.33}$$

Multiplying by $e^{i\omega t}$ to get C, substituting this in (17.18) gives

$$\Delta p(x,t) = \frac{2\rho\tilde{v}U}{ik\{K_0(ik) + K_1(ik)\}} \sqrt{\frac{a+x}{a-x}}\, e^{i\omega t}$$

and integrating this over the foil (from $-a$ to a,) to secure the lift

$$L = 2\pi\rho a\tilde{v}U\, S(k)\, e^{i\omega t} \tag{17.34}$$

where

$$S(k) = \frac{1}{ik\{K_0(ik) + K_1(ik)\}} \tag{17.35}$$

is referred to as the ***Sears function.***

Normalizing (17.34) to obtain the lift coefficient,

$$\frac{L}{\frac{1}{2}\rho(2a)U^2} = C_L = 2\pi\tilde{\alpha}\, S(k)\, e^{i\omega t} \tag{17.36}$$

where $\tilde{\alpha} = \tilde{v}/U$ is the complex amplitude of the angle of attack and we retain the real part of the right side. Hence the effect of unsteadiness of the flow is to yield an effective angle of attack, $\tilde{\alpha}S(k)$.

From the Bessel-function relations

$$K_0(ik) = -\frac{\pi}{2}\, i(J_0(k) - iY_0(k)) \quad ; \quad K_1(ik) = -\frac{\pi}{2}\,(J_1(k) - iY_1(k))$$

the Sears function can be written

$$\begin{aligned} S(k) &= -\frac{2}{\pi k[Y_1(k) - J_0(k) + i(Y_0(k) + J_1(k))]} \\ &= -\frac{2}{\pi k}\left\{\frac{Y_1(k) - J_0(k) - i(Y_0(k) + J_1(k))}{(Y_1(k) - J_0(k))^2 + (Y_0(k) + J_1(k))^2}\right\} \\ &= S_r(k) + iS_i(k) \end{aligned} \tag{17.37}$$

; S_r and S_i are the coefficients of the real and the imaginary part

Here the J_0, J_1, Y_0, Y_1 are the Bessel functions of the first and second kind of orders 0 and 1 respectively.

For the gust in the form of (17.1) we have

$$\tilde{\alpha} = \frac{\tilde{v}}{U} = \tilde{a} - i\tilde{b} \tag{17.38}$$

We want the real part of $\tilde{\alpha}\, S(k)\, e^{i\omega t}$, i.e.,

$$Re\left\{(\tilde{a} - i\tilde{b})(S_r + iS_i)(\cos\omega t + i\sin\omega t)\right\} \tag{17.39}$$

$$= (\tilde{a}S_r(k) + \tilde{b}S_i(k))\cos\omega t + (\tilde{b}S_r(k) - \tilde{a}S_i(k))\sin\omega t$$

Thus if (for simplicity) the gust varies only as $\cos\omega t$, then $\tilde{b} = 0$ and the lift coefficient is

$$C_L = 2\pi\tilde{a}\,(S_r(k)\cos\omega t - S_i(k)\sin\omega t) \tag{17.40}$$

Here we see that the responding lift is made up of both in-phase and quadrature components, and that for $k = 0$, the steady value is recovered.

An alternate form of the lift coefficient is

$$C_L = 2\pi \; Re \left\{ \tilde{\alpha} \; e^{i\omega t} \left[\frac{2i/\pi k}{H_1^{(2)}(k) + iH_0^{(2)}(k)} \right] \right\} \tag{17.41}$$

where $H_0^{(2)}$ and $H_1^{(2)}$ are the Hankel functions

$$H_\nu^{(2)}(k) = J_\nu(k) - iY_\nu(k)$$

For large k

$$H_\nu^{(2)}(k) \approx \sqrt{\frac{2}{\pi k}} \; e^{-i(k - \pi/4 - \nu\pi/2)} \tag{17.42}$$

and hence they oscillate rapidly (as sin k and cos k) as the reduced frequency is made large. A plot of the function within the square brackets in (17.41) is shown in Figure 17.3 for two origins. With the origin at mid-chord (which we have taken) the phase is seen to vary widely. Whereas for the origin taken at the leading edge the phase ($\tan^{-1}$(*Im*aginary part/*Re*al part)) varies slowly. The reason is that for the gust referred to the leading edge the gust expression would be, cf. (17.3)

$$v_g = Re \left\{ \tilde{v} \; e^{ik(\check{x}-1)+i\omega t} \right\} \tag{17.43}$$

This would produce a factor of e^{-ik} in the numerator of (17.41) which cancels the same factor in the denominator as exhibited in (17.42) as k becomes large.

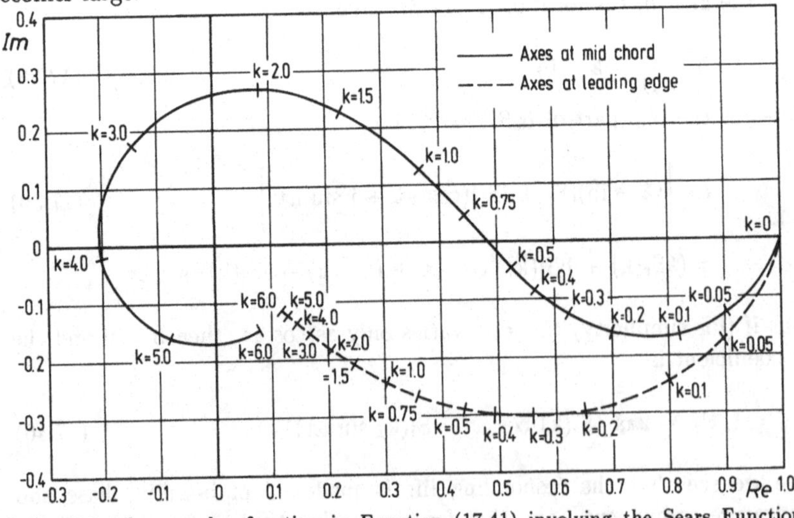

Figure 17.3 The complex function in Equation (17.41) involving the Sears Function for $|\tilde{\alpha}| = 1.0$ evaluated for axes taken at mid-chord and at the leading edge.

UNSTEADY LIFT ON HYDROFOILS OF FINITE SPAN

When we depart from the realm of two-dimensional flow we unfortunately encounter mathematical formulations which are highly resistant to analytical procedures and we must resort to numerical methods. These can be coped with by use of computers. Since there are no analytical inversion techniques for the integral equations arising in unsteady lift on foils of arbitrary planform all we can do is to formulate the problem and then exhibit results obtained via computer to see the effects of both aspect ratio and reduced frequency.

However, the steady lift on slender wings can be found approximately by analytical methods. The lift-slope of such wings is much smaller than that for large aspect ratio, and this has a bearing on interpretation of the ratio of unsteady to quasi-steady lift. It therefore seems worthwhile to digress briefly to develop the steady results for slender wings.

We consider a foil of arbitrary but slender planform as shown in Figure 17.4.

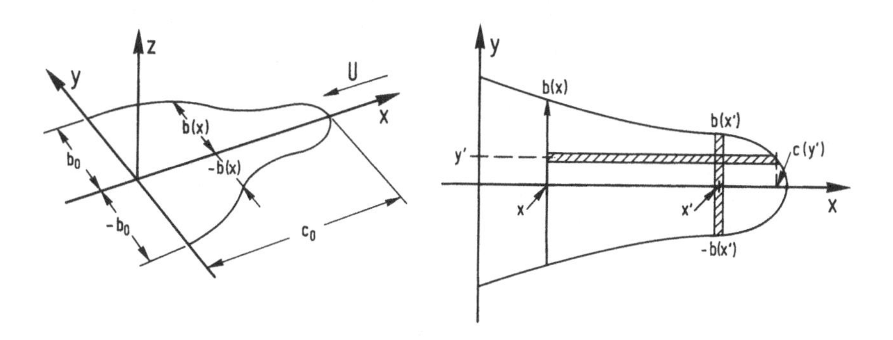

Figure 17.4 Slender foil $\frac{2b_0}{c_0} < 1$. Limits for two integration schemes indicated.

The field pressure arising from a distribution of $\Delta p(x',y')$ on this planform is, analogous to (2.23) but now in three dimensions and the pressure dipole directed in the z-direction

$$p(x,y,z) = -\frac{1}{4\pi} z \int_0^{c_0} dx' \int_{-b(x')}^{b(x')} \frac{\Delta p(x',y')}{[(x-x')^2+(y-y')^2+z^2]^{3/2}} dy' \tag{17.44}$$

The steady linearized pressure-velocity potential relationship is the same form as in two dimensions, cf. (17.9), with w and z instead of v and y.

Now suppose that the foil is beset with some specified vertical velocity w(x,y,0), then the kinematic condition is

$$w_{foil} + w(x,y,z) = 0 \;;\quad x,y \text{ in foil, } z = 0 \tag{17.45}$$

Thus we are confronted with the integral equation

$$\frac{1}{4\pi\rho U}\int_x^\infty dx''\int_0^{c_0} dx'\int_{-b(x')}^{b(x')} \frac{\Delta p(x',y')}{[(x''-x')^2+(y-y')^2]^{3/2}}\, dy' = -w(x,y,0) \tag{17.46}$$

where the kernel function is highly singular and as before of the Mangler (1951) type.

Carrying out the x''-integration, we get from that

$$\frac{1}{(y-y')^2}\left[1-\frac{x-x'}{\sqrt{(x-x')^2+(y-y')^2}}\right] \approx \frac{1}{(y-y')^2}\left[1-\frac{x-x'}{|x-x'|}\right] = \begin{cases} 0 & x' < x \\ \dfrac{2}{(y-y')^2} & x' > x \end{cases} \tag{17.47}$$

where we assume that (y–y') can be suppressed (in the radical) because the range of y and y' is much smaller than that of x and x' as the foil is slender.

With this assumption, we are faced with

$$\frac{2}{4\pi\rho U}\int_x^{c_0} dx' \,⨍_{-b(x')}^{b(x')} \frac{\Delta p(x',y')}{(y-y')^2}\, dy' = -w(x,y,0) \tag{17.48}$$

Integrating over y' by parts and retaining only the finite part we obtain

$$\frac{1}{2\pi\rho U}\int_x^{c_0} dx' \,⨍_{-b(x')}^{b(x')} \frac{\partial \Delta p(x',y')/\partial y'}{y-y'}\, dy' = w(x,y,0) \tag{17.49}$$

Interchanging the orders of integration, see Figure 17.4, we obtain

$$\frac{1}{2\pi\rho U}⨍_{-b(x)}^{b(x)} \frac{dy'}{y-y'}\int_x^{c(y')} \frac{\partial}{\partial y'}\Delta p(x',y')dx' = w(x,y,0) \tag{17.50}$$

If we now consider (following Ashley & Landahl (1977))

$$\frac{\partial}{\partial y'}\int_x^{c(y')} \Delta p(x',y')dx' = \frac{dc(y')}{dy'}\Delta p(c(y'))+\int_x^{c(y')} \frac{\partial}{\partial y'}\Delta p(x',y')dx'$$

According to Mangler (cf. p. 513, Section 4 of the Mathematical Compendium) the first term vanishes and writing the partial, chordwise lift per unit span as

$$L(x,y') = \int_x^{c(y')} \Delta p(x',y')dx'$$

we have the following integral equation of the Munk-type

$$\frac{1}{2\pi\rho U}\int_{-b(x)}^{b(x)} \frac{1}{y-y'}\frac{\partial}{\partial y'} L(x,y')\, dy' = w(x,y,0) \tag{17.51}$$

The inversion of this is (see p. 69–71)

$$\frac{\partial}{\partial y} L(x,y) = -\frac{2}{\pi}\frac{\rho U}{\sqrt{b^2(x)-y^2}}\int_{-b(x)}^{b(x)} \frac{\sqrt{b^2(x)-y'^2}}{y-y'} w(x,y')dy' + \frac{K}{\sqrt{b^2(x)-y^2}} \tag{17.52}$$

We can note that if we integrate this over y from $y = -b$ to b the left side will be $L(x,b) - L(x,-b)$, each of which is zero since $c(b) = x$ (x'–limits in L come together) and the right side involves

$$\int_{-b}^{b} \frac{1}{\sqrt{b^2(x)-y^2}}\frac{1}{y-y'} dy = 0 \quad \text{(cf. (M8.6) of the Table of Airfoil Integrals)}$$

Hence K must be zero, giving the non-circulatory solution.

Now the case of central interest is for w independent of y (or y'). Then the y'-integral in (17.52) is by (M8.6) and (M8.8) from the Table of Airfoil Integrals, πy, and hence

$$\frac{\partial}{\partial y} L(x,y) = -2\rho U w(x) \frac{y}{\sqrt{b^2(x)-y^2}}$$

or

$$\int_x^{c(y)} \Delta p(x',y)dx' = 2\rho U w(x)\sqrt{b^2(x)-y^2} \tag{17.53}$$

For the upward velocity $w(x) = U\alpha$, (i.e. foil at angle of attack α) and taking $x = 0$, $b^2(0) = b_0^2$ we have the entire chordwise lift along $x = 0$, and the total lift is simply

$$L^w = \int_{-b_0}^{b_0} dy \int_0^{c(y)} \Delta p(x',y)dx' = 2\rho U^2 \alpha \int_{-b_0}^{b_0} \sqrt{b_0^2 - y^2}\, dy$$

or

$$L^w = \pi \rho U^2 \alpha b_0^2 \tag{17.54}$$

Dividing through by $\frac{1}{2}\rho SU^2$, S being the planform area, noting that

$$\frac{b_0^2}{S} = \frac{1}{4}\frac{(\text{span})^2}{S} = \frac{A_R}{4}$$

A_R, the aspect ratio, we get finally,

$$C_L^w = \frac{\pi}{2}\alpha A_R \tag{17.55}$$

Thus the lift rate $dC_L^w/d\alpha$ at low aspect ratio is only 1/4th of that of two-dimensional sections. This formula compares well with data and more precise evaluations in the range $0 \leq A_R \leq 1$ as may be seen in Figure 17.5. We shall see next that the unsteady effects on low-aspect-ratio wings are smaller relative to this small quasi-steady lift than for two-dimensional sections.

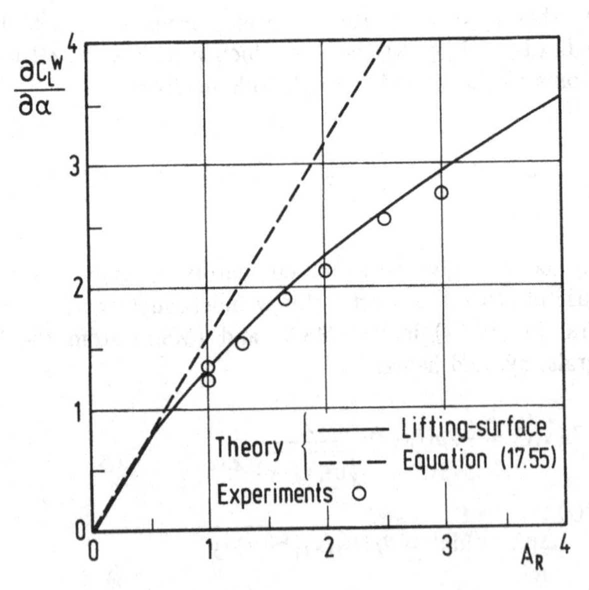

Figure 17.5 Comparison of theoretical and experimental lift-curve slopes for low-spect-ratio triangular wing. (Numerical lifting-surface theory, by courtesy of Dr. S. Widnall).

Figure 6-10, from Ashley, H. & Landahl, M. (1977). *Aerodynamics of Wings and Bodies*. Reading, Mass.: Addison-Wesley Publishing Co. By courtesy of H. Ashley & M. Landahl.

The unsteady, three-dimensional foil theory involves, by analogy with two dimensions, an integral equation of the following form

$$\frac{1}{4\pi\rho U}\int_{x}^{\infty} dx'' \int_{-b}^{b} dy' \int_{-a_t(y')}^{a_l(y')} \frac{\Delta p(x',y')\, e^{-ikx''}}{[(x''-x')^2+(y-y')^2]^{3/2}}\, dx' = -\tilde{v} \tag{17.56}$$

Rather surprisingly, according to Ashley & Landahl (1977), there have been no useful approximate solutions for either high or low aspect ratio nor for low or high reduced frequencies in range of Mach numbers between 0 and 1 except for the case of circular planform. For supersonic flow there are many analytical solutions for particular planforms.

For subsonic flows the approach has been one of numerical solution of complete lifting-surface representations. The definitive works are those of Watkins, Runyan & Woolston (1955) and Watkins, Woolston & Cunningham (1959) at the NACA (now NASA), Langley, Virginia.

Computer-generated results are shown in Figure 17.6 where we perceive the dependence of the modulus of unsteady lift relative to quasi-steady lift as a function of reduced frequency, k, and aspect ratio, A_R. It is seen that as the aspect ratio decreases, the relative effect of unsteadiness diminishes rapidly.

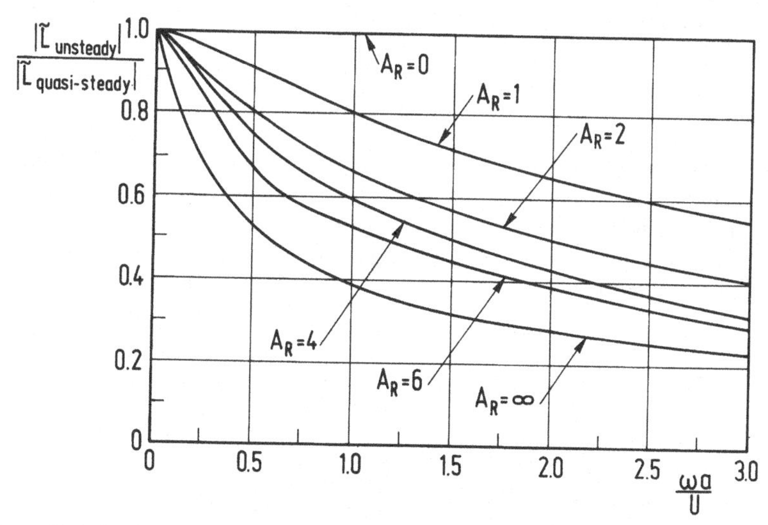

Figure 17.6 Ratio of unsteady to quasi-steady lift response of rectangular foils to sinusoidal gust of unit amplitude.

Figure 3 from Breslin, J.P. (1970). Theoretical and experimental techniques for practical estimation of propeller-induced vibratory forces. *Trans. SNAME*, vol. 78, pp. 23–40. New York, N.Y.: SNAME.

IMPLICATIONS FOR PROPELLERS

The general expression for a travelling wave of unit amplitude is

$$w = e^{(i2\pi/\lambda)(x-\nu t)}$$

where λ is the wavelength and ν is the celerity (speed of propagation). As we have seen, each propeller blade encounters each of the spatially distributed harmonics of the wake of order numbers n. Then the wavelength at any r is

$$\lambda = \frac{2\pi r}{n}$$

The relative celerity ν at this radius is effectively (at the outer radii)

$$\nu = -\omega r$$

ω being the angular velocity of the propeller. Then the incident gust takes the form

$$w = e^{inx/r}\, e^{in\omega t}$$

x being the chordwise variable. Comparing this with the gust used in the foregoing two-dimensional analysis (17.3), i.e.

$$\frac{v_g}{\tilde{v}} = e^{i(\omega a/U)(x/a)}\, e^{i\omega t}$$

We see that $k = \omega a/U$ is replaced by nc/r and the coefficient of t by $n\omega$. The propeller-section reduced frequency, nc/r, c being the expanded chord at say $r = 0.7R$, can be expressed in terms of the expanded-area ratio A_E/A_0. For a typical propeller[26] the expanded-area ratio in terms of the blade numbers and chord at 0.7 R is

$$A_E/A_0 \approx \frac{Zc_{0.7}R}{\pi R^2} \approx \frac{Zc_{0.7}}{D}\frac{1}{2}$$

Then the reduced frequency k for blades is

$$k = \frac{nc_{0.7}}{0.7R} \approx 5.7\,\frac{n}{Z}\,A_E/A_0$$

26 For the Wageningen B-series propeller (4–7 blades) $A_E/A_0 = (Zc_{0.7}/D)(1/2.144)$ (Oosterveld & van Oossanen (1975)).

For a typical $A_E/A_0 = 0.80$, and $Z = 5$ for which the relevant n are in the range $1 \lesssim n \lesssim 6$, we see that

$$0.90 < k < 5.5$$

Thus the blade loadings from the lower wake orders are incurred at low reduced frequencies and those at blade frequencies $n = Z$ at $5.7 \cdot (A_E/A_0) = 4.6$ (in this case), a high reduced frequency relative to aerodynamic experience. The aspect ratio of propeller blades is small being effectively

$$A_{R\ \mathrm{blades}} = \frac{R^2}{\pi R^2 A_E/A_0} = \frac{1}{\pi A_E/A_0} = 0.4 \quad \text{for } A_E/A_0 = 0.80$$

We surmise that the unsteady/quasi-steady thrust ratio for propeller blades may lie in the vicinity of 0.80 because of the low aspect ratio of the blades, to the extent that the aerodynamic behavior observed in Figure 17.6 can be applied to propellers.

In any event, this look at unsteady wing and section lifts should convince us that the two-dimensional theory cannot be applied to wide propeller blades because the lift-slope is too high and the unsteady reductions are too large. Thus we are led to the consideration of propellers as composed of blades which must be represented as lifting surfaces negotiating the "wavy structure" of the hull wake.

18 Lifting-Surface Theory

Ship propellers have wide blades to distribute the loading over the blades. As we have seen previously this is necessary to reduce local negative pressures to avoid, as much as possible, the generation of cavitation. We can then think of the blades as low-aspect-ratio wings that rotate and translate through the water. To describe the flow we consider the propeller blades as lifting surfaces over which we distribute singularities to model the effects of blade loading and thickness. This was done in Chapter 14 for uniform inflows and in Chapter 15 for varying blade loadings in hull wakes. But in those chapters we only considered the pressure from assumed distributions of loading and thickness over the propeller blades. To be able to find these distributions we must establish a relation between the pressure and the velocity and fulfill the kinematic boundary condition on the blades. We anticipate that the operation of the blades in the spatially varying flow, generated by the hull, will give rise to forces which vary with blade position. Hence it is necessary to treat the blades as lifting surfaces with non-stationary pressure distributions.

In this chapter an overview of the extensive literature of the past three decades is followed by a detailed development of a linear theory which is sufficiently accurate for prediction of the unsteady forces arising from only the temporal-mean spatial variations of inflow. As mentioned in Chapter 13 hull wake flows vary with time in a non-periodic manner, but lack of knowledge of their statistical characteristics has inhibited efforts to predict stochastic blade forces. Use of Fourier expansions enables identification of those harmonics of the time-averaged hull wake which give rise to the various force components acting on the shaft. The chapter concludes with an outline of a non-linear theory which is required for more accurate prediction of mean thrust, torque, efficiency and pressure distributions. Comparisons with measurements are given in Chapter 19.

OVERVIEW OF EXTANT UNSTEADY THEORY

The application of lifting-surface theory to ship propellers had to await the arrival of the digital computer because of the size and complication of the numerics involved. Ritger & Breslin (1958) devised an *ad-hoc* theory using steady lifting-line theory coupled with the unsteady-section theory of Sears which later was found to be too approximate, yet gave a good indication of the relative reduction of unsteady forces with skew. Hanaoka (1962, 1969a, 1969b) was perhaps the first to give a rational, linearized lifting-surface theory for the oscillating propeller, and also for operation in a wake, using the acceleration potential or simply the disturbance pressure as the dependent variable. A thorough, consistently linearized theory for propellers in *uniform* flow was given by Sparenberg (1959, 1960, 1984) in terms of pressure dipoles and also via vortex representations with an exhaustive elucidation of the singularities occurring in each procedure. His work was later programmed at MARIN by Verbrugh (1968) who extended it to unsteady loadings as occur in wakes. Later van Gent (1975) gave a detailed account of Verbrugh and comparisons with measurements.

Dr. Stavros Tsakonas and Miss Winnifred Jacobs at Davidson Laboratory, Stevens Institute of Technology, pioneered the application of lifting-surface theory to propellers in spatially varying wakes. They began in the late 50's with rudimentary modeling of helical blade wakes and single bound vortices at one-quarter chord with imposed kinematic condition at the three-quarter chord (à la Weissinger) in order to cope with the computer effort at that time. As these and other approximate models proved to be inadequate they were replaced by Shioiri & Tsakonas (1964) by a representation of the trailing vortices along helical surfaces determined by the convection of the undisturbed relative flow. Many other papers and reports have been given in which these procedures were improved and applied to various configurations, cf. Tsakonas, Chen & Jacobs (1967), Tsakonas, Jacobs & Rank (1968), Tsakonas & Jacobs (1969), Jacobs, Mercier & Tsakonas (1972), Jacobs & Tsakonas (1973), Tsakonas, Jacobs & Ali (1973), Jacobs & Tsakonas (1975), Tsakonas & Liao (1983). The final computer program by Tsakonas and Jacobs gave results which compared well with model measurements. That program was documented and used in Europe and in the USA.

Programs for calculation of mean thrust and torque were written by Kerwin (1961) and by Pien (1961) followed by Cox (1961), English (1962) and Nelson (1964). Later Kerwin & Lee (1978) undertook the computation of unsteady forces with a computer code called Propeller-Unsteady-Forces (PUF). It was extended by Lee (1979), at that time Kerwin's student, to include unsteady cavitation and is used extensively.

Significant contributions have been made in Germany by Isay (1970), in USSR by Bavin, Vashkevich & Miniovich (1968) and in Japan by many researchers, notably Yamazaki (1962, 1966), Koyama (1972), Tanibayashi & Hoshino (1981), Hoshino (1989a, 1989b) and Ito, Takahashi & Koyama (1977) who evaluated an early theory of Hanaoka (1969b). An excellent review of the development of mainly steady propeller theory up to 1986 has been rendered by Kerwin (1986).

Later advances concern the modeling of the force-free vortices abaft the blades and the influence of thickness distribution on steady and unsteady propeller forces. All the lifting-line and lifting-surface theorists recognized that the local trajectories of the trailing vortices were important in regard to the flow and forces induced on the blades. They were forced by the complexities of the non-linear coupling of this wake geometry with the loading to take these trajectories as being constant in radius and purely helical until the advent of sufficient computational power. An enormous literature from the "vortex chasers" has been reviewed by Hoeijmakers (1983) and by Sarpkaya (1989). The first free-wake analysis applied to ship propellers is that of Cummings (1968) who by assuming small streamwise gradients was able to transform the problem to an unsteady one in two dimensions. His calculated radial position of the tip vortex agreed well with observations. His attempt to implement a fully three-dimensional wake relaxation scheme for a lifting-line wing proved to be computationally excessive at that time.

A partially-free-wake model for propellers in peripherally uniform flow (steady loading) has been developed by Greeley & Kerwin (1982). They prescribe the radial positions of the trailing vortices and then require that the pitch of the vortices be determined by the local axial and tangential flow components. Roll-up of the vortex sheets is included through the use of empirical parameters. Leading-edge separated, vortical flow is also approximated. The non-linear behavior of the vortex wakes has been addressed by Keenan (1989) via a vortex-lattice numerical procedure. His results compare well with measured forces and vortex trajectories.

Lifting-surface methods which model the blades as laminae with zero thickness fail to give good pressure distributions along the blade leading-edge and tip regions. There the effects of thickness are substantial (as was demonstrated in Chapter 5 for the two-dimensional section). To account for both camber and thickness, *boundary-element* or *panel-methods* have been employed which make use of distributions of singularities on the actual surface. Boundary-element representations have been applied to the steady flow around marine propellers, including the hub, by Hess & Valarezo (1985) and by Hoshino (1989a, 1989b) and to the flow about propellers and ducts by Kerwin, Kinnas, Lee & Wei-Zen (1987). Unsteady flow about hydrofoils (in two dimensions) by Giesing (1968) and Basu & Hancock (1978) include large-amplitude oscillation yielding non-planar vortex trajectories. A numerical procedure for analysis of marine propellers of extreme geometries in spatially non-uniform inflow has been evolved by Kinnas & Hsin (1990) using a boundary-element method in a potential-based formulation. As this comprises a general and advanced representation of ship propellers it is outlined later at the end of this chapter.

Figure 18.1 Propeller for 4400 TEU container ship, owned by Hapag Lloyd, Germany, built by Samsung Shipyard, Korea. Propeller data: Diameter 8.250 m; 6 blades; $A_E/A_0 = 0.795$; weight: 63 000 kg; material: Alcunic 8; power: 36 500 kW at 93 RPM; manufacturer: Ostermann Metallwerke, Germany. By courtesy of Ostermann Metallwerke, Germany.

BLADE GEOMETRY AND NORMALS

Before deriving expressions for the velocities on the propeller blade we consider the blade geometry. Modern propellers are designed with blades which are skewed, as can be seen in Figure 18.1. The use of skew makes the blade sections pass through the wake "spike" or valley in a staggered fashion and hence drastically reduces the vibratory forces, as demonstrated in Chapter 12 (p. 256 and sequel). But when the blade sections are skewed, i.e., when they are displaced along the helical line at given radius and pitch, they are also moved in the axial direction (generally aft), cf. Figure 18.2. This would result in a large axial extent of the propeller and to avoid this and make the propeller fit the aperture, the blades are raked forward to compensate for skew-induced rake. We consider the general case with both skew and rake which we designate ψ_s and x_r while x_s is the rake induced by skew. Note, that in this notation ψ_s

and x_s are positive when the section is skewed in the direction opposite to the direction of rotation, while x_r is positive when the blade is raked forward (in contrast to ITTC notation).

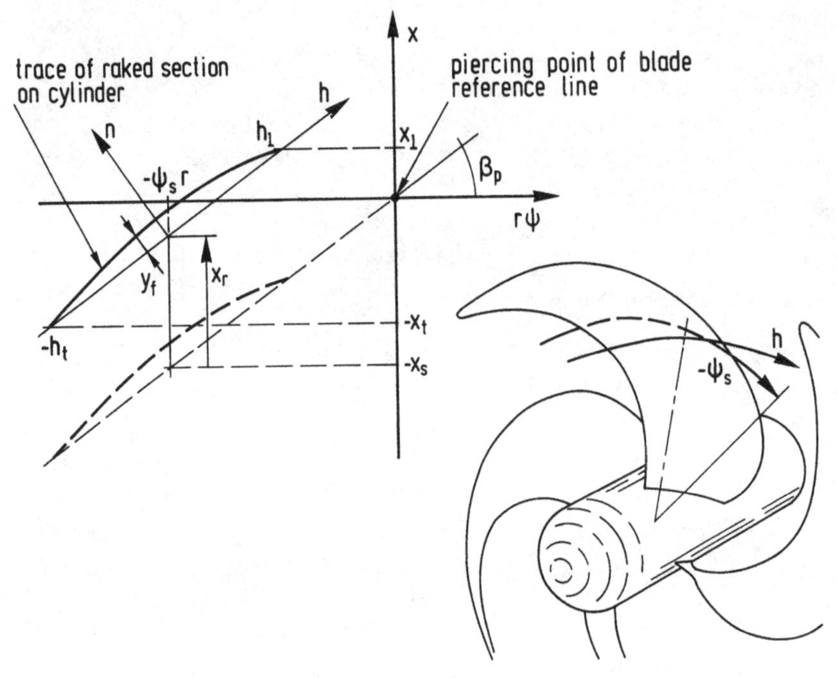

Figure 18.2 Expanded surface of a cylinder of radius r showing traces of blade camber line with skew and rake. See also Figure 14.1 and 14.2.

The parametric equations of the camber line of any section of a blade with skew and rake can be written, with respect to the blade reference line

$$x = -r\,\psi_s \tan\beta_p + x_r + h \sin\beta_p + y_f \cos\beta_p \tag{18.1}$$

$$r\psi = -r\,\psi_s + h \cos\beta_p - y_f \sin\beta_p \tag{18.2}$$

r being constant. Eliminating h between these relations gives the equation for the camber line at any fixed r as

$$F_f(x,\psi;r) = x \cos\beta_p - x_r\cos\beta_p - r\psi \sin\beta_p - y_f(h,r) = 0 \tag{18.3}$$

From the differential geometry of surfaces, as displayed in Chapter 3, p. 42 and 43, we need the derivatives of F_f to calculate the normals, cf. Equation (3.6). In the present cylindrical system they are

$$(n_a,n_r,n_t) = \left[\frac{1}{N}\frac{\partial F_f}{\partial x}, \frac{1}{N}\frac{\partial F_f}{\partial r}, \frac{1}{N}\frac{1}{r}\frac{\partial F_f}{\partial \psi}\right] \tag{18.4}$$

where $N = \sqrt{\left[\frac{\partial F_f}{\partial x}\right]^2 + \left[\frac{\partial F_f}{\partial r}\right]^2 + \left[\frac{1}{r}\frac{\partial F_f}{\partial \psi}\right]^2}$

We then have the axial derivative (from 18.3)

$$\frac{\partial F_f}{\partial x} = \cos\beta_P - \frac{\partial y_f}{\partial h}\frac{\partial h}{\partial x} = \cos\beta_P - \frac{\partial y_f}{\partial h}\frac{1}{\sin\beta_P} \tag{18.5}$$

using (14.3) to connect x and h. The camber slope $\partial y_f/\partial h$ can be considered small relative to $\cos\beta_P$. Moreover, in first order theory we replace the pitch angle β_P with the *fluid pitch angle* β ($= \tan^{-1}(U/\omega r)$) as explained in Chapter 14. We then have to first order

$$\frac{\partial F_f}{\partial x} = \cos\beta \tag{18.6}$$

For the radial derivative we have

$$\frac{\partial F_f}{\partial r} = -\left((x-x_r)\sin\beta_P + r\psi\cos\beta_P\right)\frac{\partial \beta_P}{\partial r} - \frac{\partial x_r}{\partial r}\cos\beta_P - \psi\sin\beta_P - \frac{\partial y_f(h,r)}{\partial r} \tag{18.7}$$

After eliminating ψ by using (18.3), neglecting all terms involving y_f and using (18.1) to eliminate $x-x_r$ we obtain

$$\frac{\partial F_f}{\partial r} = -\frac{2\pi r}{\sqrt{P^2 + (2\pi r)^2}}\left[\frac{\partial P}{\partial r}\left[\frac{h}{\sqrt{P^2 + (2\pi r)^2}} - \frac{\psi_s}{2\pi}\right] + \frac{\partial x_r}{\partial r}\right] \tag{18.8}$$

For constant pitch and zero rake we see that $\partial F_f/\partial r=0$ as required since we then have a regular helical surface with no radial component of the normal. For constant pitch with rake varying with radius we have

$$\frac{\partial F_f}{\partial r} = -\frac{2\pi r}{\sqrt{P^2 + (2\pi r)^2}}\frac{\partial x_r}{\partial r} = -\frac{\partial x_r}{\partial r}\cos\beta_P \tag{18.9}$$

Returning to (18.8) the physical pitch angle β_P is replaced by the fluid reference surface pitch angle β. As the terms in (18.8) can be of the same order we retain them but replace P in the coefficient of $\partial P/\partial r$ by the fluid pitch to give

$$\frac{\partial F_f}{\partial r} = -\cos\beta\left[\frac{1}{2\pi}\frac{\partial P}{\partial r}\left[\frac{h}{r\tan\beta} - \psi_s\right] + \frac{\partial x_r}{\partial r}\right] \tag{18.10}$$

From (18.3) we have for the tangential component

$$\frac{1}{r}\frac{\partial F_f}{\partial \psi} = -\sin\beta_p - \frac{\partial y_f}{\partial h}\frac{1}{\cos\beta_p} \tag{18.11}$$

For the first order orientation of the normal we approximate (18.11) by

$$\frac{1}{r}\frac{\partial F_f}{\partial \psi} = -\sin\beta \tag{18.12}$$

This completes the determination of the normal-vector derivation.

Note that the first order result (18.6) and (18.12) together with constant pitch and zero rake gives the normal

$$\mathbf{n} = (\cos\beta, 0, -\sin\beta) \tag{18.13}$$

which is the same as Equation (14.5), the normal used in Chapters 14–16 where it describes the orientation of the pressure dipoles. These dipoles were placed on the fluid reference surface, i.e., the surface parallel to the incoming (relative) flow. This situation corresponds to two-dimensional, linearized section theory, cf. p. 69, where the singularities are placed on the x-axis. If we wish to resolve details of the blade geometry, such as variation in camber, pitch and rake, we can do so by proceeding as in two-dimensional theory and calculate the flow component normal to the fluid reference surface and induced by the singularities on this surface. This flow component must be annulled by the normal component of the relative inflow velocity and here, as in two-dimensional theory, the higher-order normal should be used. This is demonstrated later when the kinematic boundary condition which the flow must meet on the blade is written out.

LINEAR THEORY

As we have not previously included blades with rake we must modify the equations of motion to include the addition of radial forces. The linearized Euler equations can be obtained as in Chapter 14, Equation (14.1), now with the radial force component included and compactly stated in vector form by

$$\left[\frac{\partial}{\partial t} - U\frac{\partial}{\partial x}\right]\mathbf{q}' + \nabla p' = \frac{1}{\rho}\mathbf{F}'(x,r,\gamma;x',r',\gamma'(t)) \tag{18.14}$$

where

$\nabla = \mathbf{i}_a\frac{\partial}{\partial x} + \mathbf{i}_r\frac{\partial}{\partial r} + \mathbf{i}_t\frac{1}{r}\frac{\partial}{\partial \gamma}$ is the gradient operator (as in (9.17))

$\mathbf{q}' = (u_a', u_r', u_t')$, the perturbation velocity due to the force $\mathbf{F}'$

$$\mathbf{F}' = -\frac{\Delta p(h',r',\gamma_0(t))}{r}\,\delta(x-x')\,\delta(r-r')\,\delta(\gamma-\gamma'(t))\,dS'\,\mathbf{n}'$$

the force from any element of the blade, acting on the fluid.

Here, as well as in the preceding Chapters 14-16, we have taken the inflow to the propeller as $-U$, the negative of the ship speed. While this can be defended from a theoretical point of view on the assumption that the velocities induced by the presence of the hull (as well as by the propeller itself) are small, any naval architect would take a pragmatic attitude and substitute U with U_A, the mean inflow velocity to the propeller disc. This is the standard procedure when using open-water propeller diagrams for propellers behind a ship, for example in a power prediction, as outlined in Chapter 12, pp. 235-238. A consistent adaption to this approach here would require further substitutions as

$$P_f \rightarrow P_A = \frac{2\pi U_A}{\omega}\ ;\ p_f^* \rightarrow p_A^* = \frac{\omega}{U_A}\ ;\ \beta \rightarrow \beta_A = \tan^{-1}\frac{U_A}{\omega r}\ ;\ \text{etc.}$$

To account for radially varying inflow, replace U_A with $U_a(r)$, the circumferentially mean inflow. Note that in that case the pitch will depend on radius. Here we avoid these additional complications and keep U and corresponding variables but remark that the above substitutions can be made.

As in Chapter 14, p. 275 we use the continuity equation to eliminate the velocity components and obtain the Poisson equation for the pressure p'

$$\nabla^2 p' = \left[\mathbf{i}_a\frac{\partial}{\partial x} + \mathbf{i}_r\left[\frac{\partial}{\partial r} + \frac{1}{r}\right] + \mathbf{i}_t\frac{1}{r}\frac{\partial}{\partial\gamma}\right]\cdot\mathbf{F}' \qquad (18.15)$$

where now the right side is modified relative to the equations of p. 277 by inclusion of the radial force component. For the Laplace operator in cylindrical coordinates, see (14.9).

It is important to note that this equation can be solved by reducing it to an ordinary differential equation by first observing that p' must be a cyclic function of the space angle γ. By inserting

$$p'(x,r,\gamma) = \sum_{-\infty}^{\infty} p'_m(x,r)\,e^{im\gamma}$$

in (18.15) the angular dependence can be removed in the equation for p'_m. Then by operation on this reduced equation by the Fourier Integral Transform

$$\tilde{p}'_m(r,\xi) = \int_{-\infty}^{\infty} p'_m(x,r)\,e^{-i\xi x}\,dx$$

it is possible to reduce (18.15) to an ordinary, inhomogeneous Bessel equation, with r as variable, whose solution is in terms of the modified Bessel functions of the first and second kind. Alternatively, one may employ the Hankel Transform

$$\hat{p}'_m(x,k) = \int_0^\infty r\, p'_m(x,r)\, J_m(kr)\, dr$$

to reduce the equation to an ordinary, inhomogeneous equation for $\hat{p}'_m$, with x as variable, whose solution involves $A\ e^{-kx} + B\ e^{kx}$ for $x > x'$ and $x < x'$. This more general approach enables meeting of boundary conditions on a cylindrical surface outside the propeller (tunnel walls) or duct of doubly infinite, x-wise extent or a cylindrical surface within the propeller radius (doubly infinite hub), or in the absence of such boundaries, solutions which vanish at infinite radius. In this latter case the field pressure is, of course, the same as that given by the solution of the Poisson equation in a boundless domain using one of the expansions of $1/R$ given in Section 5 of the Mathematical Compendium, involving either Bessel functions or modified Bessel functions, Equations (M5.22) and (M5.24). For details of the Fourier Integral Transform and the Hankel Transforms, cf. Sneddon (1951).

As in Chapter 14 the field pressure is that given by a single dipole for a boundless domain. It can be expressed in the general form (corresponding to (14.14) since $\partial/\partial n' = \mathbf{n}' \cdot \mathbf{\nabla}'$, but without the integration over the blade), as

$$p' = -\frac{1}{4\pi}\,\Delta p(h',r',t)\ \mathbf{n}'\cdot\mathbf{\nabla}'\left[\frac{1}{R}\right] \tag{18.16}$$

where $R = \sqrt{(x - x')^2 + r^2 + r'^2 - 2rr'\cos(\gamma-\gamma_0-\psi')}$ and the dummy x' and ψ' are connected by the equation of the blade nominal surface (18.3) or to first order (14.3). The gradient operator is as shown in (18.14) but operates here on the dummy variables as indicated with the '.

Equation (18.14) can now be integrated to give the induced velocity field by the method of characteristics, cf. Appendix E, p. 498. The three components of the velocity vector are integrated separately. We substitute the components of $\mathbf{q}'$ for v in Equation (E.7) and the components of $-\mathbf{\nabla}p' - \mathbf{F}'(x,r,\gamma;x',r',\gamma'(t))/\rho$, for the right side G, using (18.16) for p' operated on by $\mathbf{\nabla}$. It should be noted that the operations on $1/R$ must be carried out *before* substituting the variables in the method of characteristics which here is $x \to x''$ and $t \to t + (x - x'')/U$, cf. (E.16). Moreover, since the blade position angle of the 0th or key blade, $\gamma_0 = \omega t$, appears in the argument of the cosine in R this argument then becomes

$$\gamma - \gamma_0 - \psi' \longrightarrow \gamma - \omega\,(t + (x-x'')/U) - \psi' = \gamma - \gamma_0 - \psi' - p_f^*(x-x'') \tag{18.17}$$

We then have, after additional integration over the entire key blade,

$$\mathbf{q} = \frac{1}{4\pi\rho U}\left[\int_x^\infty dx'' \iint_S \Delta p(h',r',t+(x-x'')/U)\, \nabla\left[\frac{\partial}{\partial n'}\frac{1}{R}\right]\Bigg|_{\substack{x\to x''\\ t\to t+(x-x'')/U}} dS' \right.$$
$$- 4\pi \int_x^\infty dx'' \iint_S \frac{\Delta p(h',r',t+(x-x'')/U)}{r}\,\delta(x''-x')\,\delta(r-r')$$
$$\left. \cdot\,\delta(\gamma - \gamma_0 - \psi' - p_f^*(x-x''))\ \mathbf{n}'\,dS' \right] \tag{18.18}$$

which is the solution that satisfies the condition at upstream infinity that $\mathbf{q}(+\infty,r,\gamma,t) = \mathbf{0}$.

The gradient operation produces

$$\nabla\frac{\partial}{\partial n'}\frac{1}{R} = \frac{\mathbf{n}'}{R^3} - \frac{3(\mathbf{n}'\cdot \boldsymbol{R})\boldsymbol{R}}{R^5} = \mathbf{K}^*(x'',r,\gamma,t+(x-x'')/U) \tag{18.19}$$

where $\mathbf{K}^*$ is a kernel (vector) function.

This kernel is the same as that derived by Sparenberg (1960,1984) for a propeller in uniform inflow (with Δp independent of time and blade position). He does not include the external forces in his equations of motion, taking as the applicable solution of $\nabla^2 p = 0$, the same pressure dipole function as found here *as a consequence of their inclusion.* As may be seen by inspection of (18.18), the integral of the external force term is zero for all $x > x'_{max}$ and only contributes for $-\infty < x < x'$. It is annulled by an opposite contribution from the first term in (18.18) analogous to the two-dimensional case presented in Chapter 17. A demonstration of this is given in the next section.

The Integral Equation in Terms of Derivatives of $1/R$

To determine that distribution of pressure jump Δp over all the blades which is such that there is no flow through (normal to) the blades we enforce the kinematic condition given by Equation (3.9) in the (x,r,ψ)-system fixed to the rotating blade but restrained form translating in a superimposed flow, where $\partial F/\partial t = 0$. We must then use the relative velocity for the onset flow and have the kinematic condition

$$\mathbf{n}\cdot\mathbf{q}(r,\gamma_0 + \psi(x)) = -\,\mathbf{n}\cdot\mathbf{q}_{rel}(r,\gamma_0 + \psi(x))\ ;\ -h_t(r)\sin\beta \le x \le h_l(r)\sin\beta \tag{18.20}$$

where $\mathbf{q}_{rel}(r,\gamma_0 + \psi(x)) = (-U + u_a^w, u_r^w, u_t^w - \omega r)$ is the relative onset velocity consisting of the wake velocity, cf. Chapter 13, p. 267, plus the contribution from rotation. The velocity induced by the singularities over the blades, $\mathbf{q}$ is given by (18.18). The evaluations are made at any field point brought to the moving key blade.

In linearized theory, the evaluations in the kernel function and of the normal are made on the fluid reference surface $\psi = x\,\omega/U = p_f^* x$ and the incident flow components are taken normal to the blade surface which is considered to be everywhere close to the fluid reference surface, as mentioned on p. 340. Here, to save needless complication, we limit the treatment to blades of zero skew and zero rake. As we have seen the normal to the fluid reference surface is given by (18.13) and in the cartesian system we have (cf. Equation (9.2))

$$(n_x, n_y, n_z) = (\cos\beta, \sin\beta\cos\gamma, \sin\beta\sin\gamma)$$

Since we wish to obtain $\mathbf{n}\cdot\mathbf{q}$ we use (18.19) and obtain contributions to $\mathbf{n}\cdot\mathbf{K}^*$

$$\mathbf{n}\cdot\mathbf{n}' = \sin\beta\sin\beta'\left[\frac{1}{\tan\beta}\frac{1}{\tan\beta'} + \cos(\gamma - \gamma')\right] \quad ;$$

$$\mathbf{n}'\cdot\boldsymbol{R} = \sin\beta'\left[\frac{x - x'}{\tan\beta'} - r\sin(\gamma - \gamma')\right] \quad ;$$

$$\mathbf{n}\cdot\boldsymbol{R} = \sin\beta\left[\frac{x - x'}{\tan\beta} - r'\sin(\gamma - \gamma')\right]$$

These results are inserted into Equation (18.18) with proper substitution of variables. Note that the field point is placed on the blade reference surface of the key blade, $\gamma = \gamma_0 + \psi = \gamma_0 + p_f^* x$. The dummy point is also placed on the blade reference surface but we must integrate over all blades, hence $\gamma' = \gamma_0 + \psi' + 2\pi\nu/Z = \gamma_0 + p_f^* x' + 2\pi\nu/Z$. We then have the following integral equation

$$\sum_{\nu=0}^{Z-1}\left\{\int_{R_h}^{R} dr' \int_{-x_t(r')}^{x_l(r')} dx' \int_x^{\infty} \Delta p(h',r',\gamma_0 + p_f^*(x-x'') + 2\pi\nu/Z)\, K_\nu^*\, dx''\right.$$

$$-\frac{4\pi}{r}\int_{R_h}^{R} dr' \int_{-x_t(r')}^{x_l(r')} dx' \int_x^{\infty} \Delta p(h',r',\gamma_0+p_f^*(x-x'')+2\pi\nu/Z)$$

$$\left.\cdot f\,\delta(x''-x')\,\delta(r-r')\,\delta(\psi-\psi'-p_f^*(x-x'') - 2\pi\nu/Z)\,dx''\right\}$$

$$= -4\pi\rho U\,\frac{\mathbf{n}\cdot\mathbf{q}_{rel}}{\sin\beta} \qquad -x_t(r) \le x \le x_l(r) \qquad (18.21)$$

where $x' = h'\sin\beta'$ and $-x_t = -h_t(r')\sin\beta'$; $x_l = h_l(r')\sin\beta'$ have been used, cf. Equation (14.3), and the kernel function $K^*_\nu = \mathbf{n}\cdot\mathbf{K}^*$ is

$$K^*_\nu = \frac{f}{R^3} - 3\,\frac{g}{R^5}$$

where the coefficients f and g are

$$f = p_f^{*2}\, r\, r' + \cos(\psi - \psi' - p_f^*(x - x'') - 2\pi\nu/Z)$$

$$g = \left[p_f^* r(x'' - x') - r' \sin(\psi - \psi' - p_f^*(x - x'') - 2\pi\nu/Z)\right]$$

$$\cdot\left[p_f^* r'(x'' - x') - r \sin(\psi - \psi' - p_f^*(x - x'') - 2\pi\nu/Z)\right] \qquad (18.22)$$

and here $R = \sqrt{(x''-x')^2 + r^2 + r'^2 + 2rr'\cos(\psi - \psi' - p_f^*(x - x'') - 2\pi\nu/Z)}$.

It is obvious that all terms for $\nu \neq 0$ are regular for $\psi = \psi' = p_f^* x'$, $x'' = x' = x$, $r' = r$, as occur in the process of integration. We focus our attention on the contributions from the key blade, $\nu = 0$. We observe that for $x > x'$ this term is also regular as x'' does not encounter x' and $\delta(x'' - x') = 0$. However, for $x < x'$ the first and second terms in K^*_0 are singular as well as that from the external forces. The first term in K^*_0 is highly singular because $f(\psi = \psi', x'' = x) = 1 + p_f^{*2} r^2 \neq 0$. The second term in K^*_0 is indeterminate because in its numerator g vanishes as $x'' \to x'$, $\psi \to \psi'$ and in the x'-integral $x' \to x$. Hence, the second term in K^*_0 is Cauchy singular, i.e. as $1/(r - r')$ whereas the first term in K^*_0 is quadratically singular, i.e. as $1/(r - r')^2$. Identification of these singularities is shown in detail by Sparenberg (1984). In distinction from him we have the last term on the left hand side of (18.21) which we must eliminate.

To accomplish this we note the following similarities

$$\frac{\partial}{\partial x}\left[\frac{\partial}{\partial x'}\frac{1}{R}\right]_{\psi' = p_f^* x'} = \frac{1}{R(x - x', \gamma - \gamma_0 - p_f^* x')^3} - \frac{3(x - x')^2}{R(x - x', \gamma - \gamma_0 - p_f^* x')^5}$$

$$\left.\frac{\partial^2}{\partial x'^2}\frac{1}{R}\right|_{\substack{\psi' = p_f^* x' \\ x = x''}} = -\frac{1}{R(x'' - x', \gamma - \gamma_0 - p_f^* x')^3} + \frac{3(x'' - x')^2}{R(x'' - x', \gamma - \gamma_0 - p_f^* x')^5}$$

(18.23)

But $1/R$ satisfies the inhomogeneous Laplace equation, cf. Equation (M5.1), which is in cylindrical coordinates (cf. (14.9)) and dummy variables

$$\left[\frac{\partial^2}{\partial x'^2} + \frac{1}{r'}\frac{\partial}{\partial r'}\left[r'\frac{\partial}{\partial r'}\right] + \frac{1}{r'^2}\frac{\partial^2}{\partial \gamma'^2}\right]\frac{1}{R} = -\frac{4\pi}{r}\,\delta(x-x')\;\delta(r-r')\;\delta(\gamma-\gamma') \tag{18.24}$$

Using this together with (18.23) the first term in the kernel function (18.22) can be replaced with

$$f\left\{\frac{4\pi}{r}\,\delta(x''-x')\;\delta(r-r')\;\delta(\psi-\psi'-p_f^*(x-x'')) + \frac{3(x''-x')^2}{R(x''-x',\gamma-\gamma_0-p_f^*x')^5}\right.$$
$$\left. + \left[\frac{1}{r'}\frac{\partial}{\partial r'}\left[r'\frac{\partial}{\partial r'}\right] + \frac{1}{r'^2}\frac{\partial^2}{\partial \gamma'^2}\right]\frac{1}{R}\right|_{\substack{\psi'=p_f^*x' \\ x=x''}}\right\}$$

Insertion of this to (18.21) results in *annulling of the terms involving the delta-functions* leaving us with a new kernel function for the key blade, $\nu = 0$

$$K_0^* = \frac{3}{R\left[x''-x',\psi-\psi'-p_f^*(x-x'')\right]^5}\left\{(x''-x')^2\cos\left[\psi-\psi'-p_f^*(x-x'')\right]\right.$$
$$\left. + p_f^*\,(x''-x')(r^2+r'^2)\sin\left[\psi-\psi'-p_f^*(x-x'')\right] - rr'\sin\left[\psi-\psi'-p_f^*(x-x'')\right]\right\}$$
$$+ f\left[\frac{1}{r'}\frac{\partial}{\partial r'}\left[r'\frac{\partial}{\partial r'}\right] + \frac{1}{r'^2}\frac{\partial^2}{\partial \psi'^2}\right]\frac{1}{R}\Bigg|_{\substack{x=x'',\ \psi'=p_f^*x',\ \gamma_0\to\gamma_0+p_f^*(x-x'') \\ \gamma=\gamma_0+\psi \\ x<x'}} \tag{18.25}$$

Since the kernel functions are of different character for $x \gtrless x'$ and because K_ν^* is regular for $1 \le \nu \le Z-1$ for all x we divide the terms in the integral equation as follows

$$\sum_{\nu=1}^{Z-1}\int_{R_h}^{R} dr' \int_{-x_t(r')}^{x_l(r')} dx' \int_{x}^{\infty} \Delta p\left[h',r',\gamma_0 + p_f^*(x-x'') + 2\pi\nu/Z\right] K_\nu^*\, dx''$$

$$+\int_{R_h}^{R} dr' \int_{-x_t(r')}^{x} dx' \int_{x}^{\infty} \Delta p\left[\gamma_0 + p_f^*(x-x'')\right] K_0^*\, dx'' \qquad I_1^*(x' < x)$$

$$+\int_{R_h}^{r-\mu} dr' \int_{x}^{x_l(r')} dx' \left[\int_{x}^{x'-\varepsilon} + \int_{x'+\varepsilon}^{\infty} \Delta p\left[\gamma_0 + p_f^*(x-x'')\right] K_0^*\, dx''\right] \qquad I_2^*(x < x')$$

$$+\int_{r+\mu}^{R} dr' \int_{x}^{x_l(r')} dx' \left[\int_{x}^{x'-\varepsilon} + \int_{x'+\varepsilon}^{\infty} \Delta p\left[\gamma_0 + p_f^*(x-x'')\right] K_0^*\, dx''\right] \qquad I_3^*(x < x')$$

$$+\int_{r-\mu}^{r+\mu} dr' \int_{x}^{x_l(r')} dx' \int_{x'-\varepsilon}^{x'+\varepsilon} \Delta p\left[\gamma_0 + p_f^*(x-x'')\right] K_0^*\, dx'' \qquad I_4^*$$

$$= -\,4\pi\rho U \frac{\mathbf{n}\cdot\mathbf{q}_{rel}}{\sin\beta} \qquad (18.26)$$

where ε and μ are very small and positive.

The regions of integration over the key blade corresponding to the integrals I_1^*, I_2^*, I_3^* and I_4^* are displayed in Figure 18.3. It is only in region IV that the kernel function K_0^* becomes singular.

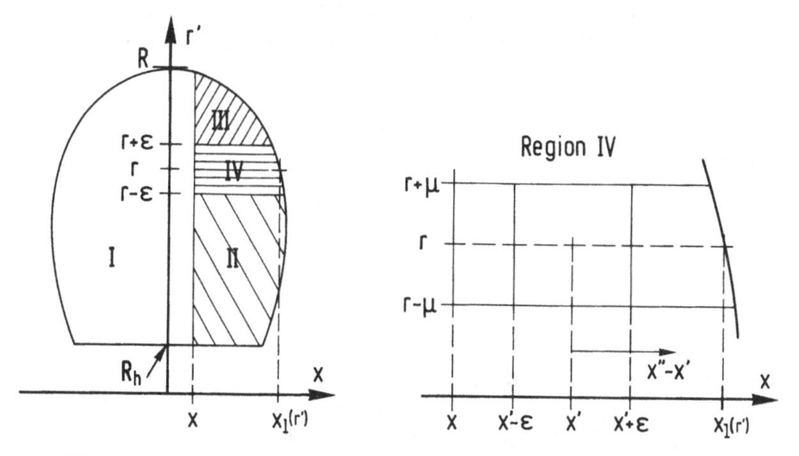

Figure 18.3 Regions of integration in the projection of the key blade.

Upon setting $\psi = p_f^* x$ and noting that $\psi' = p_f^* x'$ on the helicoidal reference surface the arguments of the trigonometric functions become $p_f^*(x'' - x')$. The term with differentiations in (18.25) then becomes

$$f\left[\frac{1}{r'}\frac{\partial}{\partial r'}\left[r'\frac{\partial}{\partial r'}\right]+\frac{1}{r'^2}\frac{\partial^2}{\partial\psi'^2}\right]\frac{1}{R}$$

$$=\left[p_f^{*2}r^2 - p_f^* r(r-r') + \cos\left[p_f^*(x''-x')\right]\right]$$

$$\cdot\left[-\frac{2}{R^3}+\frac{3(r'-r\cos(p_f^*(x''-x')))^2}{R^5}+\frac{3r^2\sin^2(p_f^*(x''-x'))}{R^5}\right]$$

Collecting similar terms and taking r' = r where permissible gives

$$K_0^* \approx -\frac{2}{R^3}\left\{\cos\left[p_f^*(x''-x')\right] + p_f^{*2}r^2\right\}$$

$$+\frac{3}{R^5}\left\{(x''-x')^2\cos\left[p_f^*(x''-x')\right] + 2p_f^* r^2(x''-x')\sin\left[p_f^*(x''-x')\right]\right.$$

$$- r^2\sin^2\left[p_f^*(x''-x')\right] + \left[\cos\left[p_f^*(x''-x')\right] + p_f^{*2}r^2\right]$$

$$\left.\cdot\left[\left[r'-r\cos\left[p_f^*(x''-x')\right]\right]^2 + r^2\sin^2\left[p_f^*(x''-x')\right]\right]\right\}$$

We expand about x" = x' to terms involving only $(x''-x')^0$ in the factor of $1/R^3$ and to terms of order $(x''-x')^2$ in the factor of $1/R^5$, ignoring terms involving r − r'. All these omitted terms give rise at most to Cauchy singularities whose integrals are zero in $x'-\varepsilon < x'' < x'+\varepsilon$ and $r-\mu < r' < r+\mu$. We are then left with the following expression for the last integral in (18.26)

$$I_4^* = \left[1+p_f^{*2}r^2\right]\int_x^{x_1(r)} dx'\Delta p(x',r,\gamma_0)\int_{r-\mu}^{r+\mu} dr'$$

$$\cdot\int_{x'-\varepsilon}^{x'+\varepsilon}\frac{1}{\left[\left[1+p_f^{*2}r^2\right](x''-x')^2+(r-r')^2\right]^{3/2}}dx'' \qquad (18.27)$$

Here we note that the integrand is positive but highly singular for x" − x' and r − r' vanishing simultaneously. Since we consider r ≈ r' we can set x' = h' sinβ and x" = h" sinβ and have $((h''-h')^2+(r-r')^2)^{3/2}$ in the denominator. It is now obvious that the integral is basically of the same structure as that of the expression for the vertical velocity of the planar wing, Equation (17.46) (where x and y correspond to h and r). This should be expected because the propeller blade is essentially a wing rotating in a superimposed flow.

Proceeding formally, we integrate over x" and obtain

$$I_4^* = 2\left[1 + p_f^{*2}r^2\right]\int_x^{x_l(r)} dx'\Delta p(x',r,\gamma_0)$$

$$\cdot\int_{r-\mu}^{r+\mu}\frac{\varepsilon}{\sqrt{\left[1+p_f^{*2}r^2\right]\ \varepsilon^2 + (r-r')^2}}\frac{1}{(r-r')^2}\,dr' \tag{18.28}$$

Holding ε fixed and $r - r'$ small we obtain for the r'-integral

$$\frac{1}{\sqrt{1+p_f^{*2}r^2}}\int_{r-\mu}^{r+\mu}\frac{1}{(r-r')^2}\,dr'$$

which as we have seen does not exist in the ordinary sense. However, the limit as $\mu \to 0$ and $\varepsilon \to 0$ of the integrals I_2^*, I_3^* and I_4^* of (18.26) exist in the sense of Hadamard (1923) by taking the formal evaluation of this last integral which is $-2/\mu$, to leave for the limit of the integrals,

$$I_2^*+I_3^*+I_4^* = \lim_{\substack{\mu \to 0\\ \varepsilon \to 0}}\left\{\int_{R_h}^{r-\mu} dr'\int_x^{x_l(r')} dx'\left[\int_x^{x'-\varepsilon} + \int_{x'+\varepsilon}^{\infty}\Delta p\left[\gamma_0+ p_f^*(x-x'')\right]K_0\, dx''\right]\right.$$

$$+\int_{r+\mu}^{R} dr'\int_x^{x_l(r')} dx'\left[\int_x^{x'-\varepsilon} + \int_{x'+\varepsilon}^{\infty}\Delta p\left[\gamma_0+ p_f^*(x-x'')\right]K_0\, dx''\right]$$

$$\left. - 4\,\frac{\sqrt{1 + p_f^{*2}r^2}}{\mu}\int_x^{x_l(r)}\Delta p(x',r,\gamma_0)\, dx'\right\}$$

The interpretation of the integral over the key blade à la Hadamard is now obvious. An outline of the evaluation of such integrals, although in one dimension, following the exposition by Mangler (1951), is given in Section 4 of the Mathematical Compendium, and by comparison with Equation (M4.10) we recognize the structure of the limit ($\mu \to 0$, $\varepsilon \to 0$), as a sum of integrals plus a singular term. Individually, both the integrals and the additional term are singular, but the limit of the sum is finite, yielding its finite part or the value obtained by ignoring the singularity.

Sparenberg (1984) obtains this result by noting that the limit processes should be to hold the field point slightly off the surface of singularities by taking $\gamma = \gamma_0 + \psi - \theta$ with θ being a very small angle. He then carries out the integrations over x" and r' and shows that the limit is the same as above. That this limit gives a finite answer, as indicated by Sparenberg (*ibid.*), is because the large negative values around $r' = r$ are annulled by the large positive values for $r' < r - \mu$ and $r' > r + \mu$.

The integral equation (18.26) is now of the form

$$\sum_{\nu=1}^{Z-1} \int_{R_h}^{R} dr' \int_{-x_t(r')}^{x_l(r')} dx' \int_{x}^{\infty} \Delta p\left[x', r', \gamma_0 + p_f^*(x - x'') + 2\pi\nu/Z\right] K_\nu^*\left[p_f^*(x'' - x'), r, r'\right] dx''$$

$$+ \int_{R_h}^{R} dr' \int_{-x_t(r')}^{x} dx' \int_{x}^{\infty} \Delta p\left[x', r', \gamma_0 + p_f^*(x - x'')\right] K_0^*\left[p_f^*(x'' - x'), r, r'\right] dx''$$

$$+ \lim_{\substack{\mu \to 0 \\ \varepsilon \to 0}} \left\{ \int_{R_h}^{r-\mu} dr' \int_{x}^{x_l(r')} dx' \left[\int_{x}^{x'-\varepsilon} + \int_{x'+\varepsilon}^{\infty} \Delta p\left[\gamma_0 + p_f^*(x - x'')\right] K_0^* \, dx'' \right] \right.$$

$$+ \int_{r+\mu}^{R} dr' \int_{x}^{x_l(r')} dx' \left[\int_{x}^{x'-\varepsilon} + \int_{x'+\varepsilon}^{\infty} \Delta p\left[\gamma_0 + p_f^*(x - x'')\right] K_0^* \, dx'' \right]$$

$$\left. - 4 \frac{\sqrt{1 + p_f^{*2} r^2}}{\mu} \int_{x}^{x_l(r)} \Delta p(x', r, \gamma_0) \, dx' \right\} = -4\pi\rho U \, \mathbf{n} \cdot \mathbf{q}_{rel} \sqrt{1 + p_f^{*2} r^2} \tag{18.29}$$

It is noted that this form of the kernel differs in form from that used by Verbrugh (1968) and van Gent (1975) but is merely a rearrangement of theirs. It is also to be noted that the evaluation of the non-singular parts in the neighborhood of the limits $r' \pm \mu$ and $x' \pm \varepsilon$ gives numerical difficulties.

We turn next to the use of Fourier expansions which enable the annulling of the external forces and yield only Cauchy-singular kernels.

The Velocity Field and Integral Equation in Fourier Series

In this section we shall use Fourier expansions of $1/R$, of the pressure difference over the blade and of the Dirac delta-function. This formulation makes it possible to recognize velocity terms in the wake of the propeller and to derive the integral equation on the blade reference surface for discrete harmonic orders. The force terms will be annulled by other terms in the expressions and the singularities will be, at most, of Cauchy type.

We return to the pressure due to the single blade element given by (18.16). In this equation we insert the Fourier expansion of $1/R$ from Equation (M5.24) and the Fourier expansion for the pressure difference over the blade (15.1). The operations carried out as indicated in (18.16) gives

$$p' = -\frac{1}{4\pi^2}\sum_{n=-\infty}^{\infty}\sum_{m=-\infty}^{\infty}\Delta p_n(h',r')\int_{-\infty}^{\infty}\left[-ikn'_a - in'_t\frac{m}{r'} + n'_r\frac{\partial}{\partial r'}\right] B_m(k,r,r')$$

$$\cdot e^{ik(x-x')}\, e^{im(\gamma-\psi')}\, e^{i(n-m)\gamma_0}\, dk \qquad (18.30)$$

where

$$B_m(k,r,r') = \begin{cases} I_{|m|}(|k|r')\ K_{|m|}(|k|r) & r > r' \\ I_{|m|}(|k|r)\ K_{|m|}(|k|r') & r < r' \end{cases} \qquad (18.31)$$

The axial velocity component induced by each patch of the blade is found by integration by the method of characteristics, in particular the same procedure as explained on p. 342. We then have

$$u'_a = \frac{1}{4\pi^2\rho U}\sum_{n=-\infty}^{\infty}\sum_{m=-\infty}^{\infty}\Delta p_n(h',r')\int_x^{\infty} dx''\int_{-\infty}^{\infty} k\left[kn'_a + n'_t\frac{m}{r'} + in'_r\frac{\partial}{\partial r'}\right] B_m(k,r,r')$$

$$\cdot e^{ik(x''-x')}\, e^{im(\gamma-\psi')}\, e^{i(n-m)(\gamma_0+p^*_f(x-x''))} dk$$

$$-\frac{n'_a}{\rho U r}\int_x^{\infty}\sum_{n=-\infty}^{\infty}\Delta p_n(h',r')\, e^{in(\gamma_0+p^*_f(x-x''))}$$

$$\cdot\ \delta(x''-x')\ \delta(r-r')\ \delta(\gamma-\gamma_0-\psi'-p^*_f(x-x''))\ dx''$$

$$(18.32)$$

The last δ-function of the last term can also be expanded in a Fourier series, cf. Equation (11.32), p. 215. Moreover, we calculate the contributions from all blades by substituting γ_0 by $\gamma_0 + 2\pi\nu/Z$ and sum $\nu = 0,..,(Z-1)$. As in (11.20) only those n and m contribute such that $n - m = qZ$; $q = 0, \pm1, \pm2,...$ This gives

$$u'_a = \frac{Z}{4\pi^2\rho U}\sum_{q=-\infty}^{\infty}\sum_{n=-\infty}^{\infty}\Delta p_n(h',r')\int_x^{\infty} dx''\int_{-\infty}^{\infty} g(k)\ B_m(k,r,r')$$

$$\cdot e^{i(k-qZp^*_f)x''}\ e^{-ikx'}\ e^{iqZp^*_f x}\ e^{i(m(\gamma-\psi')+qZ\gamma_0)} dk$$

$$-\frac{Z}{2\pi\rho U r}\sum_{q=-\infty}^{\infty}\sum_{n=-\infty}^{\infty}\Delta p_n(h',r')\ n'_a\int_x^{\infty}\delta(x''-x')\ \delta(r-r')$$

$$\cdot e^{iqZp^*_f(x-x'')}\ e^{i(m(\gamma-\psi')+qZ\gamma_0)}\ dx''$$

$$;\quad n - m = qZ \qquad (18.33)$$

where $g(k)$ is an operator given by

$$g(k) = k\left[k\, n_a' + n_t'\frac{m}{r'} + i\, n_r'\frac{\partial}{\partial r'}\right] \tag{18.34}$$

It is clear that the last double sum vanishes for $x > x'$. It is because of the different character of the velocity field for $x > x'$ and for $x < x'$ that we integrate over the blade later.

The improper x''-integrals are operators or distribution functions. They are of the type evaluated on p. 217; there, however, with 0 as the lower limit. By a similar procedure one obtains,

$$\int_x^\infty e^{i(k-k_0)x''}dx'' = \pi\,\delta(k-k_0) + \frac{i\,e^{i(k-k_0)x}}{k-k_0} \tag{18.35}$$

Use of this in (18.33) gives

$$\begin{aligned} u_a' = \frac{Z}{4\pi\rho U}\sum_{q=-\infty}^{\infty}\sum_{n=-\infty}^{\infty}\Delta p_n(h',r')\Bigg\{& g(k_0)\,B_m(k_0)\,e^{ik_0(x-x')} \\ &+ \frac{i}{\pi}\int_{-\infty}^{\infty}\frac{g(k)\;B_m(k)}{k-k_0}\,e^{ik(x-x')}dk \\ &- n_a'\,H(x'-x)\,\frac{2}{r}\,\delta(r-r')\,e^{ik_0(x-x')}\Bigg\}\,e^{i(m(\gamma-\psi')+qZ\gamma_0)} \end{aligned} \tag{18.36}$$

where $k_0 = qZp_f^*$ and H is the Heaviside step function, cf. (M2.6).

To exhibit the different behaviors of u_a' for $x \gtrless x'$ we must evaluate the k-integral for these two regions. We write (cf. Lamb (1963), p. 401 and 402)

$$\begin{aligned} &\frac{i}{\pi}\int_{-\infty}^{\infty}\frac{g(k)B_m(k)}{k-k_0}\,e^{ik(x-x')}dk = \\ &\frac{i}{\pi}\left\{\int_0^\infty\frac{g(k)B_m(k)}{k-k_0}\,e^{ik(x-x')}dk - \int_0^\infty\frac{g(-k)B_m(k)}{k+k_0}\,e^{-ik(x-x')}dk\right\} \end{aligned} \tag{18.37}$$

Observe that the modulus of k is taken in the argument of the modified Bessel functions in B_m in the above operation. We now remove the restrictions on k in the argument of $B_m(k)$ (but not on m) as we will replace k by $\zeta = \xi + i\eta$ and carry out the integrations in the complex plane. For positive k_0 the first integral has a pole at $\zeta = k_0$ (on the real axis) and the second has no pole. The numerator $g(\pm k)\;B_m(k)$ has no singularities by the properties of the modified Bessel functions of the first and second kind. The exponential can be written

$$e^{ik(x-x')} = e^{i(x-x')|\zeta|(\cos\theta + i\sin\theta)} = e^{-(x-x')|\zeta|\sin\theta}\, e^{i(x-x')|\zeta|\cos\theta}$$

and it decays to zero for $x > x'$ as $|\zeta| \to \infty$ for $0 \le \theta \le \pi/2$. For $x < x'$ the range of θ is required to be $-\pi/2 \le \theta \le 0$. This determines the contours (to be used to secure convergent values of these integrals) as indicated in Figure 18.4. For negative k_0 the first integral in (18.37) will have no pole while the second integral will have a pole at $\zeta = -k_0$ and the contours must be modified (in this case simply interchanged).

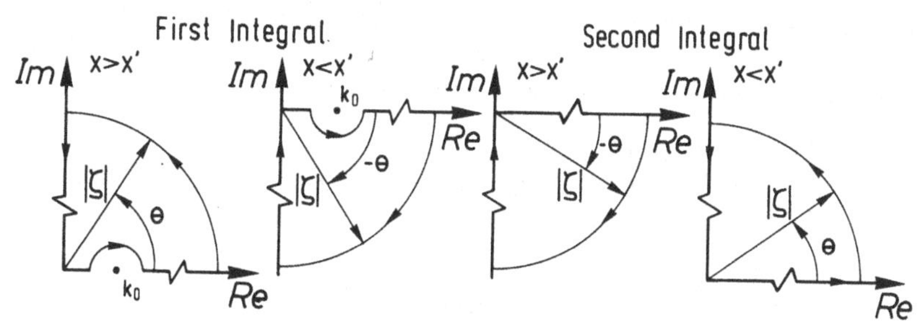

Figure 18.4 Integration contours (for $k_0 > 0$) for the integrals in Equation (18.37).

Application of Cauchy's residue formula for poles on the contour and integral theorem for non-singular functions in the limit of $|\zeta| \to \infty$ on C_∞ yields expressions with arguments of the modified Bessel functions that are purely imaginary. These modified Bessel functions can then be transformed into Bessel functions with real arguments, cf. Abramowitz & Stegun (1972), so we have after some manipulation where also η is substituted with k in the integrals of Bessel functions,

$$u_a' = \frac{Z}{4\pi\rho U} \sum_{q=-\infty}^{\infty} \sum_{n=-\infty}^{\infty} \Delta p_n(h',r') \int_0^{\infty} \frac{g(ik)\ J_m(kr')\ J_m(kr)}{k + ik_0}\, e^{-k(x-x')} dk$$

$$\cdot e^{i(m(\gamma-\psi')+qZ\gamma_0)} \quad ; \quad x > x' \qquad (18.38)$$

$$u_a' = \frac{Z}{4\pi\rho U} \sum_{q=-\infty}^{\infty} \sum_{n=-\infty}^{\infty} \Delta p_n(h',r') \left\{ 2\, g(k_0)\, B_m(k_0)\, e^{ik_0(x-x')} \right.$$

$$\left. - \int_0^{\infty} \frac{g(-ik)\ J_m(kr')\ J_m(kr)}{k - ik_0}\, e^{k(x-x')} dk - \frac{2}{r}\, n_a'\, \delta(r-r')\, e^{ik_0(x-x')} \right\}$$

$$\cdot e^{i(m(\gamma-\psi')+qZ\gamma_0)} \quad ; \quad x < x' \qquad (18.39)$$

As we now have a complete specification of u_a' for $r \gtrless r'$ and for $x \gtrless x'$ we can integrate over the key blade to obtain the total axial induced velocity in the form

$$u_a = \int_{R_h}^{r} dr' \left\{ \int_{-h_t(r')}^{h(x,r')} u_a'(x > x')\, dh' + \int_{h(x,r')}^{h_l(r')} u_a'(x < x')\, dh' \right\}$$
$$+ \int_{r}^{R} dr' \left\{ \int_{-h_t(r')}^{h(x,r')} u_a'(x > x')\, dh' + \int_{h(x,r')}^{h_l(r')} u_a'(x < x')\, dh' \right\} \qquad (18.40)$$

We next show that the third term in (18.39) which comes from the external forces on the fluid is annulled by part of the first term in the process of integration. We consider the terms which come from the integration forward of position h, i.e. from $h(x,r')$ to h_l which are

$$u_a^{fx} = 2 \left\{ \left[\int_{R_h}^{r} dr' \int_{h(x,r')}^{h_l(r')} \Delta p_n\, g(k_0)\, I_m(k_0 r')\, K_m(k_0 r)\, e^{ik_0(x-x')}\, e^{-im\psi'} dh' \right.\right.$$
$$+ \int_{r}^{R} dr' \int_{h(x,r')}^{h_l(r')} \Delta p_n\, g(k_0)\, I_m(k_0 r)\, K_m(k_0 r')\, e^{ik_0(x-x')}\, e^{-im\psi'} dh'$$
$$\left.\left. - \frac{1}{r} \int_{h(x,r')}^{h_l(r')} \Delta p_n(h',r)\, n_a'(r)\, e^{ik_0(x-x')} e^{-im\psi'} dh' \right] e^{i(m\gamma + qZ\gamma_0)} \right\}$$
$$(18.41)$$

where in the last term the complete r'-integral has been effected via $\delta(r-r')$ and the modulus of arguments and orders of the Bessel functions are assumed. With the approximation (18.13) (consistent with linearization) $n_a' = \cos\beta'$ and $n_t' = -\sin\beta'$ then $\tan\beta' = U/\omega r'$. Note, however, that we keep $n_r' \neq 0$. As $qZ = n - m$ we can manipulate the operator $g(k_0)$, (18.34) as follows

$$g(k_0) = k_0 \left[k_0\, n_a' + n_t' \frac{m}{r'} + i\, n_r' \frac{\partial}{\partial r'} \right] = n_a' \left[k_0^2 - \frac{mqZ\omega}{U} \frac{\tan\beta'}{r'} \right] + i\, k_0\, n_r' \frac{\partial}{\partial r'}$$
$$= n_a' \left[k_0^2 + \frac{m^2}{r'^2} \right] - \frac{mn}{r'^2} n_a' + i\, k_0\, n_r' \frac{\partial}{\partial r'} \quad ; \quad \left[\text{using } k_0 = qZ \frac{\omega}{U} \right]$$

But here g operates on the modified Bessel functions $I_m(k_0 r')$ and $K_m(k_0 r')$ and the term $k_0^2 + m^2/r'^2$ can be recognized as a coefficient in Bessel's differential equation, cf. Watson (1966), of which $I_m(k_0 r')$ and $K_m(k_0 r')$ are solutions. We make use of this to obtain

$$g(k_0)\, I_m(k_0 r') = \left[n_a' \frac{1}{r'} \frac{d}{dr'} \left[r' \frac{d}{dr'} \right] - \frac{mn}{r'^2} n_a' + i\, k_0\, n_r' \frac{d}{dr'} \right] I_m(k_0 r') \tag{18.42}$$

and similarly for $K_m(k_0 r')$.

We now define the *axial, radial and tangential partial-force operators*

$$\begin{Bmatrix} f_a(r') \\ f_r(r') \\ f_t(r') \end{Bmatrix} = \int_{h(x,\,r')}^{h_1(r')} \Delta p_n\,(h',r') \begin{Bmatrix} n_a'(r') \\ n_r'(r') \\ n_t'(r') \end{Bmatrix} e^{-ip_f^*(nx' - mx_r')} dh' \tag{18.43}$$

Note that $f_a(R) = 0$ and $f_a(R_h) = 0$, approximately; similarly for the other components. Insertion of the force operators together with (18.42) into (18.41) and integration by parts yields

$$u_a^{fx} = 2 \Bigg\{ f_a(r)\, k_0\, [I_m'(k_0 r)\, K_m(k_0 r) - I_m(k_0 r)\, K_m'(k_0 r)]$$

$$- \int_{R_h}^{r} \left\{ \frac{d}{dr'} \left[\frac{f_a(r')}{r'} \right] k_0\, r'\, I_m'(k_0 r') + \frac{m\,n}{r'^2} f_a(r')\, I_m(k_0 r') \right.$$

$$\left. - i\, k_0^2\, f_r(r')\, I_m'(k_0 r') \right\} dr'\, K_m(k_0 r)$$

$$- \int_{r}^{R} \left\{ \frac{d}{dr'} \left[\frac{f_a(r')}{r'} \right] k_0\, r'\, K_m'(k_0 r') + \frac{m\,n}{r'^2} f_a(r')\, K_m(k_0 r') \right.$$

$$\left. - i\, k_0^2\, f_r(r')\, K_m'(k_0 r') \right\} dr'\, I_m(k_0 r) - \frac{f_a(r)}{r} \Bigg\}\, e^{i(m\gamma + qZ\gamma_0)}\, e^{ik_0 x} \tag{18.44}$$

where $'$ indicates derivatives with respect to argument $k_0 r'$.

The first term contains the negative of the Wronskian of the applicable Bessel equation, cf. Abramowitz & Stegun (1972), which is equal to $1/k_0 r$. Hence *this term precisely annuls the last term which came from the external axial force on the fluid.* This is necessary because the q-summation of this last term is divergent. In addition, *the integration by parts gives kernel functions which are at most Cauchy singular.* This is shown later.

The "local effect" terms represented by the k-integrals in (18.38) and (18.39) are seen to decay rapidly for $|x|$ large. They can be expressed in terms of the associated Legendre function of half-integer degree and zero order as follows.

First note by Equation (M5.15), p. 516, that

$$\frac{\partial}{\partial x'}\int_0^\infty J_m(kr)\,J_m(kr')\,e^{-k|x-x'|}dk =$$

$$\pm\int_0^\infty k\,J_m(kr)\,J_m(kr')\,e^{-k|x-x'|}dk = \frac{\partial}{\partial x'}\left[\frac{1}{\pi\sqrt{rr'}}\,Q_{m-1/2}(Z)\right]$$

$$\pm \text{ for } x \gtrless x'$$

where $Z = ((x-x')^2 + r'^2 + r^2)/(2\,r\,r')$. Then from (18.34) and by manipulation of $g(ik)/(k \pm ik_0)$ we have for the k-integrals in (18.38) and (18.39)

$$C^{\pm} = \int_0^\infty \frac{g(\pm ik)\;J_m(kr)\,J_m(kr')}{k \pm ik_0}\,e^{-k|x-x'|}dk$$

$$= \pm\left\{-n_a'\left[\frac{\partial}{\partial x'} - i\,k_0\right] + in_t'\frac{m}{r'} - n_r'\frac{\partial}{\partial r'}\right\}\frac{1}{\pi\sqrt{rr'}}\,Q_{m-1/2}(Z)$$

$$+\,k_0\left\{k_0\,n_a' + n_t'\frac{m}{r'} + i\,n_r'\frac{\partial}{\partial r'}\right\}\int_0^\infty\frac{J_m(kr)\,J_m(kr')}{k \pm ik_0}\,e^{-k|x-x'|}dk$$

where the upper signs go together as do the lower.

The integral can now be written in terms of the the Legendre function by using the integral representation

$$\frac{1}{k \pm ik_0} = e^{\pm(k \pm ik_0)x}\int_{\pm x}^\infty e^{-(k \pm ik_0)x''}dx''$$

in the integrand and then integrate over k. This gives

$$C^{+} = +\left\{-n_a'\left[\frac{\partial}{\partial x'} - i\,k_0\right] + in_t'\frac{m}{r'} - n_r'\frac{\partial}{\partial r'}\right\}\frac{1}{\pi\sqrt{rr'}}\,Q_{m-1/2}(Z)$$

$$+\,k_0\left\{k_0\,n_a' + n_t'\frac{m}{r'} + i\,n_r'\frac{\partial}{\partial r'}\right\}\frac{1}{\pi\sqrt{rr'}}\int_x^\infty Q_{m-1/2}(Z''^{-})\,e^{ik_0(x-x'')}dx'' \tag{18.45}$$

$$C^{-} = -\left\{-n_a'\left[\frac{\partial}{\partial x'} - i\,k_0\right] + in_t'\frac{m}{r'} - n_r'\frac{\partial}{\partial r'}\right\}\frac{1}{\pi\sqrt{rr'}}\,Q_{m-1/2}(Z)$$

$$+\,k_0\left\{k_0\,n_a' + n_t'\frac{m}{r'} + i\,n_r'\frac{\partial}{\partial r'}\right\}\frac{1}{\pi\sqrt{rr'}}\int_{-x}^\infty Q_{m-1/2}(Z''^{+})\,e^{ik_0(x+x'')}dx''$$

where

$$\mathcal{Z}''^- = ((x''-x')^2 + r'^2 + r^2)/(2rr') \;;$$
$$\mathcal{Z}''^+ = ((x''+x')^2 + r'^2 + r^2)/(2rr')$$

Note the difference between the arguments $\mathcal{Z}$, $\mathcal{Z}''^-$ and $\mathcal{Z}''^+$ in the Legendre functions.

In the integral equation the harmonic order, n, is fixed and the sum over q must be secured which gives rise to singularities. As the q-sum of $Q_{|n-qZ|-1/2}$ has the singularity of the source function, $-1/R$, we see that we have only axial, radial and tangential dipoles which are Cauchy singular in the first lines of (18.45) for $C^{\pm}$ and the remaining terms are surely integrable because of the x''-integration.

We can now collect the foregoing results to obtain the following expression for the axial component of induced velocity at all x,r and γ

$$u_a(x,r,\gamma;\gamma_0(t)) = \frac{Z}{4\pi\rho U} \sum_{q=-\infty}^{\infty} \sum_{n=-\infty}^{\infty} e^{i(m\gamma+qZ\gamma_0)}\, e^{ik_0x}$$

$$\cdot\left\{\int_{R_h}^{R} dr' \left\{\left[\int_{-h_t(r')}^{h(x,r')} \Delta p_n\, e^{-im\psi'} C^+ dh' - \int_{h(x,r')}^{h_l(r')} \Delta p_n\, e^{-im\psi'} C^- dh'\right\}\right.\right.$$

$$-2\int_{R_h}^{r} \left\{\frac{d}{dr'}\left[\frac{f_a(r')}{r'}\right] k_0\, r'\, I_m'(k_0r') + \frac{m\,n}{r'^2} f_a(r')\, I_m(k_0r')\right.$$

$$\left. - i\,k_0^2\, f_r(r')\, I_m'(k_0r')\right\} dr'\, K_m(k_0r)$$

$$-2\int_{r}^{R} \left\{\frac{d}{dr'}\left[\frac{f_a(r')}{r'}\right] k_0\, r'\, K_m'(k_0r') + \frac{m\,n}{r'^2} f_a(r')\, K_m(k_0r')\right.$$

$$\left.\left. - i\,k_0^2\, f_r(r')\, K_m'(k_0r')\right\} dr'\, I_m(k_0r)\right\} \tag{18.46}$$

For the region forward of the leading edge of the blades $f_a = 0$ as seen from (18.43) and the second term in (18.46) also vanishes while the upper limit in the first term becomes fixed at $h_l(r')$. Downstream of the blade trailing edges, the lower limit for f_a becomes fixed at $-h_t(r')$ and the first term above vanishes. For $x \to -\infty$, actually more than two diameters or so downstream, the local term is negligible and as $|x| >> |nx' - mx'_r|$ we have the third and remaining terms of (18.46) with the partial, now the total, force operator given by

$$f_a(-\infty,r') = \int_{-h_t(r')}^{h_l(r')} \Delta p_n(h',r')\, n'_a(r')\, dh'$$

and similarly for the other components.

It is of interest to determine the mean axial component arising from the mean loading on the blade for $x \to -\infty$ and $R_h < r < R$. This term is given by the expressions for $q = 0$ and $n = 0$, hence $m = 0$ and $k_0 = qZp^*_f = 0$. As

$$\lim_{\substack{q\to 0\\ m\to 0}} k_0 I'_m(k_0 r')\, K_m(k_0 r) = \frac{1}{2r'} \; ; \; \lim_{\substack{q\to 0\\ m\to 0}} k_0 I_m(k_0 r)\, K'_m(k_0 r') = -\frac{1}{2r'}$$

which can be seen by use of (11.38) and (11.39), we then have

$$\lim_{\substack{x\to -\infty\\ n=m=q=0}} u_a = -\frac{Z}{2\pi\rho U}\left\{\frac{1}{2}\int_{R_h}^{r}\frac{d}{dr'}\left[\frac{f_a(-\infty,r')}{r'}\right]dr' - \frac{1}{2}\int_{r}^{R}\frac{d}{dr'}\left[\frac{f_a(-\infty,r')}{r'}\right]dr'\right\}$$

which is integrated directly yielding

$$u_{a0}(-\infty, r) = -\frac{Z}{2\pi\rho U r}\int_{-h_t(r)}^{h_l(r)} \Delta p_0(h',r)\, n'_a(r)\, dh' = \frac{\text{Thrust density at } r}{2\pi\rho U r} \tag{18.47}$$

a well-known result. It can be found from the linearized actuator-disc theory, p. 191. The thrust density is $\rho\Gamma\omega r$, from Equation (9.95), with $u_t \approx 0$, and the velocity far downstream is twice the value at the disc, Equation (9.102). Equation (18.46) also yields $u_{a0}(-\infty,r) = 0$ for $r > R$ as required.

This and subsequent expressions for the other components of induced velocity differ from those of Jacobs & Tsakonas (1973) who had no terms in the race for $q = 0$ for all n, nor had they distinguished the different behaviors of the kernel function within the region of the blades. Goodman (1982) derived a correction which applies in the race but not within the region between the trailing and leading edges.

Similar formulas for the radial and tangential components can be derived. In each, integration by parts annuls the external-force term and all the kernel functions are, at most, Cauchy singular.

The formula for the *tangential* component of the induced velocity is

$$u_t(x,r,\gamma;\gamma_0(t)) = \frac{Z}{4\pi\rho U r}\sum_{q=-\infty}^{\infty}\sum_{n=-\infty}^{\infty} e^{i(m\gamma+qZ\gamma_0)}\, e^{ik_0x}$$

$$\cdot\left\{\int_{R_h}^{R} dr'\left\{\int_{-h_t(r')}^{h(x,r')} dh'\left[\Delta p_n\, e^{-im\psi'}\left[imn_a'+n_t'r'\left[\frac{\partial}{\partial x'}-ik_0\right]\right]\frac{1}{\pi\sqrt{rr'}}Q_{|m|-1/2}(Z)\right.\right.\right.$$

$$+\left[\Delta p_n\, e^{-im\psi'}\left[mn_a'k_0+imn_r'\frac{\partial}{\partial r'}-k_0^2r'n_t'\right]-r'\frac{\partial}{\partial r'}\left[n_t'\Delta p_n e^{-im\psi'}\right]\frac{\partial}{\partial r'}\right]$$

$$\left.\cdot\int_x^{\infty}\frac{1}{\pi\sqrt{rr'}}Q_{|m|-1/2}(Z''^-)\, e^{-ik_0x''}dx''\right]$$

$$+\int_{h(x,r')}^{h_l(r')} dh'\left[\Delta p_n\, e^{-im\psi'}\left[imn_a'+n_t'r'\left[\frac{\partial}{\partial x'}-ik_0\right]\right]\frac{1}{\pi\sqrt{rr'}}Q_{|m|-1/2}(Z)\right.$$

$$-\left[\Delta p_n\, e^{-im\psi'}\left[mn_a'k_0+imn_r'\frac{\partial}{\partial r'}-k_0^2r'n_t'\right]-r'\frac{\partial}{\partial r'}\left[n_t'\Delta p_n e^{-im\psi'}\right]\frac{\partial}{\partial r'}\right]$$

$$\left.\left.\cdot\int_{-x}^{\infty}\frac{1}{\pi\sqrt{rr'}}Q_{|m|-1/2}(Z''^+)\, e^{ik_0x''}dx''\right]\right\}$$

$$-2\int_{R_h}^{r}\left[\left[r'\frac{df_t(r')}{dr'}-imf_r(r')\right]I_m'(k_0r')\right.$$

$$\left.+\left[k_0r'f_t(r')-mf_a(r')\right]I_m(k_0r')\right]dr'\, k_0K_m(k_0r)$$

$$-2\int_{r}^{R}\left[\left[r'\frac{df_t(r')}{dr'}-imf_r(r')\right]K_m'(k_0r')\right.$$

$$\left.\left.+\left[k_0r'f_t(r')-mf_a(r')\right]K_m(k_0r')\right]dr'k_0I_m(k_0r)\right\}$$

(18.48)

where Z, Z''^- and Z''^+ are defined in connection with (18.45).

A necessary condition for the correctness of (18.48) is that the mean tangential velocity for all h at and downstream of the trailing edge of the blades be given by

$$\frac{Z}{2\pi\rho Ur}\int_{-h_t(r)}^{h_l(r)} \Delta p_0\,(h',r)\; n_t'(r)\; dh' \quad ; \quad R_h < r < R$$

and zero outside the race ($R < r$) in analogy with the expression for the axial velocity, (18.47), and as before with disc theory. Putting $n = 0$, $m = 0$ and hence $q = 0$ and $k_0 = 0$ as on p. 358, and taking $h = -h_t$, we see that the first two h'-integrals vanish. In the next two h'-integrals all terms are zero except the underscored ones which appear to be non-zero. But these annul because they arose from the term m^2/r'. This can be seen by reversing of the integration by parts and using the Bessel equation for $m = 0$. Hence all the local terms vanish. In the remaining r'-integrals all terms vanish except

$$-\frac{Z}{2\pi\rho Ur}\int_r^R r'\frac{df_t(r')}{dr'} \lim_{\substack{m\to 0\\ k_0\to 0}} (k_0\, K_m'(k_0 r')\, I_m(k_0 r))\, dr'$$

$$= -\frac{Z}{2\pi\rho Ur}\int_{-h_t(r)}^{h_l(r)} \Delta p_0\,(h',r)\; n_t'(r)\; dh'$$

As $n_t(r) = -\sin\beta(r)$ in linear theory we obtain the correct sign and magnitude for the mean swirl velocity at all $R_h \le r \le R$ for all $h(x)$ at and abaft the trailing edge. For $R < r$ we obtain $u_{t0}(x,r) = 0$ as required from the last two integrals in (18.48).

The formula for the *radial* component of the induced velocity is

$$u_r(x,r,\gamma;\gamma_0(t)) = \frac{Z}{4\pi\rho U}\sum_{q=-\infty}^{\infty}\sum_{n=-\infty}^{\infty} e^{i(m\gamma+qZ\gamma_0)}\, e^{ik_0x}$$

$$\cdot\left\{\int_{R_h}^{R} dr'\left\{\int_{-h_t(r')}^{h(x,r')} dh'\left[\Delta p_n\, e^{-im\psi'}\left[n_a'\frac{\partial}{\partial x'} - in_t'\frac{m}{r'}\right] - \frac{\partial}{\partial r'}\left[n_r'\Delta p_n e^{-im\psi'}\right]\right]\right.\right.$$

$$\cdot\int_x^\infty \frac{\partial}{\partial r}\left[\frac{1}{\pi\sqrt{rr'}}\,Q_{|m|-1/2}(Z''^{-})\right] e^{-ik_0x''}dx''$$

$$-\int_{h(x,r')}^{h_1(r')} dh' \left[\Delta p_n\, e^{-im\psi'}\left[n_a'\frac{\partial}{\partial x'} - in_t'\frac{m}{r'}\right] - \frac{\partial}{\partial r'}\left[n_r'\Delta p_n e^{-im\psi'}\right]\right]$$

$$\cdot\int_{-x}^{\infty}\frac{\partial}{\partial r}\left[\frac{1}{\pi\sqrt{rr'}}\,Q_{|m|-1/2}(Z''+)\right] e^{ik_0x''}dx''\Bigg\}$$

$$-2\int_{R_h}^{r}\left\{ik_0f_a(r') + \frac{df_r(r')}{dr'} + i\frac{m}{r'}f_t(r')\right\} k_0\, I_m(k_0r')\, dr'\, K_m'(k_0r)$$

$$-2\int_{r}^{R}\left\{ik_0f_a(r') + \frac{df_r(r')}{dr'} + i\frac{m}{r'}f_t(r')\right\} k_0\, K_m(k_0r')\, dr'\, I_m'(k_0r)\Bigg] \quad (18.49)$$

The Integral Equation for Discrete Harmonic Orders

With the induced velocity components determined, we can now turn to the formation of the integral equation for any particular, single harmonic order, say $n = q_0Z$. For the determination of the mean loading, $q_0 = 0$ and for blade order $q_0 = \pm 1$. Because the expressions for the velocity components are unwieldy we deal with the integral equation symbolically.

The kinematic condition imposed on the key blade is given by Equation (18.20). We have the following form when taking $\gamma = \gamma_0 + \psi$, i.e. at any point on the key blade

$$\sum_{n=-\infty}^{\infty}\sum_{q=-\infty}^{\infty}\Big\{n_a\, u_{an}(x,r;\Delta p_n;n-qZ) + n_r\, u_{rn}(x,r;\Delta p_n;n-qZ)$$

$$+ n_t\, u_{tn}(x,r;\Delta p_n;n-qZ)\Big\} \cdot e^{i((n-qZ)(\gamma_0+\psi) - qZ\gamma_0)}$$

$$= -\sum_{n=-\infty}^{\infty}\left\{n_a\, u_{an}^{w}(r) + n_r\, u_{rn}^{w}(r) + n_t\, u_{tn}^{w}(r)\right\} e^{in(\gamma_0+\psi)} + n_aU + n_t\omega r$$

which reduces to

$$\sum_{q=-\infty}^{\infty}\left\{n_a\, u_{an} + n_r\, u_{rn} + n_t\, u_{tn}\right\} e^{-iqZ\psi}$$

$$= -\left\{n_a\, u_{an}^{w}(r) + n_r\, u_{rn}^{w}(r) + n_t\, u_{tn}^{w}(r)\right\} \quad ; n \neq 0 \quad (18.50)$$

$$\sum_{q=-\infty}^{\infty}\left\{n_a\, u_{a0} + n_r\, u_{r0} + n_t\, u_{t0}\right\} e^{-iqZ\psi}$$

$$= -\left\{n_a\,(-U + u_{a0}^{w}(r)) + n_r\, u_{r0}^{w}(r) + n_t\,(-\omega r + u_{t0}^{w}(r)\right\}$$

where the first equation applies for each harmonic order $n \neq 0$, and the last represents the steady case. As the amplitudes on both sides are complex (18.50) represents two equations for the real and imaginary (or the in-phase and quadrature) parts of the complex pressure jump function, Δp_n.

Had we made the preceding derivations on the basis of the mean inflow to the propeller disc U_A, with the substitutions on p. 341, the axial wake velocity in (18.50) must be defined relative to U_A, i.e., $u_{a0}^w(r)$ in (18.50) must be substituted with $u_{a0}^w(r) + U_A - U$, since $u_{a0}^w(r)$ is defined relative to the ship speed U.

We see that now the effects of all blades are included but that we must effect the q-summations which yield singular and regular parts of the kernel functions. These summations can be carried out analytically to exhibit the character of the singularities. The following is an example of one such summation.

Among the kernels in Equation (18.50) is the sum

$$S = \sum_{q=-\infty}^{\infty} I_{|n-qZ|}(|k_0|r')\, K_{|n-qZ|}(|k_0|r)\, e^{i(k_0(x-x') - qZ(\psi-\psi'))} \; ; \; r \geq r' \tag{18.51}$$

where $k_0 = qZp_f^*$ (and the tacitly assumed moduli of orders and arguments are emphasized).

Folding the q-sum (and taking the limits of the Bessel functions for $q \to 0$) we have

$$S = \sum_{q=1}^{\infty} \Big\{ I_{|n-qZ|}(k_0r')\, K_{|n-qZ|}(k_0r)\, e^{iqZ(p_f^*(x-x') - (\psi-\psi'))} + I_{|n+qZ|}(k_0r')\, K_{|n+qZ|}(k_0r)\, e^{-iqZ(p_f^*(x-x') - (\psi-\psi'))} \Big\} + \frac{1}{2n}\left[\frac{r'}{r}\right]^n$$

As we are interested in harmonic orders $n = q_0Z$ we can break up the q-sum as follows

$$S \simeq \sum_{q=1}^{q_1-1} (\,)_{|q-q_0|Z} + \sum_{q=q_1}^{\infty} (\,)_{qZ} + \sum_{q=1}^{q_1-1} (\,)_{|q+q_0|Z} + \sum_{q=q_1}^{\infty} (\,)_{qZ}$$

$$= \sum_{q=1}^{q_1-1} \Big\{ (\,)_{|q-q_0|Z} + (\,)_{|q+q_0|Z} \Big\} + S_1$$

for $q_1 >> q_0$ and where now the two infinite sums combine to

$$S_1 = 2\, Re \sum_{q=q_1}^{\infty} I_{qZ}(k_0r')\, K_{qZ}(k_0r)\, e^{iqZ(p_f^*(x-x') - (\psi-\psi'))} \tag{18.52}$$

As the order number is large here we can use asymptotic expressions for the Bessel functions, Equation (12.56), to obtain

$$S_1 = \frac{1}{\left[\left[1 + \left[p_f^* r'\right]^2 \left[1 + \left[p_f^* r\right]^2\right]\right]^{1/4}} Re \sum_{q=q_1}^{\infty} \frac{1}{qZ} X^q \qquad (18.53)$$

where

$$X = e^{\left[\sqrt{1 + \left[p_f^* r'\right]^2} - \sqrt{1 + \left[p_f^* r\right]^2}\right] Z} \left[\frac{r' \left[1 + \sqrt{1 + \left[p_f^* r\right]^2}\right]}{r \left[1 + \sqrt{1 + \left[p_f^* r'\right]^2}\right]}\right]^Z \cdot e^{iZ(p_f^*(x-x') - (\psi-\psi'))} \qquad (18.54)$$

To effect the sum in (18.53) we differentiate with respect to ψ and write

$$\frac{\partial S_1}{\partial \psi} = \frac{1}{\left[\left[1 + \left[p_f^* r'\right]^2\right] \left[1 + \left[p_f^* r\right]^2\right]\right]^{1/4}} Re \left\{\frac{-i}{1 - X} + i \sum_{q=0}^{q_1-1} X^q\right\} \qquad (18.55)$$

where we used the sum of the geometric series. As the last sum is a regular function for $X = 1$, i.e. for $x' = x$, $\psi' = \psi$, $r' = r$, we concentrate on the singular part. This can be written as

$$\frac{1}{1 - X} = \frac{1 - |X| \cos\theta + i|X| \sin\theta}{1 + |X|^2 - 2|X| \cos\theta} \quad ; \quad \theta = Z \left[p_f^*(x-x') - (\psi-\psi')\right]$$

Hence the singular part of (18.53) is by integration over ψ

$$S_{1,sing} = -\frac{1}{2Z} \frac{\ln(1 + |X|^2 - 2|X| \cos\theta)}{\left[\left[1 + \left[p_f^* r'\right]^2\right] \left[1 + \left[p_f^* r\right]^2\right]\right]^{1/4}} \qquad (18.56)$$

For $\psi' = \psi$ and $x' = x$ and letting $r - r'$ be small, we find

$$S_{1,sing} \longrightarrow -\frac{1}{Z\sqrt{1 + \left[p_f^* r\right]^2}} \ln \left[\frac{p_f^{*2}\, r\, Z}{\sqrt{1 + \left[p_f^* r\right]^2}} (r - r')\right] \qquad (18.57)$$

If we put $r = r'$

$$S_{1,sing} \longrightarrow -\frac{1}{Z\sqrt{1 + \left[p_f^* r\right]^2}} \ln \left[2 \sin \frac{Z\left[p_f^*(x-x') - (\psi-\psi')\right]}{2}\right] \qquad (18.58)$$

which for $x' = x$ is logarithmically singular at $\psi' = \psi$ and similarly for $\psi' = \psi$ in $x' - x$. Consequently the original sum (18.53) is composed of a regular part plus the singular part given by (18.56).

We can now see that all terms in the kernel functions which are single or mixed derivatives are Cauchy-singular, i.e. as $1/(r - r')$, $1/(x - x')$ or $1/(\psi - \psi')$. This is in contrast to the result in Equation (18.29) when using derivatives of $1/R$, where the kernel was found to be highly singular à la Hadamard, Section 4 of the Mathematical Compendium. From the expressions for the velocities, (18.46), (18.48) and (18.49) it can be seen that the singularities of the Hadamard type have been "softened" by use of integration by parts and the less singular Cauchy-kernels are now found in combination with derivatives of the partial loadings. When performing the summations it is advisable to multiply and divide by the Cauchy singularity to put it in evidence and to give bounded kernels.

It is to be noted that the use of Fourier expansions gives expressions for the components at the desired frequencies which are useful for specific purposes in the race and elsewhere and facilitates development of kernel functions which are integrable, in the sense of Cauchy. The wake-harmonic orders of interest are those of multiples of the blade number for thrust and torque on the shaft and those at $n = q_0 Z \pm 1$ for transverse and vertical forces and moments. This will be seen in the next section.

A remaining task is the development of a computer program for the unknown pressure-jump distribution to obtain numerical results from the foregoing. Although the approach with use of Fourier expansions of $1/R$ follows that of Tsakonas and Jacobs, cf. p. 335, the kernel developed here differs in several respects from theirs. It remains to be seen how their numerical results and those from this development compare.

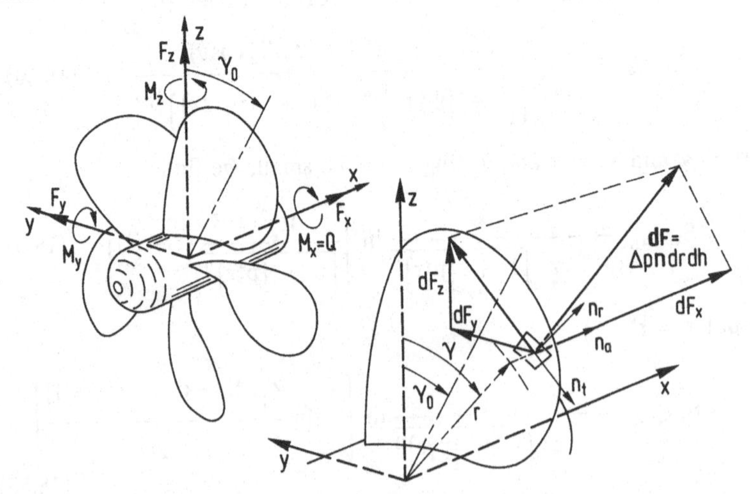

Figure 18.5 Forces and moments on propeller.

Forces and Moments

The forces and moments acting on the entire propeller are obtained by resolving the components of the forces on an element of the ν-th blade, cf. Figure 18.5, at position $\gamma = \gamma_0 + 2\pi\nu/Z$, integrate over the blade and sum over ν to obtain the total component.

Hence, the axial force or thrust on the entire propeller is

$$F_x = T = \sum_{\nu=0}^{Z-1} \sum_{n=-\infty}^{\infty} \int_{R_h}^{R} dr \int_{-h_t(r)}^{h_l(r)} \Delta p_n(h,r)\; n_a\; dh\; e^{in(\gamma_0+2\pi\nu/Z)}$$

which as we have seen (p. 212) is zero for all $n \neq qZ$, $q = 0, \pm1, \pm2, \dots$ and hence only the wake harmonics of order $n = qZ$ contribute to give

$$F_x = Z \sum_{q=-\infty}^{\infty} \int_{R_h}^{R} dr \int_{-h_t(r)}^{h_l(r)} \Delta p_{qZ}(h,r)\; n_a(h,r)\; dh\; e^{iqZ\gamma_0}$$

The axial force at any selected q is

$$(F_x)_{qZ} = Z\int_{R_h}^{R} dr \int_{-h_t(r)}^{h_l(r)} \left\{\Delta p_{qZ}(h,r)\, e^{iqZ\gamma_0} + \Delta p_{-qZ}(h,r)\, e^{-iqZ\gamma_0}\right\} n_a(h,r)\; dh$$

As Δp_{-qZ} is the complex conjugate of Δp_{qZ} we have

$$(F_x)_{qZ} = 2Z \int_{R_h}^{R} dr \int_{-h_t(r)}^{h_l(r)} n_a(h,r)\; Re\left[\Delta p_{qZ}(h,r)\, e^{iqZ\gamma_0}\right] dh \tag{18.59}$$

For the horizontal and vertical forces we need the relations between the components of the unit normals, cf. Figure 18.5

$$n_y = -(n_r \sin\gamma + n_t \cos\gamma) = -\frac{1}{2}\left[(n_t - in_r)e^{i\gamma} + (n_t + in_r)e^{-i\gamma}\right]$$

(18.60)

$$n_z = \quad n_r \cos\gamma - n_t \sin\gamma = \frac{1}{2}\left[(n_r + in_t)e^{i\gamma} + (n_r - in_t)e^{-i\gamma}\right]$$

On the ν-th blade $\gamma = \gamma_0 + 2\pi\nu/Z + \psi$ and the horizontal force is

$$F_y = -\frac{1}{2}\sum_{\nu=0}^{Z-1} \sum_{n=-\infty}^{\infty} \int_{R_h}^{R} dr \int_{-h_t(r)}^{h_l(r)} \Delta p_n(h,r) \Big\{(n_t - in_r)\, e^{i((n+1)\gamma_0+(n+1)2\pi\nu/Z+\psi)}$$

$$+ (n_t + in_r)\, e^{i((n-1)\gamma_0+(n-1)2\pi\nu/Z-\psi)}\Big\}\, dh$$

The ν-sums then give

$$F_y = -\frac{Z}{2}\sum_{q=-\infty}^{\infty}\int_{R_h}^{R} dr \int_{-h_t(r)}^{h_l(r)} \Big\{(n_t - in_r)\Delta p_{qZ-1} e^{i\psi} + (n_t + in_r)\Delta p_{qZ+1}\, e^{-i\psi}\Big\}\, dh\, e^{iqZ\gamma_0}$$

Folding the q-sum and noting that $\Delta p_{-(qZ+1)}$ is the complex conjugate of Δp_{qZ+1}, we find the sum of complex conjugates and obtain

$$F_y = -Z\, Re\sum_{q=0}^{\infty}\int_{R_h}^{R} dr \int_{-h_t(r)}^{h_l(r)} \Big\{(n_t - in_r)\Delta p_{qZ-1}\, e^{i\psi} + (n_t + in_r)\Delta p_{qZ+1}\, e^{-i\psi}\Big\}\, dh\, e^{iqZ\gamma_0} \qquad (18.61)$$

Similarly, the vertical force is

$$F_z = Z\, Re\sum_{q=0}^{\infty}\int_{R_h}^{R} dr \int_{-h_t(r)}^{h_l(r)} \Big\{(n_r + in_t)\Delta p_{qZ-1}\, e^{i\psi} + (n_r - in_t)\Delta p_{qZ+1}\, e^{-i\psi}\Big\}\, dh\, e^{iqZ\gamma_0} \qquad (18.62)$$

The differential moment about the origin at the center of the hub is $\mathbf{dM} = \mathbf{r} \times \mathbf{n}\, \Delta p\, dS$, or

$$\mathbf{dM} = (x, -r\sin\gamma, r\cos\gamma) \times (n_a, -n_r\sin\gamma - n_t\cos\gamma, n_r\cos\gamma - n_t\sin\gamma)\, \Delta p\, dS$$

$$=(rn_t, rn_a\cos\gamma - xn_r\cos\gamma + xn_t\sin\gamma, rn_a\sin\gamma - xn_r\sin\gamma - xn_t\cos\gamma)\, \Delta p\, dS$$

Integrating and summing as before we have the following expressions for the components of the moment vector:

Torque:

$$M_x = -Q = Z\left\{\int_{R_h}^{R} r\, dr \int_{-h_t(r)}^{h_l(r)} \Delta p_0\, n_t dh + Re\sum_{q=1}^{\infty}\int_{R_h}^{R} r\, dr \int_{-h_t(r)}^{h_l(r)} \Delta p_{qZ}\, n_t dh\, e^{iqZ\gamma_0}\right\} \qquad (18.63)$$

Moment about the horizontal (y-) axis:

$$M_y = Z\, Re \sum_{q=0}^{\infty} \int_{R_h}^{R} dr \int_{-h_t(r)}^{h_l(r)} \Big\{ (r\, n_a - x\, n_r - i\, x\, n_t)\, \Delta p_{qZ-1}\, e^{i\psi} + (r\, n_a - x\, n_r + i\, x\, n_t)\, \Delta p_{qZ+1}\, e^{-i\psi} dh \Big\}\, e^{iqZ\gamma_0} \tag{18.64}$$

Moment about the vertical (z-) axis:

$$M_z = Z\, Re \sum_{q=0}^{\infty} \int_{R_h}^{R} dr \int_{-h_t(r)}^{h_l(r)} \Big\{ (-i(r\, n_a - x\, n_r) - x\, n_t)\, \Delta p_{qZ-1}\, e^{i\psi} + (i(r\, n_a - x\, n_r) - x\, n_t)\, \Delta p_{qZ+1}\, e^{-i\psi} dh \Big\}\, e^{iqZ\gamma_0} \tag{18.65}$$

The foregoing identifies the various wake components which cause forces and moments at frequency $qZ\omega$. These results can be summarized as follows:

Forces and moments at frequency $qZ\omega$		Hull wake component order
Thrust	$(T)_{qZ}$	qZ
Torque	$(M_x)_{qZ}$ or $(Q)_{qZ}$	
Horizontal force	$(F_y)_{qZ}$	$qZ \pm 1$
Vertical force	$(F_z)_{qZ}$	
Horizontal moment	$(M_y)_{qZ}$	
Vertical moment	$(M_z)_{qZ}$	

It is apparent that linearized theory omits or approximates geometric details of the blade as well as higher-order terms in the equations of motion. Presently computers have sufficient capacity to enable evaluation of formulations with complete representations of blade geometry. We next give an outline of such a procedure for propellers of arbitrary geometry in spatially non-uniform inflow.

A POTENTIAL-BASED BOUNDARY-ELEMENT PROCEDURE

Here we outline a method by Kinnas & Hsin (1990) for quite exact representation of propellers of arbitrary shape in a spatially non-uniform flow of an ideal, incompressible fluid. This procedure is a recent development by Kerwin and his colleagues at MIT of computer methods for design and analysis of propellers.

The basic equation for the velocity potential on the surface of the blade S_B and that containing the shed vorticity in the blade wake S_w, cf. Figure 18.6, is obtained from Green's second identity with the use of $1/R$ as the Green function, cf. Section 3 of the Mathematical Compendium, p. 506. This equation will have the form of (M3.7) where S is the boundary surface of the fluid volume, i.e. $S = S_B + S_w^+ + S_w^-$ + a surface around the entire propeller but "far" away from it. This interpretation makes S a single closed surface. S_B is now the actual surface of the propeller blade (and not just the fluid reference surface used previously). Normals are defined positive out of the fluid as indicated in the Figure 18.6.

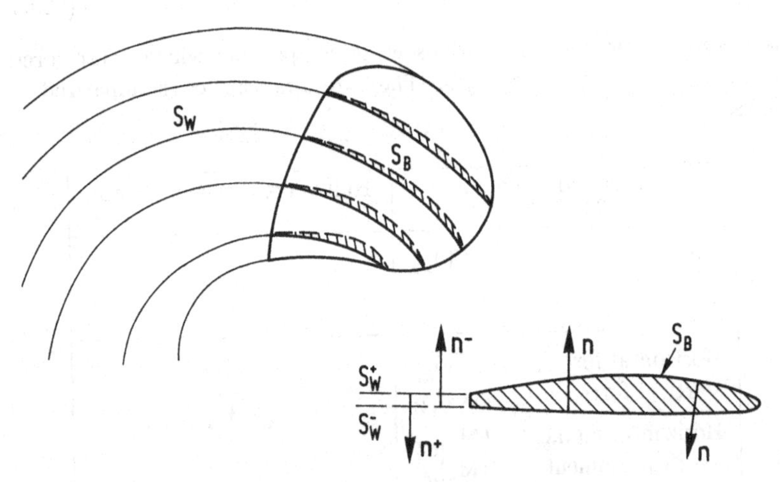

Figure 18.6 Propeller–blade and blade–wake surfaces.

The integral over the wake surfaces S_w^+ and S_w^- can be collapsed into one single surface S_w (=, say S_w^+). From (M3.7), p. 507 the contribution from the wake is

$$\phi_w = -\frac{1}{2\pi}\int_{S_w^+ + S_w^-}\left\{\phi\,\frac{\partial}{\partial n'}\frac{1}{R} - \frac{1}{R}\frac{\partial\phi}{\partial n'}\right\} dS'$$

$$= -\frac{1}{2\pi}\int_{S_w^+}\left\{\phi_+\,\frac{\partial}{\partial n'}\frac{1}{R} + \phi_-\,\frac{\partial}{\partial n'}\frac{1}{R} - \frac{1}{R}\left[\frac{\partial\phi_+}{\partial n'} + \frac{\partial\phi_-}{\partial n'}\right]\right\} dS'$$

since the distance between two points on S_w^+ and S_w^- is infinitely small. The velocity normal to the trailing-vortex sheet is continuous. To see this, note that the helical trailers will give a contribution analogous to the straight-line trailers of the wing, cf. Equation (10.5). The transverse vortices in the wake, due to the change of blade loading with time, can be interpreted as those bound to a two-dimensional wing of varying circulation and the velocity calculated according to Equation (2.38). We then have that $\partial\phi_+/\partial n_+ = -\partial\phi_-/\partial n_-$ and the last integrand is zero. The first integrand contains the jump in potential across the wake cut $\Delta\phi_W = \phi_+ - \phi_-$ which is non-zero, cf. Equation (2.60) (although this applies to the steady case). We now have the equation for the velocity potential

$$\phi = -\frac{1}{2\pi}\int_{S_B}\left\{\phi\,\frac{\partial}{\partial n'}\frac{1}{R} - \frac{1}{R}\frac{\partial\phi}{\partial n'}\right\} dS' - \frac{1}{2\pi}\int_{S_W} \Delta\phi_W \frac{\partial}{\partial n'}\frac{1}{R}\, dS' \tag{18.66}$$

We now can see that the surface potential ϕ is the superposition of potentials of distributions of sources of strength $\partial\phi/\partial n'$ and normal dipoles of strength ϕ on the face and back surfaces of the blade plus a distribution of dipoles on the wake surface S_W of strength $\Delta\phi_W$.

As in two dimensions, the history of the flow must be included. In the present formulation it appears in the wake term $\Delta\phi_W$. This can be put in evidence and an expression secured which is needed for the solution of the problem by the following argument:

First, note that the pressure jump across the wake must be zero because there is no physical barrier, such as the propeller blade, to maintain a pressure difference. The full pressure equation referred to fixed axes is given by Equation (1.27) (with $f(t) = 0$). With respect to rotating axes fixed to the blade (which is restrained from translating in a superimposed stream) for which $\gamma = \gamma_0 + \psi = \omega t + \psi$, the relative velocity vector, $\mathbf{q}_{rel} = (-U + u_a + u_a^W, u_r + u_r^W, u_t + u_t^W - \omega r)$, must be used in the pressure equation and the pressure is then given by

$$p = -\rho\left[\frac{\partial}{\partial t} + \omega\frac{\partial}{\partial\psi}\right]\phi(x,r,\omega t + \psi,t) - \frac{\rho}{2}\, q_{rel}^2 \tag{18.67}$$

The pressure jump across the wake cut will then be

$$p_- - p_+ = -\rho\left[\frac{\partial}{\partial t} + \omega\frac{\partial}{\partial\psi}\right](\phi_- - \phi_+) = \rho\left[\frac{\partial}{\partial t} + \omega\frac{\partial}{\partial\psi}\right]\Delta\phi_W \tag{18.68}$$

since the magnitudes of the velocities are equal on the lower and upper banks of the cut although there is a shearing of velocity. As now $p_- - p_+$ must be zero the solution to (18.68) must be of the form

$$\Delta\phi_w(r,\psi,t) = \Delta\phi_{te}(r,\psi_t,t-(\psi+\psi_t)/\omega) \; ; \qquad \text{for } t > |\psi+\psi_t|/\omega,\ \psi \leq -\psi_t \quad (18.69)$$

which means that the strength $\Delta\phi_w$ at current time t and given position ψ relative to the blade is equal to $\Delta\phi_{te}$ of the trailing edge at that particular time $t-(\psi+\psi_t)/\omega$ when the trailing edge was at that position, ψ_t. For steady flow

$$\Delta\phi_w(r,\psi,t) = \Delta\phi_{te}^{s}(r) \quad (18.70)$$

where $\Delta\phi_{te}^{s}(r)$ is the steady flow jump in potential at the trailing edge.

It can be seen that the value of the jump in the dipole strength at the trailing edge is given by

$$\Delta\phi_{te}(r,t) = \phi_{te}^{+}(r,t) - \phi_{te}^{-}(r,t) = \Gamma(r,t) \quad (18.71)$$

in which $\Gamma(r,t)$ is the instantaneous circulation around the blade at r and t. This is in analogy with the result of Equation (2.60) for the wing in steady flow and the condition is equivalent to the requirement that the shed vorticity be proportional to the time rate of change of the circulation around the blade section, p. 316.

As the kinematic condition on the blade is, cf. (18.20),

$$\frac{\partial\phi}{\partial n} = \mathbf{n}\cdot\mathbf{q} = -\mathbf{q}_{rel}\cdot\mathbf{n} \quad (18.72)$$

the source strength in the second term in (18.66) is a known function, giving the following Fredholm integral equation of the second kind for the unknown potential function

$$2\pi\phi + \int_{S_B}\phi\frac{\partial}{\partial n'}\frac{1}{R}\,dS' + \int_{S_W}\Delta\phi_w\frac{\partial}{\partial n'}\frac{1}{R}\,dS' = -\int_{S_B}\frac{\mathbf{q}_{rel}\cdot\mathbf{n}'}{R}\,dS' \quad (18.73)$$

The trajectory of the trailing wake surface S_w is taken to be that determined by Greeley & Kerwin (1982), with contraction and roll-up and convected by the mean inductions plus relative velocities $-U_a(r)$ and $-\omega r$.

The numerical inversion of (18.73) by Kinnas & Hsin (1990) involves discrete equal time steps, Δt, and the use of wake panels located on the prescribed wake surface at equal angular spacing $\Delta\psi_w = -\omega\,\Delta t$. The dipole and source strengths are taken as constants on each panel of the blade surface and evaluations of the kernel functions are made at the centroids of the panels at each time step, thereby converting the integral equation into a set of simultaneous algebraic equations generated by N chordwise and M spanwise blade panels on the blades and N_w panels per spanwise blade panels on the wake surface as shown in Figure 18.7.

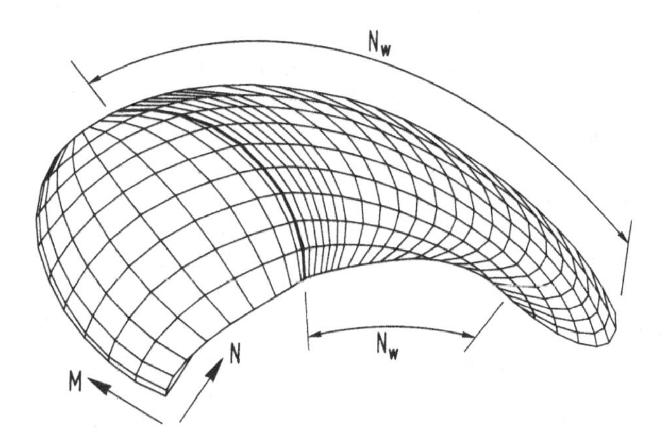

Figure 18.7 Panel arrangement for a blade and its wake, N = 20, M = 10, $\Delta\psi_W$ = 6 degrees.

Figure 2 from Kinnas, S.A. & Hsin, C.-Y. (1990). *A Boundary-Element Method for Analysis of the Unsteady Flow Around Extreme Propeller Geometries.* Cambridge, Mass.: Dept. of Ocean Eng., MIT.

By courtesy of Massachusetts Institute of Technology, USA.

Figure 18.8 Discretized geometry of highly skewed propeller.

Figure 10 from Kinnas, S.A. & Hsin, C.-Y. (1990). *A Boundary-Element Method for Analysis of the Unsteady Flow Around Extreme Propeller Geometries.* Cambridge, Mass.: Dept. of Ocean Eng., MIT.

By courtesy of Massachusetts Institute of Technology, USA.

The solution is carried out iteratively. For $t < 0$ the steady case is considered and the unsteady inflow is then "turned on" at $t = 0$. At each time step the propeller is rotated and (18.73) is solved using the solution of the previous step until convergence. Further details of the numerical procedures can be found in Kinnas & Hsin (*ibid.*) as well as results showing consistency of their boundary-element method with analytical results for forces of a flat plate.

Among the results reported were calculated unsteady forces on a highly skewed propeller shown in Figure 18.8. The wake was of a special type with large amplitudes of the 5th, 10th and 15th harmonics 5 to 15 times larger than those of typical ship wakes and each being 10 per cent of the mean axial velocity. This wake was used to determine the number of panels required at four frequencies to secure convergent results. It was found that 40 chordwise and 30 spanwise panels give converged results after about three rotations of the propeller.

Harmonic analysis of the total unsteady thrust on this 5-bladed propeller revealed amplitudes at each of the harmonics of 1, 5, 10 and 15 times, which was to be expected, according to linear theory, cf. p. 367. But additional amplitudes arising from the sum and difference of the spatial orders present in the wake were found. These amplitudes are explained as arising from the non-linear leading-edge suction forces giving rise to forces at twice the inflow harmonic orders as well as contributions from the sum and difference of the wake orders. The results at frequencies other than the four in the imposed wake gave amplitudes varying from 5 to 15 per cent of those at the basic orders. We surmise that these fractions are unduly large because of the non-typical large values of the imposed amplitudes at the 5th, 10th and 15th orders.

This highly sophisticated procedure gives very close approximation of the pressure distribution for use in cavitation inception and for close prediction of mean thrust, torque and efficiency. As hull-wake harmonics at blade order and multiples thereof are small, it is sufficient to use the linearized theory for unsteady forces. Correlations with some measurement, which are displayed in the next chapter, show that the predictions of unsteady forces are reasonable to good.

We may conclusively remark that although this procedure, as outlined here, may seem simple, in particular when compared with the detailed formulas for the velocity field using Fourier series, p. 350–361, the numerical aspects such as generation of grids over the blades and placing of collation points are complicated. With the high accuracy and generality of this and similar methods it can be considered as a "numerical cavitation tunnel", and like the real cavitation tunnel the results obtained, e.g., field pressures or blade forces, contain the whole "signal" and must be further analyzed to give the harmonics. Moreover, in the opinion of the authors, computer programs, despite their completeness, do not provide as much physical insight to the various aspects of propeller flows as do the more lengthy, albeit approximate, mathematical analyses also presented in this chapter.

Several aspects of propeller hydrodynamics have not been addressed in the foregoing. Model measurements of unsteady shaft forces have revealed rapidly varying modulations in amplitudes by large factors frequently within 1/Z-th revolution, always at blade frequency (and integer multiples). This requires a theory to account for flows having spatial harmonics with amplitudes varying stochastically with time.

Another problem area is presented by the coupling of propeller inductions with hull-wake vorticity. Calculations of the alteration of nominal-to-effective wake by this mechanism have been given, cf. Huang & Groves (1980), Goodman (1979), Falcão de Campos & van Gent (1981) and Breslin (1992). A single calculation by Van Houten (see Breslin, Van Houten, Kerwin & Johnsson (1982)) showed large alterations of the harmonics of importance to the prediction of unsteady forces as well as the extent of transient cavitation.

Moreover, the effects of viscosity and turbulence, which are particularly significant in off-design operation, have been virtually untouched. Undoubtedly these aspects will be attacked in the future by computer-effected representations analogous to those currently being applied to turbulent hull flows, cf. Stern, Toda & Kim (1991) who are attempting to account for the alteration of stern flows by the propeller. This and other recent developments can be followed in the latest reports of the Committees of the International Towing Tank Conferences and the bibliographies published in the ITTC proceedings, (e.g. ITTC (1990a, 1990b)).

19 Correlations of Theories with Measurements

The number of comparisons that have been made of calculated and measured blade-frequency thrust, torque and other force and moment components are very few because of the paucity of data. In this chapter comparisons of theoretical predictions with experimental data will be given. Results obtained by various theories will also be compared. The chapter concludes with presentation of a simple procedure, based upon the K_T–J curve of the steady case, for a quick estimate of the varying thrust at blade frequency.

The measurement of blade-frequency forces on model propellers requires great care in the design of the dynamometer which must have both high sensitivity and high natural frequencies well above the model blade frequency. After a number of failures a successful blade-frequency propeller dynamometer capable of measuring six components (three forces, three moments) was evolved at David Taylor Research Center (DTRC) about 1960.

Measurements were made with a triplet of three-bladed propellers of different blade-area ratio designed to produce the same mean thrust. This set was tested in the DTRC 24-inch water tunnel alternately abaft three- and four-cycle wake screens which produced large harmonic amplitudes of the order of $0.25 \cdot U$ in order to obtain strong output-to-"noise" levels! The three-cycle screens give rise to blade-frequency thrust and torque whereas the four-cycle wake produces transverse and vertical forces and moments about the y- and z-axes which in general come from the fourth and second harmonic orders of blade loading on a three-bladed propeller, as was demonstrated on p. 367.

We shall present the forces and moments in dimensionless form in the usual way

$$K_F = \frac{F}{\rho N^2 D^4} \quad ; \quad K_M = \frac{M}{\rho N^2 D^5} \tag{19.1}$$

The direction of forces and moments cf. Figure 18.5 are indicated by subscript where we in accordance with tradition call x-direction thrust and torque T and Q, instead of F_x and M_x. Hence

$$K_T = \frac{T}{\rho N^2 D^4} \qquad K_Q = \frac{Q}{\rho N^2 D^5} \tag{19.2a}$$

$$K_{F_y} = \frac{F_y}{\rho N^2 D^4} \qquad K_{M_y} = \frac{M_y}{\rho N^2 D^5} \tag{19.2b}$$

$$K_{F_z} = \frac{F_z}{\rho N^2 D^4} \qquad K_{M_z} = \frac{M_z}{\rho N^2 D^5} \tag{19.2c}$$

$$K_{F_{ys}} = \frac{F_{ys}}{\rho N^2 D^4} \tag{19.2d}$$

F_{ys}: horizontal force on single blade (index s)

$$K_N(r) = \frac{F_N(r)}{\rho N^2 D^4} \tag{19.2e}$$

$F_N(r)$: blade normal force (local loading)

A further index indicates multiple of shaft frequency, so

$(K_{F_y})_0$	steady, usually index omitted and written K_{F_y}
$(K_{F_y})_1$	shaft frequency
$(K_{F_y})_2$	twice shaft frequency
:	
$(K_{F_y})_Z$	blade frequency
$(K_{F_y})_{2Z}$	twice blade frequency
:	

Figure 19.1 displays the blade-frequency thrust variation with expanded-area ratio as measured and calculated by various means. It is clear that the quasi-steady approximation, using the slope of the $K_T - J$ curve, see p. 380, (indicated as Quasi-Steady + $\partial K_T / \partial J$) fails to reveal the decreasing $(K_T)_Z$ as the expanded-area ratio increases or, as we have seen earlier (cf. p. 332), this is equivalent to increasing the reduced frequency. It is also clear that using the two-dimensional Sears function without induced effects grossly overestimates the data or upon an *ad-hoc* allowance for induced effects (indicated as Quasi-Steady + 2D Unsteady, à la Ritger & Breslin (1958)) seriously underestimates. The theory of Jacobs & Tsakonas (1973) (designated as unsteady lifting-surface theory in Figure 19.1) tracks the data well. Their correlation with blade-frequency torque amplitude in Figure 19.2 is not as good but far better than all other displayed approximations. The correlations of transverse forces and moments are shown in Figure 19.3 and Figure 19.4. Here agreement with data is somewhat less close.

Comparisons of calculations and data for an unskewed propeller over a large range of advance ratios, as shown in Figure 19.5, show remarkably close agreement while in the case of skew the trend is different. It was later found that the blades of the skewed propeller twisted under load at reduced J.

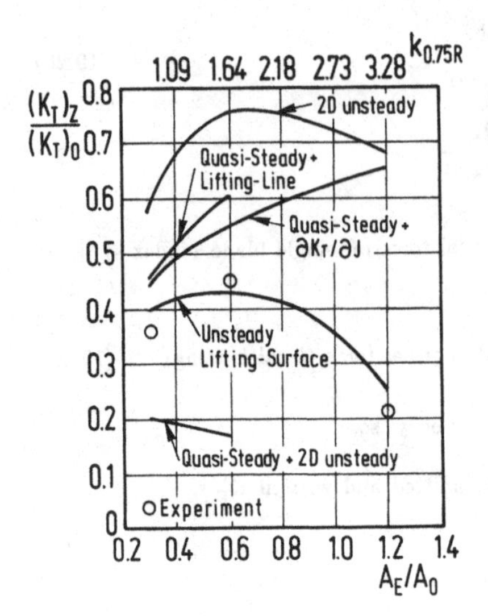

Figure 19.1 Correlation of blade-frequency thrust over range of expanded-area ratio at design K_T.

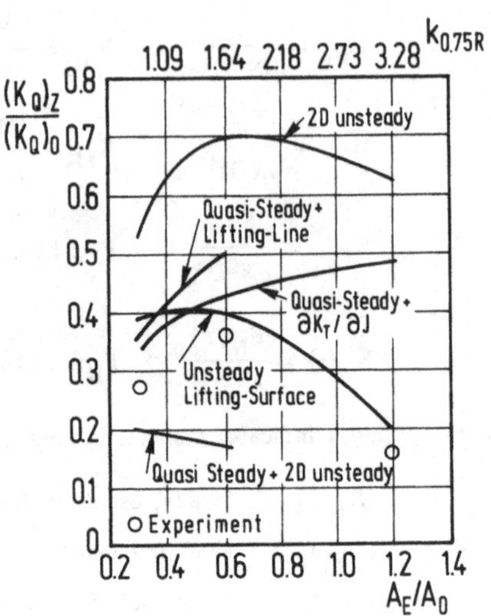

Figure 19.2 Correlation of blade-frequency torque over range of expanded-area ratio at design K_T.

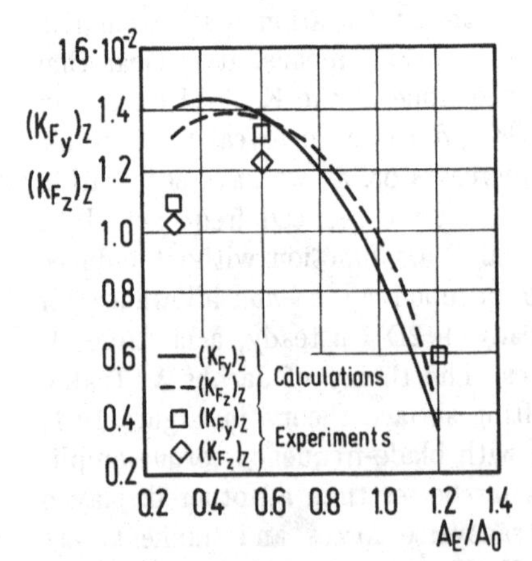

Figure 19.3 Correlation of experimental blade-frequency transverse forces with lifting-surface theory at design K_T.

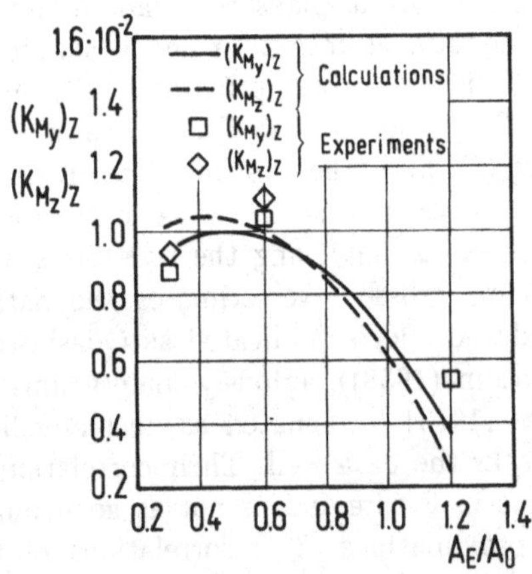

Figure 19.4 Correlation of experimental blade-frequency bending moments with lifting-surface theory at design K_T.

Figures 6, 7 and 8 from Breslin, J.P. (1970). Theoretical and experimental techniques for practical estimation of propeller-induced vibratory forces. *Trans. SNAME*, vol. 78, pp. 23–40. New York, N.Y.: SNAME.

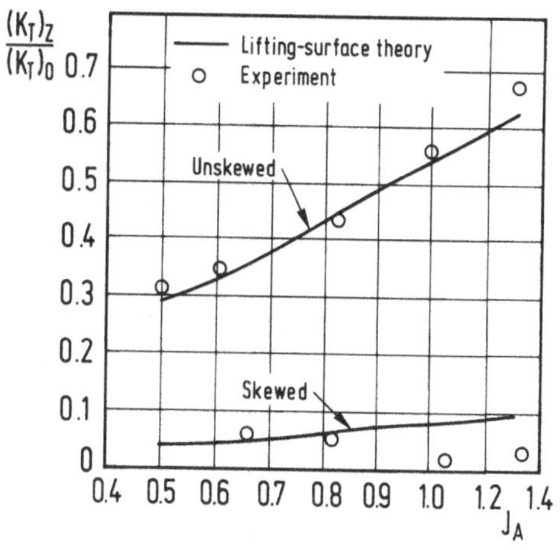

Figure 19.5 Correlation of experimental blade-frequency thrust with lifting-surface theory considering skew ($A_E/A_0 = 0.6$).

Figure 9 from Breslin, J.P. (1970). Theoretical and experimental techniques for practical estimation of propeller-induced vibratory forces. *Trans. SNAME*, vol. 78, pp. 23–40. New York, N.Y.: SNAME.

Force and Moment Coefficients	Davidson Laboratory	Massachusetts Institute of Technology (MIT)
$(K_T)_3$	1.05	0.90
$(K_Q)_3$	1.15	0.96
$(K_{F_y})_3$	1.14	1.20
$(K_{M_y})_3$	0.95	0.97
$(K_{F_z})_3$	1.14	1.36
$(K_{M_z})_3$	0.99	1.04

Table 19.1 Ratio of predicted to experimental results for DTRC propeller 4118. Thrust and torque induced by three-cycle wake screen; remaining forces and moments by four-cycle wake screen.

Table 6 from Tsakonas, S. (1978). Discussion of Kerwin, J.E. & Lee, C.-S.: Prediction of steady and unsteady marine propeller performance by numerical lifting-surface theory. *Trans. SNAME*, vol. 86, pp. 247–8. New York, N.Y.: SNAME.

Highly favorable comparisons of the Davidson Laboratory program with predictions made at MIT by Kerwin & Lee (1978) are given in the following Table 19.1 and in Figures 19.6, 19.7 and 19.8. The Davidson Laboratory program is based on the work of Tsakonas and colleagues, cf. p. 335, using Fourier expansions of $1/R$, i.e. the approach outlined on p. 350 and

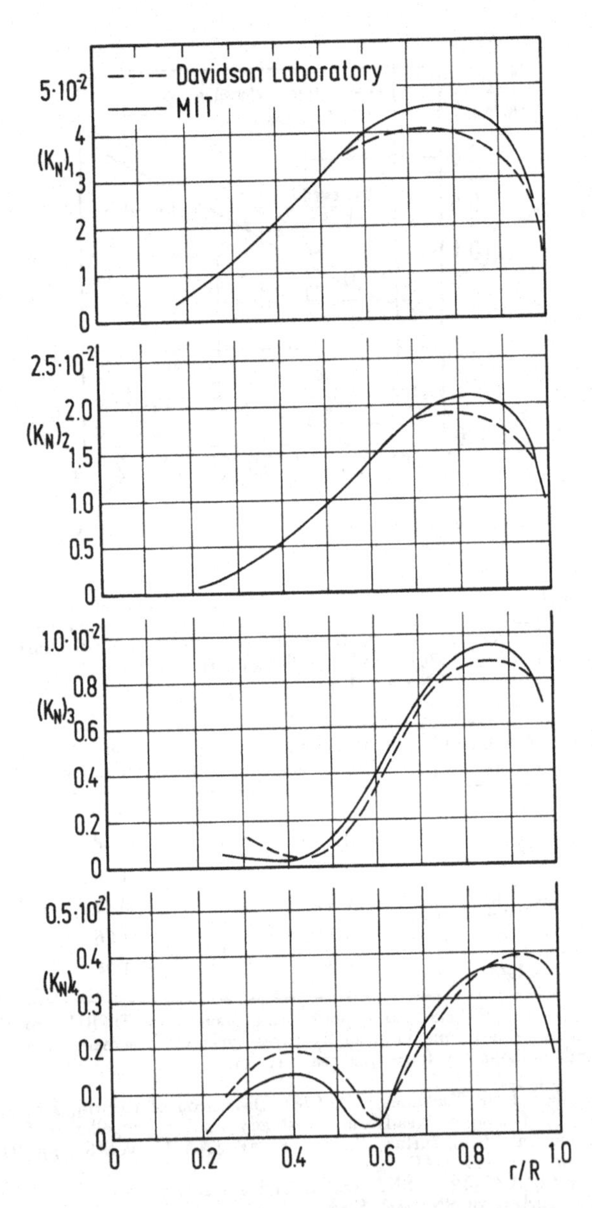

Figure 19.6 Comparison of theoretical results for radial distribution of blade normal force. Note the difference in scale of the force coefficients.

Figure 31 from Tsakonas, S. (1978). Discussion of Kerwin, J.E. & Lee, C.-S.: Prediction of steady and unsteady marine propeller performance by numerical lifting-surface theory. *Trans. SNAME*, vol. 86, pp. 247–8. New York, N.Y.: SNAME.

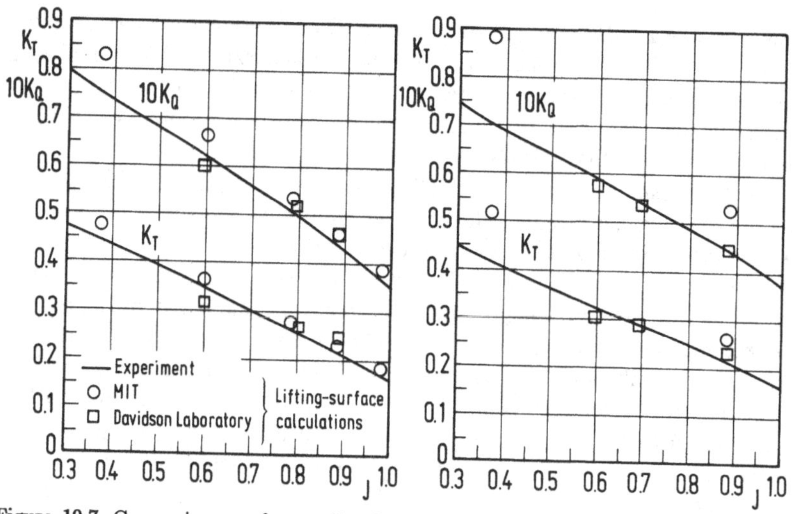

Figure 19.7 Comparison of predicted performances by theories with open-water characteristics of DTRC propeller model 4382 (36-deg. skew).

Figure 19.8 Comparison of predicted performance by theories with open-water characteristics of DTRC propeller model 4384 (108-deg. skew).

Figures 32 and 33 from Tsakonas, S. (1978). Discussion of Kerwin, J.E. & Lee, C.-S.: Prediction of steady and unsteady marine propeller performance by numerical lifting-surface theory. *Trans. SNAME*, vol. 86, pp. 247-8. New York, N.Y.: SNAME.

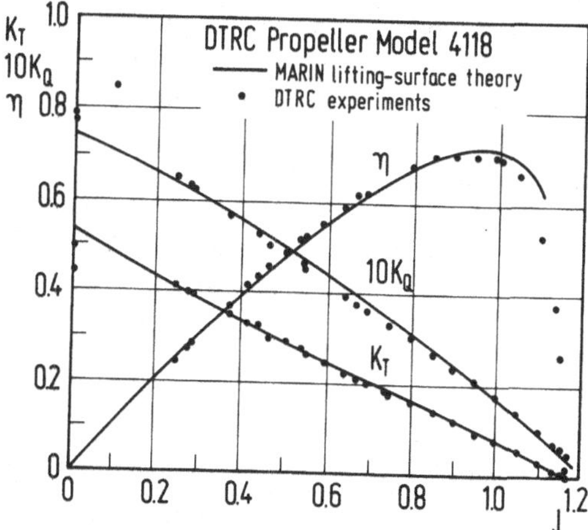

Figure 19.9 Comparison of open-water characteristics of DTRC model propeller 4118; experiment and MARIN theory.

Figure 34 from van Gent, W. (1978). Discussion of Kerwin, J.E. & Lee, C.-S.: Prediction of steady and unsteady marine propeller performance by numerical lifting-surface theory. *Trans. SNAME*, vol. 86, p. 249. New York, N.Y.: SNAME.

sequel, although differing as noted. The MIT method uses a discrete-vortex/source-line representation of the blades and can be considered as a predecessor of the method outlined on p. 368 and sequel. Here the results of the Davidson Laboratory program generally compare more closely with the measurements than MIT's.

Favorable results by van Gent (1978) are given in Figure 19.9 for steady propeller performance which displays his ability to calculate open-water performances over a wide range of advance ratios.

An interesting comparison of the amplitudes of alternating thrust on a single blade for the first eight shaft harmonics as calculated by many investigators, was compiled by Schwanecke (1975). Kerwin & Lee (1978) added their results in 1978. The sources of the calculations are identified in the Table 19.2 and the comparison is shown in Figure 19.10 and 19.11. In Figure 19.10 the bar graphs grouped to the left are those identifed as "exact" theories in the sense that they account for unsteady effects in a three-dimensional way. Those grouped to the right are "approximate" theories which utilize a variety of quasi-steady or two-dimensional unsteady approaches. It is seen that the "exact" or most rational theories give results which fall in a rather narrow band, whereas the approximate methods yield answers which vary widely. Of the "exact" methods based on unsteady lifting-surface theory those numbered 4 and 6 use derivatives of $1/R$ while those numbered 2, 3 and 5 use Fourier expansions. Number 7 uses a discrete-vortex/source-line representation of the blade. These three approaches are outlined in Chapter 18 on p. 343, 350, 368 and sequel. For the last two, more recent developments of these basic approaches are presented there. This applies in particular to the third (number 7) where a potential formulation with boundary elements is outlined instead of discrete vortices.

An attempt was made by Breslin (1969) to collapse the unsteady thrust calculated at Davidson Laboratory for propellers in widely different wakes generated by a screen at DTRC and by vanes at MARIN. The results are shown in Figure 19.13 for the wake blade-frequency amplitude variations with radius displayed in Figure 19.12. It is seen that the collapse of these results is remarkable.

The basis of this normalization is that at "zero" reduced frequency (quasi-steady) the fluctuating thrust can be approximated by the product of the slope of the steady K_T vs. J curve as demonstrated by the following:

The K_T–J curve is approximated by a straight line in the neighborhood of the design J_0, i.e.,

$$K_T = K_{T_0} + \left[\frac{\partial K_T}{\partial J}\right]_0 (J - J_0) \tag{19.3}$$

Number in Figure	Organization	Method
1	Ship Research Inst., Japan	Hanaoka (1969b)
2	Stevens Inst. Tech., USA	Tsakonas *et. al.* (1968), (1969)
3	DTRC, USA	Tsakonas *et. al.* (1968), (1969)
4	Univ. of Hamburg, Germany	Isay (1970), Klingsporn (1973), Bauschke & Lederer (1974)
5	Centrum Techn. Okretowej, Poland	Tsakonas *et. al.* (1968), (1969)
6	MARIN, The Netherlands	Verbrugh (1968), van Gent (1975)
7	MIT, USA	Kerwin & Lee (1978)
8	HSVA, Germany, Tech. Univ. Wien, Austria	Schwanecke & Laudan (1972)
9	Admiralty Res. Lab., Great Britain	Murray & Tubby (1973)
10	CETENA, Italy	–
11	Krylov Shipbldg. Inst., USSR	Voitkunsky *et. al.* (1973)
12	SSPA, Sweden	Johnsson (1968)
13	Mitsubishi Heavy Ind., Japan	McCarthy (1961)

1 — 7 Unsteady lifting-surface methods
8 — 11 Approximation methods
12 — 13 Quasi-steady approximation methods

Table 19.2 Illustration of Numbers in Figure 19.10 and 19.11.
The method of Tsakonas *et. al.* (1968), (1969), (i.e. Tsakonas, Jacobs & Rank (1968), Tsakonas & Jacobs (1969)), has been programmed by various programmers. For details cf. Schwanecke (1975).

Figures 19.10, 19.11 and Table 19.2, from Schwanecke, H. (1975). Comparative calculation of unsteady propeller blade forces. In *Proc. Fourteenth International Towing Tank Conference (ITTC)*, vol. 3, pp. 357–97. Ottawa: National Research Council Canada.
By courtesy of ITTC and H. Schwanecke.

The excursion of K_T due to a variation in the effective J because of wake harmonics at blade frequency is then

$$\delta K_T = \left[\frac{\partial K_T}{\partial J}\right]_0 \delta J \qquad (19.4)$$

As J = U/ND and the principle variation is due to the axial harmonic (as shown in Chapter 13, p. 267), then, assuming single-screw ship,

$$\delta J = \frac{\delta U}{ND} \approx \frac{u\,\S_n}{ND} = J_0\left[\frac{u\,\S_n}{U}\right] \quad ; \ n = Z \qquad (19.5)$$

as the wake harmonic is given in fraction of the model speed.

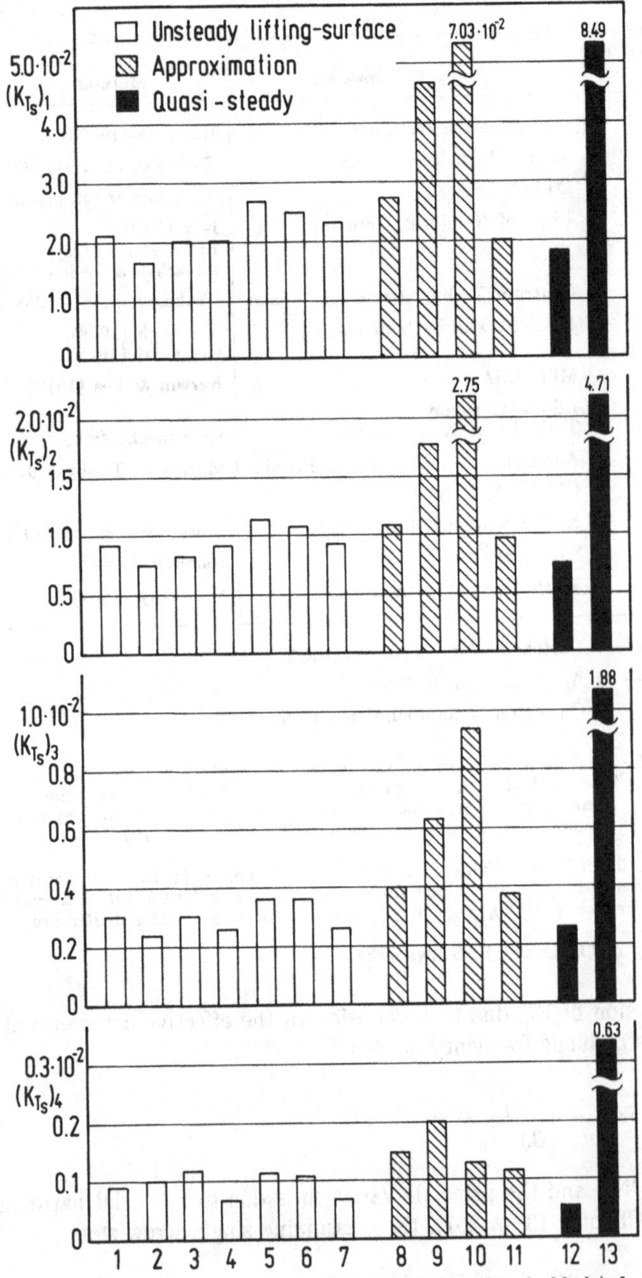

Figure 19.10a Comparison of results for blade axial force (single blade) from various theories in the same wake. Note the difference in scale of the force coefficient. For the sources of calculation, cf. Table 19.2.

(Original figure is reproduced from Schwanecke (1975) with results of Kerwin & Lee (1978), added).

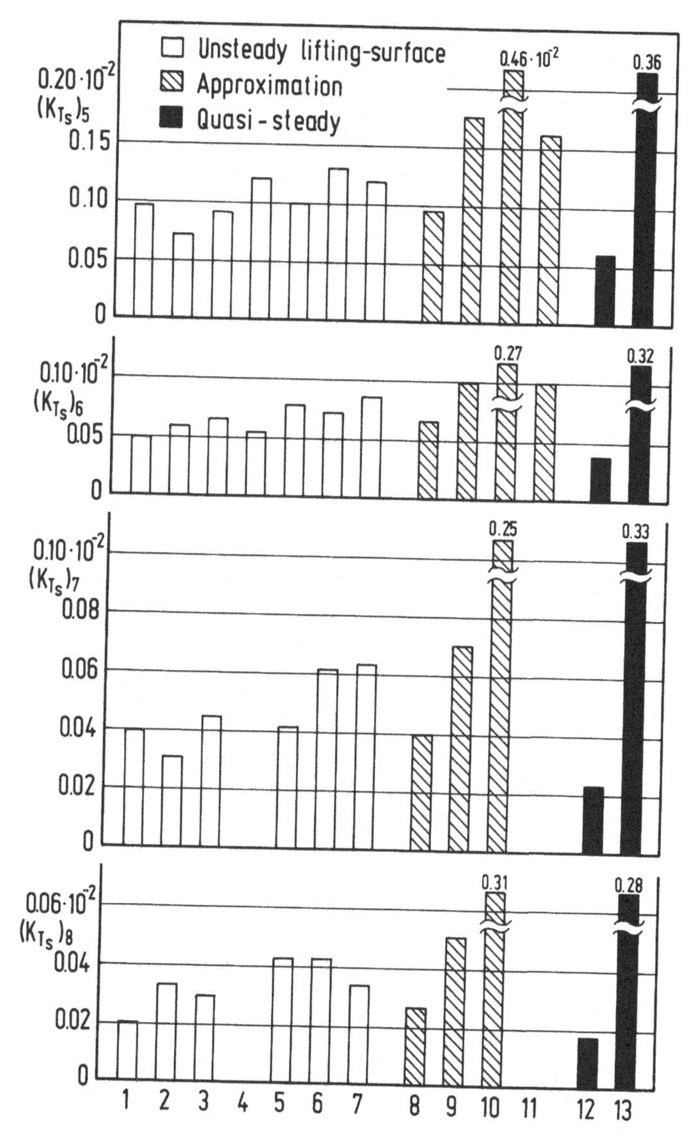

Figure 19.10b Comparison of results for blade axial force (single blade) from various theories in the same wake. Note the difference in scale of the force coefficient. For the sources of calculation, cf. Table 19.2.

(Original figure is reproduced from Schwanecke (1975) with results of Kerwin & Lee (1978), added).

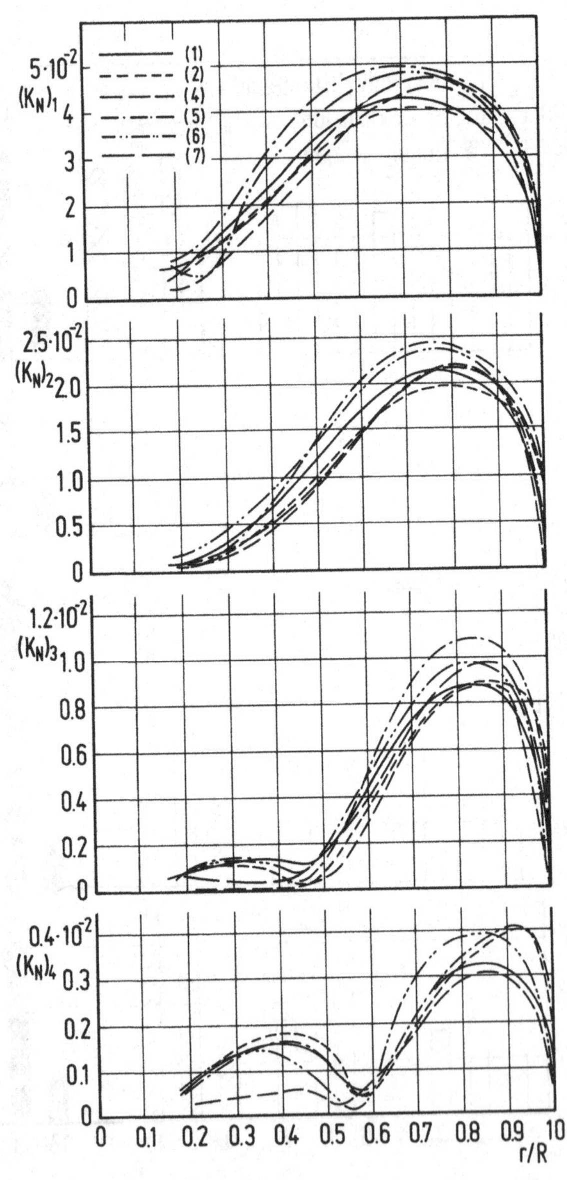

Figure 19.11 Radial distribution of blade normal force by lifting-surface procedures. Note the difference in scale of the force coefficients. For the sources of the calculations, cf. Table 19.2.

(Original figure reproduced from Schwanecke (1975) with results by Kerwin & Lee (1978) added).

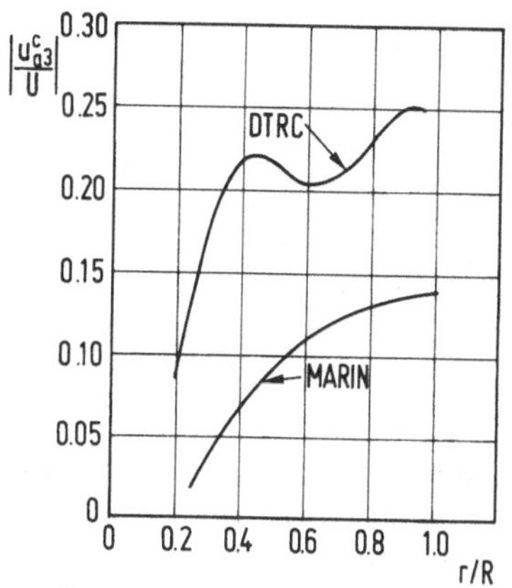

Figure 19.12 First blade harmonic of axial wake velocity u^c_{a3}/U produced by 3-cycle wake generators at DTRC and MARIN.

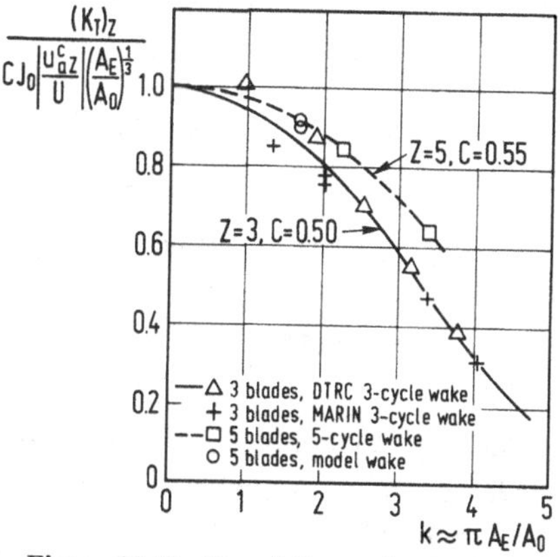

Figure 19.13 Correlation of theoretical results for two wakes and various propellers, cf. Equation (19.6)

Figures 22 and 23 from Breslin, J.P. (1969). Report on vibratory propeller, appendage and hull forces and moments. In *Proc. Twelfth International Towing Tank Conference (ITTC)*, ed. Istituto Nazionale per Studi ed Esperienze di Architettura Navale, pp. 280–90. Rome: Consiglio Nazionale delle Ricerche.
By courtesy of ITTC.

Then with (19.5), (19.4) becomes

$$\delta K_T \approx (K_T)_Z = J_0 \left[\frac{\partial K}{\partial J}\right] \frac{u^c_{aZ}(0.8R)}{U} \tag{19.6}$$

where the harmonic amplitude is taken at the 0.8 radius. This requires that the wake harmonic at this radius is representative for the whole wake variation as felt by the propeller. Hence the variation of the wake and the radius must not change sign.

At the time (1969) the K_T-slope was considered to vary as $(A_E/A_0)^{1/3}$ (A_E/A_0 being the expanded area ratio) as determined by Breslin from the Gawn series and thus $\partial K_T/\partial J$ was replaced by Breslin by $C(A_E/A_0)^{1/3}$ where the coefficient C was assumed to be dependent on the number of blades and adjusted to make the normalized results pass through unity at a reduced frequency of zero. Thus the form of the normalized ordinate values in Figure 19.13 is based on this quasi-steady approximation of the McCarthy (1961) type. Subsequent analysis of the Gawn and Wageningen propeller series data has yielded the empirical results for the K_T–J curve slope near optimum efficiency collected in Table 19.3. The method can be used for obtaining reasonable engineering estimates of unsteady thrust under the aforementioned conditions of the wake.

Number of Blades Series	$\frac{\partial K_T}{\partial J}$
Gawn-3	$-0.57\ (A_E/A_0)^{1/2}$
Wag'n-4	$-0.50\ (A_E/A_0)^{1/5}$
Wag'n-5	$-0.51\ (A_E/A_0)^{1/8}$

Table 19.3 Expressions for $\frac{\partial K_T}{\partial J}$ deduced from propeller data.

Attempts to apply this quasi-steady method of collapsing calculations made on various propellers in ship wakes were not successful whenever the axial wake harmonic changes sign along the radius as usually is exhibited by odd-order harmonics. Then the selection of the effective value is not clear. This lead the senior author to devise an operator technique in which the unsteady-propeller-force computer program was run eight times, to provide the partial derivative of the unsteady loading with respect to the radius at eight radial locations instead of only in one radius. For details, such as the application of this procedure, see Breslin (1971).

20 Outline of Theory of Intermittently Cavitating Propellers

Cavitation on ship propellers has been the bane of naval architects and ship operators since its first discovery on the propellers of the British destroyer *Daring* in 1894. Primary interest in propeller-blade cavitation was, for many years, centered upon the attending blade damage and the degradation of thrust arising from extensive, steady cavitation. It was not until the advent of the rapid growth in the size of merchant ships in the past three decades (with concurrent marked increases in blade loading) that extensive, *intermittent* or *unsteady cavitation* appeared and was indicted as the cause of large forces exciting highly objectionable hull vibration. Efforts in the modeling of hull wakes in water tunnels date back to about 1955 (cf. van Manen (1957b)) when tests of propeller models in fabricated axially non-uniform flows were being conducted at Maritime Research Institute Netherlands (MARIN), National Physical Laboratory (NPL) and Hamburgische Schiffbau-Versuchsanstalt (HSVA). Non-stationary blade cavities were observed then but there seem to have been no notice or measurement of unsteady near-field pressures attending unsteady cavitation until the experimental work of Takahashi & Ueda (1969). They measured pressures at one point above a propeller in a water tunnel in uniform and non-uniform flow and gave a brief contribution to the 12th International Towing Tank Conference (ITTC) in Rome in 1969. Their principal results are shown in Figure 20.1, where it is seen that the pressure amplitudes increased dramatically with reduced cavitation number. Subsequently there have been many experimental studies in water tunnels, vacuum tanks and on ships which have verified the connection between hull pressures and changing cavitation patterns as the blades sweep through the high-wake (low-velocity) regions.

The pressure amplitudes generated by intermittent cavitation are generally four to six times those from the sum of blade loading and thickness by non-cavitating blades. Moreover, the forces are about an order of magnitude larger because of the relatively slow spatial decay or attenuation with distance from the cavities. Another significant departure is that the pressures and forces at twice and sometimes three times blade frequency are strong from intermittent cavitation whereas at these frequencies the non-cavitating propeller effects are generally negligible. The reasons for these dramatic differences will become apparent from the analysis which follows.

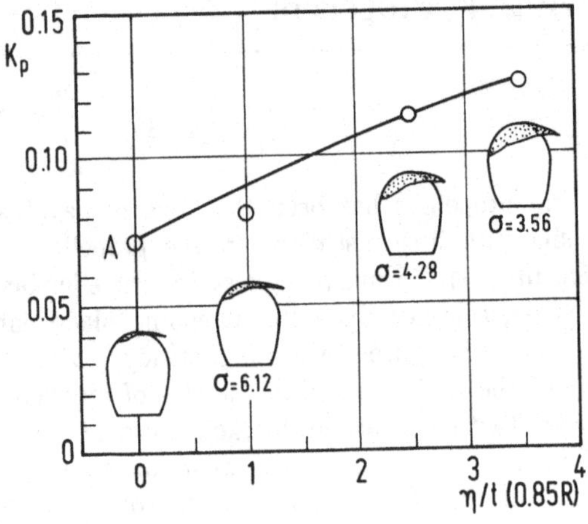

Figure 20.1 Fluctuating-pressure amplitude on flat plate above cavitating propeller in non-uniform inflow. Cavity thicknesses estimated from visual observation; point A considered as non-cavitating.

Figure 5 from Takahashi, H. & Ueda, T. (1969). An experimental investigation into the effect of cavitation on fluctuating pressures around a marine propeller. In *Proc. Twelfth International Towing Tank Conference (ITTC)*, ed. Istituto Nazionale per Studi ed Esperienze di Architettura Navale, pp. 315–7. Rome: Consiglio Nazionale delle Ricerche. By courtesy of ITTC and H. Takahashi and T. Ueda.

When cavitation ensues over an area of the blade, an interface between the volume in which the water is vaporized and the outer fluid, is developed, as explained in Chapter 8. All along this interfacial surface the vaporization process is taking place at a pressure close to the vapor pressure specific to the fluid temperature. As the outer pressure field continues to change, as with increasing angle of attack, the interface volume grows, its downstream boundary extending and the radial extent penetrating to smaller radii. One may ask: How can this give rise to large field pressures since everywhere on the vapor-fluid interfacial surface the pressure is constant? The answer is that a volume change takes place and it is as though the outer fluid sees a changing blade thickness and camber which are effectively altering the outer flow pattern and hence the potential. We shall see this in the following.

A BASIC ASPECT OF THE PRESSURE FIELD GENERATED BY UNSTEADY CAVITATION

The growth and recession of large transient vapor cavities on ship propeller blades, as they transit the region of low inflow in the wake abaft the hull, is a complicated phenomenon. As a prelude to the detailed developments which follow, it is thought to be helpful to portray an elementary time-varying, cavity-generated flow to motivate and provide insight to the mathematical representations employed later.

To demonstrate the physics of a cavity-generated flow, we may consider a spherically symmetric, irrotational, incompressible, radial flow emanating from a single point (taken as the origin of coordinates) in a boundless fluid. As the flow is irrotational, it is described by the velocity-potential function of a time-dependent point source, cf. Equation (1.50), which models the cavity,

$$\phi_c = -\frac{M(t)}{4\pi R} \tag{20.1}$$

from which the radial velocity is

$$u_r(R,t) = \frac{\partial \phi_c}{\partial R} \tag{20.2}$$

where R is the distance between the field point and the source point.

We have only radial velocity and can use Equation (1.27) to secure the relationship between velocity and pressure

$$\frac{\partial \Phi}{\partial t} + \frac{1}{2}\left[\frac{\partial \Phi}{\partial R}\right]^2 = \frac{p_a - p(R,t)}{\rho} \tag{20.3}$$

where p_a is the ambient total pressure at large R. For a cavity driven or developed by the action of a flow having a potential ϕ_d, we take the total potential to be

$$\Phi = \phi_c + \phi_d \tag{20.4}$$

The kinematic condition on the cavity surface $R = r_c$ is

$$\frac{\partial \Phi}{\partial R} = \frac{\partial r_c}{\partial t} \tag{20.5}$$

or

$$\frac{M}{4\pi r_c^2} + \frac{\partial \phi_d}{\partial R} = \frac{\partial r_c}{\partial t} \qquad \text{on } R = r(t) \tag{20.6}$$

We assume that the driving potential provides only a weak distortion to the spherical surface $R = r_c(t)$ and can be neglected in (20.6) to give

$$M \simeq 4\pi r_c^2 \frac{\partial r_c}{\partial t} \tag{20.7}$$

and hence (20.1) becomes

$$\phi_c \simeq -\frac{r_c^2 \dfrac{\partial r_c}{\partial t}}{R} \tag{20.8}$$

The use of (20.8) in (20.4) and thence to (20.3) produces

$$-\left[\frac{2r_c\left[\dfrac{\partial r_c}{\partial t}\right]^2 + r_c^2\left[\dfrac{\partial^2 r_c}{\partial t^2}\right]}{R}\right] + \frac{\partial \phi_d}{\partial t} + \frac{1}{2}\frac{r_c^4\left[\dfrac{\partial r_c}{\partial t}\right]^2}{R^4} = \frac{p_a - p(R,t)}{\rho} \tag{20.9}$$

where $\phi_d = \phi_d(R,t)$; neglecting $\partial\phi_d/\partial R$.

When evaluating (20.9) on the cavity surface we may take ϕ_d as $\phi_d(0,t)$ since the bubble radius is very small compared with the spatial scale of the driving potential ϕ_d (which may be considered to be that arising from the changing blade loading during transit of the wake peak). The pressure on the cavity surface is taken to be the vapor pressure (ignoring surface tension) so $p(r,t) = p_v$, a constant, and then (20.9) on $R = r_c$ yields the relation

$$-\left[2\left[\frac{\partial r_c}{\partial t}\right]^2 + r_c\frac{\partial^2 r_c}{\partial t^2}\right] + \frac{1}{2}\left[\frac{\partial r_c}{\partial t}\right]^2 = \frac{p_a - p_v}{\rho} - \frac{\partial \phi_d(0,t)}{\partial t}$$

or

$$\frac{1}{2r_c^2\dfrac{\partial r_c}{\partial t}}\frac{\partial}{\partial t}\left[r_c^3\left[\frac{\partial r_c}{\partial t}\right]^2\right] = \frac{p_v - (p_d + p_a)}{\rho} = \frac{p_{cav}}{\rho} \tag{20.10}$$

where $p_d = -\rho\, \partial\phi_d(0,y)/\partial t$ = the driving pressure.

In order for the cavity to expand, the right side of (20.10) must be positive. This occurs when the driving pressure becomes more negative than the ambient pressure minus the vapor pressure. The ambient pressure is that composed of the atmospheric pressure plus the static pressure of the fluid at the depth of the cavity.

Integrating (20.10) under the assumption that p_{cav} does not vary significantly in the very short time for the cavity to grow yields

$$r_c^3\left[\frac{\partial r_c}{\partial t}\right]^2 \simeq \frac{2p_{cav}}{3\rho}\int_0^t \frac{d}{dt'} r_c^3\, dt' \quad ; \quad \frac{\partial r_c(0)}{\partial t} = 0$$

whence

$$\frac{\partial r_c}{\partial t} = \sqrt{\frac{2p_{cav}(0)}{3\rho}\left\{1 - \left[\frac{r_c(0)}{r_c(t)}\right]^3\right\}} \tag{20.11}$$

After the initial period, then, $r_c(0)/r_c(t) << 1$ and we see that the rate of growth approaches a constant, namely, $\sqrt{2p_{cav}(0)/3\rho}$. Then the first-order pressure increment, p_c (not to be confused with p_c used in Chapter 8 as the pressure in the cavity, here assumed to be the vapor pressure p_v) at any field radius R due to the expanding cavity, is, from (20.8)

$$p_c = -\rho\,\frac{\partial \phi_c}{\partial t} = \rho\,\frac{\left[2r_c\left[\frac{\partial r_c}{\partial t}\right]^2 + r_c^2\,\frac{\partial^2 r_c}{\partial t^2}\right]}{R} \tag{20.12}$$

and for $\partial r_c/\partial t \simeq$ a constant, $\partial^2 r_c/\partial t^2 \simeq 0$,

$$p_c \approx \frac{2\rho}{R}\left[\frac{2}{3}\,\frac{p_{cav}(0)}{\rho}\right]^{3/2} t \tag{20.13}$$

or

$$p_c \approx \frac{2\rho}{R}\left[\frac{2}{3\rho}\,(p_v - (p_d + p_a))\right]^{3/2} t \tag{20.14}$$

Thus, at a distance from an expanding spherical cavity, the pressure signature is proportional to the 3/2-power of the net pressure causing vaporization, that is, the amount that p_d is more negative than the cavitation-inhibiting pressure, $p_a - p_v$.

Returning to (20.7), we see that the source strength $M(t)$ can be written

$$M(t) = \frac{4}{3}\,\pi\,\frac{\partial {r_c}^3}{\partial t} \quad \text{or} \quad M(t) = \frac{\partial V_c}{\partial t}$$

where V_c is the cavity volume. Then the potential of a spherical cavity is

$$\phi_c = -\frac{1}{4\pi R}\,\frac{\partial V_c}{\partial t}$$

and the incremental pressure is

$$p_c = -\rho\,\frac{\partial \phi}{\partial t} = \frac{\rho}{4\pi R}\,\frac{\partial^2 V_c}{\partial t^2} \tag{20.15}$$

a pressure source whose strength is proportional to the acceleration of the cavity volume.

On propeller blades, cavities are clearly far from spherical, being stretched out into sheets because of the high tangential velocity of the blade. Consequently, such cavities are modelled by area distribution of sources (and sinks) as done in Chapter 8. Nevertheless, a formula quite analogous to (20.15) is found for the pressure at relatively large distances from the propeller, as will be shown later.

It is important to note that, when cavitation occurs near the water surface, the effect of the presence of that surface is to reduce the pressures at field points. As the motion is rapid, gravity is negligible in the near field, cf. Appendix F, p. 500. The pressure change due to cavitation would then be given by

$$p_c = \frac{\rho}{4\pi}\frac{\partial^2 V_c}{\partial t^2}\left[\frac{1}{R} - \frac{1}{R_i}\right]$$

where $R = \sqrt{x^2 + y^2 + z^2}$, $R_i = \sqrt{x^2 + y^2 + (2d-z)^2}$ and the surface is at $z = d$, cf. Figure 20.2.

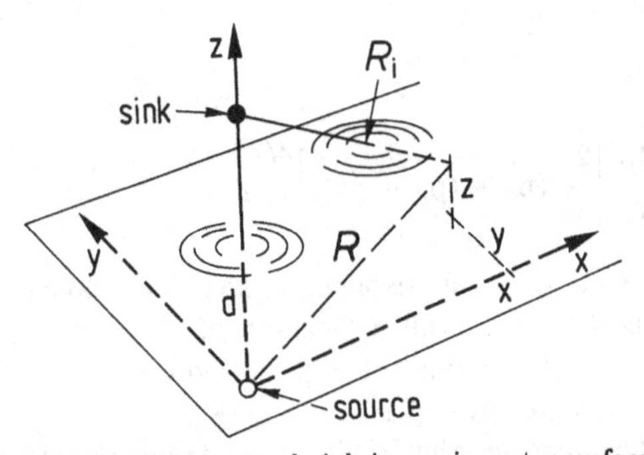

Figure 20.2 Source and sink image in water surface.

Clearly, along the locus $z = d$, $p_c = 0$. Now if we ask how the pressure varies at $x >> d$ for $y = 0$, say, then expanding each radical by the binomial theorem gives

$$p_c \approx \left\{\frac{1}{x}\left[1 - \frac{1}{2}\left[\frac{z}{x}\right]^2 + \ldots\right] - \frac{1}{x}\left[1 - \frac{1}{2}\left[\frac{2d-z}{x}\right]^2 + \ldots\right]\right\}$$

or

$$p_c \approx \frac{2d(d-z)}{x^3}$$

This decay with distance is to be contrasted with that from the source in the absence of the surface, namely, as $1/x$.

Thus, the pressure "signature" at distance from unsteady cavitation can be rapidly quenched when the free surface is present.

It is hoped that the foregoing rather lengthy development provides an explanation as to how intermittent cavitation can produce large field pressures even though the pressure on the expanding (or contracting) cavity surface is only the vapor pressure of water (or close to that pressure) at the ambient temperature.

Another question arises from the fact that experimentally observed pressures from intermittently cavitating propellers have frequently been found to have components at twice and thrice blade frequency that are comparable to those at blade frequency. How can this be when it has never been the case for non-cavitating propellers? We attempt to answer this question next.

PRESSURE FIELD DUE TO CAVITATING PROPELLER

With the insight obtained up to now it is not difficult to establish a (linearized) model of the flow over the propeller as due to combination of blade loading and blade and cavity thickness. We focus on the time-varying contribution from unsteady cavitation and make the representation approximate by ignoring the harmonic contributions of $1/R$ which is possible under the following restrictions:

i. The cavities are short in length.

ii. The cavity is limited radially so its center is at some outer effective radius R_c.

iii. The angular duration of the cavitation is small, i.e., $\gamma_c - \gamma_i$ is not large; γ_c = blade angle at cavity collapse; γ_i = inception angle (< 0 for single screw).

The general velocity potential generated by a single intermittently cavitating blade (in the absence of boundaries other than the blades) can be expressed by

$$\phi = -\frac{1}{4\pi}\int_{S_c(t)} \frac{m_c}{R}\, dS + \frac{1}{4\pi\rho U}\int_x^\infty p_\ell\left[x'', r, \gamma, t - \frac{x - x''}{U}\right] dx'' + \frac{1}{2\pi}\int_S \frac{V\tau'}{R} dS \tag{20.16}$$

where the first term is a source distribution of density m_c over the instantaneous area of the blade, $S_c(\gamma_0(t))$, on which the cavitation extends, Equation (2.39). The second term is the contribution from the pressure field generated by the distribution of the jump in pressure (or loading) across the blade elements. The potential is obtained from the pressure due to unsteady loading, p_ℓ, Equation (15.2) by use of the pressure-potential relation, Equation (1.30), and by integration by the method of characteristics, cf. Appendix E. The third term is the potential induced by the blade semi-thickness τ, Equation (4.43).

We shall focus only on the source or cavity term which is designated as ϕ_c. Since R is the straight-line distance between any field point (x,r,γ) and any dummy point $(x',r',\gamma_0+\psi')$, we have explicitly

$$\phi_c(x,r,\gamma;\gamma_0(t)) = -\frac{1}{4\pi}\int_{r_c}^{R} dr' \int_{h_t^c}^{h_l^c} \frac{m_c(r',h',t)}{\sqrt{(x-x')^2+r^2+r'^2-2rr'\cos(\gamma-\gamma_0-\psi')}}\,dh' \tag{20.17}$$

[27]

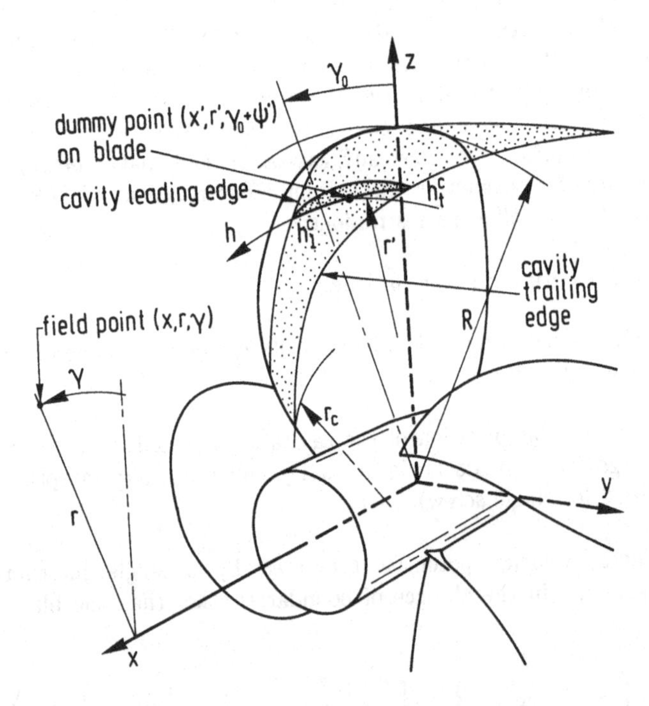

Figure 20.3 Schematic of instantaneous intermittent cavitation on single blade.

[27] While we have measured the distance h_t from the blade reference axis to the trailing edge assuming that the trailing edge is abaft the blade reference axis (cf. Figure 14.1) the cavity trailing edge may well be both fore or abaft this axis. We then indicate this by orienting the cavity trailing ordinate h_t^c as positive along the helix arc (positive fore).

Here, as may be observed in Figure 20.3, $r_c = r_c(\gamma_0)$ is the lowest instantaneous radial extent of the cavity at any blade angular position $\gamma_0 = \omega t$; R is the blade radius; $h^c_t = h^c_t(r',\gamma_0)$ is the cavity trailing edge along the helix arc; $h^c_l = h^c_l(r',\gamma_0)$ is the leading-edge location; r' is the dummy radius along the surface (a ruled surface), see also Figure 14.1 and 14.2 p. 274.

Now it is known that the source density m_c is given by the jump in the velocity normal to the source sheet. It is also known that the cavity ordinate η is connected to this jump in the normal velocity by

$$\frac{\partial\eta(r',h';t)}{\partial t} - V\,\frac{\partial\eta(r',h';t)}{\partial h'} = m_c \tag{20.18}$$

where $V = \sqrt{U_a(r)^2 + (\omega r')^2}$ is the resultant relative velocity at radius r'. To see this relation we digress to examine in detail the kinematics on the cavity and on the blade in way of the cavity.

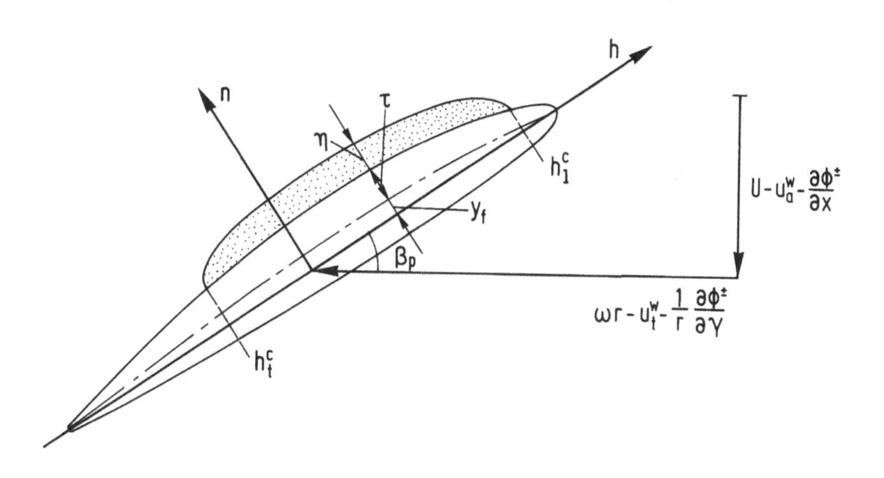

Figure 20.4 Schematic of blade section at radius r displaying relative velocities; u^w_a, u^w_t hull axial and tangential wake; $\partial\phi/\partial x$, $(1/r)(\partial\phi/\partial\gamma)$, axial and tangential induction from loading, thickness and cavity.

Source-Density Cavity-Ordinate Relation

The equation of the cavity surface is $F^+(h,n,r,t) = 0$ in the local blade section coordinate system (h,n,r), cf. Figure 20.4. It is seen that the explicit composition of F^+ is

$$F^+(h,n,r,t) = n - (y_f + \tau + \eta(h,r,t)) = 0 \tag{20.19}$$

We recognize here Equation (3.12), p. 43, from Chapter 3 where now $y_f+\tau+\eta$ is substituted for f (in Chapter 3), h for x, n for z and r for y. The kinematic boundary condition requires the partial derivatives (Equation (3.13))

$$\frac{\partial F^+}{\partial t} = -\frac{\partial \eta}{\partial t}$$

$$\frac{\partial F^+}{\partial h} = -\left[\frac{\partial y_f}{\partial h} + \frac{\partial \tau}{\partial h} + \frac{\partial \eta}{\partial h}\right] ; \frac{\partial F^+}{\partial n} = 1 ; \frac{\partial F^+}{\partial r} = -\left[\frac{\partial y_f}{\partial r} + \frac{\partial \tau}{\partial r} + \frac{\partial \eta}{\partial r}\right] \tag{20.20}$$

The boundary condition, similar to Equation (3.14) (with U = 0) then becomes

$$\frac{\partial \eta}{\partial t} = -\left[\frac{\partial y_f}{\partial h} + \frac{\partial \tau}{\partial h} + \frac{\partial \eta}{\partial h}\right] V_h^+ + (1)V_n^+ - \left[\frac{\partial y_f}{\partial r} + \frac{\partial \tau}{\partial r} + \frac{\partial \eta}{\partial r}\right] V_r^+ \tag{20.21}$$

where the subscripts on V^+ indicate the component.

The equation of the lower surface of the section in way of the cavity is

$$F^-(h,n,r) = n - (y_f - \tau) = 0 \tag{20.22}$$

The boundary condition corresponding to (20.21) on the lower surface is

$$0 = -\left[\frac{\partial y_f}{\partial h} - \frac{\partial \tau}{\partial h}\right] V_h^- + (1)V_n^- - \left[\frac{\partial y_f}{\partial r} - \frac{\partial \tau}{\partial r}\right] V_r^- \tag{20.23}$$

As $\partial y_f/\partial r$, $\partial \tau/\partial r$ and $\partial \eta/\partial r$ are very small, and the radial velocity component, V_r, is also small, we may readily neglect the last terms in (20.21) and (20.23). As (20.21) and (20.23) hold over the same range of h (in way of the cavity), we may subtract (20.23) from (20.21) to secure

$$\frac{\partial \eta}{\partial t} + \frac{\partial \eta}{\partial h} V_h^+ = -\frac{\partial y_f}{\partial h}(V_h^+ - V_h^-) - \frac{\partial \tau}{\partial h}(V_h^+ + V_h^-) + V_n^+ - V_n^- \tag{20.24}$$

Now, from Figure 20.4

$$V_h^{\pm} = \left[-r\omega + u_t^W + \frac{1}{r}\frac{\partial \phi^{\pm}}{\partial \gamma}\right]\cos\beta_p + \left[-U + u_a^W + \frac{\partial \phi^{\pm}}{\partial x}\right]\sin\beta_p \tag{20.25}$$

Then

$$V_h^+ - V_h^- = \frac{1}{r}\frac{\partial}{\partial \gamma}(\phi^+ - \phi^-)\cos\beta_p + \frac{\partial}{\partial x}(\phi^+ - \phi^-)\sin\beta_p \tag{20.26}$$

and as $(V_h^+ - V_h^-)\partial y_f/\partial h$ is seen to involve products of only second-order terms, no contribution is secured. In contrast

$$\frac{\partial \tau}{\partial h}(V_h^+ + V_h^-) \simeq 2\frac{\partial \tau}{\partial h}\{-r\omega\cos\beta_p - U_a\sin\beta_p\} \simeq -2\sqrt{U_a^2+(r\omega)^2}\,\frac{\partial \tau}{\partial h} \tag{20.27}$$

again neglecting products of $\partial \tau/\partial h$ and perturbations.

The jump in the n-component $V_n^+ - V_n^-$ can arise only from the source distribution m_τ modeling the thickness and the cavity, $m_C(h,r,t)$. As shown by Equation (2.45), p. 35, the jump in normal velocity across a source sheet is equal to the source density, in this case $m_\tau + m_C$. Consequently (20.24) reduces to

$$\frac{\partial\eta}{\partial t} - (r\omega\cos\beta_P + U_a\sin\beta_P)\frac{\partial\eta}{\partial h} = 2\frac{\partial\tau}{\partial h}\sqrt{U_a^2 + (r\omega)^2} + m_\tau + m_c$$

As we know from thin-section theory that $m_\tau = -2\,\partial\tau/\partial h\,V$, then we obtain finally

$$\frac{\partial\eta}{\partial t} - V\frac{\partial\eta}{\partial h} = m_c \tag{20.28}$$

Returning to (20.17) and inserting (20.28) yields

$$\phi_c = -\frac{1}{4\pi}\int_{r_c}^{R} dr' \int_{h_t^c}^{h_l^c} \frac{\left[\dfrac{\partial\eta}{\partial t} - V\dfrac{\partial\eta}{\partial h'}\right]}{\sqrt{(x-x')^2+r^2+r'^2-2rr'\cos(\gamma-\gamma_0-\psi')}}\,dh' \tag{20.29}$$

Extracting the time derivative (while accounting for the variation of the radical with time, as $\gamma_0 = \omega t$), and integrating the second term by parts over h (using that $\eta\left[h_l^c\right] = 0$, $\eta\left[h_t^c\right] = 0$), yields

$$-4\pi\phi_c = \frac{\partial}{\partial t}\int_{r_c}^{R} dr' \int_{h_t^c}^{h_l^c} \frac{\eta}{R}\,dh'$$

$$-\omega\int_{r_c}^{R} dr' \int_{h_t^c}^{h_l^c} \frac{\eta r r'\sin(\gamma-\gamma_0-\psi')}{R^3}\,dh' + \int_{r_c}^{R} dr' V \int_{h_t^c}^{h_l^c} \eta\frac{\partial}{\partial h'}\frac{1}{R}\,dh' \tag{20.30}$$

Carrying out the h'-differentiation in the last term and combining terms gives

$$-4\pi\phi_c = \frac{\partial}{\partial t}\int_{r_c}^{R} dr' \int_{h_t^c}^{h_l^c} \frac{\eta}{R}\,dh' + \int_{r_c}^{R} dr' \frac{V\sqrt{P_f^{*2}+r'^2}}{P_f^*}\int_{h_t^c}^{h_l^c} \frac{\eta(x-x')}{R^3}\,dh'$$

$$+ r\int_{r_c}^{R} dr' \int_{h_t^c}^{h_l^c} \frac{1}{R^3}\left[-\omega r' + \frac{r'}{\sqrt{P_f^{*2}+r'^2}}V\right]\eta\sin(\gamma-\gamma_0-\psi')\,dh' \tag{20.31}$$

where we have used the relations between the dummy variables, h', x' and ψ', Equation (14.3), and together with (14.20), $h' = \sqrt{P_f^{*2} + r'^2}\,\psi'$, with $P_f^* = P_f/2\pi = U/\omega$, P_f the fluid pitch at radius r', and $x' = P_f^*\psi'$. (We have not allowed for blade skew or rake for simplicity; to do so substitutions as displayed in Chapter 18, p. 338, must be made).

Now using assumption ii. then $\sqrt{P_f^{*2} + r'^2} \approx r'$ (for pitch-diameter ratios about unity), and $V = \sqrt{U_a(r)^2 + (\omega r')^2} \approx \omega r'$ for $r' = R_c \approx R$. Thus the last term vanishes or, in reality, is weak. For $r' >> x'$ (the x-wise extent of the cavity is relatively small especially under assumption i., then

$$-4\pi\phi_c = \frac{\partial}{\partial t}\int_{r_c}^{R}\frac{A_c(r',\gamma_0)}{R}\,dr - \frac{\omega x}{P_f^*}\int_{r_c}^{R}\frac{r'^2 A_c(r,\gamma_0)}{R^3}\,dr' \qquad (20.32)$$

where $A_c(r',\gamma_0)$ is the cavity cross-section area

$$A_c(r',\gamma_0) = \int_{h_t^c}^{h_1^c}\eta(r',h',t)\,dh' \qquad (20.33)$$

Under the assumption iii. that the variation of R throughout the cavitation interval $\gamma_c - \gamma_i$ is small and makes little harmonic contribution (in addition the space angle γ range is limited to the vicinity of zero) and moreover assuming that $A_c(r',\gamma_0)$ is concentrated at the outer radii so the last term in (20.32) can be approximated by replacing r' by R_c (the effective cavitation radius), we have

$$\phi_c = -\frac{1}{4\pi R_0}\frac{\partial V_c(\gamma_0)}{\partial t} + \frac{\omega R_c^2 x}{P_f^*}\frac{V_c(\gamma_0)}{R_0^3} \qquad (20.34)$$

where $R_0 = \sqrt{x^2 + (r - R_c)^2}$ and V_c is the cavity volume

$$V_c(\gamma_0) = \int_{r_c}^{R} A_c(r',\gamma_0)\,dr'$$

Thus, an educated seal could say, when swimming along with a strange ship having only a single-bladed, intermittently cavitating propeller – "Aha! At large distances, I see a point source whose strength is the time derivative of the cavity volume – and, behold!, this transforms to a time-varying line source and a dipole as I move closer and then, when very close, there is, in addition, a distributed mess of chordwise dipoles!"

Finally, expressing the volume as a complex Fourier series in blade angle γ_0,

$$V_c = \sum_{n=-\infty}^{\infty} (V_c)_n \, e^{in\gamma_0}$$

we find the total contribution *from all blades* by the substitution $\gamma_0 \to \gamma_0 + 2\pi\nu/Z$ and sum for all blades $\nu = 0, .., Z-1$, cf. p. 212. This yields

$$\phi_c \approx \frac{Z}{4\pi} \sum_{q=-\infty}^{\infty} \left[\frac{iqZ\omega}{R_0} - \frac{\omega R_c^2 x}{P_f^* R_0^3} \right] (V_c)_{qZ} \, e^{iqZ\gamma_0} \tag{20.35}$$

The total pressure $p_c \approx -\rho \, \partial\phi_c/\partial t = -\rho\omega \, \partial\phi_c/\partial\gamma_0$ is approximately

$$p_c = \frac{\rho}{4\pi} \sum_{q=-\infty}^{\infty} \left[\frac{q^2 Z^3 \omega^2}{R_0} + \frac{i\omega^2 q Z^2 R_c^2 x}{P_f^* R_0^3} \right] (V_c)_{qZ} \, e^{iqZ\gamma_0} \tag{20.36}$$

Instead of making assumption iii. that the variation of R throughout the cavitation interval is small we can return to (20.32) and exhibit the harmonic dependence of the spatial propagation function $1/R$, by expressing the $1/R$ in a series of Legendre functions, cf. Mathematical compendium, Section 5, with which the harmonic components of $1/R$ are individually secured.

At sufficient distances where the source term of (20.32) dominates, (20.32) can then be written, using (M5.22), p. 518

$$\phi_c \simeq -\frac{1}{4\pi} \frac{\partial}{\partial t} \sum_{m=-\infty}^{\infty} \int_{r_c(t)}^{R} \frac{A_c(r',t)}{\pi\sqrt{rr'}} \, Q_{|m|-\frac{1}{2}}(Z) \, e^{-im\psi_1^c(r',t)} dr' \, e^{im(\gamma-\gamma_0)} \tag{20.37}$$

where we have assumed that the cavity is concentrated at the leading edge $\psi' \approx \psi_1^c$. Note that (20.37) gives the potential for a single blade.

Now we see that there is harmonic or cyclic variation imbedded in the limit $r_c(t)$, in $A_c(r',t)$ and $\psi_1^c(r',t)$. This cavity-generated harmonic contribution can be placed in evidence by writing (20.37) in a Fourier series

$$\phi_c = -\frac{1}{4\pi} \frac{\partial}{\partial t} \sum_{m=-\infty}^{\infty} \sum_{n=-\infty}^{\infty} C_n^{(m)}(x,r) \, e^{im\gamma} \, e^{i(n-m)\gamma_0(t)} \tag{20.38}$$

where the "cavitation amplitude function" is

$$C_n^{(m)} = \frac{1}{2\pi} \int_0^{2\pi} d\gamma'' \, e^{-in\gamma''} \int_{r_c(\gamma'')}^{R} \frac{A_c(r',\gamma'')}{\pi\sqrt{rr'}} \, Q_{|m|-\frac{1}{2}}(Z) \, e^{-im\psi_1^c(\gamma'')} dr' \tag{20.39}$$

Equation (20.38) is analogous to that for the pressure field of a propeller in a wake. The total cavitation potential ϕ_c for a Z-bladed propeller is

$$\phi_c = -\frac{Z}{2\pi}\frac{\partial}{\partial t} Re\left\{\frac{1}{2}C_0^0 + \sum_{n=1}^{\infty} C_n^{(n)} e^{in\gamma} + \sum_{q=1}^{\infty} C_0^{(qZ)} e^{iqZ(\gamma-\gamma_0)}\right.$$

$$\left. + \sum_{q=1}^{\infty}\sum_{n=1}^{\infty}\left[C_n^{(n-qZ)} e^{i[(n-qZ)\gamma+qZ\gamma_0]} + C_n^{(n+qZ)} e^{i[(n+qZ)\gamma-qZ\gamma_0]}\right]\right\} \tag{20.40}$$

In this "menagerie" of terms we can recognize the source-like term given by n = qZ. This constituent is

$$(\phi_c)_{qZ}^{(qZ)} = -\frac{Z}{2\pi}\frac{\partial}{\partial t}\left[Re\, C_{qZ}^{(0)} e^{iqZ\gamma_0}\right]$$

$$= -\frac{Z}{2\pi}\frac{\partial}{\partial t}\left[\frac{Re}{2\pi}\int_0^{2\pi} d\gamma'' e^{-iqZ\gamma''}\int_{r_c(\gamma'')}^{R}\frac{A_c(r',\gamma'')}{\pi\sqrt{rr'}}\,Q_{-\frac{1}{2}}(\mathcal{Z})dr' e^{iqZ\gamma_0}\right] \tag{20.41}$$

which is source-like because it is independent of γ and decays like $\sqrt{x^2+r^2}$ for large x and r as can be seen in Section 6 of the Mathematical Compendium. For sufficiently large $\mathcal{Z}$ we can use the approximation (M6.6), p. 520, and beyond the sphere $x^2+r^2 = (2R)^2$, (20.41) can be taken independent of r' to first order. For such regions the radial integral gives the cavity volume, and the result is

$$(\phi_c)_{qZ}^{(qZ)} \simeq -\frac{Z}{2\pi}\frac{\partial}{\partial t} Re\left[(V_c)_{qZ} e^{iqZ\gamma_0}\right]\frac{1}{R_0} \tag{20.42}$$

where $(V_c)_{qZ}$ is the complex amplitude of the qZ-th harmonic of the cavity volume and $R_0 = \sqrt{x^2+r^2}$. As $(V_c)_{qZ} = \left[a_{qZ} - ib_{qZ}\right]/2$ we have, as $\gamma_0 = \omega t$,

$$(\phi_c)_{qZ}^{(qZ)} = \frac{1}{4\pi} qZ^2\omega\,\frac{a_{qZ}\sin\, qZ\gamma_0 - b_{qZ}\cos\, qZ\gamma_0}{R_0} \tag{20.43}$$

The dominant pressure is $-\rho\partial\phi/\partial t$

$$(p_c)_{qZ}^{(qZ)} = -\frac{\rho}{4\pi}(qZ\omega)^2\, Z\,\frac{a_{qZ}\cos\, qZ\gamma_0 + b_{qZ}\sin\, qZ\gamma_0}{R_0}$$

$$= \frac{\rho Z}{4\pi}\frac{(qZ\omega)^2}{R_0}\left|(V_c)_{qZ}\right|\cos(qZ\gamma_0-\alpha)\ ;\quad \alpha = \tan^{-1}\frac{b_{qZ}}{a_{qZ}} \tag{20.44}$$

We can see from this that unless the qZ-th harmonic of the cavity volume decreases faster than $1/q^2$, higher harmonics of the pressure can be larger than that of the first order. This is the answer (at long last!) to the question posed earlier (p. 393).

It is to be noted that first q-term in (20.40) contains terms which involve the zeroth harmonic of (20.39) which vanish rapidly with increasing R_0. However, these and other terms can be of significance near to the propeller.

Some idea of the harmonics of the blade-cavity volume can be gleaned from measurements made from photographs of intermittent cavitation on a five-bladed propeller on a twin-screw tanker by Det Norske Veritas, reported by Søntvedt & Frivold (1976). Figure 20.5 shows variation of the cavity volume as a function of blade-position angle referred to the wake peak.

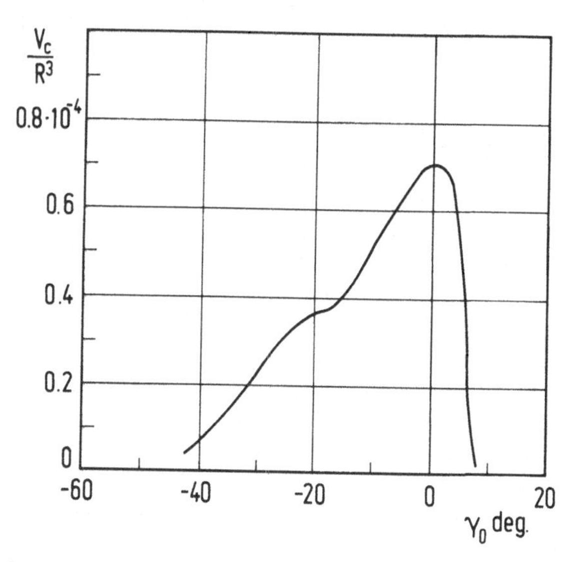

Figure 20.5 Cavity volume on a propeller blade as a function of blade angle. Harmonics of Fourier-series approximation shown in Figure 20.6.

Figure 4 from Breslin, J.P. (1977) (adapted from Figure 3, Søntvedt & Frivold (1976)). A theory for the vibratory forces on a flat plate arising from intermittent propeller-blade cavitation. In *Proc. Symp. on Hydrodynamics of Ship and Offshore Propulsion Systems*, paper 6, session 3. Høvik: Det Norske Veritas.
By courtesy of Det Norske Veritas, Norway.

An approximate fit of a 20-term Fourier series to this variation is also shown. The amplitudes of the various harmonic orders of shaft frequency are displayed in Figure 20.6. Here we see that the mean value of the cavity volume is about equal to the blade-rate (5th shaft order) harmonic and that the decrease of the higher order amplitudes is "slow".

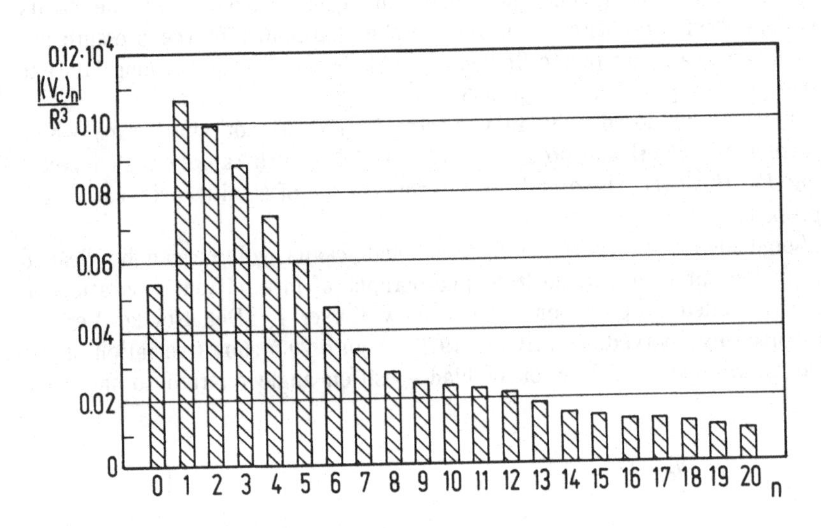

Figure 20.6 Harmonic amplitudes $|(V_c)_n|$ of cavity volume variations shown in Figure 20.5.

It is necessary to point out that the coupling of the blade loadings and the propagation functions and those couplings of the "cavitation function" harmonics with its propagation functions are not completely analogous. In the case of the loadings, Δp_n, there is a one-to-one relationship to the harmonic orders of the wake whereas cavitation ensues because of the total angle-of-attack variation contributed by the sum of the wake harmonics (the first few being the dominant ones). Thus the harmonics of the cavity-generated pressures are not related to any single specific wake harmonic. This is so because the cavity extent is a non-linear function of the incidence angle.

The behavior exhibited by (20.44) could have been extracted immediately from (20.32) by taking

$$R = R_0 = \sqrt{x^2 + (r - R_c)^2}$$

without all the preceding detail. However, we would have had no measure of the region in which the space point must lie for the validity of the approximation, nor would we see the multiplicity of terms which can contribute in the near field. All other terms can be interpreted as weighted integrals of the cavity area and ordinates, the weighting function being multipoles of various orders. As in the case of the spherically expanding cavity volume, the magnitude of $|(V_c)_{qZ}|$ and the other cavity dimensions involved depend upon the orchestrated interaction between the loading Δp (and the blade thickness) and the cavity sources which evolve from the simultaneous satisfaction of the kinematic condition on the wetted portions of the blades and the spatial constancy of pressure on the cavitating region in the presence of all other blades. An overview of methods available for the solution of this problem, involving as it does simultaneous integral equations, is given below.

Finally, it is to be noted that this double Fourier expansion procedure of p. 399 and 400 could be applied to the last two "dipole" integrals in (20.31) to secure analogous structures involving weighted integrals of the partial-cavity areas and the cavity ordinates. As this would not add further insight and would burden the reader with more unpalatable mathematics, we omit additional elaboration.

NUMERICAL SOLUTION OF THE INTERMITTENTLY-CAVITATING-PROPELLER PROBLEM

There are several computer-effected solutions for non-cavitating propellers based on unsteady lifting-surface theory which are of such demonstrated accuracy as to be used for practical design purposes. Such is not the case for the intermittently cavitating propeller. In the past, Johnsson & Søntvedt (1972) and Noordzij (1976) have employed the steady, partially cavitating, two-dimensional section theory of Geurst (1959) in a parametric sense with the variations of local angle of attack being looked after via lifting-line theory. Kaplan, Bentson & Breslin (1979) improved upon Noordzij by employing the equivalent unsteady angle of incidence as provided by the Davidson Laboratory unsteady-propeller program, developed by Tsakonas, Jacobs & Ali (1976). Comparison with measured cavity outlines and hull pressures predicted by the methods of Johnsson–Søntvedt and Noordzij have been fairly good, but they have not been adopted as reliable predictors.

Evaluations of the method of Kaplan *et al.* (1979), when compared with several measured model and ship hull pressures as carried out by Lloyd's Register of Shipping, have shown remarkable agreement.

To obtain hull forces, Vorus, Breslin & Tien (1978) have applied the reciprocity procedure, which relies entirely on an arbitrary estimate of cavity thickness. Stern & Vorus (1983) have advanced a non-linear theory in which the cavity cross section is constrained to be a semi-ellipse. However, none of the cited formulations have addressed the complete, three-dimensional, unsteady, mixed-boundary-condition problem, posed by an intermittently cavitating blade in the presence of all other blades.

A very numerically oriented method, developed at MIT, for the solution of this problem uses a representation of the propeller blades and cavities by lattices of vortices and line sources. The method was extended by Lee (1979), (1980) from the non-cavitating procedure (cf. Kerwin & Lee (1978)), a predecessor of the surface-panel method by Kinnas & Hsin (1990), outlined in Chapter 18, p. 368 and sequel. Applications are presented by Breslin, Van Houten, Kerwin & Johnsson (1982) where also a description of the method is given, and by Kerwin, Kinnas, Wilson & McHugh (1986) and later refinements in the cavitation model by Kinnas & Fine (1989).

As in two dimensions, cf. Chapter 8, p. 142 and sequel, the sources represent the blade and cavity thicknesses while the vortices model the pressure jump over the blade including the disturbance from the cavity. The leading-edge correction by Kinnas (1985, 1991) is included. The procedure is to solve the steady, non-cavitating problem first, then turn on the unsteadiness of the inflow and let the solution proceed until the steady-state, oscillatory, non-cavitating solution is achieved. Finally, the cavity solution is turned on and the solution process continued until a steady-state, oscillatory cavity solution has been achieved (cf. Kerwin *et al.* (1986)).

Results are shown next, together with results of other procedures. Further results where the mirror effects of the hull and water surfaces are taken into account are shown in Chapter 22.

COMPARISON OF CALCULATED AND OBSERVED TRANSIENT BLADE CAVITATION AND PRESSURES

There are several comparisons of observed and calculated instantaneous cavity geometries to be found in the literature since the mid 1970s. Here we limit ourselves to recounting the correlations made by Lee (1980) who used the surface-panel method and by Lloyd's Register using a procedure developed by Kaplan, Bentson & Breslin (1979).

Figure 20.7 shows a comparison of steady cavity geometry from the Kerwin-Lee theory (as presented by Lee (1980)) with observations made at DTRC. Here we see that the locus of the trailing edge of the cavity as calculated compares well with the mean locus drawn by Boswell (1971), particularly in view of the large variations from blade to blade.

The same propeller model was operated in the DTRC 24-inch water tunnel abaft a wake screen which produced the iso-wake contours shown in Figure 20.8.

The radial and chordwise extents of cavitation at various angular positions in this wake are shown in Figure 20.9. Cavity geometries calculated by Lee with this wake input are shown in Figure 20.10. Unfortunately direct overlaps of these calculated outlines and those observed were not made by Lee who states that agreement was found to be excellent.

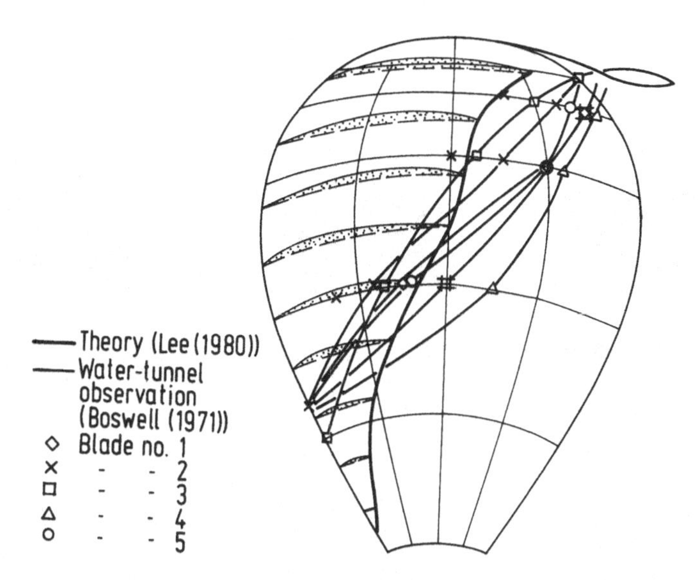

Figure 20.7 Comparison of predicted and observed extent of steady cavitation on DTRC propeller model 4381, at advance ratio $J_A = 0.7$ and cavitation index $\sigma_N = 1.92$.

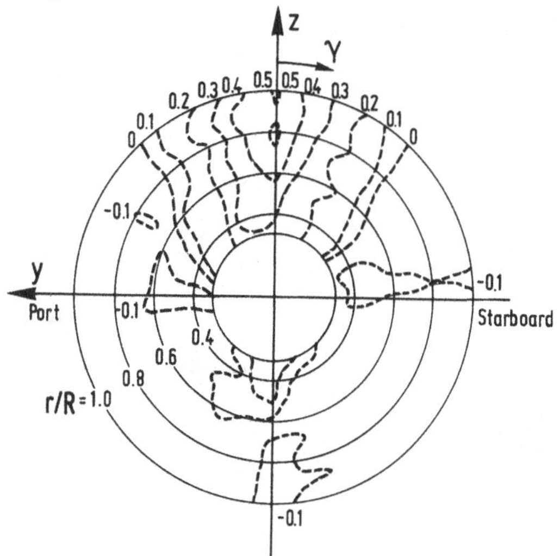

Figure 20.8 Wake field (u_a^W/U) measured behind screen. This wake field is used for the results shown in Figure 20.9 and 20.10.

Figures 20.7 and 20.8 Figures 5 and 6 from Lee, C.-S. (1980). Prediction of the transient cavitation on marine propellers by numerical lifting-surface theory. In *Proc. Thirteenth Symp. on Naval Hydrodynamics*, ed. T. Inui, pp. 41–64. Tokyo: The Shipbuilding Research Association of Japan.

By courtesy of Massachusetts Institute of Technology, USA, and Office of Naval Research, USA.

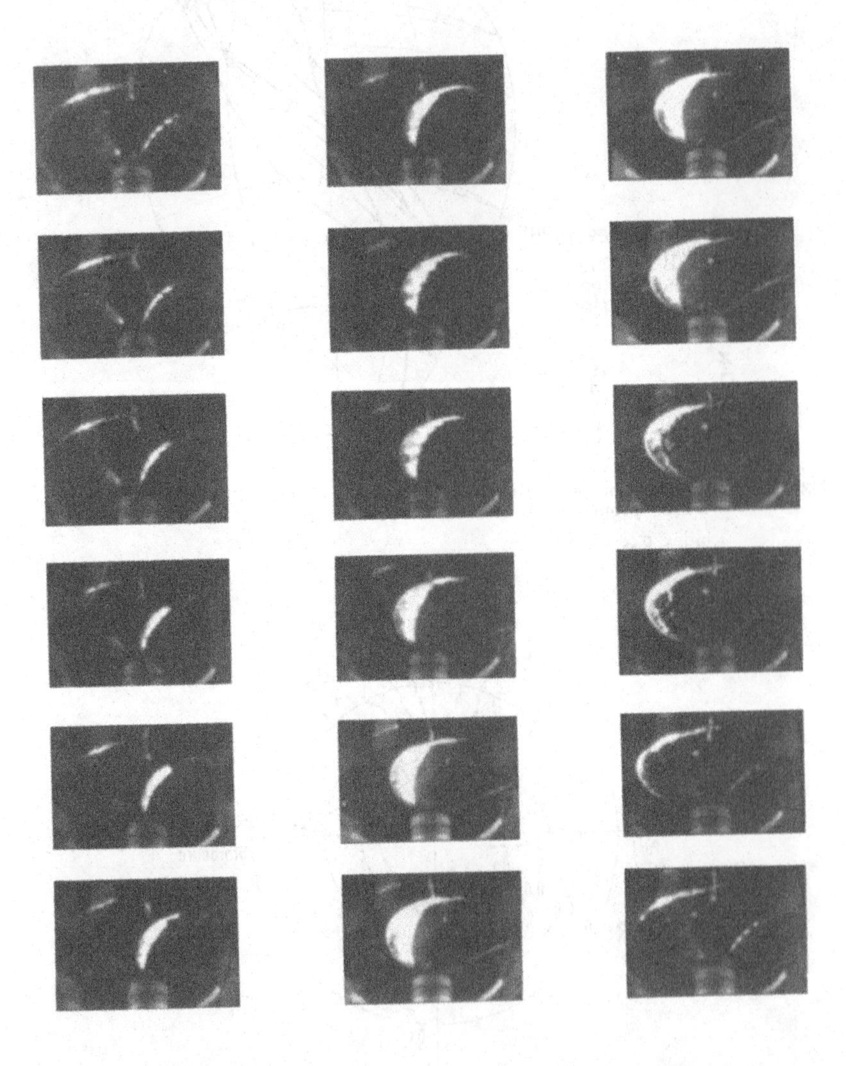

Figure 20.9 DTRC propeller model 4381 pictured at $J_A = 0.889$ with $\sigma_N = 3.53$.

Figures 20.9 and 20.10 figures 7 and 8 from Lee, C.-S. (1980). Prediction of the transient cavitation on marine propellers by numerical lifting-surface theory. In *Proc. Thirteenth Symp. on Naval Hydrodynamics*, ed. T. Inui, pp. 41–64. Tokyo: The Shipbuilding Research Association of Japan.

By courtesy of Massachusetts Institute of Technology, USA, and Office of Naval Research; USA.

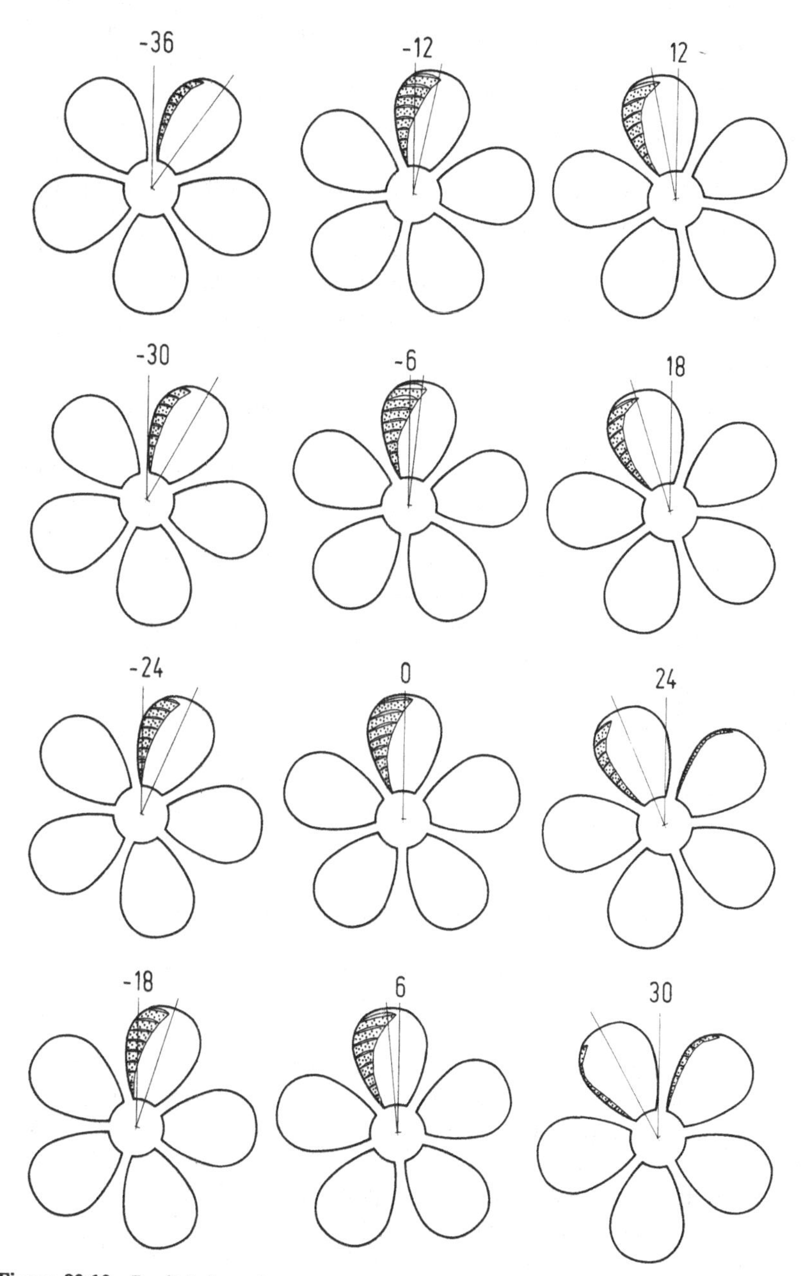

Figure 20.10 Predicted cavity extent and thickness distribution of DTRC propeller model 4381 under same conditions as pictured in Figure 20.9.

Lee also calculated the unsteady cavitation on a propeller of the *M/S Berge Vanga*, a twin-screw oil/ore carrier, for which stereo-photographic observations were reported by Holden & Øfsti (1975). Reasonable agreement was found from the maximum thickness of the cavities; 24 cm calculated; 10 to 20 cm for the stereo-photographs. But the predicted extent of the cavitation was found to be much larger than observed resulting in a maximum cavity volume some 12 times larger than deduced from the pictures. The lack of correlation was ascribed by Lee to the use of model wake data with no correction for scale effects. Subsequent studies by Kerwin have shown that his cavities are extremely sensitive to the wake input and he now employs an estimated effective wake as input, derived from the nominal model wake by an approximate procedure. Moreover, Kinnas (1985) has demonstrated pronounced shortening and reduced cavity area due to leading-edge thickness as was detailed in Chapter 8.

Remarkably successful comparisons of hull pressures (measured in several European cavitation facilities and on three ships) with computed values from a composite theory developed by Kaplan, Bentson & Breslin (1979) are displayed in Figures 20.11 through 20.13. Their mathematical model uses an adaptation of the steady, partially and fully cavitating section theory of Geurst (1959), (1961) together with an unsteady-angle-of-attack input from the Davidson Laboratory (Tsakonas, Jacobs & Ali (1976)) program for unsteady non-cavitating propellers.

Close correlation between calculated, model and full scale pressures are seen in Figure 20.11 at blade frequency ($q = 1$) and generally not as close for the second and third orders of blade frequency. Excellent agreements with measurements can be observed in Figure 20.12; model-ship correlation is beyond belief!

In Figure 20.13 the comparisons are excellent to good at blade frequency, not as good at twice blade frequency and poor at thrice blade frequency. Kaplan *et al.* (*ibid.*) sought to ascribe the poor correlation at thrice blade frequency to the lack of precise knowledge of the third harmonic of the wake. This explanation does not take into account that the cavitation arises dominantly from the sum of the first few wake harmonics of shaft frequency and as described earlier produces significant higher harmonics by itself which are not keyed to any specific wake harmonic.

The Kaplan *et al.* (*ibid.*) formulation cannot be defended as entirely rational as it does not include the dynamics of the cavitation and the hull-boundary effect was approximated by applying a factor of 2 to the free-field values. Unfortunately there have not been sufficiently many correlations to ascertain the validity of any mathematical model in a statistical sense. We can note that the various correlations to data provide encouragement that the effects attributed to intermittently cavitating blades can be quite well approximated by both semi- and more fully rational models when the wake input is modified to give the effective wake.

Number of blades	4
Diameter	5.18 m
P/D ratio at 0.7 R	1.065
Rate of revolution	124 RPM
Ship speed	18 knots
Power (shaft)	12 000 HP

Table 20.1 Refrigerated ship propeller characteristics.

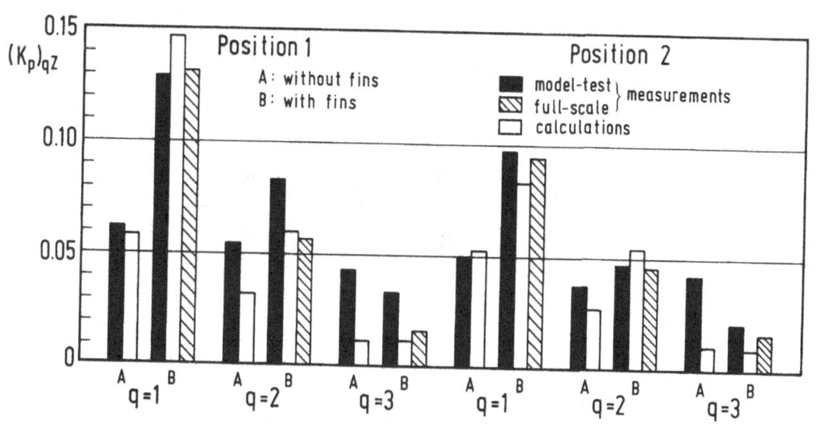

Figure 20.11 Pressure amplitudes at two positions on a refrigerated ship without and with fins.

Number of blades	4
Diameter	4.2 m
P/D ratio at 0.7 R	0.9
Rate of revolution	187 RPM
Ship speed	21 knots
Power (shaft)	10 400 HP

Table 20.2 RO–RO carrier propeller characteristics.

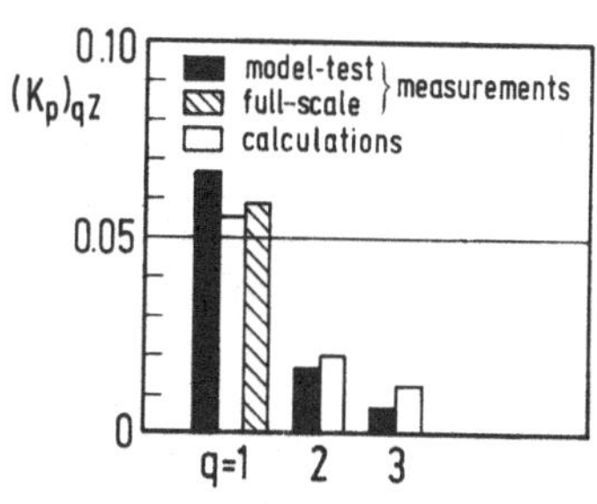

Figure 20.12 Pressure amplitudes for three harmonics for RO–RO carrier.

Number of blades	4
Diameter	6.3 m
P/D ratio at 0.7 R	0.881
Rate of revolution	114 RPM
Ship speed	19.1 knots
Power (shaft)	14 364 HP

Table 20.3 Cargo liner propeller characteristics.

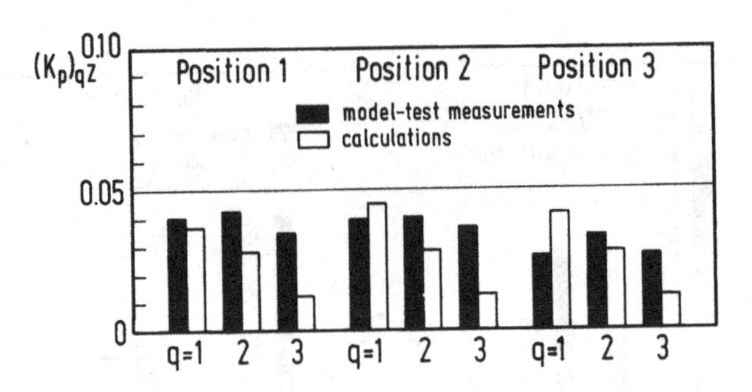

Figure 20.13 Pressure amplitudes at three positions, cargo liner, for three harmonics.

Figures 20.11, 20.12, 20.13 and Tables 20.1, 20.2 and 20.3 from Kaplan, P., Bentson, J. & Breslin, J.P. (1979). Theoretical analysis of propeller radiated pressure and blade forces due to cavitation. In *Proc. Symp. on Propeller Induced Ship Vibration*, pp. 133–45. London: RINA.

21 Forces on Simple Bodies Generated by Intermittent Cavitation

The velocity potential and the pressures generated at field points in the vicinity of an intermittently cavitating propeller have been seen to be complicated. In order to deal with the forces induced on bodies by "hand-turned" mathematics not only must the bodies be of simple form but also the velocity field of the intermittently cavitating propeller must be simplified. In the process of this simplification we shall have to impose restrictions.

Let us first reconsider the linearized pressure-velocity-potential relation in regard to application to a long body whose length is say, L. If we adopt new variables such that

$$p = \rho U^2 \breve{p} \quad ; \quad x = L\breve{x} \quad ; \quad \phi = LU\breve{\phi} \quad ; \quad t = \frac{1}{q\omega Z}\breve{t}$$

where the quantities with ˘ are dimensionless, then (1.30) transforms to

$$\breve{p} = \frac{\partial \breve{\phi}}{\partial \breve{x}} - \frac{q\omega ZL}{U}\frac{\partial \breve{\phi}}{\partial \breve{t}} \tag{21.1}$$

where q is the blade-frequency order, Z the number of blades, ω the angular velocity and U the free-stream or body speed. It is clear that

$$\frac{q\omega ZL}{U} = 2\pi \frac{qZ}{J}\frac{L}{D} >> 1$$

for all hull forms of interest, so that

$$\breve{p} \approx -\frac{q\omega ZL}{U}\frac{\partial \breve{\phi}}{\partial \breve{t}} \quad \text{(very nearly)} \tag{21.2}$$

which says that the convective pressure is negligible and the unsteady forces on hulls are purely inertial and independent of historical effects arising from changes in circulation and shed vorticity.

We then impose the restrictions displayed in the previous chapter (p. 393 and sequel) and approximate the potential of the propeller as that of a concentrated point source, cf. (20.34). This is used in the following.

The Flat Plate

Let us consider at first a "hull" of negligible draft (as a planing boat) having a water plane in the form of a rectangle (for simplicity) extending aft and forward of the propeller which is replaced by a pulsating source at some effective radius R_c from the propeller axis at a depth, d, below the water plane as shown in Figure 21.1. Moreover we shall let the plate be stretched longitudinally to $\pm\infty$ which makes it possible to reduce the problem to one in two dimensions provided one is content to find only the vertical force.

To make the further approximation that the pressure induced on the plate is twice that induced in the same position but in a boundless fluid, would give an infinite vertical force. Hence this rudimentary model is entirely untenable because no account of the presence of the free surface is taken.

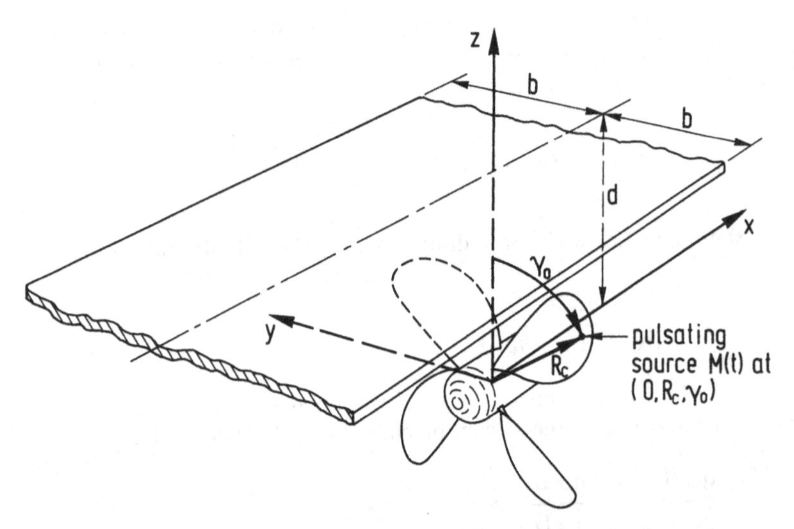

Figure 21.1 Flat plate in water surface in presence of a pulsating source.

Instead, we then consider the plate in the plane of the free surface. The free-surface condition at the high frequencies, qZω, is cf. Appendix F (Equation (F.13), p. 502) that the potential of the total motion is effectively zero there. We take this total potential ϕ as

$$\phi(x,y,z,t) = \phi_P(x,y,z,t) + \phi_S(x,y,z,t) - \phi_S(x,y,2d-z,t) \qquad (21.3)$$

where the last term is the negative image of the source (i.e. the cavitating propeller) in the plane $z = d$ and ϕ_P is the potential of the plate. Note that the contribution to ϕ from the source and its image vanishes for $z = d$.

Consequently, the vertical velocity induced on the plate by the source and its image is

$$2\left.\frac{\partial\phi_S(x,y,z)}{\partial z}\right|_{z=d} = 2w_S(x,y,d) \tag{21.4}$$

or twice the vertical velocity induced by the source alone on that locus. We now have the following boundary-value problem:

Find ϕ_P such that

$$\frac{\partial^2\phi_P}{\partial x^2} + \frac{\partial^2\phi_P}{\partial y^2} + \frac{\partial^2\phi_P}{\partial z^2} = 0 \quad ; \quad z \leq d \tag{21.5}$$

and

$$\frac{\partial\phi_P}{\partial z} = -\,2w_S(x,y,d) \qquad ; \qquad z = d, \quad |y| \leq b \tag{21.6}$$

$$\phi_P = 0 \qquad ; \qquad z = d, \quad |y| > b \tag{21.7}$$

$$\phi_P \longrightarrow 0 \qquad ; \qquad x,y \longrightarrow \infty, \quad z \longrightarrow -\infty \tag{21.8}$$

We are only interested in the force which is obtained by integrating the linearized pressure (1.30) over the plate or

$$F_z = -\rho\int_{-\infty}^{\infty}dx\int_{-b}^{b}\left[\frac{\partial\phi_P}{\partial t} - U\,\frac{\partial\phi_P}{\partial x}\right]dy \tag{21.9}$$

As the plate is distributed symmetrically fore and aft of the propeller plane ($x \equiv 0$), the second term (the convection pressure) is observed to integrate to zero. This is immediately seen to be unrealistic as the ship surface extends aft only to the order of a propeller diameter. We may defend this configuration from two aspects. Firstly, the convective term is weak in comparison to $\partial\phi_P/\partial t$ as demonstrated on p. 411. Secondly, $\partial\phi_P/\partial x$ vanishes more rapidly with x than ϕ_P. Thus the force under this gross representation is taken to be

$$F_z = -\rho\,\frac{\partial}{\partial t}\int_{-\infty}^{\infty}dx\int_{-b}^{b}\phi_P(x,y,d,t)dy \quad ; \; \phi_S(x,y,d) \equiv 0 \tag{21.10}$$

and the force is known when the integral

$$I^*(y,z,t) = \int_{-\infty}^{\infty}\phi_P(x,y,z,t)dx \tag{21.11}$$

can be specified on z = d. Performing the following operation

$$\left[\frac{\partial^2}{\partial y^2} + \frac{\partial^2}{\partial z^2}\right] I^* = -\int_{-\infty}^{\infty} \frac{\partial^2 \phi_P}{\partial x^2} dx = 0$$

as

$$\phi_P = 0 \quad ; \quad \frac{\partial \phi_P}{\partial x} = 0 \qquad \text{at } x = \pm\infty$$

Thus $I^*(y,z)$ satisfies the Laplace Equation in two dimensions; then, the boundary conditions on I^* are and from Equations (21.11) and (21.6), (21.7) and (21.8)

$$\frac{\partial I^*(y,d)}{\partial z} = -2\int_{-\infty}^{\infty} w_S(x,y,d)dx \quad ; \qquad |y| < b \tag{21.12}$$

$$I^*(y,d) = 0 \quad ; \qquad |y| > b \tag{21.13}$$

and

$$I^*(y,d) \longrightarrow 0 \quad ; \qquad y \longrightarrow \pm\infty, \quad z \longrightarrow -\infty$$

To solve this mixed-boundary-condition problem one observes the analogy with the section on vertical dipole distributions in Chapter 2 (p. 29–31), in particular Equations (2.16)–(2.17), which correspond to (21.12) and (21.13). Thus we see that $I^*(y,z)$ can be represented by a distribution of vertical dipoles over the region of the plate cross section

$$I^*(y,z) = -\frac{(z-d)}{\pi}\int_{-b}^{b} \frac{\sigma_z(y',t)}{(y-y')^2 + (z-d)^2} dy' \tag{21.14}$$

Here we see that this construction yields $I^*(y,0) = 0$, $|y| > b$. It may be shown (as displayed in Equation (2.16)) that the limit as $z \to d$ *from below* is

$$\lim_{z \to d+0_-=d_-} I^*(y,z,t) = \sigma_z(y,t) \quad ; \qquad |y| \le b \tag{21.15}$$

Thus once σ_z is found then I^* on the lower side of the plate is known. The kinematic condition presented by (21.12) yields the integral equation for σ_z

$$-\frac{1}{\pi}\int_{-b}^{b} \frac{\sigma_z(y',t)}{(y-y')^2} dy' = -2\int_{-\infty}^{\infty} w_S(x'',y,d,t)dx''$$

The left hand side can be transformed into

$$\frac{1}{\pi}\frac{\partial}{\partial y}\int_{-b}^{b}\frac{\sigma_z}{y - y'}\,dy' = -2\int_{-\infty}^{\infty} w_s(x'',y,d,t)dx''$$

and by integration over y

$$\int_{-b}^{b}\frac{\sigma_z}{y-y'}\,dy' = -2\pi\int_{-\infty}^{\infty}dx''\int^{y} w_s(x'',y,d,t)dy' + C_1 = W(y) + C_1 \tag{21.16}$$

where C_1 is a constant.

The inversion of this equation is (cf. Equation (A.19) in Appendix A)

$$\sigma_z(y) = -\frac{1}{\pi^2\sqrt{b^2 - y^2}}\left[\int_{-b}^{b}\frac{\sqrt{b^2 - y'^2}}{y - y'}(W(y')+C_1)dy' + C_2\right] \tag{21.17}$$

As ϕ_P and hence σ_z must be zero at $y = \pm b$, we obtain, by use of Equation (M8.14) from the Table of Airfoil Integrals, p. 525,

$$\int_{-b}^{b}\frac{\sqrt{b^2 - y'^2}}{b - y'}W(y')dy' + \pi b C_1 + C_2 = 0$$

$$-\int_{-b}^{b}\frac{\sqrt{b^2 - y'^2}}{b + y'}W(y')dy' - \pi b C_1 + C_2 = 0$$

C_1 and C_2 can be found from these conditions and inserted back again into (21.17). After use of (M8.14) again and some reduction we achieve

$$\sigma_z(y) = -\frac{1}{\pi^2}\sqrt{b^2 - y^2}\int_{-b}^{b}\frac{W(y')}{\sqrt{b^2 - y'^2}}\frac{1}{y - y'}dy' = I^*(y,d,t) \tag{21.18}$$

The Equations (21.18), (21.15), (21.11) and (21.10) give the force as

$$F_z = \frac{\rho}{\pi^2}\frac{\partial}{\partial t}\int_{-b}^{b}dy\sqrt{b^2 - y^2}\int_{-b}^{b}\frac{W(y')}{\sqrt{b^2 - y'^2}}\frac{1}{y - y'}dy' \tag{21.19}$$

The y-integral yields $-\pi y'$ (Table of Airfoil Integrals, Equation (M8.14)). Using the definition of W from Equation (21.16), we have

$$F_z = 2\rho \frac{\partial}{\partial t} \int_{-b}^{b} dy' \frac{y'}{\sqrt{b^2 - y'^2}} \int_{-\infty}^{\infty} dx'' \int^{y'} w_s(x'',y'',d,t)dy''$$

and integrating by parts over y' yields

$$F_z = 2\rho \frac{\partial}{\partial t} \int_{-b}^{b} dy' \sqrt{b^2 - y'^2} \int_{-\infty}^{\infty} w_s(x'',y',d,t)dx'' \tag{21.20}$$

Equation (21.20) is quite general in that the vertical velocity w_s can be that induced by any appropriate singularity, isolated or distributed. The temporal acceleration $\partial w_s/\partial t$ is, to the order of the approximation, cf. (1.8)

$$-\frac{1}{\rho}\frac{\partial p_s}{\partial z}\bigg|_{z=d}$$

i.e., proportional to the pressure gradient produced by the singularity or its aggregate distribution at any point on the plate.

We now specify w_s as that due to a point source of strength $M(\gamma_0(t))$ located at $x' = 0$, $y' = -R_c \sin\gamma_0$ and $z' = R_c \cos\gamma_0$, cf. p. 411 and Figure 21.1. This specialization yields

$$\int_{-\infty}^{\infty} w_s(x'',y',d,t)dx'' =$$

$$\frac{M(\gamma_0(t))}{4\pi}(d-R_c\cos\gamma_0)\int_{-\infty}^{\infty} \frac{1}{[x''^2+(y'+R_c\sin\gamma_0)^2+(d-R_c\cos\gamma_0)^2]^{3/2}}dx'' =$$

$$\frac{M(\gamma_0(t))\ (d-R_c\cos\gamma_0)}{2\pi((y'+R_c\sin\gamma_0)^2+(d-R_c\cos\gamma_0)^2)} \tag{21.21}$$

Here we see that harmonic dependence on time or blade-position angle is inherent in both M and its geometric coefficient. In keeping with the approximation that the effect of distributed intermittent cavitation may (under the conditions derived in Chapter 20, p. 393) be modelled by a concentrated, stationary, pulsating source and an axial dipole at $\gamma_0 \approx 0$ we simplify Equation (21.21) by setting $\gamma_0 = 0$ thereby taking only the *wash* from a fixed source. (The axial dipole will not contribute here because it is odd in x).

With $\gamma_0 = 0$, in Equation (21.21), the remaining y'-integral in (21.20) can be effected, with the substitution $y' = b \sin\theta$, to give

$$F_z = \rho \frac{\partial M}{\partial t} \frac{b}{\sqrt{1 + \left[\frac{d - R_c}{b}\right]^2} + \frac{d - R_c}{b}} \tag{21.22}$$

We see that since the relative clearance $(d-R_c)/b$ (see Figure 21.1) is very small the force virtually increases linearly with the plate semi-width b, and, hence, becomes infinite as $b \to \infty$. It is interesting that the force is bounded even when the effective clearance $d - R_c$ vanishes. As expected, the force vanishes as the clearance tends to infinity for fixed b vanishing as $b^2/2d$, i.e., slowly. Clearly, the effective source strength will vary sharply with clearance because of the change in blade cavitation with varying penetration of the hull boundary layer or wake within a critical clearance. In addition, the decay with increasing effective clearance is not applicable for the case of a hull water plane which terminates shortly abaft the propeller due to the relieving effect of the free surface. This emphasizes the necessity of solving the problem of a flat hull of limited length which must be done by numerical methods. Thus although our hand-turned result is simple we cannot regard it as realistic. We attempt to remove this shortcoming by showing results for an "ellipsoidal" ship later.

To complete Equation (21.22), the source strength is taken to be the time-rate-of-change of cavity volume which varies cyclically with blade position cf. (20.34). Then for a Z-bladed propeller, we use the source term of (20.35) which is

$$M(t) = -Z \sum_{q=-\infty}^{\infty} iqZ\omega(V_c)_{qZ}\, e^{iqZ\gamma_0(t)}$$

where $(V_c)_{qZ}$ is the amplitude of the q-th harmonic of the cavity volume. The amplitude of the force on the plate at frequency qZ is then

$$(F_z)_{qZ} = \rho Z^3(q\omega)^2\, (V_c)_{qZ} \frac{b}{\sqrt{1 + \left[\frac{d - R_c}{b}\right]^2} + \frac{d - R_c}{b}} \tag{21.23}$$

The foregoing analysis to secure the potential or the force on the simplest of geometries is seen to be lengthy. This is because we approached the problem via the classical diffraction or scattering technique wherein the body is "constructed" in the presence of the external, disturbing or exciting singularities. A much more adroit procedure exists which avoids the solution of the diffraction problem when one seeks only particular force components induced on the body by nearby singularities. As this method is general it can be applied to both simple and actual ship forms. We turn to an outline of the derivation of this procedure.

HULL FORCES WITHOUT SOLVING THE DIFFRACTION PROBLEM

Vorus (1974) advanced a generalization of a procedure due to Chertock (1965) for finding forces on a body by invoking Green's theorem which avoids the necessity of solving the diffraction or boundary-value problem posed by creating the body in the presence of any ensemble of singularities. Here the derivation is limited to the vertical force associated with a rigid-body mode of motion. Vorus's landmark paper should be studied to appreciate the generality of his theory.

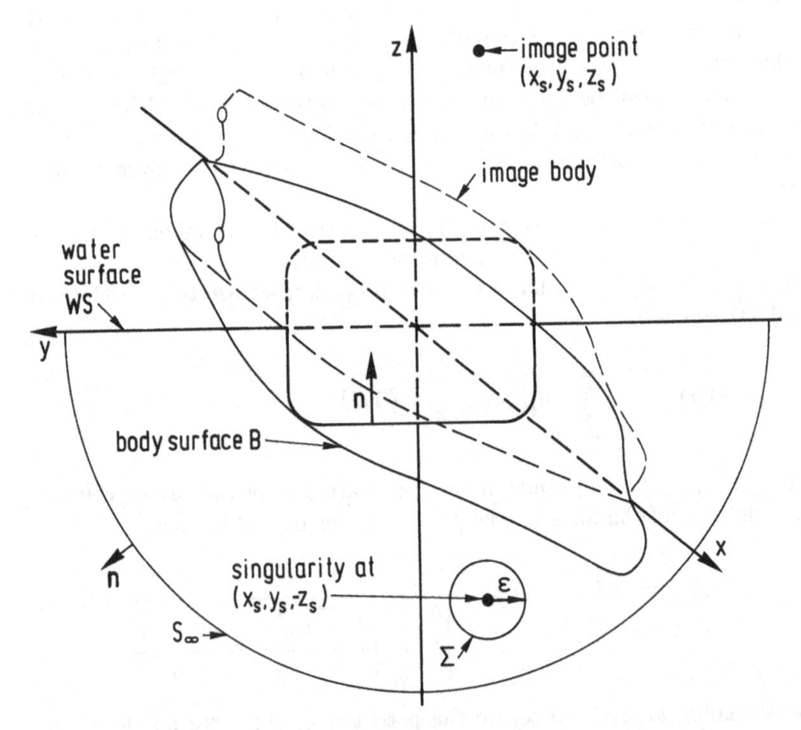

Figure 21.2 Schematic of hull fluid domain and location of singularities. Note that here we deviate from our previous standard and place the origin of the coordinate system in the free surface.

Consider the fluid of concern to be in the lower half of space bounded by the body surface B, the water surface WS, a spherical boundary at infinity S_∞, and a sphere of radius ε and surface Σ around a singularity at (x_s, y_s, $-z_s$), as depicted in Figure 21.2. The character of this point singularity is left unspecified at this juncture.

We wish to use Green's second identity and for this purpose need two potentials ϕ and ψ, cf. Section 3 of the Mathematical Compendium, p. 506. The total potential ϕ due to the disturbance is due to the singularity, its negative image in the upper half space and to diffraction in the body. The other potential ψ is suitably constructed by considering the motion of the body B, and its reflection B_i in the fluid region extended to the upper half space. Both B and B_i are given a motion of unit velocity in the j-th direction, where j = 1 corresponds to x–, j = 2 to y– and j = 3 to z–direction. The boundary condition in the motion potential ψ_j can be seen from Equation (3.19) if we take U, V and W = 1 respectively since we have unit motion of the bodies and remember that **n** is now positive out of the fluid. We then have on the body

$$\frac{\partial \psi_j}{\partial n} = n_j \qquad j = 1, 2, 3 \tag{21.24}$$

where n_j are the components of **n**.

To find the force, say in the j-th direction, we integrate the pressure projected in that direction, over the body

$$F_j = \int_B p n_j \, dS$$

We use the linearized relation between pressure and velocity potential and neglect the convective term, as explained previously on p. 411, to obtain

$$F_j = \rho \frac{\partial}{\partial t} \int_B \phi n_j \, dS \tag{21.25}$$

in which we take

$$\phi = \phi_b + \phi_s + \phi_{si} \tag{21.26}$$

ϕ_b is the body or diffraction potential;

ϕ_s is the potential of the singularity at (x_s, y_s, $-z_s$);

ϕ_{si} is the potential of the negative of the singularity at the image point in the upper half space, i.e., at (x_s, y_s, z_s).

The potentials in (21.26) must satisfy

$$\frac{\partial \phi}{\partial n} = 0 \qquad \text{on B} \tag{21.27}$$

$$\phi = 0 \qquad \text{on WS} \tag{21.28}$$

$$\phi_s + \phi_{si} = 0 \qquad \text{on WS} \tag{21.29}$$

$$\phi \longrightarrow 0 \qquad \text{on } S_\infty \tag{21.30}$$

and to employ the motion potential ψ_j, we must also require

$$\psi_j(x,y,0) = 0 \qquad \text{on WS}$$

outside of the body, and $\psi_j \to 0$ on S_∞.

Limiting our attention to the vertical force then $n_j = n_3 = n_z$, we write out Green's second identity, cf. (M3.3), explicitly for all surfaces bounding the fluid, taking $\psi_3 = \psi$ for convenience, and using (21.24) on the body

$$\int_B \phi\, n_z\, dS + \int_{WS} \phi \frac{\partial \psi}{\partial z} dS + \int_\Sigma \phi \frac{\partial \psi}{\partial n} dS + \int_{S_\infty} \phi \frac{\partial \psi}{\partial n} dS =$$

$$\int_B \psi \frac{\partial \phi}{\partial n} dS + \int_{WS} \psi \frac{\partial \phi}{\partial z} dS + \int_\Sigma \psi \frac{\partial \phi}{\partial n} dS + \int_{S_\infty} \psi \frac{\partial \phi}{\partial n} dS \tag{21.31}$$

The second and fourth integrals on the left side vanish as well as the first, second and fourth integrals on the right side as a consequence of the construction of ϕ and ψ, i.e. the conditions that these two potentials fulfil on WS and S_∞. Hence, we have

$$\int_B \phi\, n_z\, dS = \int_\Sigma \left\{ \psi \frac{\partial}{\partial n} (\phi_b + \phi_s + \phi_{si}) - (\phi_b + \phi_s + \phi_{si}) \frac{\partial \psi}{\partial n} \right\} dS \tag{21.32}$$

Now on the surface Σ, ψ, ϕ_b and ϕ_{si} are regular and as the surface Σ is shrunk to zero, they will not contribute. For a source,

$$\phi_s = -\frac{M}{4\pi R}$$

where $R = \sqrt{(x-x_s)^2+(y-y_s)^2+(z+z_s)^2}$ is the distance from the source to any field point. Then $\phi_{si} = M/4\pi R_i$ where R_i is the distance to the image point, i.e. the same as R with z_s replaced by $-z_s$. Then in the first integral in the right side the normal is directed inward to the sphere Σ, and hence $\partial/\partial n = -\partial/\partial r$ where r is the local radial direction. We then have

$$\int_\Sigma \psi \frac{\partial}{\partial n}(\phi_b+\phi_s+\phi_{si})\,dS = -\int_\Sigma \psi \frac{\partial}{\partial r}\left[\phi_b - \frac{M}{4\pi r} + \frac{M}{4\pi R_i}\right]\Bigg|_{r=\varepsilon} dS$$

As the sphere Σ is shrunk to zero, the potential function ψ takes its value $\psi(x_s, y_s, -z_s)$ at the center of the sphere and as the functions $\partial\phi_b/\partial r$ and $\partial(1/R_i)/\partial r$ are regular at this point their contributions vanish with the surface area $4\pi\varepsilon^2 \to 0$. We are then left with (upon effecting the differentiation)

$$\int_\Sigma \psi \frac{\partial}{\partial n}(\phi_b+\phi_s+\phi_{si})dS = -\,\psi(x_s, y_s, -z_s)\frac{M}{4\pi\varepsilon^2}\cdot 4\pi\varepsilon^2 = -\psi M(t) \tag{21.33}$$

In the second integral on the right side of (21.32) we see again that ϕ_b and ϕ_{si} contribute nothing and as $\phi_s = -\,M/4\pi\varepsilon$ on Σ and the surface area of the sphere is $4\pi\varepsilon^2$ and $\partial\psi/\partial n = -\,\partial\psi/\partial r$ is bounded, this integral vanishes. Then the vertical force on the body is, using (21.33) and (21.32) in (21.25) with $j = 3$, the vertical direction

$$F_z = \rho \frac{\partial}{\partial t}\left\{M(t)\ \psi\Big[x_s,\ y_s(\gamma(t)),\ -z_s(\gamma(t))\Big]\right\} \tag{21.34}$$

or in general for the disturbance due to a point source the force in the j-th direction is

$$F_j = \rho \frac{\partial}{\partial t}\left\{M(t)\ \psi_j(x_s,\ y_s,\ -z_s)\right\} \tag{21.35}$$

Hence the vertical force is obtained by evaluating the unit vertical-motion potential of the double body at the location of the source performing the operation indicated by Equation (21.35). We note that the vertical motion of the double body induces velocities which are even functions of z and hence the potential is odd in z and therefore vanishes at $z = 0$ as required by the high-frequency condition.

Although (21.35) is a simple formula it must be realized that for each force and moment which is desired one must solve the boundary-value problems posed by the unit motions of the body in each motion which meets the kinematic condition on the body and the pressure condition $\psi_j(x,y,0) = 0$ on the water surface. Thus, for example, if one wishes the transverse force on the hull it is necessary to find ψ_2 which satisfies $\partial\psi_2/\partial n = n_2 = n_y$ on the wetted hull and to meet the free-surface condition the image hull has to move in the negative or opposite direction. This problem is not simple.

The vertical force due to a dipole at $(x_S, y_S, -z_S)$ can be obtained by replacing $(x_S, y_S, -z_S)$ by $(x_S-x', y_S-y', -(z_S-z'))$ and applying the operator $-\{n_x\, \partial/\partial x' + n_y\, \partial/\partial y' + n_z\, \partial/\partial z'\}$ to (21.34) thus modified (and thereafter setting $x' = y' = z' = 0$). The source strength M is replaced by the dipole strength $\Sigma(t)$. n_x, n_y, n_z are the direction cosines of the dipole axis. The formula for the potential is given in Equation (1.68).

It is most interesting to note that the vertical force on the wetted lower half of the double model in the presence of the water surface is the same as the force on the double model in the presence of the singularity when the whole is immersed in a boundless fluid.

Let us consider the double hull in a boundless fluid (no free surface) in the presence of a single isolated singularity. Then the foregoing development applies with the diffraction potential for this problem, satisfying $\partial\phi/\partial n = 0$ on B which is now the double hull and, of course, integrals over the water surface are absent. The same formula is obtained and we see that the force on this entire form is identical to that on the wetted hull in the presence of the free surface and the singularity plus its negative image. Thus, both experimentally and theoretically, the vertical force on a floating body can be secured by testing or solving the problem of the double body in the presence of the singularity in a boundless fluid or effectively deeply submerged. This has been shown in a less adroit fashion by Vorus (see Cox, Vorus, Breslin & Rood (1978)). A more detailed demonstration is given in the next chapter (p. 430 and sequel).

We next demonstrate the efficacy of Vorus's procedure by applying it to the circular cylinder of infinite length.

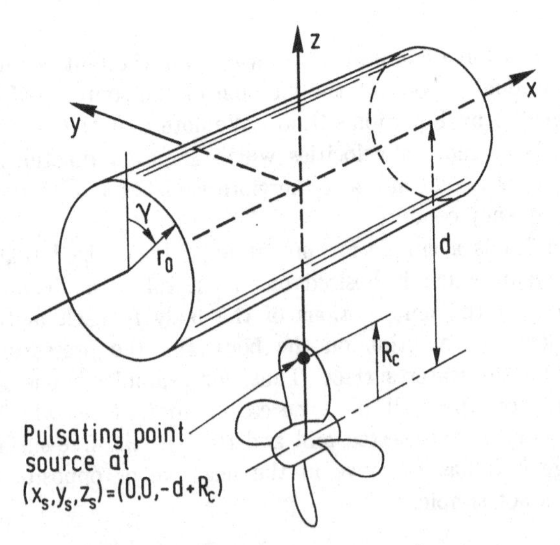

Figure 21.3 Cylinder in presence of source.

The Circular Cylinder

Consider a cylinder of radius r_0 which extends axially from $x = -\infty$ to ∞ in the presence of a point source at $x_s = 0$, $y_s = 0$, $z_s = -d + R_c$, cf. Figure 21.3. The cylinder is semi-submerged with the center lying in the waterplane. The flow around the cylinder in unit velocity vertical motion is clearly two-dimensional and can be described as the flow around vertically directed dipoles along the cylinder axis, or from Equation (1.46)

$$\psi_3 = -\frac{\Sigma_z}{2\pi}\frac{z}{y^2 + z^2} \tag{21.36}$$

We use cylindrical coordinates $z = r\cos\gamma$, $y = -r\sin\gamma$ and use (21.24) to determine the dipole strength Σ_z. Then $n_z = -\cos\gamma$ and we have

$$\left.\frac{\partial\psi_3}{\partial n}\right|_{r=r_0} = -\left.\frac{\partial\psi_3}{\partial r}\right|_{r=r_0} = -\frac{\partial}{\partial r}\left[-\frac{\Sigma_z}{2\pi}\frac{\cos\gamma}{r}\right]\Bigg|_{r=r_0} = -\frac{\Sigma_z}{2\pi}\frac{\cos\gamma}{r_0^2} = -\cos\gamma \tag{21.37}$$

and hence

$$\Sigma_z = 2\pi r_0^2$$

Insert this into (21.36) to obtain

$$\psi_3 = -\frac{r_0^2}{r}\cos\gamma \tag{21.38}$$

Now, from (21.35), the force generated on this floating cylinder due to a pulsating source is

$$F_z = -\rho\frac{\partial}{\partial t}\left[M(t)\frac{r_0^2}{d - R_c}\cos\pi\right] = \rho\frac{\partial M}{\partial t}\frac{r_0^2}{d - R_c} \tag{21.39}$$

As for the flat plate, the force diverges as the cylinder radius tends to infinity. It is also seen that the force vanishes slowly as the clearance between source and cylinder $d - r_0 - R_c \to \infty$, although again the blocking effect of the cylinder abaft the source contributes to this slow decay. Referring to the force on the plate given by Equation (21.22), we see that for the effective clearance there $d - R_c << b$, the forces on the plate and the cylinder are of identical structure as to be expected.

Ellipsoid of Three Unequal Axes

The ellipsoid (length L, breadth B, draft T) is the most ship-like of all the mathematical forms. The vertical force induced on such a general ellipsoid by a pulsating point source can be calculated from the velocity potential deduced from Lamb (1963) together with the method of Vorus as explained above. Although this derivation is not difficult it is rather lengthy and only the results will be presented here. The calculations have been carried out with the proportions length/breadth = 5, breadth/draft = 3, propeller diameter/length = 0.0375 (diameter/draft = 0.56). These proportions put the propeller 0.023·L fore of the aft end of the "ship" with its tips touching the base line at a clearance of 0.25 D, as outlined in Figure 21.4. From this one sees that the form is very open fore of the propeller. But in this region a ship would have a very narrow and deep form with normals directed mainly horizontally so this portion of the hull would not contribute greatly to the vertical force. The source representing the cavity volume has been placed at the upper tip of the propeller.

The ratio of the force coefficient at given clearance ratio c/D to that at zero clearance is graphed in Figure 21.4, showing a gradual but small decay with increased clearance on the assumption that the cavity volume harmonic does not vary with tip clearance. This is not the case since the hull boundary layer varies with clearance, hence also cavity volume. *This result indicates that the main importance of tip clearance is its rôle in the development of the blade cavitation which will be sensitive to the depth of penetration of the hull wake or boundary layer by the blade-tip region.*

Further expositions of the effect of ellipsoid proportions are left for interested readers. We now turn to arbitrary hull forms of practical geometry.

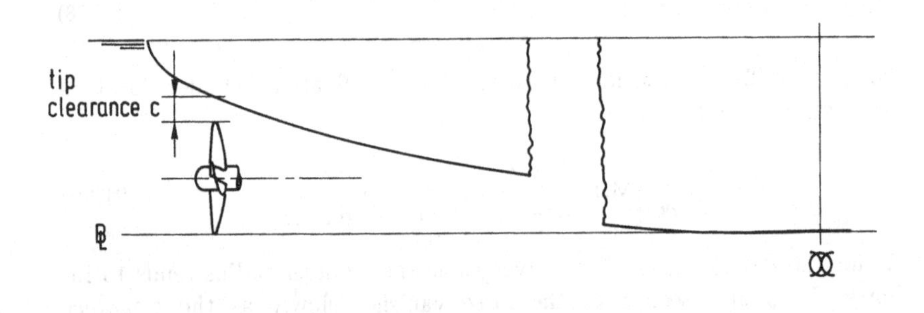

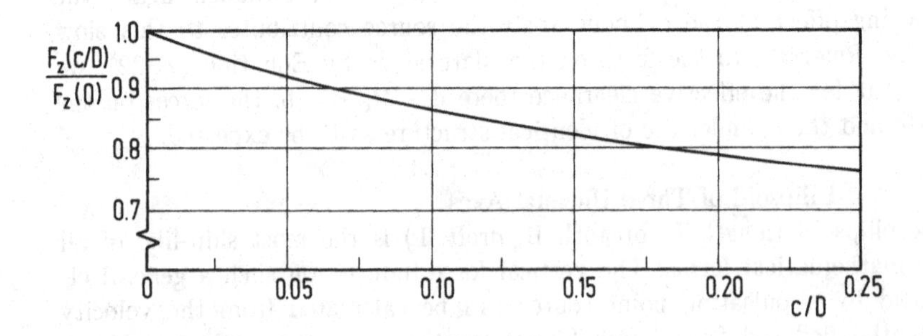

Figure 21.4 Variation of vertical force with tip clearance on an ellipsoidal hull.

22 Pressures on Hulls of Arbitrary Shape Generated by Blade Loading, Thickness and Intermittent Cavitation

In the preceding chapter we calculated the force generated by intermittent cavitation on simple forms. It was necessary to model not only the ship as a simple form but also the cavitating propeller. Although these simplifications gave useful information it was also obvious that the results could not be used for practical purposes. This was the price we had to pay for being able to obtain results by "hand-turned mathematics".

When we now wish to obtain results for actual, intermittently cavitating propellers behind real ship forms of given (arbitrary) shape we must expect the mathematics to be too complicated to be manipulated into closed expressions for forces and pressures. Instead, we shall describe a general, computer-effected method for solving the problem for a propeller behind a ship and we shall also present results of this theory for cavitating propellers. Furthermore, we shall correlate these results with those obtained by model experiments.

REPRESENTATION OF HULLS OF ARBITRARY SHAPE IN THE PRESENCE OF A PROPELLER AND WATER SURFACE

It has been demonstrated in the foregoing that a propeller operating in a temporally uniform but spatially varying hull wake produces, through the concerted action of all the blades, a potential flow and pressure field composed of many components, all of which are at frequencies $qZ\omega$. As these frequencies are large compared with those which can give rise to wave generation on the free water surface, the appropriate linearized boundary condition, imposed by the presence of the water surface, is that the total velocity potential in the undisturbed locus of that surface must be zero. A proof of this is presented, for sake of completeness, in Appendix F, p. 500.

Thus, if $\phi(x,y,d+z,t)$ is the potential of a propeller at depth d below the plane $z = 0$ in an infinite fluid with the origin of coordinates above the center of the propeller (x forward, y to port and z vertically upward)[28], then the potential of the propeller in the presence of the water surface at $z = 0$ is

[28] In this section we once more as we did in the last chapter deviate from our standard and put the origin of the coordinate system in the free surface.

$$\phi_{pr}(x,y,z,t) = \phi(x,y,d+z,t) - \phi(x,y,d-z,t) \tag{22.1}$$

since, at $z = 0$ (the plane of the undisturbed water surface), we have $\phi_{pr}(x,y,0,t) \equiv 0$ for all (x,y,t). Thus (22.1) is constructed to satisfy the high-frequency boundary condition. The second term (with sign) in (22.1) is referred to as the negative image of the propeller in the water surface.

If we wish to insert a ship hull into this flow field, we may first employ a surface distribution of sources on the wetted hull plus their negative images over the reflected hull. The determination of the source strength densities is effected by solving an integral equation generated from the requirement that the normal velocity imposed by the propeller and its negative image must be balanced or annulled at each and every hull element by the normal velocity induced by the concerted action of all the hull sources. This was actually the procedure used for the case of a flat plate, p. 412–418, although there the simple geometry of propeller and ship (point source and plate) made the analytical solution possible. But for real ship and propeller geometries this process was found to require excessive computer effort since three components of the propeller-induced velocity must be calculated for the scalar product with the normal vector. This effort is reduced by two-thirds by solving this exterior Neumann problem by exploiting the interior Dirichlet problem through the use of normal dipoles whose concerted potential evaluated on the *interior side of the hull surface* is required to annul the propeller potential (one function) at each and every element. A proof of this equivalent procedure follows.

Replacement of the Exterior Neumann Problem by an Interior Dirichlet Problem[29]

As we shall see later the problem posed by the wetted hull, propeller and free surface can be solved by finding the dipole density over the double hull in the presence of the propeller. The equation for the dipole densities follows from solving a Dirichlet problem on the interior side of the hull.

The problem posed by the generation (or creation) of a double hull in the presence of a single propeller requires a hull potential ϕ_B which meets the Neumann condition

$$\frac{\partial \phi_B}{\partial n} = -\frac{\partial \phi}{\partial n} \quad ; \quad \text{on } S = S_B + S_I \tag{22.2}$$

where ϕ is the potential of the disturbance (in our case the propeller, Equation (22.1)), $\mathbf{n}$ the outward normal from S, S_B the hull surface, and S_I is the image or reflected hull; thus S is the double hull, Figure 22.1. The difficulty with solving the hull generation problem to meet (22.2) directly is that the right side is composed of three terms

$$\frac{\partial \phi}{\partial n} = n_x \frac{\partial \phi}{\partial x} + n_y \frac{\partial \phi}{\partial y} + n_z \frac{\partial \phi}{\partial z}$$

[29] This proof is due to Arthur S. Peters, Professor Emeritus, Courant Institute, New York University, New York, New York, USA.

and this calculation is highly consumptive of computer effort. Fortunately, the same Neumann condition can be met by solving the *interior* Dirichlet problem requiring within the interior of S that $\phi_B = -\phi$ (a single scalar function).

Let S denote a bounded, simple, smooth, closed surface, cf. Figure 22.1. Let D_i denote the bounded interior domain of S, and let D_e indicate the unbounded exterior of S. Define **n** as the unit normal to S directed from D_i to D_e. Designate

$$R_s = \sqrt{(x-x')^2 + (y-y')^2 + (z-z')^2}$$
$$R = \sqrt{(X-x')^2 + (Y-y')^2 + (Z-z')^2} \qquad (22.3)$$

where Q(x',y',z') and P(x,y,z) are points on S and T(X,Y,Z) is a point in either D_i or D_e.

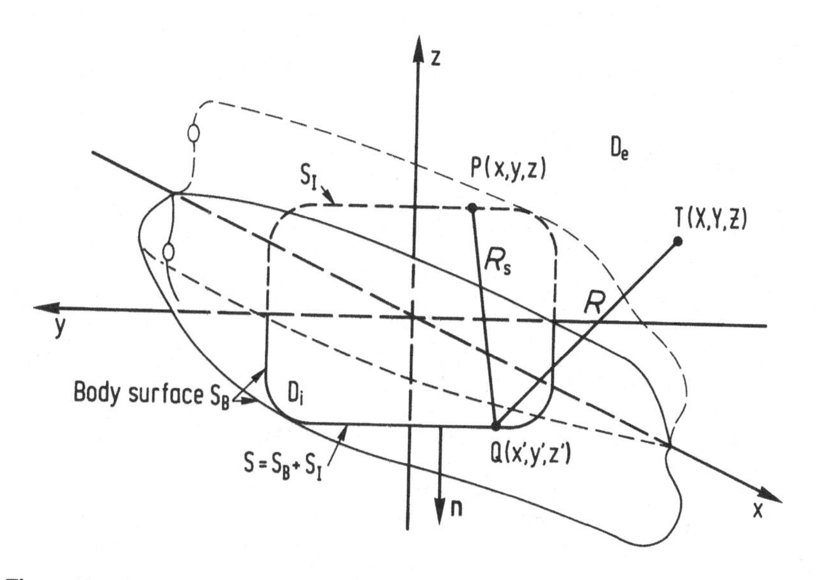

Figure 22.1 Definition interior and exterior domains and points P and Q on S and T a field point.

Consider the potential by a distribution of normal dipoles, cf. Equation (2.46)

$$\phi_B(T) = -\frac{1}{4\pi}\int_S \sigma(Q)\,\frac{\partial}{\partial n'}\left[\frac{1}{R}\right] dS \qquad (22.4)$$

which satisfies the Laplace equation

$$\nabla^2\phi_B(T) = 0$$

for T in either D_i or D_e. The potential (22.4) is discontinuous at S (cf. Equation (2.49)). If

$$\phi_B^+(P) = \lim_{T \to P} \phi_B(T) \qquad \text{T in } D_e \tag{22.5}$$

$$\phi_B^-(P) = \lim_{T \to P} \phi_B(T) \qquad \text{T in } D_i \tag{22.6}$$

then

$$\phi_B^+(P) = -\frac{\sigma(P)}{2} - \frac{1}{4\pi}\int_S \sigma(Q)\, \frac{\partial}{\partial n'}\left[\frac{1}{R_s}\right] dS \tag{22.7}$$

$$\phi_B^-(P) = +\frac{\sigma(P)}{2} - \frac{1}{4\pi}\int_S \sigma(Q)\, \frac{\partial}{\partial n'}\left[\frac{1}{R_s}\right] dS \tag{22.8}$$

where

$$\sigma(P) = -\left[\phi_B^+(P) - \phi_B^-(P)\right] \tag{22.9}$$

On the other hand, it is shown (Kellogg (1967)) that, if $\sigma(Q)$ is continuous on S and if either of the limits

$$\lim_{T \to P} \mathbf{n}\cdot\nabla\phi_B = \left[\frac{\partial \phi_B}{\partial n}\right]_e \qquad \text{T in } D_e$$

or

$$\lim_{T \to P} \mathbf{n}\cdot\nabla\phi_B = \left[\frac{\partial \phi_B}{\partial n}\right]_i \qquad \text{T in } D_i \tag{22.10}$$

exists, then the other does and

$$\left[\frac{\partial \phi_B}{\partial n}\right]_e = \left[\frac{\partial \phi_B}{\partial n}\right]_i = \left[\frac{\partial \phi_B}{\partial n}\right]_P \tag{22.11}$$

that is, the normal derivative or velocity is continuous through the dipoled surface. Now suppose that

$$\phi_B(T) = -\frac{1}{4\pi}\int_S \sigma(Q)\, \frac{\partial}{\partial n'}\left[\frac{1}{R}\right] dS \qquad \text{T in } D_e \tag{22.12}$$

is to be such that

$$\left[\frac{\partial \phi_B}{\partial n}\right]_e = -\left[\frac{\partial \phi}{\partial n}\right]_e = -\frac{\partial \phi}{\partial n} \tag{22.13}$$

where, in our case, ϕ is the potential of the propeller which is, of course, a regular harmonic function in the domain containing the double hull surface S and its interior.

In accordance with the properties of surface dipole distributions cited in the foregoing, this Neumann condition (22.13) can be satisfied if we demand that the dipole density $\sigma(Q)$ be such that

$$\phi_B(T) = -\phi(T) = -\frac{1}{4\pi}\int_S \sigma(Q)\, \frac{\partial}{\partial n'}\left[\frac{1}{R}\right] dS \tag{22.14}$$

for all points T in D_i. This ensures the satisfaction of

$$\left[\frac{\partial\phi_B}{\partial n}\right]_i = \left[\frac{\partial\phi_B}{\partial n}\right]_e = -\left[\frac{\partial\phi}{\partial n}\right]_e = -\frac{\partial\phi}{\partial n}$$

and we see that the exterior Neumann problem is thus reduced to an interior Dirichlet problem for, as T → P (on the interior of S), we see from (22.8) that σ must satisfy

$$-\phi(P) = \frac{\sigma(P)}{2} - \frac{1}{4\pi}\int_S \sigma(Q)\,\frac{\partial}{\partial n'}\left[\frac{1}{R_S}\right] dS \tag{22.15}$$

a Fredholm integral equation of the second kind of the form identical to that for which procedures are available for its numerical inversion.

Since $\phi_B^-(P) = -\phi(P)$, then, from the jump relation (22.9), we obtain

$$\sigma(P) = -\left[\phi_B^+(P) + \phi(P)\right] \tag{22.16}$$

Hence, once σ is found on the hull, the total pressure on the hull can be calculated from

$$p = -\rho\left[\frac{\partial}{\partial t} - U\frac{\partial}{\partial x}\right]\left[\phi_B^+ + \phi\right] = \rho\left[\frac{\partial}{\partial t} - U\frac{\partial}{\partial x}\right]\sigma(w,s,t) \tag{22.17}$$

where w is arc length along a waterline and s is arc length along a section or frame line, Figure 22.2. If the half-breadths of the ship are expressed as $b = b(x,z)$, then the pressure at frequency qZ can be deduced from

$$(p)_{qZ} = \rho\left[-iqZ\omega\,(\sigma)_{qZ}(w,s) - U\,\frac{\partial(\sigma)_{qZ}}{\partial w}\,\frac{\partial w}{\partial x}\right]e^{iqZ\gamma_0} \tag{22.18}$$

As $(\sigma)_{qZ}$, the complex amplitude, is known only numerically, the evaluation of the second term requires numerical differentiation. Alternatively, $(\sigma)_{qZ}$ can be inserted into (22.14) and the quadrature presented by $\partial\phi_B/\partial x$ on S evaluated by converting the integral to a sum.

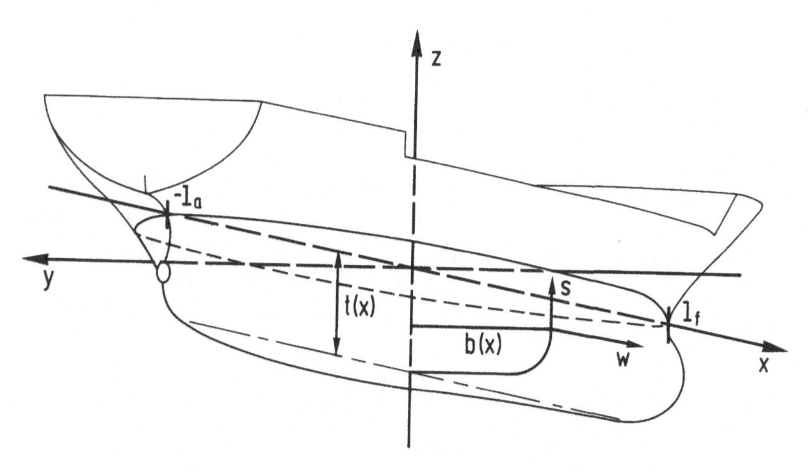

Figure 22.2 Definition of s and w on ship hull.

The potential of the hull, ϕ_B, as represented by a distribution of normal dipoles and their negative free-surface images of the wetted surface, is expressed by

$$\phi_B(x,y,z,t) = -\frac{1}{4\pi}\int_{-l_a}^{l_f} dx' \int_{-t(x')}^{0} \sigma(x',y')\frac{\partial}{\partial n'}\left[\frac{1}{R} - \frac{1}{R_i}\right]\frac{\partial w}{\partial x'}\frac{\partial s}{\partial z'}\, dz' \tag{22.19}$$

where $-l_a$ is the aft extent of waterplane, l_f is the forward extent, t(x) the depth of the keel along the centerplane; σ the as-yet unknown normal dipole density, w and s the waterline and sectional arc lengths, respectively, cf. Figure 22.2, and

$$\frac{1}{R} = \frac{1}{\sqrt{(x-x')^2 + (y-y')^2 + (z-z')^2}} = \frac{1}{R(x,y,z;x',y',z')} \tag{22.20}$$

$$\frac{1}{R_i} = \frac{1}{\sqrt{(x-x')^2 + (y-y')^2 + (z+z')^2}} = \frac{1}{R_i(x,y,z;x',y',z')} \tag{22.21}$$

It is clear that $\phi_B(x,y,0,t) \equiv 0$ ($R_i = R$ on $z = 0$). On the inner side of the hull, satisfaction of the condition $\phi_B^- = -\phi_{Pr}$ requires that σ be such that

$$\frac{\sigma(x,z)}{2} - \frac{1}{4\pi}\int_{-l_a}^{l_f} dx' \int_{-t(x')}^{0} \sigma(x',y')\frac{\partial}{\partial n'}\left[\frac{1}{R} - \frac{1}{R_i}\right]\frac{\partial w}{\partial x'}\frac{\partial s}{\partial z'}\, dz'$$
$$= -\phi_{Pr}(x,b(x,z),z) = -\left[\phi(x,b(x,z),d+z) - \phi(x,b(x,z),d-z)\right] \tag{22.22}$$

where $-\infty < z \le 0$, and y has been replaced by

$$y = \pm b(x,z) \quad ; \quad + \text{ for port;} \quad - \text{ for starboard}$$

the equation defining the hull half-breadths at each (x,z), and where ϕ_{Pr} is the potential of the propeller and its negative image in the water surface (cf. 22.1).

This integral equation is not of the form available from the previous work of Hess & Smith (1962), who developed an inversion procedure for the same kind of Fredholm equation, but having a kernel $\partial(1/R)/\partial n'$. However, it will now be shown that the problem posed by (22.22) can indeed be solved by converting to an equation having the Hess-Smith kernel.

The solution of (22.22) can be obtained in terms of dipole strengths ν required to solve the problem posed by the *double* hull in the presence of a *single* propeller as depicted schematically in Figure 22.3.

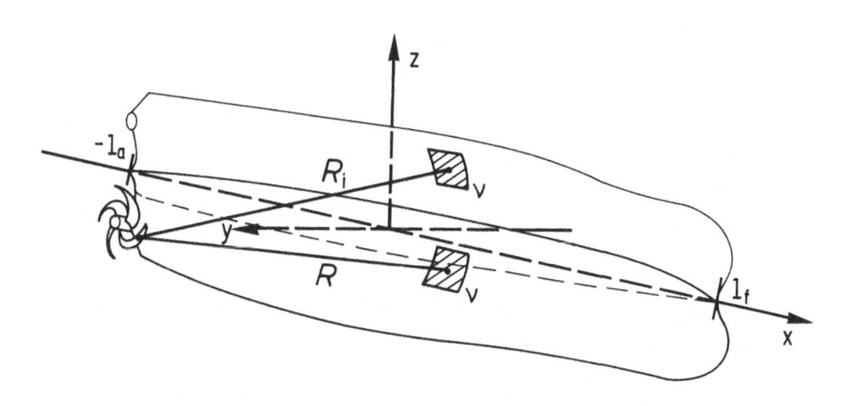

Figure 22.3 Schematic of double hull in the presence of a single propeller.

The integral equation for this problem is

$$\frac{\nu(x,z)}{2} - \frac{1}{4\pi}\int_{-l_a}^{l_f} dx' \int_{-t(x')}^{t(x')} \nu(x',z') \frac{\partial}{\partial n'}\left[\frac{1}{\sqrt{(x-x')^2+(b-y')^2+(z-z')^2}}\right]\frac{\partial w}{\partial x'}\,\frac{\partial s}{\partial z'}\,dz'$$
$$= -\phi(x,b,d+z) \qquad (22.23)$$

To achieve a right-hand side identical to that in (22.22), replace z by $-z$ throughout and subtract the result from (22.23) to secure

$$\frac{\nu(x,z)-\nu(x,-z)}{2} - \frac{1}{4\pi}\int_{-l_a}^{l_f} dx' \int_{-t(x')}^{t(x')} \nu(x',z') \frac{\partial}{\partial n'}\left[\frac{1}{R} - \frac{1}{R_i}\right]\frac{\partial w}{\partial x'}\,\frac{\partial s}{\partial z'}\,dz'$$
$$= -(\phi(x,b,d+z)-\phi(x,b,d-z)) \qquad (22.24)$$

where $1/R$ and $1/R_i$ are defined by (22.20) and (22.21).

Now write the z'-integral in symbolic form

$$\int_{-t(x')}^{t(x')} F(z')dz' = \int_{-t(x')}^{0} F(z')dz' + \int_{0}^{t(x')} F(z')dz' = \int_{-t(x')}^{0} (F(z') + F(-z'))dz'$$

As both $\mathbf{n}(x',y',z')$ and $\mathbf{V}(x',y'z')$ are odd functions of z', $\partial/\partial n' = \mathbf{n}\cdot\mathbf{V}$ is even and, as (from (22.20) and (22.21))

$$\frac{1}{R(-z')} - \frac{1}{R_i(-z')} = -\left[\frac{1}{R(z')} - \frac{1}{R_i(z')}\right]$$

then (22.24) becomes

$$\frac{\nu(x,z)-\nu(x,-z)}{2} - \frac{1}{4\pi}\int_{-l_a}^{l_f} dx' \int_{-t(x')}^{0} (\nu(x',z')-\nu(x',-z'))\frac{\partial}{\partial n'}\left[\frac{1}{R} - \frac{1}{R_i}\right]\frac{\partial w}{\partial x'}\frac{\partial s}{\partial z'} dz'$$
$$= -(\phi(x,b,d+z)-\phi(x,b,d-z)) \qquad (22.25)$$

Comparison of (22.25) and (22.22) shows that

$$\sigma(x,z) = \nu(x,z) - \nu(x,-z) \qquad (22.26)$$

Thus,

- ***the dipole strength distributed over the wetted hull and used in describing the flow over the hull in the presence of the propeller and its high-frequency image in the free surface can be obtained by considering the problem of the flow past the double hull in the presence of only the propeller.***

If a rigid-surface condition is imposed on the locus of the water surface (the zero-frequency condition as defined in Appendix F, p. 502), then the analogous result is

$$\sigma_{rs} = \nu(x,z) + \nu(x,-z) \qquad (22.27)$$

Equation (22.26) implies that tests of non-cavitating and cavitating propeller-hull configurations can be carried out using a double model well submerged in an open-to-atmosphere towing tank. Of course, for cavitation to ensue, sufficient speed must be provided to match the ship cavitation number. The speed required will be somewhat less than the ship speed.

It is important to note that, as shown above (cf. (22.16)), the dipole density ν gives, in our case, the sum of the double-body potential ϕ_B and the propeller potential taken on the ship surface, i.e.,

$$\nu(x,z) = -\left[\phi_B(x,b(x,z),z) + \phi(x,b(x,z),z)\right] \qquad (22.28)$$

and, for the hull in the joint presence of the free surface and the propeller (with its negative image), we have

$$\sigma(x,z) = -\left[\phi_B(x,b(x,z),z) - \phi_B(x,b(x,z),-z)\right]$$
$$-\left[\phi(x,b(x,z),z) - \phi(x,b(x,z),-z)\right]$$
$$= -\left[\phi_B(x,b(x,z),z) - \phi_B(x,b,(x,z),-z)\right] - \phi_{pr}(x,b(x,z),z)$$
$$(22.29)$$

Note that all potentials as well as dipole strengths also depend on time.

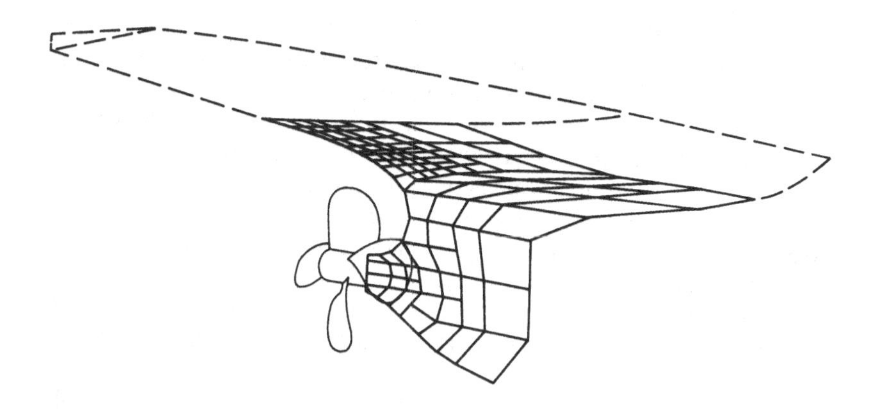

Figure 22.4 Panel isometric representation of SSPA 2148-A hull in vicinity of the propeller; only starboard side shown (dotted lines indicate extent of hull to load waterline and to transom).

Figure 3 from Breslin, J.P., Van Houten, R.J., Kerwin, J.E. & Johnsson, C.-A. (1982). Theoretical and experimental propeller-induced hull pressures arising from intermittent blade cavitation, loading and thickness. *Trans. SNAME*, vol. 90, pp. 111-51. New York, N.Y.: SNAME.

By courtesy of SNAME, USA.

To secure the numerical inversion of (22.24) using the procedure of Hess & Smith (1962), the hull is divided into quadrilateral elements as shown in Figure 22.4 and the coordinates of their corners are supplied to a modified version of the Hess-Smith program. The *right-hand* side is the single propeller potential evaluated at the hull panels. This potential must include the contributions from intermittent cavitation, blade-loading variations and blade thickness, cf. (20.16). A numerical method as applied at MIT to secure this required input propeller potential is referred to in Chapter 20 p. 403-404.

Selection of panel size is dictated by the behavior of the propeller potential in the longitudinal and athwartship directions with respect to the propeller. Forward of the propeller, all the various terms attenuate monotonically with longitudinal distance, x. The strongest loading and cavity harmonics yield spatial signatures which decay very "rapidly". For the loading the decay is as $\partial Q_{|n-qZ|-\frac{1}{2}}/\partial x$, from (15.4), with maximum loading for n = 0. For the cavity the decay is as $Q_{qZ-\frac{1}{2}}$ ($C_0^{(qZ)}$ in (20.39)). This yields, by use of (M6.6), decays as $1/x^{2qZ+2}$ and $1/x^{2qZ+1}$, respectively. The component generated by blade-rate loading on the blade falls off as $1/x^2$ (Equation (15.9)) and that from the blade-frequency harmonic of the cavity volume decays as $1/x$ (Equation (20.44)). Thus, beyond one diameter the panel lengths can be increased substantially.

Downstream of the propeller, however, the potentials have similarly attenuating components *plus* non-decaying constituents arising from the helically distributed vortex wake. The non-vanishing parts vary with axial distance as $e^{\pm iqZ\omega x/U}$ as can be seen from the velocity expression, Equations (18.46), (18.48) or (18.49) (where $k_0 = qZ\omega/U$). These spatial oscillations pass through a complete cycle for an x-variation given by $\Delta x/D = J/qZ$. When, for example, $q = 1$ and $Z = 4$ and as J is the order of 0.6, then $\Delta x = 0.15 \cdot D$ and to have four panels in each of these intervals yields a panel length of $0.0375 \cdot D$.

It is envisioned that the net effect of these "rapid" x-wise undulations will produce small contributions to forces from residuals arising from the changes in hull sections with x.

The athwartship variations of the propeller potentials are, in general, decaying, but with oscillations varying as $e^{i(n-qZ)\gamma}$ (cf. (18.46), (18.48), (18.49) for loading, and (20.40) for cavity). The space angle γ is given by $-\tan^{-1}(y/z)$, y being athwartship distance and z the vertical distance of any hull point from the propeller axis. The value $y = y_1$ at which $\cos(n-qZ)\gamma$ first changes sign for $n = 4$, $q = -1$ and $Z = 4$ is $y_1 = 0.20\, z_1$. As z_1 is of the order of 1.5 R, the buttock at which $\cos 8\gamma$ changes sign is $y_1 = 0.30$ R. To secure a minimum panel distribution over the domain wherein this onset potential changes sign, one may choose two panels of width 0.15 R. As we move out a frame line, the amplitude of the potential decreases and the panel width can be increased. For the strong local contribution arising from mean propeller loading, for which $n = 0$, and say, $qZ = -4$ (for blade-rate contribution of a four-bladed propeller), we note that the complete cycle of

$$\begin{matrix}\sin \\ \cos\end{matrix}\left\{4\tan^{-1}\left[\frac{y}{z}\right]\right\}$$

is achieved only for $y \to \pm\infty$. It is also to be noted that, for the contributions arising from the blade-rate harmonics of the cavity and the blade-rate harmonics of the blade loading given by $n = \pm qZ$, there are no variations with space angle γ so that the panels are beset with monotonically descending onsets which would permit the use of relatively large panels.

As in all panel methods, a compromise has to be struck between the estimated requirements for accuracy and the computing capacity and expense. In the application which follows, 512 panels over the double hull are used, with a distribution of panel size which is thought to be adequate.

CORRELATION OF THEORY AND MEASUREMENTS

Measurements of propeller-induced pressures on a model of a RO-RO ship have been made in the SSPA large variable-pressure water tunnel. Characteristics of this facility are summarized in Breslin, Van Houten, Kerwin & Johnsson (1982) as well as in SSPA publications. However, it is necessary to point out a highly significant feature of this facility in which the water surface is covered by marine plywood between the model and the steel work of the tunnel as indicated in Figure 22.5. The space above the plywood is virtually filled with water. The plywood is supported by

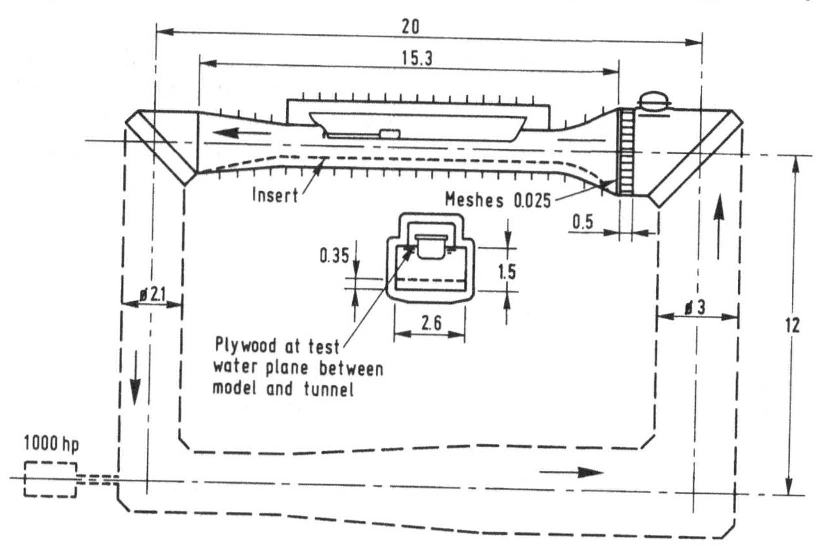

Main data for test sections of SSPA cavitation tunnel no. 2

	High Speed	Low-Speed Section	
		Original	With Insert
Length, m	2.5	9.6	9.6
Area B · H, m^2	diam. = 1 m	2.6 · 1.5	2.6 · 1.15
Maximum speed m/s	23	6.9	8.8
Minimum cavitation number	0.06[a]	1.45[b]	0.50[b]

[a] Empty tunnel; in propeller tests cavitation on right-angle gear dynamometer sets $\sigma = 0.15$ as the lower limit.
[b] At propeller shaft in a position 0.2 m below roof.

Figure 22.5 The SSPA large cavitation tunnel with low-speed test section in place. Dimensions in meters.

Figure 7 from Breslin, J.P., Van Houten, R.J., Kerwin, J.E. & Johnsson, C.-A. (1982). Theoretical and experimental propeller-induced hull pressures arising from intermittent blade cavitation, loading and thickness. *Trans. SNAME*, vol. 90, pp. 111–51. New York, N.Y.: SNAME.

vertical posts from the cap over the test section. This barrier or "wood surround" prevents the generation of large-amplitude waves since the model is tested at high speeds to achieve much larger Reynolds numbers than when modeling Froude numbers. This wood with water backing is considered by SSPA to be non-reflecting because of the acoustic properties of water-saturated wood. This aspect will be discussed later.

The afterbody sections and pressure-transducer locations are displayed in Figure 22.6. Propeller geometry is given in Figures 22.7a and 22.7b. The measured wake fractions as a function of angular position at four radii are displayed in Figure 22.8. The ship and propeller dimensions and operating conditions are summarized in Table 22.1.

When mounting the model in the tunnel, the draft aft was adjusted to level with the stern wave at a draft of 11.15 m full-scale. (Measurements conducted later in the towing tank showed that the height of the stern wave was overestimated by about 0.5 m full-scale). The cavitation numbers cited in the table were, however, used as the increase of static pressure does not fully correspond to the height of the stern wave.

Thrust and Efficiency

Open-water tests for determining the propeller characteristics were conducted in the towing tank and the corresponding characteristics in the "behind" condition at atmospheric conditions in the cavitation tunnel. At the design advance ratio, the following comparisons between calculated and measured thrust coefficients and efficiency were secured:

	Computed	Open Water	Behind Condition
Thrust coefficient	0.181	0.185	0.187
Efficiency	0.613	0.600	0.608

Cavitation Patterns

Cavitation tests in the tunnel were carried out at different combinations of advance ratio and cavitation number. Cavitation patterns were TV-filmed, from which the patterns can be sketched. Pressure fluctuations at blade and twice blade frequencies were registered. Cavitation patterns at $J_A = 0.588$ and $\sigma_{0.7R} = 0.301$ are shown in Figure 22.9 for SSPA propeller model P1841.

In Figure 22.10, amplitudes of the pressure fluctuations, measured by Gage 3, are given as curves of K_P versus $\sigma_{0.7R}$ for propeller model P1841. These results confirm earlier experience that there is very little influence of advance ratio on the pressure magnitude variation when plotted in this way. Table 22.2 (p. 441) compares amplitudes obtained at three transducers with theoretical results.

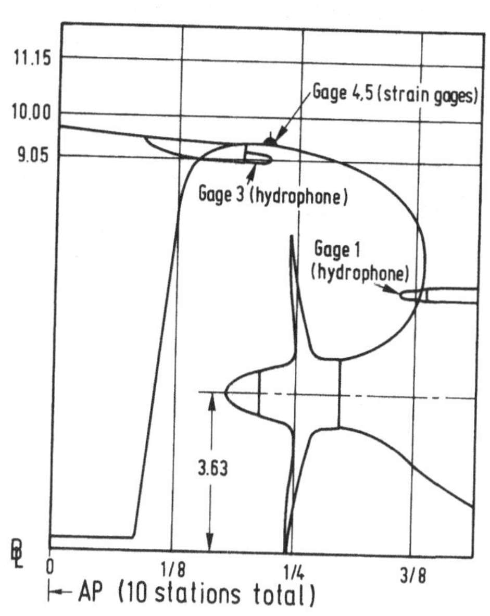

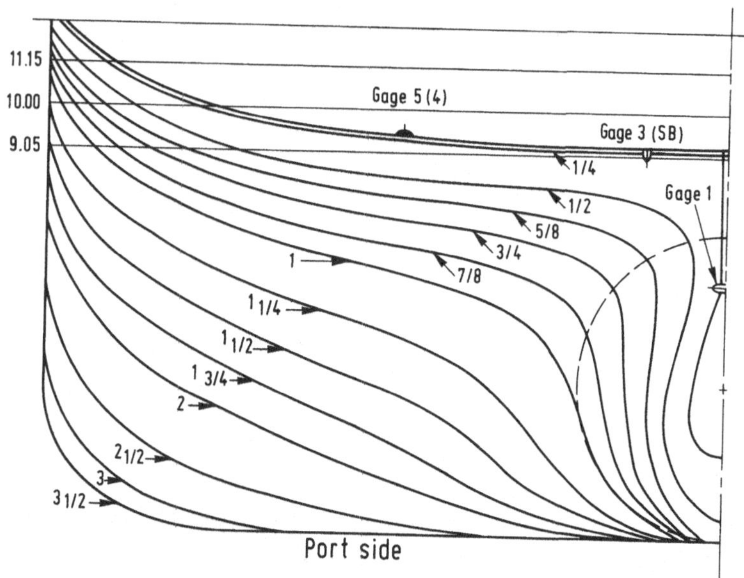

Figure 22.6 RO–RO ship model; stern shape and gage positions.

Figure 8 from Breslin, J.P., Van Houten, R.J., Kerwin, J.E. & Johnsson, C.-A. (1982). Theoretical and experimental propeller-induced hull pressures arising from intermittent blade cavitation, loading and thickness. *Trans. SNAME*, vol. 90, pp. 111–51. New York, N.Y.: SNAME.

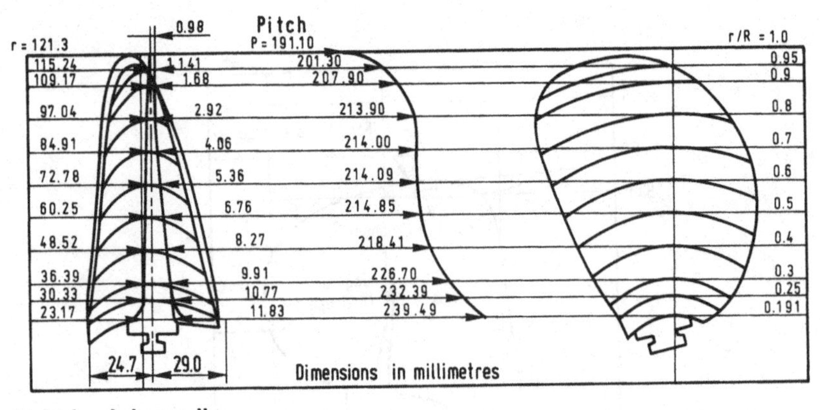

Right-handed propeller			
Number of blades	$Z = 5$	Pitch ratio at 0.7 R	$P_{0.7}/D = 0.848$
Diameter	$D = 242.6$ mm	Expanded-area ratio	$A_E/A_0 = 0.80$

Figure 22.7a SSPA Propeller Model P1841: Blade form and pitch distribution.

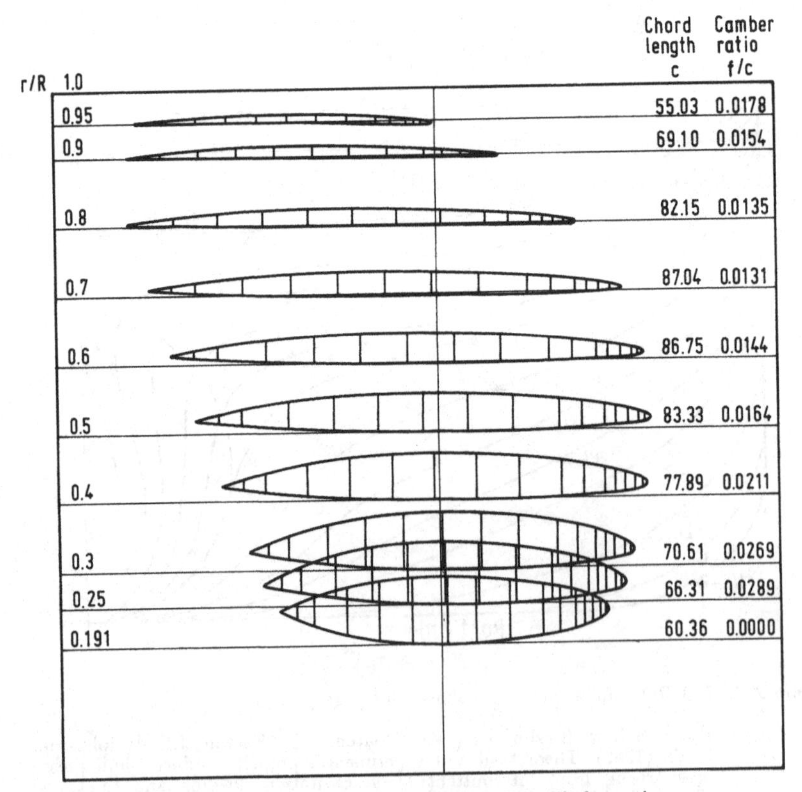

Figure 22.7b SSPA Propeller Model P1841: Blade sections.

By courtesy of SSPA, Sweden.

Ship Dimensions

Length	210.00 m
Breadth	31.16 m
Scantling draft	10.80 m (fore & aft)
Corresponding displacement	50 330 m^3
Normal draft	9.05 m (fore & aft)
Corresponding displacement	40 400 m^3

The ship stern is shown in Figure 22.6.

Propeller Operating Conditions

Ship speed	22.0 knots
Propeller speed	126 RPM
Full-scale mean effective-wake coefficient	0.290
Thrust	1560 kN
Height of waterline above propeller at zero speed (design draft 9.05 m)	5.42 m
Advance ratio	$J_A = 0.580$
Thrust coefficient	$K_T = 0.182$
Thrust-loading coefficient	$C_{Th} = 1.378$
Cavitation numbers	$\sigma_{U_A} = 4.64$
	$\sigma_N = 1.56$
	$\sigma_{0.7R} = 0.302$

Dimensions of Prototype Propeller

Diameter	6.6 m
No. of blades	5
Expanded-area ratio	0.80
Hub ratio	0.191
Blade-section mean line	NACA, **a** = 0.80
Blade thickness	NACA 16

	Pitch	Section Lengths		Camber	Thickness
r/R	P/D	h_l/D	h_t/D	f/c	t/D
0.191	0.987	0.128	0.121	0.0	0.0488
0.25	0.958	0.139	0.134	0.0290	0.0444
0.3	0.935	0.146	0.144	0.0269	0.0408
0.4	0.900	0.157	0.163	0.0211	0.0341
0.5	0.886	0.160	0.182	0.0164	0.0278
0.6	0.883	0.154	0.200	0.0144	0.0221
0.7	0.882	0.138	0.218	0.0131	0.0168
0.8	0.882	0.105	0.232	0.0135	0.0120
0.9	0.857	0.048	0.236	0.0153	0.0078
0.95	0.830	0.0003	0.226	0.0178	0.0058
1.0	0.788	−0.136	0.136	...	0.0040

The model propeller is shown in Figure 20.7.

Table 22.1 RO-RO ship dimensions, propeller dimensions and operating conditions.

From Breslin, J.P., Van Houten, R.J., Kerwin, J.E. & Johnsson, C.-A. (1982). Theoretical and experimental propeller-induced hull pressures arising from intermittent blade cavitation, loading and thickness. *Trans. SNAME*, vol. 90, pp. 111-51. New York, N.Y.: SNAME.

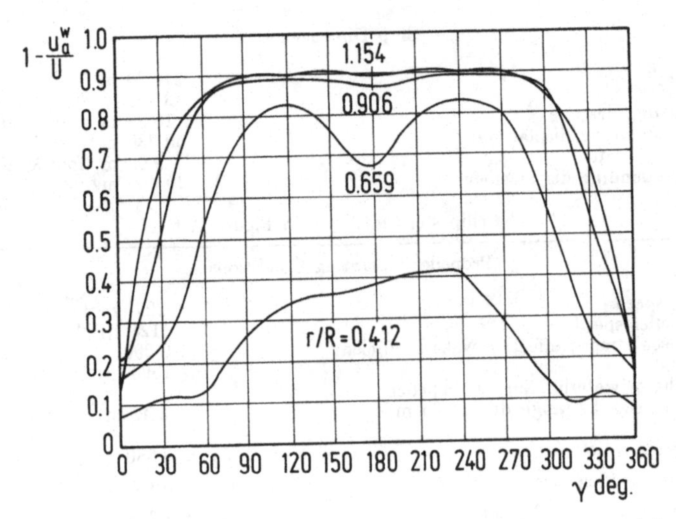

Figure 22.8 RO–RO ship model axial wake velocity distribution measured in towing tank with Prandtl tubes.

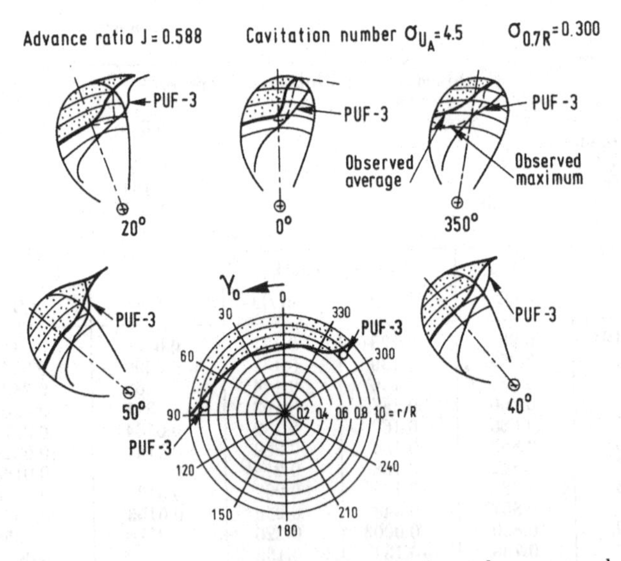

Figure 22.9 Radial extent of cavities at various blade angles, measured and computed, on SSPA propeller model P1841 operating aft of RO–RO hull. Positive blade angle is counterclockwise from 12 o'clock looking aft, i.e. positive in direction of rotation. PUF–3 indicates results calculated by MIT program for intermittently cavitating propellers.

Figures 22.8 and 22.9: Figures 12 and 13 from Breslin, J.P., Van Houten, R.J., Kerwin, J.E. & Johnsson, C.-A. (1982). Theoretical and experimental propeller-induced hull pressures arising from intermittent blade cavitation, loading and thickness. *Trans. SNAME*, vol. 90, pp. 111–51. New York, N.Y.: SNAME.

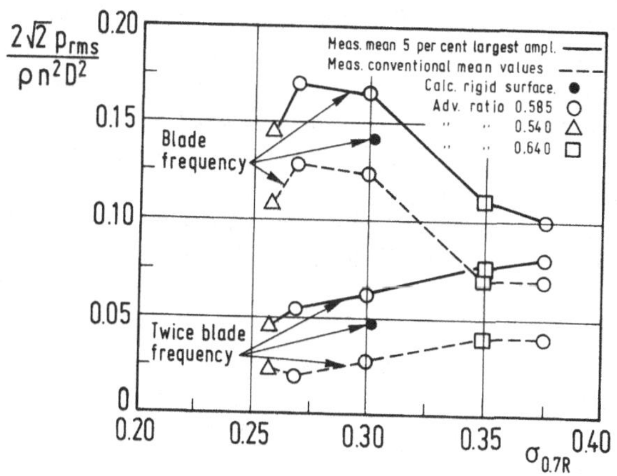

Figure 22.10 Normalized amplitudes of pressure fluctuations — Gage 3 (centerline, above SSPA propeller model P1841).

	$\lvert(K_p)_{q5}\rvert = \frac{\lvert(p)_{q5}\rvert}{\rho N^2 D^2}$					
Gage	Meas'd	Calculated		Ratio Meas'd/Calc'd		Ratio Calc'd Free Surface / Rigid Surface
		Free	Rigid	Free	Rigid	
	Blade Frequency, q = 1					
3 C.L.	0.083	0.062	0.071	1.34	1.17	0.87
4 St'bd.	0.053	0.020	0.033	2.69	1.59	0.59
5 Port	0.035	0.016	0.028	2.23	1.25	0.56
	Twice Blade Frequency, q = 2					
3 C.L.	0.033	0.021	0.024	1.52	1.33	0.87
4 St'bd.	0.012	0.008	0.012	1.60	1.00	0.63
5 Port	0.008	0.005	0.009	1.60	0.89	0.56

Note: Calculated blade angles for maximum positive pressures for Gages 3, 4, 5, are 15.5°, 16.5°, and 13.6°, respectively, at blade frequency and 1.8°, 2.5° and 1.7° at twice blade frequency. This indicates the dominance of the source behavior of the cavity. Positive blade angle is counterclockwise from 12 o'clock looking aft, i.e. positive in direction of rotation.

Table 22.2 Comparison of measured and computed dimensionless pressure coefficient amplitudes for two conditions on locus of water surface. (For gage positions, see Figure 22.6).

Figures 22.10 and Table 22.2 adapted from Figure 10 and Table 3 from Breslin, J.P., Van Houten, R.J., Kerwin, J.E. & Johnsson, C.-A. (1982). Theoretical and experimental propeller-induced hull pressures arising from intermittent blade cavitation, loading and thickness. *Trans. SNAME*, vol. 90, pp. 111–51. New York, N.Y.: SNAME.

The method used to calculate the flow from the intermittently cavitating propeller was developed at MIT. It uses a representation of the propeller blades by lattices of vortices and sources, the latter also modelling cavities. A description of this method can be found in Breslin *et al.* (*ibid.*). The corresponding computer program designated as PUF-3, was initially applied to SSPA propeller P1842 using the nominal model wake. Cavity outlines were found to correlate poorly with SSPA observations. A second calculation was made using an effective wake derived from the nominal wake in the manner described below. Again the cavity geometry did not comport with measurements. Examination of SSPA cavity outlines showed that, in the collapsing phase, the leading edge of the cavity retreats from the leading edge of the blade. As PUF-3 at that time could accommodate only cavities which extend from the locus of the leading edges of the blades, it was necessary to secure data in which this aspect of cavity behavior obtained throughout the episode. Fortunately, SSPA model P1841 performed in this fashion and the geometry of that design was used.

The nominal axial wake, as defined at four radii from SSPA measurements, shown in Figure 22.8, is interpolated by spline cubics to give the wake at 10 radii as displayed in Figure 22.11 (curve at $r/R = 0.191$ not shown). The harmonics of this nominal wake, however, are not an appropriate wake to use as input to PUF-3. It is well known that an operating propeller causes incoming vortex lines to move relative to each other, rather than merely to convect at advance velocity. This relative motion changes the corresponding wake velocities. In PUF-3, however, this nonlinear effect was not accounted for, so that the distortion of the nominal wake by the propeller must be estimated, and the resulting "effective wake" used as input to PUF-3. Unfortunately, there is no rigorous method available for the prediction of effective wakes in the general, non-axisymmetric case, and the method of Huang & Groves (1980), which predicts effective wakes in the axisymmetric case, is used as the basis for an approximate scheme, described in some detail by Van Houten (see Breslin *et al.* (1982)). This method produces the modified or effective wake curves graphed in Figure 22.12. The corresponding harmonics of the axial component are provided in Table 22.3.

The use of these effective-wake values in the PUF-3 program gave the cavity outlines labeled PUF-3 in Figure 22.9. Although the predictions of cavity extents are greater (effects of blade thickness were not accounted for) than those mean loci observed at SSPA, the discrepancies in the velocities of the cavity volumes are not nearly as great since Figure 22.13 shows that the predicted cavities are quite thin in those areas of the blade where no cavitation was observed at SSPA. Consequently, the computed pressures are not expected to be seriously affected by the presence of these excess but very thin cavities.

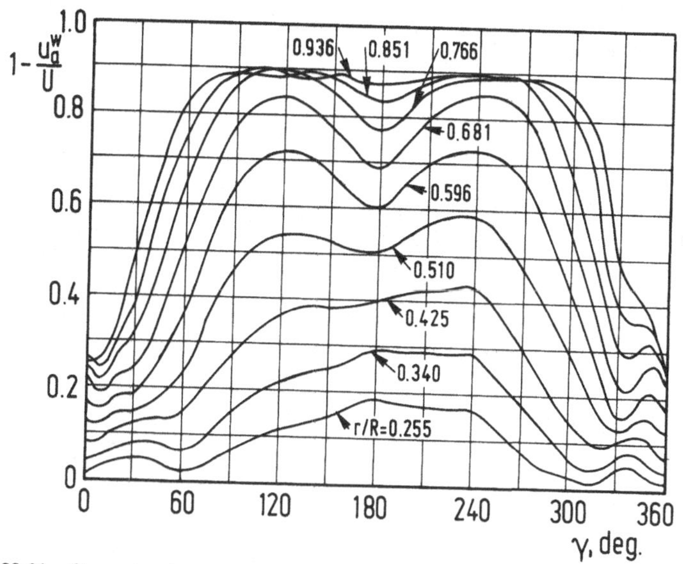

Figure 22.11 Computer-interpolated axial velocity distribution, nominal wake velocity, of RO-RO ship model.

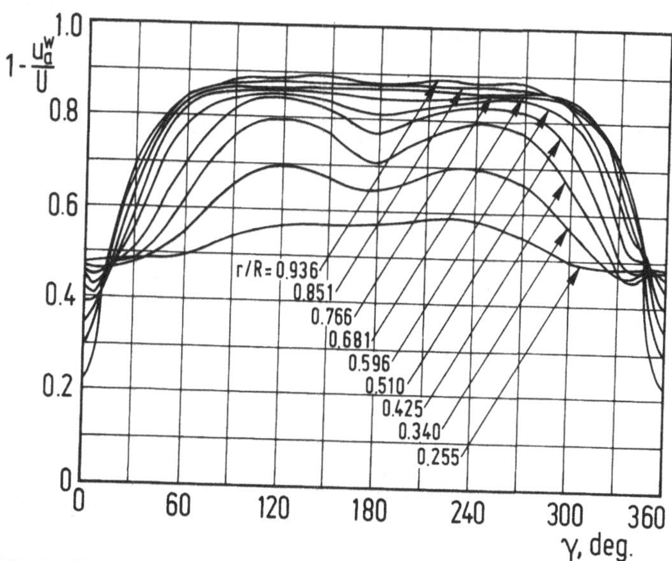

Figure 22.12 Computed axial velocity distribution, effective wake velocity, for RO-RO ship model.

Figure 22.11 and 22.12 from Breslin, J.P., Van Houten, R.J., Kerwin, J.E. & Johnsson, C.-A. (1982). Theoretical and experimental propeller-induced hull pressures arising from intermittent blade cavitation, loading and thickness. *Trans. SNAME*, vol. 90, pp. 111–51. New York, N.Y.: SNAME.

NOMINAL

Axial Wake Harmonics

r/R	A_0	A_1	A_2	A_3	A_4	A_5	A_6	A_7
0.191	−0.951	−0.046	0.014	0.008	−0.002	−0.007	−0.002	−0.003
0.300	−0.874	−0.102	0.015	0.018	−0.000	−0.011	−0.004	−0.005
0.400	−0.769	−0.162	−0.009	0.028	0.003	−0.008	−0.004	−0.004
0.500	−0.630	−0.224	−0.065	0.037	0.010	0.002	−0.003	−0.001
0.600	−0.484	−0.270	−0.132	0.035	0.015	0.015	−0.001	0.004
0.700	−0.363	−0.283	−0.174	0.011	0.009	0.020	0.003	0.008
0.800	−0.285	−0.261	−0.180	−0.034	−0.010	0.013	0.005	0.008
0.900	−0.240	−0.228	−0.166	−0.076	−0.031	−0.000	0.004	0.005
0.950	−0.226	−0.214	−0.159	−0.089	−0.039	−0.008	0.000	0.001
1.000	−0.216	−0.205	−0.152	−0.095	−0.045	−0.015	−0.005	−0.003

r/R		B_1	B_2	B_3	B_4	B_5	B_6	B_7
0.191	...	−0.006	0.009	−0.001	−0.003	−0.000	−0.002	−0.002
0.300	...	−0.012	0.014	−0.001	−0.004	−0.000	−0.003	−0.004
0.400	...	−0.017	0.015	−0.001	−0.002	−0.000	−0.004	−0.005
0.500	...	−0.021	0.008	0.000	0.004	−0.001	−0.003	−0.005
0.600	...	−0.022	−0.003	0.000	0.009	−0.001	−0.002	−0.005
0.700	...	−0.021	−0.011	−0.002	0.009	−0.002	−0.002	−0.005
0.800	...	−0.017	−0.014	−0.005	0.002	−0.004	−0.004	−0.005
0.900	...	−0.011	−0.011	−0.008	−0.005	−0.004	−0.005	−0.006
0.950	...	−0.009	−0.009	−0.007	−0.007	−0.004	−0.005	−0.005
1.000	...	−0.006	−0.007	−0.006	−0.006	−0.003	−0.004	−0.004

Tangential Wake Harmonics

r/R	A_0	A_1	A_2	A_3	A_4	A_5	A_6	A_7
0.191	0.000	0.000	0.000	0.000	0.000	0.000	0.000	0.000
0.300	−0.003	0.001	0.001	0.001	0.001	0.000	0.000	0.000
0.400	−0.005	0.002	0.002	0.002	0.002	0.001	0.000	0.000
0.500	−0.007	0.004	0.004	0.003	0.003	0.001	0.001	0.000
0.600	−0.010	0.005	0.005	0.004	0.003	0.001	0.001	0.001
0.700	−0.012	0.006	0.006	0.005	0.004	0.002	0.001	0.001
0.800	−0.009	0.004	0.003	0.003	0.003	0.001	0.001	0.001
0.900	−0.001	0.001	−0.001	−0.001	−0.001	−0.000	0.000	0.000
0.950	−0.000	0.000	−0.002	−0.002	−0.001	−0.001	0.000	0.000
1.000	−0.000	0.000	−0.002	−0.002	−0.001	−0.001	0.000	0.000

r/R		B_1	B_2	B_3	B_4	B_5	B_6	B_7
0.191	...	0.000	0.000	0.000	0.000	0.000	0.000	0.000
0.300	...	−0.024	−0.004	0.001	0.001	0.000	0.000	0.000
0.400	...	−0.046	−0.007	0.002	0.003	0.001	0.000	0.000
0.500	...	−0.068	−0.011	0.003	0.004	0.001	0.000	0.000
0.600	...	−0.090	−0.014	0.004	0.005	0.002	0.000	0.000
0.700	...	−0.112	−0.018	0.005	0.006	0.002	0.000	0.001
0.800	...	−0.123	−0.026	0.000	0.004	0.001	0.000	0.000
0.900	...	−0.128	−0.036	−0.008	−0.001	−0.000	0.000	−0.000
0.950	...	−0.129	−0.038	−0.009	−0.002	−0.001	0.000	−0.001
1.000	...	−0.129	−0.038	−0.009	−0.002	−0.001	0.000	−0.001

EFFECTIVE

Axial Wake Harmonics

r/R	A_0	A_1	A_2	A_3	A_4	A_5	A_6	A_7
0.191	−0.505	−0.003	0.001	0.000	−0.000	−0.000	−0.000	−0.000
0.300	−0.425	−0.088	−0.028	0.014	0.004	0.001	−0.001	−0.000
0.400	−0.340	−0.141	−0.078	0.014	0.007	0.009	0.000	0.003
0.500	−0.278	−0.161	−0.107	−0.009	−0.001	0.010	0.003	0.005
0.600	−0.241	−0.156	−0.113	−0.043	−0.017	0.003	0.003	−0.004
0.700	−0.221	−0.151	−0.112	−0.066	−0.030	−0.008	−0.001	−0.000
0.800	−0.211	−0.154	−0.113	−0.074	−0.040	−0.020	−0.011	−0.008
0.900	−0.206	−0.166	−0.120	−0.075	−0.048	−0.032	−0.025	−0.018
0.950	−0.204	−0.173	−0.124	−0.074	−0.052	−0.038	−0.031	−0.023
1.000	−0.203	−0.177	−0.127	−0.074	−0.055	−0.044	−0.037	−0.027

r/R		B_1	B_2	B_3	B_4	B_5	B_6	B_7
0.191	...	0.000	−0.001	0.000	0.000	0.000	0.000	0.000
0.300	...	0.008	−0.003	−0.000	−0.002	0.000	0.001	0.002
0.400	...	0.011	0.004	0.000	−0.005	0.001	0.001	0.002
0.500	...	0.011	0.008	0.002	−0.003	0.002	0.002	0.003
0.600	...	0.009	0.008	0.005	0.002	0.003	0.003	0.004
0.700	...	0.006	0.006	0.005	0.005	0.002	0.003	0.003
0.800	...	0.003	0.002	0.002	0.003	0.001	0.002	0.002
0.900	...	0.001	−0.000	−0.001	0.000	0.001	0.002	0.001
0.950	...	0.001	−0.000	−0.002	−0.001	0.002	0.003	0.002
1.000	...	0.002	0.000	−0.002	−0.001	0.003	0.004	0.003

Note: Nominal values assumed for tangential and wake harmonics.

Axial harmonics: $A_n = -\dfrac{u^c_{an}}{U}, \qquad B_n = -\dfrac{u^s_{an}}{U}$

Tangential harmonics: $A_n = +\dfrac{u^c_{tn}}{U}, \qquad B_n = +\dfrac{u^s_{tn}}{U}$

cf. Chapter 13, Equation (13.3), (13.4).

Table 22.3 Nominal axial and tangential wake harmonics; effective axial wake harmonics.

Adapted from Tables 4 and 5 from Breslin, J.P., Van Houten, R.J., Kerwin, J.E. & Johnsson, C.-A. (1982). Theoretical and experimental propeller-induced hull pressures arising from intermittent blade cavitation, loading and thickness. *Trans. SNAME*, vol. 90, pp. 111–51. New York, N.Y.: SNAME.

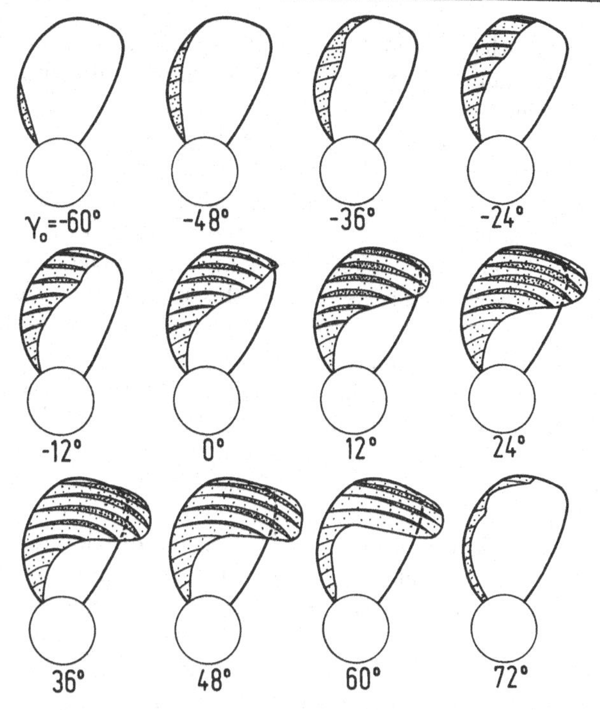

Figure 22.13 Computed PUF-3 cavity outlines and thickness distributions for SSPA propeller model P1841 on RO-RO ship model based on computer effective wake. Positive blade angle is counterclockwise from 12 o'clock looking aft, i.e. positive in direction of rotation.

Figure 17 from Breslin, J.P., Van Houten, R.J., Kerwin, J.E. & Johnsson, C.-A. (1982). Theoretical and experimental propeller-induced hull pressures arising from intermittent blade cavitation, loading and thickness. *Trans. SNAME*, vol. 90, pp. 111–51. New York, N.Y.: SNAME.

Comparison of Computed and Measured Pressures

To calculate the pressures on the hull the MIT vortex-lattice program PUF-3 was used to calculate the field potentials arising from propeller blade loading, thickness and intermittent cavitation. These potentials were then used as input to a computer program developed at Davidson Laboratory which calculates the hull potential in the presence of the propeller using the theory described here on p. 426 and sequel. The calculated pressure amplitudes are compared with SSPA determinations of the means of the upper 5 per cent largest amplitudes in Table 22.2. Computed results are provided for two conditions imposed on the locus of the undisturbed water surface, one being the high-frequency boundary condition (zero potential), the other for the rigid-surface condition (zero vertical derivative of the potential) as described previously (cf. p. 432 and sequel).

Concentrating on the 5th and 6th columns in Table 22.2, we see that the comparison is generally good for the use of the theoretical values based on the rigid-wall condition imposed on the locus of the water surface. Consistently the measured-calculated ratios are larger when the actual free-surface approximation is used which is to be expected since the calculated values are smaller for that condition due to the relieving effect of the water surface. It is most noteworthy that the measured-calculated ratios for Gage 3 (see Figure 22.6 for gage positions) are not markedly different for these two imposed conditions because this gage is very close to the propeller. This is highly consistent because the hull pressures at this point close to the propeller are dominated by the blade cavities, whereas the contributions from either of the negative or positive images of the propeller (imposed by the free and rigid conditions) are weak, being distant from Gage 3. Consistently, even at Gage 3, the calculated pressure with the rigid-surface condition is somewhat higher (15 per cent) than that with the free surface. In contrast, the relative effects of these surface conditions are marked for Gage 4 and 5 which are distant from the propeller and hence relatively close to the negative and positive images (or to the water plane). In addition, they are consistently different in the correct sense.

While this correlation suggests that the plywood surround and the presence of the laterally disposed steel work of the SSPA tunnel act as rigid, fully-reflecting boundaries it is the contention of C.-A. Johnsson and colleagues at SSPA that the plywood barrier with water above and below is non-reflecting because the product of the mass density of the wood and the speed of sound in the wood is close to that product for the water. This view can be reinforced by considering reflection and transmission of plane acoustic waves incident on such barriers backed by water, including that the thickness of the wood is very small compared to the length of the pressure waves.

Assuming this is true and ignoring the reflecting effects of the tunnel steel work, the SSPA test configuration may be considered to be approximated by the double model and a single propeller all of which is deeply submerged in a boundless sea. This is the very configuration which was treated on p. 431 and sequel.

The computed pressures are then proportional to the dipole density ν alone found from the calculations at the gage locations. The tabulated values in Table 22.2 are then proportional to $\nu(x,y,-z) - \nu(x,y,z)$ for the free-surface (high-frequency) case, cf. (22.26) and to $\nu(x,y,-z) + \nu(x,y,z)$ for the rigid surface, cf. (22.27). We may then calculate the pressures for the condition that the waterplane is non-reflecting as follows

$$k(\nu(x,y,-z) + \nu(x,y,z)) = p_{rs} \qquad \text{(rigid surface)} \qquad (22.30)$$

$$k(\nu(x,y,-z) - \nu(x,y,z)) = p_{fs} \qquad \text{(free surface)} \qquad (22.31)$$

where k is a constant and upon adding[30]

$$k\nu(x,y,-z) = \frac{p_{rs} + p_{fs}}{2} = p_{ns} \tag{22.32}$$

p_{ns} being the pressure with non-reflecting water plane.

Using the values in columns 3 and 4 in Table 22.2 we can deduce the values of p_{ns} and compare them with the measured values as displayed in Table 22.4.

	$\lvert(K_{p_{ns}})_{q5}\rvert = \frac{\lvert(p_{ns})_{q5}\rvert}{\rho N^2 D^2}$		
Gage	Measured	Calculated	Meas'd / Calc'd
	Blade Frequency, q = 1		
3 C.L.	0.083	0.066	1.25
4 St'bd.	0.053	0.027	1.98
5 Port	0.035	0.022	1.60
	Twice Blade Frequency, q = 2		
3 C.L.	0.033	0.023	1.44
4 St'bd.	0.012	0.010	1.20
5 Port	0.008	0.007	1.14

Table 22.4 Comparison of measured and computed dimensionless pressure coefficient amplitudes for the assumption that the SSPA test facility provides a non-reflecting barrier in the locus of the waterplane. For gage positions, cf. Figure 22.6.

Here we see that the comparison between measured and calculated values for a non-reflecting surface condition is not as close as for the rigid-surface assumption, as shown in Table 22.2, but is better than for the free-water-surface condition which applies to the ship and to tests in a vacuum tank. The average ratio of measured-calculated is 1.6 at blade frequency and 1.26 at twice blade frequency indicating that the theory underestimates but does quite well on this basis. Subsequent developments revealing an error in Kerwin's computer code having the effect of overpredicting of cavity volumes as well as the more recent work of Kinnas referred to in Chapter 8 cast doubt on this correlation of theory and experiment. However, the values used from the SSPA measurements being the mean of the 5 per cent highest (due to temporal variations in the model wake) are some 50 to 100 per cent higher than the average of all the

[30] This ignores the phases of these components but as these were found to be small (see note, Table 22.2) we proceed to add amplitudes.

values and were not suitable for a comparison with the theory which used spatial harmonics of the time-averaged wake as input. In any event it is believed that the most reasonable basis for evaluation of the theory is that which regards the surface about the hull in the SSPA facility to be rigid. Clearly more correlations of theory and experiment are needed.

It is of interest to observe the calculated cavity length and cavity volume on a blade as a function of blade position. Figure 22.14 shows that the maximum cavity length lags the blade angle at the wake minimum ($\gamma = 0^\circ$) by 45° at 0.851 radius and displays inception at −45° and collapse at 80° as defined there because of the finite size of the blade element at the leading edge. These angles are seen to comport very well with the observations as shown in Figure 22.9. The lag of the cavity growth is a feature of the dynamics of the cavity which is not included in quasi-steady formulations. The variation of the calculated cavity volume (normalized by cube of the propeller radius) displays, in Figure 22.15, a phase lag relative to the wake minimum of 30°.

As we have seen, it is the blade-frequency harmonics of the cavity-volume acceleration which are important and these in effect have been also rendered by Professor Kerwin in Table 22.5, Breslin *et al.* (*ibid.*), by presenting the harmonic amplitudes of the cavity-volume velocity. Here we see that the amplitude of $\partial V_c/\partial t$ at blade frequency is 0.013 (5th shaft harmonic) and that at twice blade frequency is 0.00328. From equation (20.36) we have that the pressures vary as $(qZ)^2$ times the cavity-volume harmonic. Consequently the pressures vary as (qZ) times the cavity-volume *velocity* harmonic and as this contribution is thought to dominate the pressure signature we may expect the ratio of the pressures at twice blade frequency to those at blade frequency to be

$$\frac{|(p)_{10}|}{|(p)_5|} = \frac{(2(5))}{(5)}\,\frac{(0.00328)}{(0.013)} = 0.50 \tag{22.33}$$

Hence we may conclude that the pressures at twice blade frequency will be smaller than those at blade frequency and indeed that is the case as may be seen in Table 22.2. The extent to which the measured and calculated pressures at twice blade frequency and at blade frequency do not conform to this simplistic rule may be seen in Table 22.6. It should be expected that the results for the rather distant locations at Gage 4 and 5 would conform better (to the ratio 0.50) than for Gage 3 but that is not the case for the experimental values. Yet the cited phase angles from the computed results as given in the note in Table 22.2 are seen to be virtually the same for all gage locations indicating the strong coherence expected from the approximate point-source-like behavior of the cavitation.

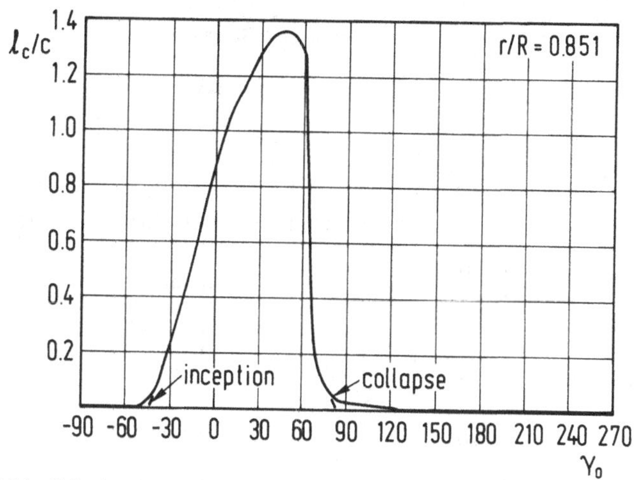

Figure 22.14 Calculated variation of cavity length with blade rotation angle at 0.851 radius for SSPA propeller model P1841 on RO–RO ship model.

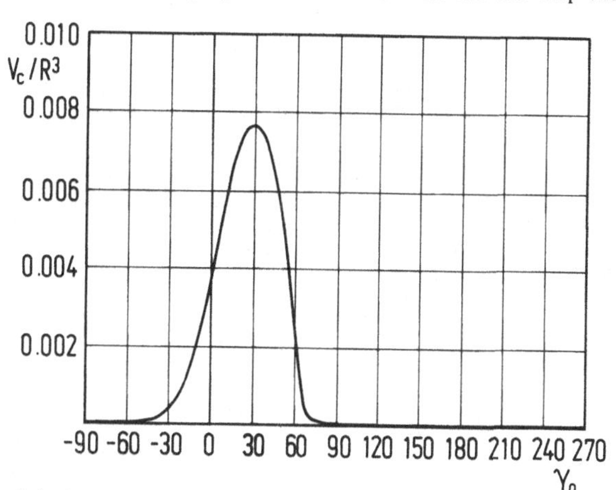

Figure 22.15 Calculated cavity volume variation with blade rotation angle for SSPA propeller model P1841 in effective wake of RO–RO ship model.

Figure 22.14 and 22.15 Figures 18 and 19 from Breslin, J.P., Van Houten, R.J., Kerwin, J.E. & Johnsson, C.-A. (1982). Theoretical and experimental propeller-induced hull pressures arising from intermittent blade cavitation, loading and thickness. *Trans. SNAME*, vol. 90, pp. 111–51. New York, N.Y.: SNAME.

In summary, this single, fairly successful correlation of calculated and measured pressures is encouraging but we must keep in mind the old saying "*one swallow does not make a summer*" (*Proverbs*, John Heywood, 1546). It is clearly necessary to make many correlations of this theory and controlled experiments before one can conclusively decide upon the applicability of this cavitation and hull-reflection theory.

Cacity–Volume Velocity Harmonics				
Shaft Harmonic Number q	A_q	B_q	Amplitude	Phase deg.
0	0.00000	0.00000	0.00000	0
1	0.00590	–0.01274	0.01404	155
2	0.01752	–0.01438	0.02267	129
3	0.02356	–0.00449	0.02399	101
4	0.01817	0.00741	0.01962	68
5	0.00602	0.01156	0.01303	28
6	–0.00332	0.00649	0.00792	333
7	–0.00442	–0.00106	0.00455	257
8	–0.00002	–0.00410	0.00410	180
9	0.00342	–0.00154	0.00375	114
10	0.00242	0.00222	0.00328	48
11	–0.00102	0.00277	0.00295	340
12	–0.00263	0.00015	0.00264	273
13	–0.00104	–0.00219	0.00242	205
14	0.00153	–0.00171	0.00229	138
15	0.00206	0.00067	0.00217	72

A_q, B_q and Amplitudes non–dimensionalized on NR^3

Table 22.5 Harmonic amplitude of cavity velocity for SSPA propeller model P1841 in effective wake of RO–RO ship model (PUF–3 Output).

Table 6 from Breslin, J.P., Van Houten, R.J., Kerwin, J.E. & Johnsson, C.–A. (1982). Theoretical and experimental propeller–induced hull pressures arising from intermittent blade cavitation, loading and thickness. *Trans. SNAME*, vol. 90, pp. 111–51. New York, N.Y.: SNAME.

	$\frac{\lvert(p)_{10}\rvert}{\lvert(p)_5\rvert}$			
Gage	Measured	Calculated		
		Free Surf.	Rigid Surf.	Non–Refl'g Surf.
3 C.L.	0.39	0.34	0.34	0.34
4 St'bd.	0.23	0.38	0.36	0.38
5 Port	0.23	0.32	0.33	0.33

Table 22.6 Ratio of pressure amplitudes at twice blade frequency to those at blade frequency for measured and calculated values.

CORRELATIONS OF THEORY AND MEASUREMENTS FOR NON-CAVITATING CONDITIONS

As demonstrated in the preceding chapters the presence of cavitation adds a high degree of complexity to the flow. This complexity is reflected not only in the theory and calculations but also in the experiments. For the simpler (yet still complicated) cases of non-cavitating propellers there are several correlations of induced forces and pressures which reflect the applicability of the theory, presented earlier in this chapter, of taking into account the presence of the hull. For the sake of completeness we give a very brief summary of these.

A number of experiments to determine the blade-frequency vertical and horizontal forces and torsional moments on two Series 60 models (C_B = 0.60, V-stern and parent form) were carried out by the late Professor Emeritus Frank M. Lewis at MIT during the 1960's (Lewis (1969)). Breslin (1981) presented a comparison of the computed vertical forces. Tsakonas & Liao (1983) using the same computer program as Breslin (*ibid.*) deduced the axial, horizontal and vertical forces.

Lateral or transverse forces on a spheroid generated by offset three-bladed propellers were measured at DTRC and the results reported by Cox, Vorus, Breslin & Rood (1978). The forces were measured at blade frequency and compared with theoretical predictions made by Vorus and by Breslin, cf. Cox *et al.* (*ibid.*). While Breslin used source panels in his representation of the spheroid (the "hull") Tsakonas & Liao (1983) employed the same panel distribution, but invested with normal dipoles.

Huse (1968) measured blade-frequency propeller forces and hull pressures at 34 points on the stern of a model of a tanker self-propelled in a towing tank. These results were compared with theoretically predicted values by Breslin (1981) who used a subdivision of the hull into 480 dipoled panels for calculations of the pressures.

In summary, the status of these available comparisons lends great credence to the usefulness of the theory and programs developed this far. Surely further detailed successful correlations must be achieved particularly between theory and full scale measurements before the highly conservative community of naval architects will adopt these methods as part of the design procedure.

SUMMARY AND CONCLUSION

As we have seen in the previous pages the propeller, when working in the confused flow behind the ship not only experiences time-varying pressure distributions on the blades but it also creates time-varying pressure distributions on the ship hull. The former pressures integrate to forces on the propeller and hence on the shaft and bearings while the latter integrate to hull forces. All these forces will, because of their time dependence, give rise to vibration and noise.

When considering the entire ship the exciting sources are propeller(s) diesel engines and sea waves. The propeller is the major source of vibration and noise which affects

- crew habitability
- fatigue of structural members and machinery parts
- navigational electronics
- gun and missile directors on naval ships
- acoustical signatures, particularly of submarines.

For deeply submerged submarines, the vibration and acoustical signatures are generated by generally non-cavitating propellers as well as by machinery.

But for many classes of surface vessels the dominant effects are those arising from intermittently cavitating propellers. While not a problem for years for most ships, the intermittent-cavitation aspect began with the growth in the size of tankers with the concurrent increases in blade loading and tip speed (larger propellers at somewhat lower RPM's) in the presence of large local wake fractions on full forms. Finer forms such as container ships, express cargo liners and naval auxiliaries also suffer from intermittent cavitation because of reduced cavitation index in combination with sharp wake "valleys".

For the designer of ships it is both natural and important to seek answers to the following questions:

- What are the dominant characteristics of propeller-hull configurations which portend serious hull sub-structure vibrations?

- Can these properties be identified from available preliminary design features at a sufficiently early stage to enable effective changes which may be expected to reduce the likelihood of objectionable vibrations?

These are indeed difficult questions since we know that the vibratory responses of hulls and sub-structures (deck houses, engine foundations, shafting etc.) are not only a function of the magnitudes of the exciting forces and moments but also of the coincidence (or near coincidence) of excitation and natural frequencies of the hull and sub-assemblies. Thus there have been cases in which the excitations have been considered to be large but the vibratory response acceptable because of fortunate mismatch of exciting and significant resonant frequencies and conversely relatively small excitations which have been attended by highly objectionable vibration and noise because of coincidence of one or more structural frequencies with multiples of blade frequency.

As the structural characteristics of a hull are generally not known early in the design stage, the various and myriad frequencies cannot be determined. Indeed, even when the complete structure has been decided it is a difficult problem to predict with accuracy the resonant frequencies in the neighborhood of the significant multiples of blade frequency. As with many features of ship design the naval architect must seek guidance from the experience gained with vessels of similar properties in regard to both the possible level of excitations and the resonant frequencies of the hull and sub-structures.

Useful guidance can be found in the studies of extensive measurements on many ships, made by Holden, Fagerjord & Frostad (1980) and by Johannessen & Skaar (1980) who have given a plan of action for vibration assessment in several stages, seeking to provide answers to the foregoing questions. These papers as well as other to be found in the literature contain a wealth of information gleaned from experience compiled over the past twenty-five years.

It is very fortunate, that the experience gained has been analyzed, as for example in the references given above, by several researchers who have endeavored to correlate excitations and responses to extract criteria for avoidance or mitigation of vibrations. But it is unfortunate, in the opinion of the authors, that the methodologies employed in some of such analyses have their origin more in physical intuition than in the rational physics of the phenomena involved. While one may question some of their methods we must applaud their aims to reduce this complex problem to simple formulae and criteria. It is one thing to criticize their most considerable efforts but quite another to replace their empirically derived results by other formulas reflecting what we may consider more rational mechanics. This would require an extensive research project which would involve many other topics than propeller hydrodynamics.

23 Propulsor Configurations for Increased Efficiency

Great increases in the cost of fuel and the advent of very large tankers and bulk carriers have focused the attention, during the last decades, on means to enhance the efficiency of ship propulsion. An obvious way of obtaining an efficiency increase is to use propellers of large diameter driven by engines at low revolutions, as can be deduced from the developments in Chapter 9. Such a solution is, however, in many cases not practically possible. This has then given impetus to the study and adoption of unconventional propulsion arrangements, consisting, in general, of static or moving surfaces in the vicinity of propellers.

A distinct indication of the serious and extensive activity in the development and use of unconventional propulsors may be seen in the report of the Propulsor Committee of the 19th ITTC (1990b) which lists seven devices including large diameter, slower turning propellers. Here we summarize the hydrodynamic characteristics of six of these devices omitting larger-diameter propellers. The six devices are: Coaxial contrarotating propellers, propeller with vane wheel, with pre-swirl stators, with post-swirl stators, ducted propellers and propellers operating behind flow-smoothing devices.

Emphasis is given in the following to a variational procedure which, in a unified fashion enables nearly optimum design of several of these configurations.

Propulsive Efficiency

Propulsive efficiency is conventionally thought of as the product of the open-water efficiency of the propulsor, the hull efficiency and a factor termed the relative-rotative efficiency. This was emphasized by Glover (1987) in his assessment of various unconventional propulsors, and can be seen from any book on ship propulsion, cf. for example Harvald (1983). The relative-rotative efficiency is considered as a correction to the propulsor (propeller) efficiency arising from the alteration in the radial distribution of loading when operating in the hull viscous and potential wake. The hull efficiency is given in terms of the wake fraction and the thrust-deduction fraction as $(1-t)/(1-w)$. Although it is necessary to consider the total propulsive efficiency the propeller designer has mainly influence on the propulsor efficiency to which we limit our considerations here.

The principal losses are those of the propeller which induces mean axial and swirl or tangential components in the race comprising a loss of kinetic energy. In addition a very significant loss is caused by the viscous drag of the blades. An indication of the relative importance of these three losses in efficiency has been provided by Glover (1987) for representative vessels with conventional propellers over a wide range of disc loadings as displayed in Table 23.1.

C_{Th}	Losses, per cent			Total Efficiency per cent
	axial	rotational	blade drag	
0.56	15.5	6.7	16.4	61.4
1.43	22.7	5.6	13.9	57.8
3.44	32.1	4.8	14.3	48.8
5.98	40.6	6.9	10.7	41.8

Table 23.1 Values of propeller efficiency and its components for a range of thrust-loading values.

From Glover, E.J. (1987). Propulsive devices for improved propulsive efficiency. In *Trans. Institute of Marine Engineers*, vol. 99, paper 31, pp. 23–9. London: The Institute of Marine Engineers.
By courtesy of the Institute of Marine Engineers, Great Britain.

It is clear that the addition of auxiliary surfaces which can reduce the propeller loading by producing thrust and by reducing the rotational losses, although these are relatively small, provide the possibility of higher efficiency than the propeller alone with several provisos. These are:

i. The thrust loading of the optimum propeller is sufficiently high.

ii. The added viscous and induced drag of the added surfaces is sufficiently less than their production of thrust.

iii. The cost of the installation can be recovered within (generally) five years.

iv. There are no or little "side effects" requiring costly maintenance or failure of reliable operation such as cavitation or vibration.

In the following we shall discuss the hydrodynamic aspects of unconventional propulsors, i.e. points i. and ii. and only touch very briefly on iv. taking into account cavitation and vibration.

A PROCEDURE FOR OPTIMUM DESIGN OF PROPULSOR CONFIGURATIONS

A unified procedure for determining the optimum distributions of loadings on interacting lifting surfaces has been developed by Kerwin, Coney & Hsin (1986). The configurations embraced are: a propeller alone; a contra-rotating set of propellers; a propeller forward of a freely rotating vane wheel; a propeller in the presence of stator arrays forward (pre-swirl) or aft (post-swirl). This is accomplished by providing these various auxiliary blades with arbitrary rotations both in magnitude and direction. The same general procedure involving application of the variational calculus has been applied to propeller-duct configurations by Kinnas & Coney (1988) and by Coney (1989) in his doctoral dissertation.

Kerwin *et al.* (1986) make use of lifting-line theory, reminding that, in spite of the latter day development of elaborate lifting-surface representations, lifting-line theory remains as an essential tool for propeller design, particularly for the determination of optimum loading distributions of the propeller and auxiliary surfaces. Kerwin's procedure is essentially a numerical vortex-lattice version of the traditional lifting-line method, as presented here in Chapters 11 and 12, with all the restrictions imposed for lifting-line representations. One important feature is the replacement of the continuous distribution of circulation along the bound vortices by discrete lattices of concentrated straight-line elements of piecewise constant strength, thereby introducing as many unknowns Γ_i as there are segments, cf. Figure 23.1. From each segment trailing vortices are convected whose strengths are determined from the difference in circulation across each segment.

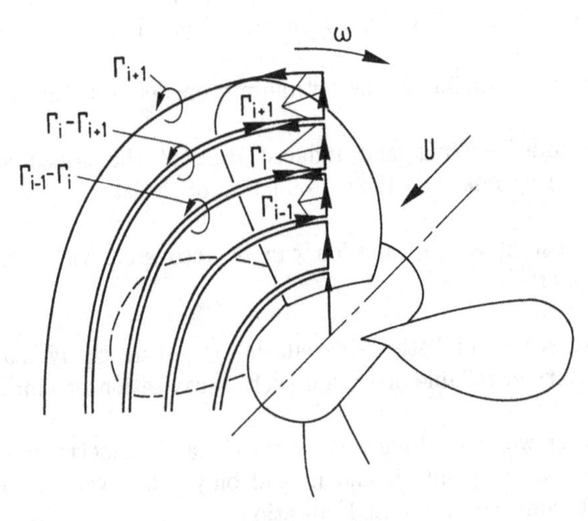

Figure 23.1 Lifting-line lattice of bound and free, trailing vortices. (Compare with Figure 11.1)

Expressions for the thrust and torque of a single propeller in the absence of other bodies are given by (11.74) and (11.75). There, $V\cos\beta_i = \omega r - u_t$ and $V\sin\beta_i = U_a - u_a$ are inserted and instead of integrals we use sums over the discrete segments. To the induced axial and tangential components we must add the corresponding induced components from the bound and trailing vortices of all the blades of the other propulsor components. We subdivide the thrust and torque into inviscid and viscid parts, with the total thrust $T = T_I + T_V$, and then have for the j-th component

$$T_I^{(j)} = \rho\, z^{(j)} \sum_{i=1}^{M^{(j)}} \left[\omega^{(j)} r_i - u_t^{(j)}(r_i) - \bar{u}_t^{(j)}(r_i)\right] \Gamma_i^{(j)}\, \Delta r_i \tag{23.1}$$

$$Q_I^{(j)} = \rho\, z^{(j)} \sum_{i=1}^{M^{(j)}} \left[U_a(r_i) - u_a^{(j)}(r_i) - \bar{u}_a^{(j)}(r_i)\right] \Gamma_i^{(j)} r_i\, \Delta r_i \tag{23.2}$$

and the viscous thrust and torque by

$$T_V^{(j)} = -\frac{1}{2}\rho z^{(j)} \sum_{i=1}^{M^{(j)}} c(r_i) C_D(r_i) \left[U_a(r_i) - u_a^{(j)}(r_i) - \bar{u}_a^{(j)}(r_i)\right] V^{(j)}(r_i)\Delta r_i \tag{23.3}$$

$$Q_V^{(j)} = \frac{1}{2}\rho z^{(j)} \sum_{i=1}^{M^{(j)}} c(r_i) C_D(r_i) \left[\omega^{(j)} r_i - u_t^{(j)}(r_i) - \bar{u}_t^{(j)}(r_i)\right] V^{(j)}(r_i) r_i \Delta r_i \tag{23.4}$$

where V is the resultant inflow velocity

$$V^{(j)} = \sqrt{\left[U_a(r_i) - u_a^{(j)}(r_i) - \bar{u}_a^{(j)}(r_i)\right]^2 + \left[\omega^{(j)} r_i - u_t^{(j)}(r_i) - \bar{u}_t^{(j)}(r_i)\right]^2}$$

and

$u_a^{(j)}(r_i)$, $u_t^{(j)}(r_i)$ are the axial and tangential velocities induced by the j-th component on itself;

$\bar{u}_a^{(j)}(r_i)$, $\bar{u}_t^{(j)}(r_i)$ are the axial and tangential velocities induced by all other components on the j-th component.

The velocities induced by the helicoidal trailers of the j-th component on itself are calculated by use of the law of Biot-Savart, as presented in Chapter 11. In analogy with Equations (11.30) or (11.34) the velocities can be expressed as products of the vortex strength and an influence function so we can write formally, summing over all trailer segments

$$u^{(j)}_{a,t}(r_i) = \sum_{k=1}^{M^{(j)}} u^{(j)}_{a,t\ k}(r_i)\ \Gamma^{(j)}_k \tag{23.5}$$

For the linear case the trailers are aligned with the onset flow $U_a(r_i)$, $\omega^{(j)}r_i$, while for the moderately loaded case it is aligned with the onset flow plus the induced velocities. This means that $u^{(j)}_{a,t\ k}(r_i)$ also depends on the induced angle β_i and hence on $\Gamma^{(j)}_k$. For the numerical calculation of the induced velocities Kerwin *et al.* (*ibid.*) use the efficient formulas by Wrench, (1957).

Similar to the above the velocities induced on the j-th component of the propulsor by all the other components is written

$$\bar{u}^{(j)}_{a,t}(r_i) = \sum_{m \neq j} \sum_{k=1}^{M^{(m)}} \bar{u}^{(m)}_{a,t\ k}(x^{(j)},r_i)\ \Gamma^{(m)}_k \tag{23.6}$$

Since in the design for optimum efficiency only the steady case is of interest, we consider the mean velocities induced on one propulsor component from the others. This corresponds to velocities induced by an actuator disc and expressions for the velocity corresponding to Equations (9.68) and (9.93) (linear disc) with proper discretization of the circulation can be used.

Application of Variational Optimization

Instead of deriving optimum criteria corresponding to Betz or Lerbs, cf. p. 227-235 a numerical version of that derivation using calculus of variations and Lagrange multipliers, but working with the unknown circulations, $\Gamma^{(j)}_i$ is used. This is now demonstrated for a two-unit propulsor system.

Here the purpose is to find discrete circulation values, $\Gamma^{(1)}_i$, $\Gamma^{(2)}_j$, $i = 1,..,M^{(1)}$, $j = 1,..,M^{(2)}$, such that the total power $P = \omega^{(1)}Q^{(1)} + \omega^{(2)}Q^{(2)}$ absorbed by the propulsor is minimized subject to the constraints that $T^{(1)} + T^{(2)} = T_r$, the required thrust and that the torque $Q^{(2)} = qQ^{(1)}$, with the ratio q specified. We then have a function to be minimized under two constraints and when using the calculus of variations, as outlined in Section 7 of the Mathematical Compendium, p. 522-523, we must have two Lagrange multipliers. The auxiliary function H, (M7.11) is then

$$H = \omega^{(1)}Q^{(1)} + \omega^{(2)}Q^{(2)} - \ell_T\left[T_r - T^{(1)} - T^{(2)}\right] - \ell_Q\left[qQ^{(1)} - Q^{(2)}\right] \tag{23.7}$$

The optimization requirement imposes that the partial derivative of H with respect to each of the unknown circulation strengths must vanish (an extension of (M7.10) to more variables). The following set of equations is obtained

$$\frac{\partial H}{\partial \Gamma_i^{(j)}} = \left[\omega^{(1)} - q\ell_Q\right]\frac{\partial Q^{(1)}}{\partial \Gamma_i^{(j)}} + \left[\omega^{(2)} + \ell_Q\right]\frac{\partial Q^{(2)}}{\partial \Gamma_i^{(j)}} + \ell_T\left[\frac{\partial T^{(1)}}{\partial \Gamma_i^{(j)}} + \frac{\partial T^{(2)}}{\partial \Gamma_i^{(j)}}\right] = 0$$

$$\text{for } j = 1,2 \text{ and } i = 1, 2, .., M^{(j)} \qquad (23.8)$$

We must also use the partial-derivative condition with respect to ℓ_T and ℓ_Q (which was implicitly used in the similar but simpler optimizations carried out in Chapters 9 and 12). This gives

$$T^{(1)} + T^{(2)} - T_r = 0 \qquad (23.9)$$

$$qQ^{(1)} - Q^{(2)} = 0 \qquad (23.10)$$

The derivatives in (23.8) are obtained by use of (23.1)-(23.6). We then have for the derivative of the inviscid thrust of component number 1 with respect to its circulation number i, (23.1)

$$\frac{\partial T_I^{(1)}}{\partial \Gamma_i^{(1)}} = \rho\, z^{(1)}\left[\left[\omega^{(1)} r_i - \sum_{k=1}^{M^{(1)}} u_{t\,k}^{(1)}(r_i)\, \Gamma_k^{(1)} - \sum_{k=1}^{M^{(2)}} \bar{u}_{t\,k}^{(2)}(x^{(1)}, r_i)\, \Gamma_k^{(2)}\right]\Delta r_i \right.$$

$$\left. - \sum_{k=1}^{M^{(1)}} u_{t\,i}^{(1)}\left[r_k\right]\, \Gamma_k^{(1)}\Delta r_k\right] \qquad (23.11)$$

The other components will have similar expressions. It is now seen that Equations (23.8)-(23.10) are a non-linear system of $M^{(1)}+M^{(2)}+2$ equations for the $M^{(1)}$ unknown circulations on propulsor component 1 and for $M^{(2)}$ unknowns on component 2, plus the two unknown Lagrange multipliers ℓ_T and ℓ_Q. This set of equations can be replaced by a linear set by regarding the Lagrange multipliers as known constants where they multiply the unknown circulations and in addition by assuming the induced velocities in the constraint equations to be known. The equations are then solved iteratively, using the ℓ's from the prior iteration as well as the induced velocities in (23.5) and (23.6). Coney (1989) reports that initial estimates of $\ell_Q = 0$; $\ell_T = -1$ and zero for the induced velocities in (23.9) and (23.10) were suitable to get the iterative process started. Generally, less than 10 iterations were required to get converged solution to the set of non-linear equations.

The trajectories of the propeller shed vortices are frozen until convergence in the iteration over the ℓ's is obtained. Then, by evaluating the hydrodynamic pitch angle β_i at each control segment, new trajectories are used and a new iteration carried out. Convergence tests showed that 10 segments on the key blade of each of the composed propellers were sufficient. Interaction velocities induced by propeller trailing vortices in the downstream region are calculated from the trailers set at 0.80 of the radius from which they are shed at the propeller blade as an allowance for contraction. For points upstream of the propeller the uncontracted radii are used.

Kerwin *et al.* (1986) compared their variational procedure for optimum distributions of loading with those found by use of Lerbs' method. The two methods were found to give nearly identical distributions of circulation over the lifting line, confirming that for light loading their procedure gives the correct optimum.

Optimized Loadings for a Single Propeller

A further exposition of results obtained using Kerwin's optimization procedure will be given here. The parametric dependence of optimum loading upon advance ratio J for a free-running propeller at fixed speed U (and hence for different rates of revolution) as determined by Coney (1989) is displayed in Figure 23.2. At each J the propeller was required to deliver the same thrust and hence the same thrust loading coefficient. The result for J = 0.10 (largest rate of revolution) is reminiscent of that of the elliptic loading on a wing which is known to give minimum induced drag. As J increases, N decreases; the circulation increases (to develop the thrust) and the peak value moves outward because added circulation at outer radii gives more effective increments to thrust. However, higher loading "outboard" gives increasing torque as may be seen in Equation (23.2) and hence an increasing penalty for thrust increments at outer radii. These conflicting effects are accounted for in the optimization procedure yielding the distributions shown for each advance ratio. The results shown are in agreement with the findings for optimum loadings for the linear actuator disc, as given in Equation (9.106). There a small advance ratio ($J = U/(ND) = U/\omega/(2\pi D)$) gives nearly constant loading over the radius whereas high advance ratios give loadings increasing at the tip.

The efficiency without viscous losses for this propeller is shown in Figure 23.3. It is seen that it has the same behaviour as for the actuator disc, cf. Figure 9.16, p. 193 (given C_{Th}, varying J), and that the efficiency approaches the ideal efficiency for $J \rightarrow 0$ as it should. Inclusion of frictional drag produces the large degradation at low advance ratios because of the high rate of revolutions giving high tip velocities. We see an optimum advance ratio as the parasitic drag decreases and hence there is an optimum combination, with respect to efficiency, of thrust loading and advance ratio.

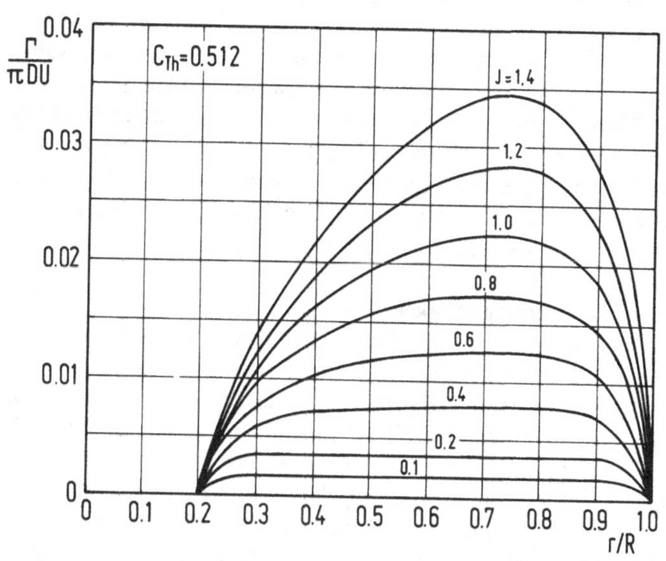

Figure 23.2 Optimum radial circulation distributions for a free–running hubless propeller for varying values of advance ratio. Each circulation distribution was required to generate the same value of thrust loading C_{Th} = 0.512. Viscous effects are ignored. The propeller is 5–bladed.

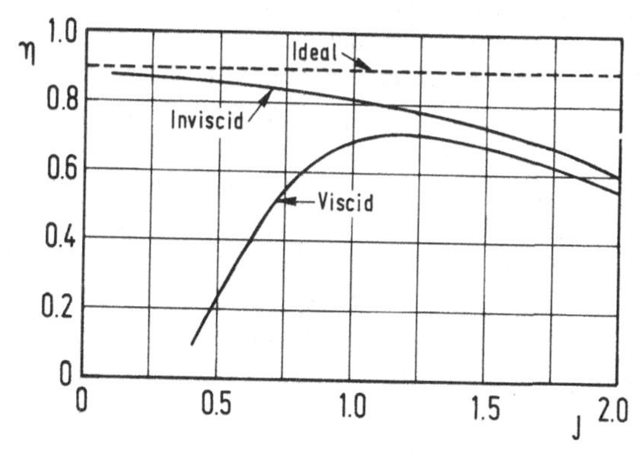

Figure 23.3 Efficiencies of 5–bladed, free–running propeller for inviscid and viscid conditions; operating specification as given under Figure 23.2.

Figure 23.2 and 23.3 from Coney, W.B. (1989). *A Method for the Design of a Class of Optimum Marine Propulsors.* Ph.D. thesis. Cambridge, Mass.: Dept. of Ocean Eng., MIT.

OPTIMIZED LOADINGS ON COMPOUND PROPULSOR CONFIGURATIONS

The optimization procedure outlined in the previous section is suitable for optimization of propulsor configurations consisting of a propeller and interacting lifting surfaces. The flow must be hydrodynamically of the same character as for a single propeller, i.e., it is due to loading and thickness of lifting surfaces and can be basically modeled as seen previously by combinations of sources and vortices. The fundamental characteristics of such propulsors are described next.

Contrarotating Propellers

This concept appears to date back to the Swedish pioneer naval architect John Ericsson who applied contrarotating propellers to shallow-draught craft to overcome directional instability arising from unbalanced forces produced by single propellers. Ericsson patented his design in 1836 and applied it to the steam boat *Robert F. Stockton* in 1839. The principal application of contrarotating propellers has been to torpedoes since the invention of Whitehead in 1864. In the torpedo, cancellation of the torque is essential to prevent spinning and to secure directional stability. Until very recently there has been no application to large surface ships because of the cost and complication of the required epicyclic gearing. A single application to submarines in the case of the USS *Jack* in 1965 is recalled by its designer J.G. Hill, in his discussion of a paper by Cox & Reed (1988) on contrarotating propellers.

Over the past several decades model tests and computer evaluated theories have yielded predicted power reductions in the range of 6 to 15 per cent for application of contrarotating propellers to surface ships. The reasons for these reductions appear to be varying. Glover (1987) recalls that model tests in Britain in the late 1960's showed a gain of 10.9 per cent in open water propeller efficiency and of 11 per cent in propulsive efficiency. But for a 200 000 dwt tanker the propeller efficiency increased by only 3.5 per cent and the propulsive efficiency decreased by 2.5 per cent because of a very large increase in thrust deduction and a large decrease in relative rotative efficiency! In contrast, tests in Japan displayed gains of 7 and 12 per cent in propeller and propulsive efficiency, respectively, for the case of a 97 000 dwt tanker and model predictions cited by Dashnaw, Forrest, Hadler & Swensson (1980) gave power reductions as much as 12 per cent for a high-speed RO-RO ship with contrarotating propellers. Model test results reported by Hadler (1969) showed reductions of about 10 per cent.

Two epochal applications of contrarotating propellers to full-scale ships were made in Japan in 1988. A 37 000 dwt bulk carrier *Juno* was retrofitted with a contrarotating propeller system. Details are given by Nishiyama *et al.* (1990). The other application was made to the existing car carrier *Toyofuji No. 5*, Japan Ship Exporters' Association (1991). Test results show reduction of power as compared to the performances of the conventional propellers originally fitted to these ships of 13.5 to 15 per cent. Mechanical inspections after a year's operation revealed no mechanical problems. Should prolonged operation of these contrarotating propeller systems prove their reliability, these installations must be regarded as a marine-engineering break-through.

The reader who desires detailed information on design methods for contrarotating propellers up to the mid 1980's is advised to study the review and extensions of theory by Cox & Reed (1988) and the extensive literature to which they refer.

We end this section on contrarotating propellers by showing an application of the optimum circulation procedure by Kerwin *et al.* (1986), outlined on p. 456–460. The propellers were of the same diameter with equal and opposite torques, hence $q = -1$, $\omega^{(1)} = -\omega^{(2)}$, p. 458–459. The total thrust was the same as for the comparative single propeller, yielding the low thrust loading $C_{Th} = 0.69$. Distributions of optimum circulation for each of the pair of 5-bladed contrarotating propellers are shown in Figure 23.4. The increase of 5.55 per cent in efficiency over the single propeller is produced by the increase in ideal efficiency of each propeller because of decreased loadings and the near cancellation of the swirl of the forward propeller by the after propeller which is also benefited by the axial flow from the forward wheel.

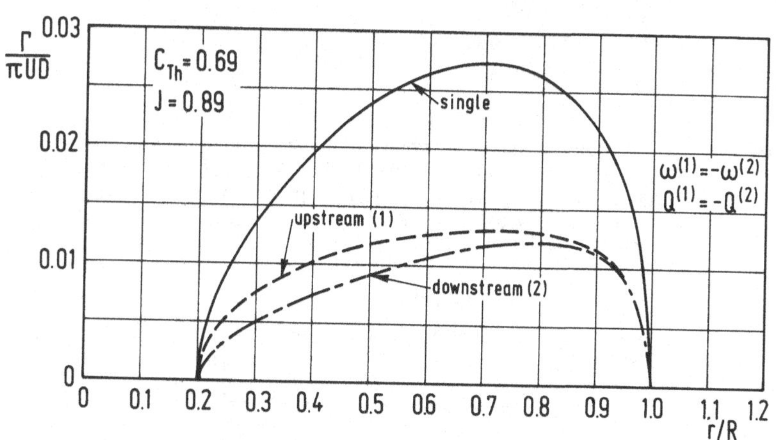

Figure 23.4 Optimum distribution of circulation on single propeller and on forward and aft propellers of a contrarotating pair.

From Kerwin, J.E., Coney, W.B. & Hsin, C.-Y. (1986). Optimum circulation distributions for single and multi-component propulsors. In *Proc. Twenty-First American Towing Tank Conference (ATTC)*, ed. R.F. Messalle, pp. 53–62. Washington, D.C.: National Academy Press. By courtesy of ATTC, USA.

Figure 23.5 Propeller and Grim vane wheel. Building yard: Bremer Vulkan, Germany. Propeller: Diameter: 6.100 m; 4 blades; $A_E/A_0 = 0.655$; weight: 18 900 kg; material: Alcunic 8; power: 12 180 kW at 117 RPM. Vane wheel: Diameter 7.400 m; 9 blades, material: Alcunic 8. Manufacturer: Ostermann Metallwerke, Germany.
By courtesy of Ostermann Metallwerke, Germany.

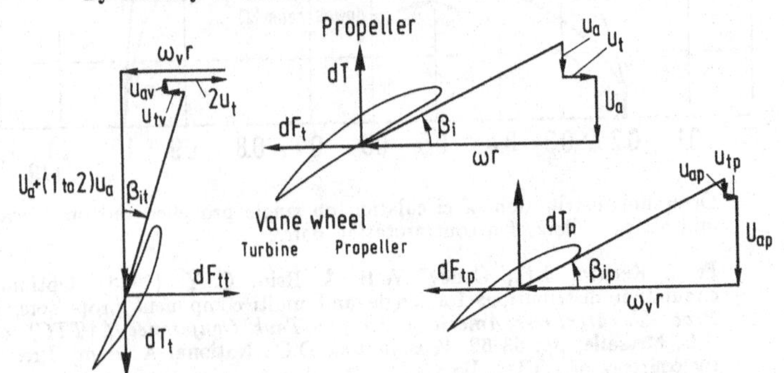

Figure 23.6 Velocity diagrams of propeller and vane wheel.

Propeller-Vane-Wheel Propulsors

A vane wheel is composed of a set of equally-spaced narrow blades mounted on a hub which freely rotates around the propeller shaft which is extended abaft the propeller, cf. Figure 23.5. This device, invented by Professor Dr. Otto Grim (1966,1980) is characterized by a diameter of some 25 per cent greater than that of the propeller, with pitch distributions such that the region in the propeller race performs as a turbine and in the portion outside the race as a propeller, as can be seen in Figure 23.6.

It is clear that the vane wheel which rotates in the same direction as the propeller at an angular velocity in the range of 35 to 50 per cent of that of the propeller makes use of both the axial and the tangential momentum losses of the propeller.

The propeller and vane wheel has a larger diameter than the propeller alone, entraining a greater volume of water and hence an increment in efficiency as dictated by momentum considerations. A similar increment can, of course, be secured by using a single propeller of the same, larger diameter, but as Blaurock (1990) points out, this would require lower shaft speed and in addition, increased vibratory excitation attending reduced tip clearance. In contrast, the lightly loaded vanes can extend much nearer to the hull without producing significant vibratory forces.

The principal parameter indicative of the gain in propulsor efficiency is the thrust-loading coefficient C_{Th}, the gain rising with increasing C_{Th}. The minimum C_{Th} for a zero gain in efficiency is 0.50 according to Blaurock (*ibid.*). In practice the minimum C_{Th} must be greater than 1.0 to obtain a reasonably short pay-back period of the installation costs.

It is interesting to note that the first model tests carried out with the vane-wheel design displayed much lower performance than predicted by Grim's calculations. This was later understood to be caused by the low Reynolds numbers at which open-water and particularly self-propulsion tests are normally conducted. Later experiments by Blaurock & Jacob (1987) using large models were conducted in the cavitation tunnel at Hamburgische Schiffbau-Versuchsanstalt (HSVA) over a range of Reynolds numbers. The results displayed in Figure 23.7 show the dominating effect of Reynolds number on the gain in efficiency over a wide range in thrust-loading coefficients.

Model tests at HSVA have since then been made without the vane wheel. The full-scale prediction of the effect of the wheel is then made by use of Grim's theory which has been verified by many sea trials. These measurements have shown gains in propulsive efficiency in the range of 7 to 10 per cent depending upon C_{Th} as summarized in Figure 23.7.

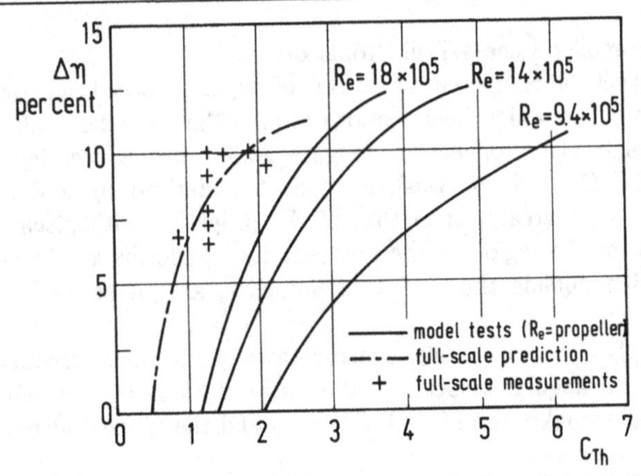

Figure 23.7 Gain in efficiency by means of a vane wheel predicted by model tests and achieved in full scale.

Figure 6 and 25 from Blaurock, J. (1990). An appraisal of unconventional aftbody configurations and propulsion devices. *Marine Technology*, vol. 27, no. 6, 325–36.

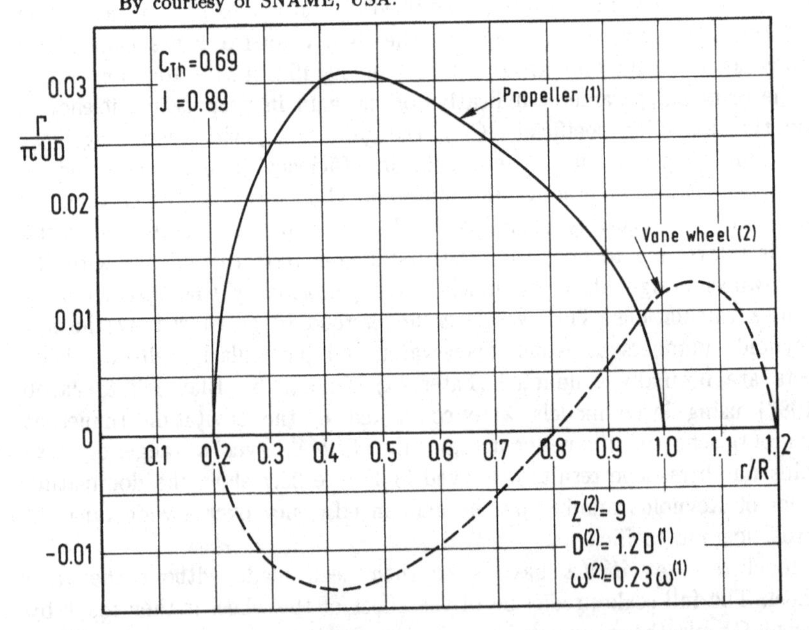

Figure 23.8 Optimum circulation distribution for propeller and vane–wheel combination.

From Kerwin, J.E., Coney, W.B. & Hsin, C.–Y. (1986). Optimum circulation distributions for single and multi–component propulsors. In *Proc. Twenty-First American Towing Tank Conference (ATTC)*, ed. R.F. Messalle, pp. 53–62. Washington, D.C.: National Academy Press.

Blaurock relates that measurements of fluctuating pressures on sister hulls in way of the propeller without and with vane wheel showed amplitudes with the wheel to be halved at blade frequency and some 30 per cent less at twice blade frequency. This is ascribed to reduced propeller cavitation attending the reduction in propeller loading.

In application of the unified variational procedure by Kerwin *et al.* (1986), described here on p. 456–460, to determine optimum propeller and vane-wheel loading distributions this propulsor is analogous to a pair of tandem propellers of unequal diameter with the aft unit rotating at a rate of revolutions at a fraction of that of the forward element and as the torque on the vane wheel is zero, then $Q^{(2)} = 0$, giving the constraint parameter $q = 0$. The optimum radial distributions of circulations for a propeller and vane wheel are given in Figure 23.8. The operation conditions are the same as shown previously for the single and contrarotating propeller, p. 463. It is seen that the propeller component is more lightly loaded, near the tip, relative to the single propeller.

Propellers with Stators

Arrangements consisting of an array of fixed blades forward of a propeller are referred to as pre-swirl stators, and as post-swirl stators for arrays fitted downstream. The basic purpose of pre-swirl stators is to produce a swirling flow opposed to the direction of rotation of the propeller, thereby annulling the swirl induced by the propeller and at the same time increasing the relative tangential velocity of the propeller blades.

These combinations are special cases of two-component propulsors and optimization can be carried out using the method of Kerwin *et al.* (1986) as outlined here on p. 456–460. The optimization procedure is greatly simplified since there is no rotation and no power applied to the stators and the ratio of torques of propeller and stator is not constrained. The optimum radial distributions for a propeller operating behind a pre-swirl stator are shown in Figure 23.9. The operating conditions are the same as for the single propeller and contrarotating propellers for which results are shown in Figure 23.4. It is seen that the circulation is distributed more uniformly yielding higher efficiency in combination with partial cancellation of swirl downstream resulting in reduced rotational losses. Note also that the stator circulation is opposite to that of the propeller. The increase in efficiency was 3 per cent over that of the single propeller. This gain is relatively small due to the light loading of the propulsor.

The number of equally-spaced stator blades is never taken equal to the number of propeller blades since this would give rise to large vibratory shaft thrust and torque. The number of stator blades should be chosen so that the least common multiple of the number of propeller blades and stator blades is large; for example 5 and 9. Also the diameter of the stator array should be about 15 per cent greater than the propeller to avoid impingement of stator-blade tip vortices on the propeller blades.

We shall finally mention the potential for reduction of unsteady blade cavitation and hence vibratory hull forces by the dual use of pre-swirl stators for enhancement of efficiency and reduction of wake harmonics. This has been demonstrated by Coney (1989) who used non-uniformly spaced stator blades to reduce the first-harmonic variation in tangential velocity in the case of an inclined shaft.

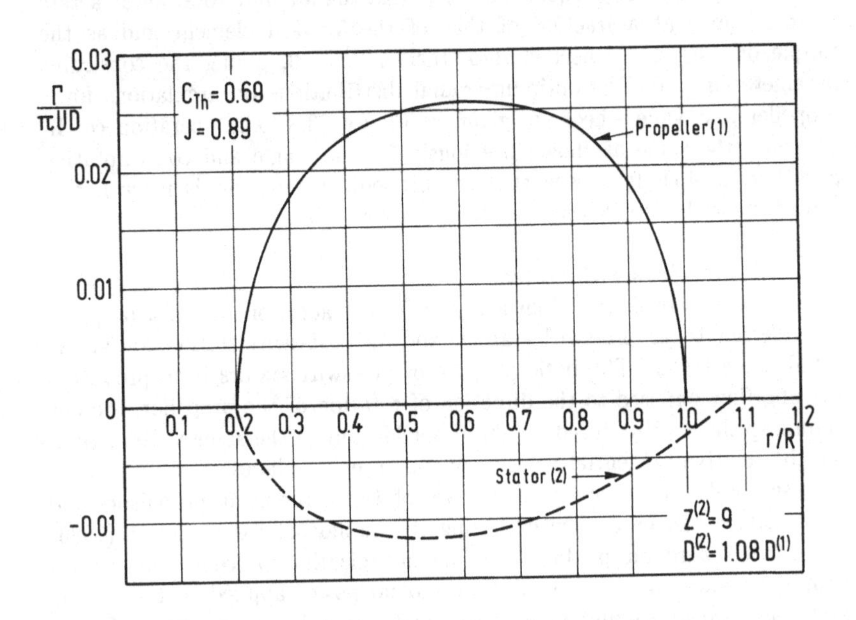

Figure 23.9 Optimum circulation distribution for propeller behind pre-swirl stator.

From Kerwin, J.E., Coney, W.B. & Hsin, C.-Y. (1986). Optimum circulation distributions for single and multi-component propulsors. In *Proc. Twenty-first American Towing Tank Conference (ATTC)*, ed. R.F. Messalle, pp. 53–62. Washington, D.C.: National Academy Press. By courtesy of ATTC, USA.

Ducted Propellers

The concept of a propeller within a nozzle or duct was developed experimentally by Stipa (1931) and by Kort (1934). Their results showed that very significant increases of efficiency could be achieved for heavily loaded propellers. Ducted propellers have been widely used since the 1930's on tugs, push-boats, trawlers and torpedoes. Later they have been applied to large tankers and bulk carriers.

Figure 23.10 Propeller and duct propulsor of multipurpose anchor–handling tug/supply vessel, owned by A.P. Møller, Denmark. Two propulsion units; controllable pitch propeller; diameter: 4.00 m; power per shaft, appr.: 4400 kW at 99/141 RPM; bollard pull, maximum, appr.: 1500 kN. Manufacturer: KaMeWa.
By courtesy of A.P. Møller, Denmark.

The design of propellers and ducts as interacting bodies was for many years based on extensive experimentation with models at MARIN, as summarized by van Manen (1957a) and by van Manen & Oosterveld (1966) and also at SSPA, cf. Dyne (1973). Extensive reviews have been given by Sachs & Burnell (1962) and by Weissinger & Maass (1968).

Theoretical analyses of increasing sophistication have been advanced by Tachmindji (1958), Morgan (1962), Morgan & Caster (1968), Ryan & Glover (1972), Tsakonas & Jacobs (1978), Falcão de Campos (1983), Van Houten (1986), Kerwin, Kinnas, Lee & Wei-Zen (1987), Coney (1989), Kinnas & Coney (1988) and Kinnas, Hsin & Keenan (1990). For light loading, Sparenberg (1969) showed that the representation of a propeller by an actuator disc in a duct in axisymmetric flow yields the efficiency of the actuator disc alone regardless of the duct shape. Any effect of acceleration or deceleration of the flow by the duct is of second order. This is because the duct is beset by only axially symmetric radial flow induced by the disc and hence has peripherally uniform loading. Consequently there is no induced drag.

Earlier momentum analysis by Küchemann & Weber (1953) for ducted propellers showed that the ideal efficiency of a propulsor composed of a uniformly loaded actuator disc in a duct is given by

$$\eta^{d}_{ideal} = \frac{2}{1 + \sqrt{1 + \tau_d C_{Th}}} \tag{23.12}$$

and the additional axial velocity at the disc induced by the duct is

$$\frac{u^{d}_{a}}{U} = \frac{\tau_d - 1}{\tau_d}\left(1 + \sqrt{1 + \tau_d C_{Th}}\right) \tag{23.13}$$

where

$$\tau_d = \frac{T_P}{T_P + T_d}$$

T_P is the propeller thrust;

T_d is the thrust of the duct resulting from the induction of the actuator disc;

$T = T_P + T_d$, the total thrust of the propulsor, upon which C_{Th} is based.

This result is a specialization of that which can be obtained from momentum analysis for actuator discs as given in many texts, cf. for example Harvald (1983). For $\tau_d = 1$ ($T_d = 0$) (23.12) is the same as the ideal efficiency for a disc, Equation (9.108), and from the results there (Chapter 9) it can be deduced that (23.12) in general is approximate since non-linear effects and rotation of the propeller slipstream are not taken into account although these effects are of importance in typical duct applications.

For ducts developing positive thrust, $\tau_d < 1$, and for the same required C_{Th}, the ducted propeller yields a higher ideal efficiency than the propeller alone for which $\tau_d = 1$. Equation (23.13) shows increased (more negative) velocity for $\tau_d < 1$. This is the case of the so-called accelerating duct shown in Figure 23.11 in which an increase of flux of fluid mass flow is induced by the inward-directed loading on the duct and its attendant vorticity. Conversely, the decelerating duct shown in Figure 23.11 yields negative T_d and hence $\tau_d > 1$ with consequent lower efficiency than an open propeller at the same C_{Th}. This reduces the mass flux of water while raising the pressure at the rotor. Accelerating ducts, often called Kort nozzles, are used on slow-moving vessels with large C_{Th} such as tugs, push-boats, trawlers and large tankers and bulk carriers. Decelerating ducts are employed on fast bodies such as torpedoes and some naval combatants to inhibit cavitation because of higher pressure at the rotor. They are often referred to as pumpjets.

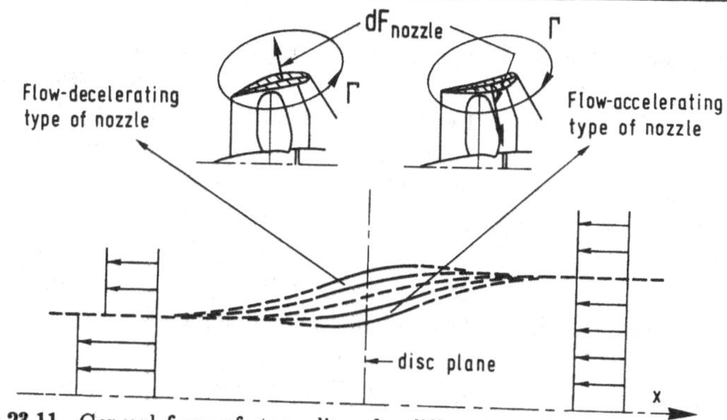

Figure 23.11 General form of streamlines for different nozzle types

From van Manen, J.D. & Oosterveld, M.W.C. (1966). Analysis of ducted-propeller-design. *Trans. SNAME*, vol. 74, pp. 522–62. New York, N.Y.: SNAME.

All propulsors develop thrust by imparting axial acceleration to a mass of water. The highest efficiency is secured by absorbing the largest mass flow and imparting the smallest change in velocity while keeping the product of the mass flux and acceleration constant. An accelerating duct acts to increase the mass flow and the rotor is the agent which produces the boost in velocity to its ultimate value in the race downstream.

A more realistic estimate of the efficiency than given previously is obtained by accounting for the profile drag force of the duct which typically has relatively large surface area and maximum thickness-chord ratio. A reasonable approximation is to multiply the ideal efficiency by a factor

$$\eta_f = \frac{T - D_d}{T}$$

where D_d is the duct profile drag which can be expressed by

$$D_d = \frac{1}{2} \rho U^2 C_{D_d} 2\pi R c_d$$

where

C_{D_d} is the duct profile drag coefficient;

$2\pi R c_d$ is the duct planform area, c_d being the chord length of the duct.

Using this factor η_f an estimate of the efficiency of a ducted propulsor in a real fluid is

$$\eta = \left[1 - \frac{2 c_d}{R} \frac{C_{D_d}}{C_{Th}}\right] \frac{2}{1 + \sqrt{1 + \tau C_{Th}}} \tag{23.14}$$

This shows that the importance of the loss in efficiency from duct drag becomes less as the total required thrust loading, C_{Th}, increases, i.e., at higher C_{Th} the induced losses by the rotor predominate. Clearly, the duct chord must be minimized to achieve a maximum efficiency. However, the duct length cannot be made arbitrarily small since the duct must achieve a certain level of radial force to develop the thrust or the necessary pressure to inhibit rotor cavitation. If the duct sectional lift coefficient becomes too large, boundary-layer separation will ensue with severe degradation in performance. Evaluations of ducted-propeller efficiencies for the MARIN criterion that the duct sectional radial force coefficient should not exceed unity are given in Figure 23.12. This criterion gives a correlation between duct length-diameter ratio c_d/D and τ. In the results shown in the figure the formula $C_{D_d} = 2C_f(1 + (t_d/c_d)^2)$ is used where C_f is the skin friction coefficient, a function of Reynolds number based on duct chord length c_d, and t_d/c_d is the duct-section-thickness-chord-length ratio. From Figure 23.12 it is seen that for accelerating ducts the performance is very sensitive to duct length for small total loading coefficients but beyond a $C_{Th} = 1.5$ the efficiency is insensitive to duct viscous drag. Note that as T_d rises with increased C_{Th} the product $\tau_d C_{Th}$ varies at least quadratically with C_{Th}; consequently for $C_{Th} << 1.0$ the efficiency becomes that of the actuator disc as found by Sparenberg (1969).

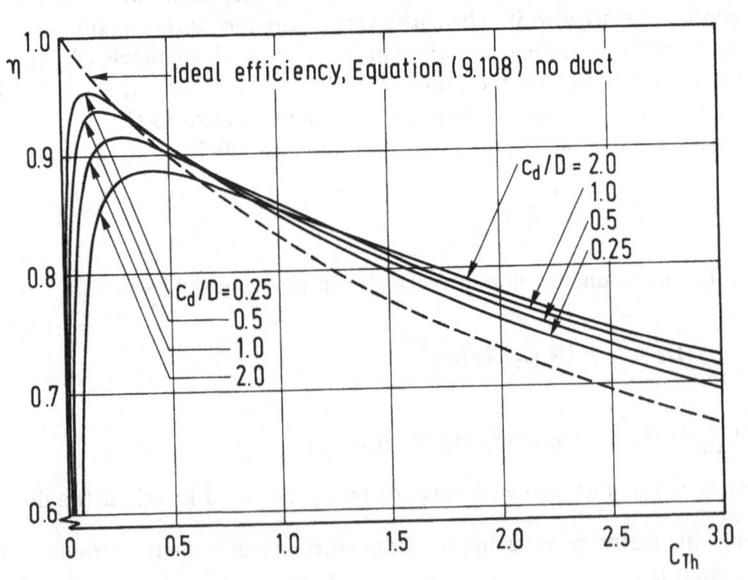

Figure 23.12 Optimum efficiencies with frictional losses of ducted propellers.

From Oosterveld, M.W.C. (1970). *Wake Adapted Ducted Propellers.* Publ. no. 345. Wageningen: Netherlands Ship Model Basin.
By courtesy of Maritime Research Institute Netherlands, The Netherlands.

Axial momentum theory can only yield some overall trends in performance of ducted propulsor systems. A more complete treatment can be obtained via a computer-evaluated mathematical model employing singularity distributions together with variational methods as has been introduced by Kerwin, Coney & Hsin (1986) (as outlined here on p. 456–460). This model has been extended by Kinnas & Coney (1988) who employ a hybrid representation involving detailed paneling of the duct surface and its wake with dipole distributions. The propeller blades are modelled as lifting lines by Green functions which are composed of free-space dipoles plus their "generalized" image function in the duct which is forced to satisfy the kinematic condition on the duct. The lifting lines are discretized by a lattice of vortex elements of constant strength. The surface distributions on the duct are also discretized. These sets of unknown singularity strengths are then determined by employing the calculus of variations to find the blade loading yielding required total thrust for minimum torque analogous to the procedure described earlier.

By this model it is possible to include or predict:

i. the non-symmetrical loading induced by the finite number of propeller blades which gives rise to induced drag on the duct;

ii. propeller-blade profile drag;

iii. the optimum distribution of blade loading;

iv. the dependence of duct thrust upon blade loading and rotation rate or advance ratio;

v. the losses due to propeller and duct-induced swirl;

vi. the mechanics for determining the propulsor geometry with or without constraints such as cavitation index.

Results are given by Kinnas & Coney (1990) for a simple duct shown in Figure 23.13. Comparison of the optimum blade circulations for the open and ducted propeller (with zero tip clearance) is also displayed in this figure. The variations of propeller and duct thrust-loading coefficients as functions of total thrust-loading coefficient are shown in Figure 23.14. It is seen that for small loadings the duct thrust is slightly negative, being made up of viscous drag and small duct-induced drag. As thrust is increased the percentage of the total thrust generated on the duct rises as the flow is now so modified that net thrust is generated on the duct and the duct is accelerating the flow. This is the typical operation of most ducted propulsor applications.

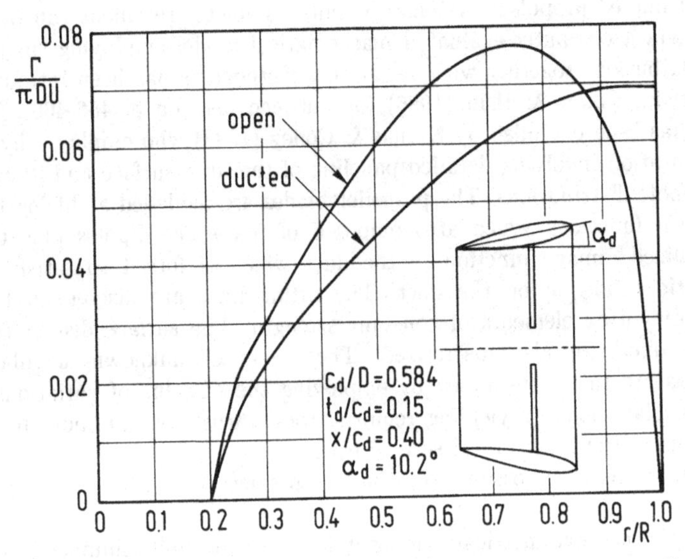

Figure 23.13 Optimum circulation distributions for open and ducted propeller; J = 1.14 and C_{Th} = 1.04. The effect of the hub is not included.

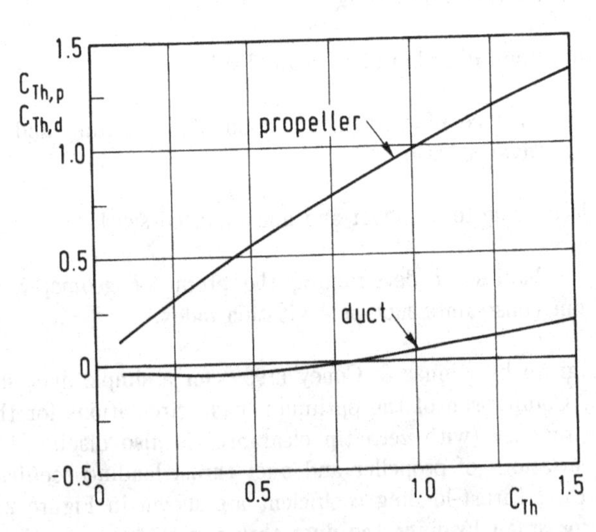

Figure 23.14 Duct and propeller thrust–loading coefficients as a function of the total (duct plus propeller) thrust–loading coefficient.

Figure 23.13 and 23.14 from Kinnas, S.A. & Coney, W.B. (1990). *The Generalized Image Model – an Application to the Design of Ducted Propellers.* Cambridge, Mass.: Dept. of Ocean Eng., MIT.

The propulsor efficiencies for optimized propellers operating in the duct are shown in Figure 23.15 as a function of total thrust. Compared with that for the optimized propeller alone the propulsor efficiency is inferior for $C_{Th} < 0.7$. The gain of efficiency is seen to grow rapidly for $C_{Th} > 1.0$. It should be noted, however, that this example demonstrating the method of Kinnas & Coney (*ibid.*) and the most important of the hydrodynamic characteristics of ducted propulsors is calculated for light loadings. In typical applications for tankers the thrust loading is much higher ($C_{Th} \sim 2.5 - 5.0$) and trawlers and tugs, except at free-sailing condition, operate at $C_{Th} > 3$, as indicated in Figure 9.17, p. 194.

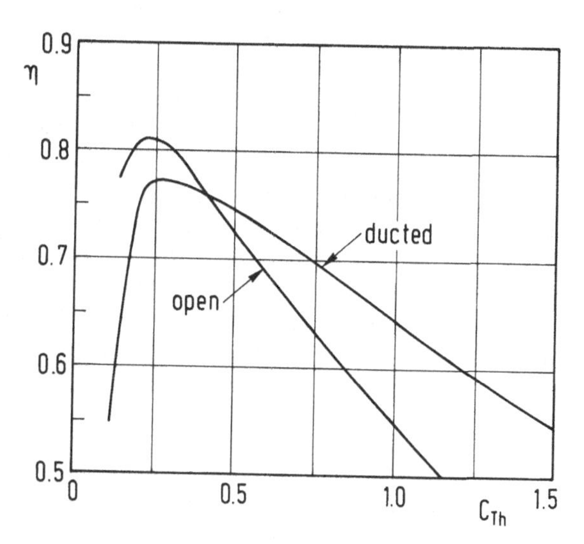

Figure 23.15 Efficiency of open and ducted propellers.

From Kinnas, S.A. & Coney, W.B. (1990). *The Generalized Image Model – an Application to the Design of Ducted Propellers.* Cambridge, Mass.: Dept. of Ocean Eng., MIT.

This optimization procedure at present does not give optimum loading on the duct nor a realistic accounting for the viscous flow effects for small blade-duct gap. A multistep design procedure is described by Coney (1989) which includes the potential-flow effects for a range of tip gaps or clearances. He shows that the calculations can be simplified by using an image system for the trailing vortices which is based on the two-dimensional image of a vortex in a circle.

Application of accelerating ducts to large tankers and bulk carriers has been frequent since the mid 1960's yielding observed gains in propulsive efficiencies in the range of 5 to 12 per cent. However, unexpected damage due to blade-tip cavitation imploding on the duct inner surface in the neighborhood of 12-o'clock region has been a problem causing eventual cracking and objectionable noise and vibration. As Glover (1987) notes, the application of "conventional" ducts to large ships came to an abrupt halt with the cancellation of orders for large tankers following the onset of the oil crisis in the mid 1970's.

Modification of blade-loading distributions can be obtained, instead of operating the propeller inside a duct, by altering the geometry of the blade tips making it possible for them to carry radially directed loading. This provides better spreading of trailing vorticity thus giving decreased induced drag. Gomez (1976) applied end plates to the propeller blades, as related by Glover (1987). Highly sophisticated mathematical lifting-surface procedures for propellers with and without end plates have been developed by Klaren & Sparenberg (1981), de Jong & Sparenberg (1990) and de Jong (1991). Bladelets fitted to the propeller blade tips like winglets on airplane wings were examined by Goodman & Breslin (1980), by Itoh, Tagori, Ishii & Ide (1986) and by Itoh (1987) while blade tips bent towards the suction side of the propeller were examined theoretically by Andersen & Andersen (1987). Applications have not been reported except by Gomez (Glover (1987)) so these attempts at improving the propeller efficiency should be considered as in the developing phase.

Summary

As a conclusion to these sections on unconventional propulsive devices consisting of a propeller and added lifting surfaces we give Table 23.2 with a summary of Coney's (1989) comparative calculations for optimum single propellers at three disc loadings with five compound propulsors. Here we see that all propulsor alternatives display decreased efficiency with increased thrust loading. Because the induced losses increase we see that the percentage gains of efficiency are greater for larger thrust loadings, as expected. Contrarotating propellers show the highest gains of all the unducted propulsors. The efficiencies of the post-swirl stator are about the same for light loadings as the vane wheel. At the highest loading the vane wheel shows a very significant advantage. The ducted propeller with post-swirl stator vanes shows the greatest gain at $C_{Th} = 2.65$. The ducted-propeller results at all values of C_{Th} clearly reflect the losses due to parasitic drag of the duct.

A further summary of the possible power savings of these and other propulsors including flow-equalizing devices of which three are described in the next section is shown in Table 23.3, p. 483.

Propulsor	C_{Th} = 0.6625		C_{Th} = 1.325		C_{Th} = 2.65	
	η	$\Delta\eta$ p.c.	η	$\Delta\eta$ p.c.	η	$\Delta\eta$ p.c.
Single propeller	0.733	–	0.632		0.508	–
Vane-wheel propulsor	0.779	6.3	0.705	11.6	0.607	19.5
Contrarotating propellers	0.803	9.6	0.724	14.6	0.614	20.9
Propeller/Stator	0.779	6.3	0.696	10.1	0.591	16.3
Ducted propeller	0.746	1.8	0.679	7.4	0.576	13.4
Ducted propeller/Stator	0.797	8.7	0.737	16.6	0.636	25.2
Ideal efficiency	0.874	19.2	0.792	25.3	0.687	35.3

Duct thrust specified 15 per cent of the total thrust; duct length taken to be the propeller radius; propeller tip gap was zero.

Table 23.2 Summary of performances for various propulsor alternatives.

Table 6.7 from Coney, W.B. (1989). *A Method for the Design of a Class of Optimum Marine Propulsors.* Ph.D. thesis. Cambridge, Mass.: Dept. of Ocean Eng., MIT.

By courtesy of Massachusetts Institute of Technology, USA.

FLOW-CONDITIONING DEVICES

A number of devices as well as alteration of the stern lines have been developed for affecting the flow to the propeller. Among these are: asymmetric sterns; the Mitsui-integrated-duct system; the Schneekluth wake-equalizing duct; Grothues-Spork guide vanes attached to the hull forward of the propeller. These devices operate in the highly complex flow over the stern of the ship where separated-boundary-layer flows, cross-flows and bilge vortices may exist. For this reason the development of these means of flow conditioning depends upon model tests. As we have only briefly mentioned such complicated flows we limit this presentation to a very short outline of flow-conditioning devices and instead refer to the literature. Overviews of these devices have been provided by Blaurock (1990) and Glover (1987) and further references can be found in the report of the Propulsor Committee of the ITTC (1990b).

Asymmetric Sterns

Asymmetric sterns are so designed that they create a swirl ahead of the propeller in the direction opposite to the rotation of the propeller. In combination with the swirl introduced by the propeller this should reduce the rotational losses in the slipstream. According to Blaurock (1990) asymmetric sterns when compared to symmetric ones give smaller mean axial wake, smaller turbulence intensities, and larger irregularity of the overall flow field. However, the physical mechanisms that produce the power savings have so far not been completely elucidated, despite widely extended research.

It is, of course, necessary that the increased efficiency obtained by introducing swirl to the propeller is not annulled by an increase in resistance of the ship, but this has not been found by the ship models tested; on the contrary, decreases in resistances were found. Blaurock (*ibid.*) conjectures that a further explanation of the effect of asymmetry is that for a right-handed propeller it is the region on the port side of the hull above the propeller shaft where the pressure gradient is high that is critical with respect to flow separation. For an asymmetric stern the angles of run are smaller in this region yielding improved flows. Future developments in numerical techniques for calculation of the flow over ship hulls and further model tests will no doubt give more insight into the hydrodynamics of asymmetric sterns. Until then, design of the lines of these sterns depends on the experience of the designer and the hull form must always be tested in the model basin.

Power savings of 6 to 7 per cent are considered as realistic. Asymmetric sterns may be used in combination with wake-equalizing ducts.

Wake–Equalizing Duct

The wake-equalizing or Schneekluth duct is attached to the hull ahead of the propeller, above the shaft, the duct diameter being approximately half of that of the propeller, cf. Figure 23.16. The duct can be fitted behind the stern as a whole ring or as two semi-rings on both sides of the stern. For asymmetric sterns or for relatively fast ships only one semi-duct may be used.

According to Schneekluth (1985) the duct equalizes the wake field by accelerating the flow in the upper region of the propeller disc where the flow with no duct would be slower and it decelerates it in the other regions where the inflow would be faster. Moreover, it creates more axial inflow to the propeller.

The wake-equalizing duct has the following features:

i. Improvement in propeller efficiency by a more favorable inflow field to the propeller.

ii. Reduction of flow separation.

iii. Generation of thrust on the duct, as by the Kort nozzle, by lifting forces with net forward components.

iv. Reduction in unsteady propeller forces and cavitation due to more uniform inflow to the propeller.

Figure 23.16 Schneekluth wake-equalizing duct and propeller of 71 375 t tanker, owned by Deutsche Shell Tanker, Germany, built by Deutsche Werft, Germany. Duct diameter: 3.00 m; length: 1.5 m. Power: 11 769 kW. By courtesy of Schneekluth Hydrodynamik, Germany.

For full form ships the main advantage with this type of duct is the efficiency increase while for finer ships it is the flow equalizing resulting in reduced propeller-induced vibrations. Gains in overall efficiency up to 14 per cent have been mentioned by Schneekluth (*ibid.*) whereas Blaurock (1990) shows 4 to 5 per cent.

Figure 23.17 Mitsui–integrated–duct propeller of 250 000 dwt tanker *Esso Copenhagen*. Propeller diameter: 8.90 m; 6 blades; power appr.: 24 000 kW at 80 RPM.
By courtesy of Mitsui Engineering & Shipbuilding Co., Ltd., Japan.

Figure 23.18 Grothues spoiler, retrofitted to improve efficiency, of container ship, owned by H.W. Janssen, Germany, built by Thyssen Nordseewerke, Germany. Propeller diameter: 5.600 m; 5 blades; A_E/A_0 = 0.675; weight: 14 200 kg; material: CuAlNi F640: power: 10 600 kW at 130 RPM.
By courtesy of Ostermann Metallwerke, Germany.

Mitsui Duct

The Mitsui-Integrated-Duct Propeller was initially developed at Mitsui Engineering & Shipbuilding Co., Ltd. and Exxon International Company, cf. Narita, Yagi, Johnson & Breves (1981). It consists of a non-axisymmetric duct placed forward of the propeller such that the trailing edge of the duct sections are aligned with the tips of the propeller blades, cf. Figure 23.17.

On the basis of model tests Narita *et al.* (*ibid.*) explain the features of the duct system as follows:

i. The Mitsui duct makes the stern flow into the propeller disc more homogeneous and stabilizes the flow around the stern. The decrease of the hull resistance is believed to be attributable to these effects.

ii. Inward radial flow around the stern creates a positive thrust on the duct.

iii. The propeller efficiency is increased due to acceleration of the flow to the propeller and to reduced required thrust because of lower thrust-deduction fraction.

The gains in efficiencies are 5 to 10 per cent with the greatest gains for slow, full-form ships. In addition other benefits may be obtained, such as reduced propeller cavitation and hull vibration level.

Grothues Spoiler

This device, named after its inventor, Grothues-Spork (1988), consists of a set of fins attached to the hull just ahead of the propeller, cf. Figure 23.18. Its development was based on observations of flows over the stern of ship models where it was found that for U-shaped models cross flows strengthen bilge vortices.

The Grothues-spoiler fins are designed to affect this flow with expected benefits, as summarized by Grothues-Spork (*ibid.*):

i. By the diversion of high-energy cross flows entering and reinforcing the bilge vortices rotational energy is recovered which would otherwise be lost after separation.

ii. Due to the flow deflection at the fins the forces on them have net forward thrust components.

iii. Propeller efficiency is increased both by the re-orientation of the flow and the resulting acceleration of the axial flow components.

These combined effects reduce angular variations of the velocity field to the propeller and hence reduce propeller generated vibration.

Model tests with many models carried out at the Berlin Model Basin – VWS showed increases of efficiency of 3 to 7 per cent and decreases of resistance of 1 to 3 per cent. This was accompanied by slight increases in propeller efficiency and reductions of thrust and torque fluctuations due to improvements in the wake distributions.

SUMMARY AND CONCLUSION

A collection of reported improvements in efficiency obtained by various unconventional propulsors is given in Table 23.3. It is seen that while some of the predicted gains are very big most of them are in the order of magnitude of some (few) per cent.

With the relatively small, yet still important, efficiency increases that can be obtained it is necessary with complex computer programs and high-accuracy measurements in model tests to secure optimum design where the benefit obtained by introducing additional surfaces or flow-conditioning devices are not annulled by detrimental effects in the form of profile drag, vortex shedding etc..

In each instance of comparison of the effectiveness of application of unconventional propulsors it is essential to determine if the conventional propeller design is the optimum which can be achieved subject to the constraints. It must be remembered that a succesful design is always a compromise and is characterized not only by high efficiency but also by low levels of vibration and noise as well as by little or no cavitation.

There is little doubt that the future will bring computer methods and model-testing techniques of a sophistication and completeness that we can only dream of today. To the hydrodynamicist as well as to the naval architect they will be powerful tools which will provide deeper insight into the problems of hydrodynamic propulsion and with which higher levels of optimization will be achieved in design. To anyone working in this field the development and use of these tools will be a challenge!

Propulsor Type		Power reductions, per cent		Number of applications at sea
		Prediction, calculations or model tests	Full scale data	
Low-RPM propeller		5 – 18	–*	many
Coaxial contrarotating propellers		7 – 20	15 – 16	2** many
Propeller with vane wheel		8 – 12	6 – 8.5	59
Ducted propeller	Axisymmetric duct	5 – 20	–	many
	Asymmetric duct	less than above	–	many
	Duct in front of propeller	5 – 12	5	–
Pre-swirl devices	Radial reaction fins in front of the propeller	3 – 8	7 – 8	–
	Asymmetric stern	1 – 9	–	30, 42**
Post-swirl devices	Additional thrusting fins at the rudder	1 – 8	8 – 9	–
	Rudder-bulb system with fins	1 – 3	–	–
	Fins on propeller fairwater	3 – 7	4	40
Flow-smoothing device	Wake-equalizing duct	6 – 7	–	many
	Guide vanes in front of the propeller	2 – 10	5 – 10	30

* Data not given indicated by – ; ** Both numbers given

Table 23.3 Comparison of unconventional propulsor performance with respect to efficiency. For more details, cf. ITTC (1990b).

From (simplified) Table 8 of ITTC (1990b). Report of the Propulsor Committee. In *Proc. 19th International Towing Tank Conference*, vol. 1, pp. 109–60. El Pardo: Canal de Experiencias Hidrodinamicas.
By courtesy of ITTC.

Appendix A Inversion of the Airfoil Integral Equations

The integral Equation (5.13) can be inverted by several different procedures. A heuristic development is given by Newman (1977) which circumvents much detailed mathematics but requires a knowledge of complex-variable theory. We shall employ a method involving a Fourier series which leads to an integral whose value is included as a special case of a more general formula. This choice shortens the derivation here.

The integral equation to be inverted is

$$\int_{-a}^{a} \frac{\gamma(x')}{x - x'} dx' = h(x) \tag{A.1}$$

where γ is unknown and h is given. Here we omit conditions which h(x) must meet. Let

$$x = -a\cos\theta \quad ; \quad x' = -a\cos\theta' \quad ; \quad dx' = a\sin\theta'\, d\theta' \tag{A.2}$$

Then we have

$$\int_{0}^{\pi} \frac{\gamma(-a\cos\theta')\sin\theta'}{\cos\theta - \cos\theta'} d\theta' = -h(-a\cos\theta) \tag{A.3}$$

We now assume that $\gamma(-a\cos\theta')\sin\theta'$ can be expanded in a Fourier series of the following form

$$\gamma(-a\cos\theta')\sin\theta' = a_0 + \sum_{n=1}^{\infty} a_n \cos n\theta' \tag{A.4}$$

Substituting in (A.3) we obtain

$$a_0 \int_{0}^{\pi} \frac{d\theta'}{\cos\theta - \cos\theta'} + \sum_{n=1}^{\infty} a_n \int_{0}^{\pi} \frac{\cos n\theta'}{\cos\theta - \cos\theta'} d\theta' = -h(-a\cos\theta) \tag{A.5}$$

The integrals

$$J_n^* = \int_0^{\pi} \frac{\cos n\theta'}{\cos\theta - \cos\theta'} d\theta' \tag{A.6}$$

are so-called Glauert (1948) integrals. They must be taken as principal-value integrals as the denominator vanishes at $\theta' = \theta$. Their reduction by him involves development of a recursion formula whose detailed evaluation is omitted (because it is lengthy) and moreover his result is given for a single order parameter, n, which is positive. In the general case we must distinguish the results for $0 < n$ and $0 > n$. A direct evaluation can be effected as follows.

In (A.6) we ensure that this is an even function of n by taking $|n|$. This enables us to replace θ' by $-\theta'$ and upon reversing the limits

$$J_n^* = \int_{-\pi}^{0} \frac{\cos|n|\theta'}{\cos\theta - \cos\theta'} d\theta'$$

Upon adding this to the foregoing we see that

$$J_n^* = \frac{1}{2}\int_{-\pi}^{\pi} (....)d\theta' = \frac{1}{2}\int_0^{2\pi} (....)d\theta'$$

which can be written as

$$J_n^* = \frac{1}{2}\, Re \left\{ \int_0^{2\pi} \frac{e^{i|n|\theta'}}{\cos\theta - \cos\theta'} d\theta' \right\} \tag{A.7}$$

where *Re* indicates that only the real part is retained.

Now we can write

$$\cos\theta' = \frac{e^{i\theta'} + e^{-i\theta'}}{2} = \frac{\zeta + \zeta^{-1}}{2}$$

where the complex variable $\zeta = e^{i\theta'}$ describes the unit circle in the complex ζ-plane as θ' varies from 0 to 2π. Introducing the function ζ, noting that $d\theta' = -id\zeta/\zeta$, we transform (A.7) to

$$\begin{aligned} J_n^* &= \frac{Re}{2}\left\{ 2i \int \frac{\zeta^{|n|}}{\zeta^2 - (2\cos\theta)\zeta + 1} d\zeta \right\} \\ &= Re \left\{ i \int \frac{\zeta^{|n|}}{(\zeta - \cos\theta)^2 + \sin^2\theta} d\zeta \right\} \end{aligned} \tag{A.8}$$

Factoring the denominator we see that the integrand has poles at

$$\zeta = \zeta_1 = \cos\theta + i \sin\theta \text{ and } \zeta = \zeta_2 = \cos\theta - i \sin\theta \qquad (A.9)$$

which lie on the unit circle as shown in Figure A.1. The path of integration as θ' varies from 0 to 2π is along the circle and the small semi-circles around the poles as shown in the figure. Equation (A.8) is now of the form

$$J_n^* = Re\left\{i \int \frac{\zeta^{|n|}}{(\zeta - \zeta_1)(\zeta - \zeta_2)} d\zeta\right\} \qquad (A.10)$$

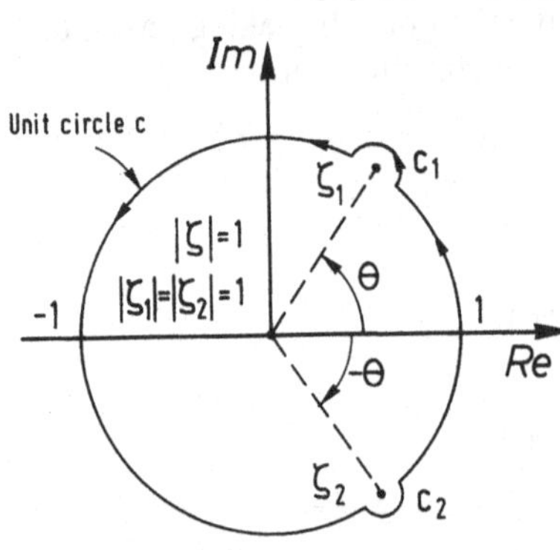

Figure A.1

where ζ_1 and ζ_2 are defined in (A.9). We now can apply Cauchy's Integral Formula which states that such integrals have the value given by $2\pi i \cdot$ (sum of the residues at all the poles within the contour), cf. for example Phillips (1947). Here the residues at $\zeta = \zeta_1$ and $\zeta = \zeta_2$ are

$$\frac{\zeta_1^{|n|}}{\zeta_1 - \zeta_2} \text{ and } \frac{\zeta_2^{|n|}}{\zeta_2 - \zeta_1}$$

respectively. As ζ_1 and ζ_2 lie on the unit circle we must include contributions from the semi-circles c_1 and c_2 which can be shown to produce $\pi i\ \Sigma$(residues) at ζ_1 and ζ_2. Hence by Cauchy

$$\oint_c \frac{\zeta^{|n|}}{(\zeta - \zeta_1)(\zeta - \zeta_2)} d\zeta + \int_{c1} (...)d\zeta + \int_{c2} (...)d\zeta = 2\pi i \sum (\text{residues})$$

or from the preceding statement concerning the contributions from c_1 and c_2 we have

$$\oint_c \frac{\zeta^{|n|}}{(\zeta - \zeta_1)(\zeta - \zeta_2)} d\zeta = \pi i\left[\frac{\zeta_1^{|n|}}{\zeta_1 - \zeta_2} + \frac{\zeta_2^{|n|}}{\zeta_2 - \zeta_1}\right]$$

and upon insertion of the values of ζ_1 and ζ_2, (A.9)

$$= \frac{\pi i}{2i \sin\theta}\left[e^{i|n|\theta} - e^{-i|n|\theta}\right] = \frac{\pi i \sin|n|\theta}{\sin\theta} \qquad (A.11)$$

(from the definition of $\sin|n|\theta$ in terms of exponentials). Use of this evaluation of the integral in (A.10) yields, finally, the even function of n and θ

$$J_n^* = -\frac{\pi \sin|n|\theta}{\sin\theta} \tag{A.12}$$

We now see that had we not ensured that the result must be an even function of n we would have secured an odd function as sin $n\theta$ is odd in n. Glauert's development does not make this distinction as he considered only $n \geq 1$.

When we now use (A.12) in (A.5) we see that the first term in (A.5) vanishes and we obtain

$$\sum_{n=1}^{\infty} a_n \sin n\theta = \frac{1}{\pi} h(-a \cos\theta) \sin\theta \tag{A.13}$$

We may now calculate the coefficients a_n by multiplying both sides of (A.13) by $1/\pi \sin m\theta$ and integrating over θ from $-\pi$ to π

$$\frac{1}{\pi} \sum_{n=1}^{\infty} a_n \int_{-\pi}^{\pi} \sin n\theta \sin m\theta \, d\theta = \frac{1}{\pi^2} \int_{-\pi}^{\pi} h(\theta) \sin\theta \sin m\theta \, d\theta \tag{A.14}$$

By the product into sum identity the left side of (A.14) integrates to

$$\frac{1}{2\pi} \sum_{n=1}^{\infty} a_n \left[\frac{\sin(m-n)\theta}{m-n} - \frac{\sin(m+n)\theta}{m+n} \right]_{-\pi}^{\pi} = 0 \qquad m \neq n$$

For $m = n$

$$\int_{-\pi}^{\pi} \sin n\theta \sin n\theta \, d\theta = \left[\frac{\theta}{2} - \frac{1}{4n} \sin 2n\theta \right] \Bigg|_{-\pi}^{\pi} = \pi$$

Hence (A.14) yields

$$a_n = \frac{1}{\pi^2} \int_{-\pi}^{\pi} h(\theta) \sin\theta \sin n\theta \, d\theta \tag{A.15}$$

Using this in (A.4) we can write for γ as a function of the variable θ

$$\gamma(-a \cos\theta)\sin\theta = a_0 + \frac{1}{\pi^2} \sum_{n=1}^{\infty} \left\{ \left[\int_{-\pi}^{\pi} h(\theta')\sin\theta' \sin n\theta' \, d\theta' \cos n\theta \right\} \tag{A.16}$$

where θ in the integral has been changed to θ', being a dummy variable.

We now interchange the orders of summation and integration and calculate the sum

$$S = \sum_{n=1}^{\infty} \sin n\theta' \cos n\theta = \frac{1}{2}\sum_{n=1}^{\infty} (\sin n(\theta'+\theta) + \sin n(\theta'-\theta))$$

Consider for brevity

$$S_0 = \sum_{n=1}^{\infty} \sin n\alpha = Im\left\{\sum_{n=1}^{\infty} \left[e^{i\alpha}\right]^n\right\}$$

where *Im* indicates retention of only the coefficient of the imaginary part $e^{i\alpha} = \cos\alpha + i\sin\alpha$.

Upon writing $\zeta = e^{i\alpha}$ we see that S_0 is the sum of a geometric series which we can write

$$\sum e^{i\alpha n} = \zeta + \zeta^2 + \zeta^3 + ... + \zeta^{n-1} + \zeta^n + ...$$

$$\zeta\sum e^{i\alpha n} = \zeta^2 + \zeta^3 + \zeta^4 + ... + \zeta^{n-1} + \zeta^n + \zeta^{n+1} + ...$$

whence in the limit $n \to \infty$ for $\zeta < 1$ we obtain by subtraction

$$\sum_{n=1}^{\infty} e^{i\alpha n} = \frac{\zeta}{1-\zeta} = \frac{e^{i\alpha}}{1-e^{i\alpha}} = \frac{i}{2}\left\{\frac{\cos\frac{\alpha}{2} + i\sin\frac{\alpha}{2}}{\sin\frac{\alpha}{2}}\right\}$$

Hence

$$S_0 = \frac{1}{2}\cot\frac{\alpha}{2}$$

noting that for $\alpha = \theta' + \theta$ and $\theta' - \theta$, the original sum S is now

$$S = \frac{1}{4}\left\{\cot\left[\frac{\theta'+\theta}{2}\right] + \cot\left[\frac{\theta'-\theta}{2}\right]\right\}$$

$$= \frac{\sin\theta'}{4\sin\left[\frac{\theta'+\theta}{2}\right]\sin\left[\frac{\theta'-\theta}{2}\right]} = \frac{1}{2}\frac{\sin\theta'}{\cos\theta - \cos\theta'} \tag{A.17}$$

With this result (A.16) reduces to

$$\gamma(-a\cos\theta) = \frac{1}{\sin\theta}\left[a_0 + \frac{1}{\pi^2}\int_0^{\pi} \frac{\sin^2\theta'\ h(-a\cos\theta')}{\cos\theta - \cos\theta'}\, d\theta'\right] \quad (A.18)$$

where the θ'-integral has been folded making use of the evenness of the integrand.

Returning to the original variables x and x' we see that (A.18) takes the form

$$\gamma(x) = -\frac{1}{\pi^2\sqrt{a^2 - x^2}}\left[\pi K + \int_{-a}^{a} \frac{\sqrt{a^2 - x'^2}}{x - x'}\, h(x')\, dx'\right] \quad (A.19)$$

where $K = -\pi a a_0$.

We observe that this solution contains an arbitrary constant K. This indicates that there is an infinity of solutions to this integral equation of the first kind. The constant K has a physical interpretation as shown in Chapter 5, p. 70–71.

For limits which are not symmetrically disposed, i.e. for trailing edge at $x = a_1$ and leading edge at $x = a_2$, (A.19) is

$$\gamma(x) = -\frac{1}{\pi^2\sqrt{(x-a_1)(a_2-x)}}\left[\pi K + \int_{a_1}^{a_2} \frac{\sqrt{(x'-a_1)(a_2-x')}}{x - x'}\, h(x')dx'\right] \quad (A.20)$$

Appendix B The Kutta-Joukowsky Theorem

The following development is taken from Milne–Thomson[31] (1955) with some changes. As the derivation given there involves the use of complex-variable theory some preliminary relations are first given.

The complex potential of the two-dimensional flow of an inviscid irrotational incompressible fluid $w(z)$, $z = x + iy$, $i = \sqrt{-1}$ is expressed as

$$w(z) = \phi(x,y) + i\psi(x,y) \tag{B.1}$$

The complex velocity is, because w is analytic

$$\frac{dw}{dz} = \frac{\partial w}{\partial x}\frac{\partial x}{\partial z} = \frac{\partial \phi}{\partial x} + i\,\frac{\partial \psi}{\partial x} \tag{B.2}$$

where ϕ is the velocity potential and ψ is the stream function. As

$$u = \frac{\partial \phi}{\partial x} \tag{B.3}$$

then as the stream function is orthogonal to the potential function and

$$v = -\frac{\partial \psi}{\partial x} \tag{B.4}$$

according to the sign convention in Figure B.1. (The sign convention is opposite to that used by Milne–Thomson (*ibid.*)).

$$\delta\psi = \psi_1 - \psi_2 = u\delta y$$

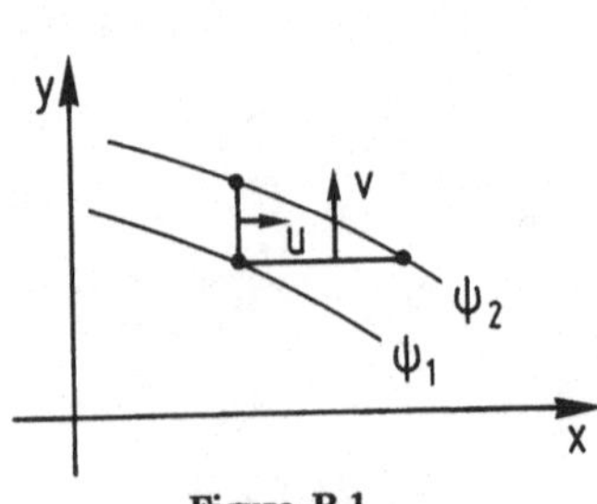

Figure B.1

hence $\quad u = \dfrac{\partial \psi}{\partial y}$,

+ being flux from left to right looking along increasing y,

and $\quad v = -\dfrac{\partial \psi}{\partial x}$

31 By courtesy of Macmillan & Co. Ltd., Great Britain.

Then the complex velocity q is

$$q = \frac{dw}{dz} = u - iv \tag{B.5}$$

The theorem developed originally by ***Kutta*** in 1902 and independently by ***Joukowsky*** in 1906 is as follows:

- ***An airfoil at rest in a uniform wind of speed* V*, with circulation* Γ *round the airfoil, undergoes a lift* $\Gamma\rho V$ *perpendicular to the wind. The direction of the lift vector is got by rotating the wind velocity vector through a right angle in the sense opposite to that of the circulation.***

Proof: Since there is a uniform wind, the velocity at a great distance from the airfoil must tend simply to the wind velocity, and therefore if $|z|$ is sufficiently large, we may write

$$\frac{dw}{dz} = -Ve^{i\alpha} + \frac{A}{z} + \frac{B}{z^2} + \ldots. \tag{B.6}$$

where α is the *incidence* or *angle of attack*, Figure B.2. Thus

$$w = -Ve^{i\alpha}z + A\log z - \frac{B}{z} + \ldots.$$

As the circulation is, by definition, the line integral of the scalar product of the velocity and the differential arc length about any closed curve about the foil, then

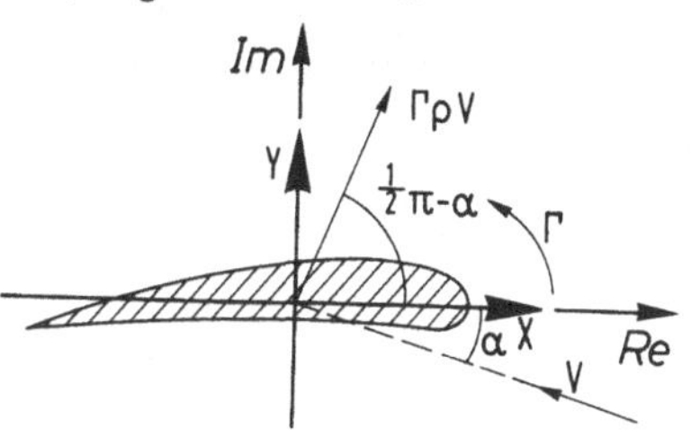

Figure B.2

$$\int_c \frac{dw}{dz} \cdot ds = \int_c \left[-Ve^{i\alpha} + \frac{A}{z} + \frac{B}{z^2}\right] dz = 2\pi i A = \Gamma$$

as the other terms produce nothing while the logarithm increases by $2\pi i$ when one goes round the foil in the positive sense. Therefore we have that $A = -i\,\Gamma/2\pi$ and hence

$$\left[\frac{dw}{dz}\right]^2 = V^2e^{2i\alpha} + \frac{i\Gamma Ve^{i\alpha}}{\pi z} - \frac{\Gamma^2 + 8\pi^2 BVe^{i\alpha}}{4\pi^2 z^2} + \ldots \tag{B.7}$$

If we now integrate this round a circle whose radius is sufficiently large for the expansion (B.7) to be valid, the Theorem of Blasius (see below) gives

$$X - iY = \frac{i\rho}{2}\int_c \left[\frac{dw}{dz}\right]^2 dz = \frac{1}{2}\, i\rho(2\pi i)\left[\frac{i\Gamma V e^{i\alpha}}{\pi}\right]$$

all the other terms yielding zero (from the calculus of residues) or

$$X - iY = -i\rho\Gamma V e^{i\alpha}$$

This may be written, by replacing i by –i

$$X + iY = \rho V\Gamma e^{i(\pi/2-\alpha)} \tag{B.8}$$

using $i = e^{i\pi/2}$. We see from Figure B.2 that this force is perpendicular to the incident velocity V and has all the properties stated in the theorem.

This theorem has the following ***corrollaries***:

- ***The lift is independent of the form of the profile.***
- ***The drag (force along the wind axis) is zero as there is no account taken of viscosity or energy dissipation.***
- ***If the airfoil is regarded as moving in air otherwise at rest, the lift is got by rotating the velocity vector of the airfoil through a right angle in the same sense as the circulation.***
- ***The theorem of Blasius applied to (B.7) gives the moment about the origin.***

$$M = Re\left\{2\pi i\rho B V e^{i\alpha}\right\} \tag{B.9}$$

Theorem of Blasius

Let a fixed cylinder be placed in a liquid which is moving steadily and irrotationally. Let X, Y and M be the components along the axes and the moment about the origin of the pressure thrusts on the cylinder. Then, neglecting external forces,

$$X - i\,Y = \frac{1}{2}\, i\, \rho \int_c \left[\frac{dw}{dz}\right]^2 dz \tag{B.10}$$

$$M = Re\left\{-\frac{1}{2}\,\rho\int_c z\left[\frac{dw}{dz}\right]^2 dz\right\} \tag{B.11}$$

where w is the complex potential, ρ the density, and the integrals are taken round the contour of the cylinder.

<u>*Proof*</u> For the action on the arc ds at P we have, cf. Figure B.3

$$dX = -\,p\,dy \quad ; \quad dY = p\,dx \quad ; \quad dM = p(x\,dx + y\,dy)$$

Thus

$$d(X - iY) = -\,ip d\bar{z} \;;$$

$$dM = Re\,\{pz d\bar{z}\}$$

(where $^{-}$ indicates complex conjugate).

From the pressure equation, (Equation (1.21))

$$p = p_0 - \frac{1}{2}\,\rho(u^2 + v^2)$$

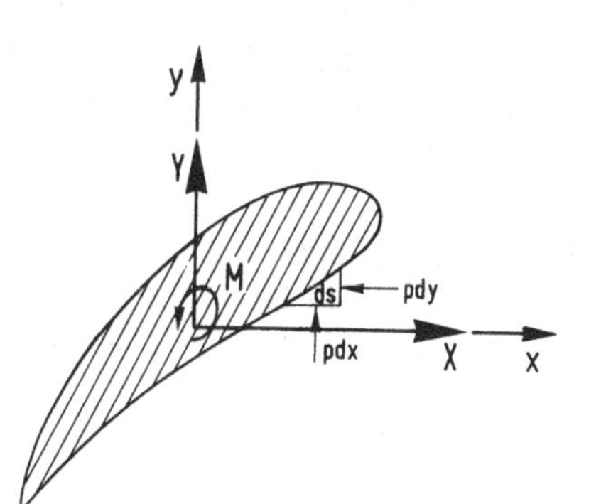

Figure B.3

where p_0 is a constant.

Since a constant pressure can have no resultant effect, we can take

$$p = -\frac{1}{2}\,\rho q^2 = -\frac{1}{2}\,\rho\,\frac{dw}{dz}\,\frac{d\bar{w}}{d\bar{z}}$$

so that

$$d(X - iY) = \frac{1}{2}\,i\,\rho\,\frac{dw}{dz}\,d\bar{w}$$

$$dM = Re\left\{-\frac{1}{2}\,\rho z\,\frac{dw}{dz}\,d\bar{w}\right\}$$

But on c, ψ = constant and therefore $d\bar{w} = dw$, so that

$$d(X - iY) = \frac{1}{2}\,i\rho\left[\frac{dw}{dz}\right]^2 dz$$

$$dM = Re\left\{-\frac{1}{2}\,\rho z\left[\frac{dw}{dz}\right]^2 dz\right\}$$

and the theorem follows by integration round c.

The contour c can be enlarged to any contour provided that we do not encompass any additional singularities of the integrand.

Appendix C The Mean Value of the Radial Velocity Component Induced by a Helical Vortex at Downstream Infinity

The radial component u_r^{he} of a single helical vortex, generated at $x = 0$, $\gamma = 0$, $r = r'$ can be readily composed from Equation (11.28), p. 214, by noting that this component (making use of (9.2)) can be expressed by

$$u_r^{he}(x,r,\gamma) = - v^{he} \sin\gamma + w^{he} \cos\gamma$$

Then from (11.28)

$$u_r^{he}(x,r,\gamma) = \frac{V}{4\pi}\Gamma(r_0)\int_{-\infty}^{0}\frac{r_0 \sin\beta \sin(\gamma-\gamma') - (x-Ut')\cos\beta\cos(\gamma-\gamma')}{R^3}dt' \tag{C.1}$$

where $R = \sqrt{(x-Ut')^2 + r^2 + r_0^2 - 2rr_0\cos(\gamma-\gamma')}$; $V = \sqrt{U^2 + (r_0\omega)^2}$.
From the definition of R we have

$$\frac{\partial}{\partial x}\frac{1}{R} = -\frac{x - Ut}{R^3} \quad ; \quad \frac{\partial}{\partial \gamma}\frac{1}{R} = -\frac{rr_0 \sin(\gamma-\gamma')}{R^3}$$

so (C.1) can be written

$$u_r^{he}(x,r,\gamma) = \frac{V}{4\pi}\Gamma(r_0)\int_{-\infty}^{0}\left[-\frac{\sin\beta}{r}\frac{\partial}{\partial\gamma} + \cos\beta\cos(\gamma-\gamma')\frac{\partial}{\partial x}\right]\frac{1}{R}dt' \tag{C.2}$$

Now, substitute the Fourier expansion of $1/R$, (M5.24), p. 518, and carry out the differentiations to obtain for $r \geq r_0$

$$u_r^{he}(x,r,\gamma) = \frac{V}{4\pi^2}\Gamma(r_0)\sum_{m=-\infty}^{\infty} i\int_{-\infty}^{0}dt'\int_{-\infty}^{\infty}\left[-m\frac{\sin\beta}{r} + \cos\beta\cos(\gamma-\gamma')\,k\right]$$

$$\cdot\, I_m(kr_0)K_m(kr)e^{ik(x-Ut')}e^{im(\gamma-\gamma')}dk \tag{C.3}$$

where absolute m's and k's are used on I and K and r and r_0 are interchanged in Bessel functions for $r \leq r_0$.

On the helical vortex, $\gamma' = \omega t'$ and substituting $x' = x - Ut'$ in (C.3) and using the complex definition of $\cos(\gamma - \gamma')$ yields, after some manipulations,

$$u_r^{he}(x,r,\gamma) = \frac{1}{4\pi^2}\Gamma(r_0)\sum_{m=-\infty}^{\infty} i\int_x^{\infty} dx' \int_{-\infty}^{\infty}\Bigg[-\frac{m}{r}\,e^{im(\gamma - p_f^* x)}\,e^{i(k+mp_f^*)x'} +$$

$$\frac{1}{2}\cot\beta\, k\Big[e^{i(m+1)(\gamma - p_f^* x)}e^{i(k+(m+1)p_f^*)x'} + e^{i(m-1)(\gamma - p_f^* x)}e^{i(k+(m-1)p_f^*)x'}\Big]\Bigg]$$

$$\cdot\, I_m(kr_0)K_m(kr)dk \qquad \text{(C.4)}$$

where $p_f^* = \omega/U$.

To find the velocity at downstream infinity we let $x \to -\infty$ and use

$$\lim_{x \to -\infty}\int_x^{\infty} e^{ikx'}dx' = 2\pi\delta(k)$$

We can then carry out the k-integration and (C.4) becomes

$$u_r^{he}(x,r,\gamma) = \frac{1}{2\pi}\Gamma(r_0)\sum_{m=-\infty}^{\infty} i\Bigg[-\frac{m}{r}\,e^{im(\gamma - p_f^* x)}\, I_m(kr_0)K_m(kr)\Big|_{k=-mp_f^*}$$

$$-\frac{1}{2}\cot\beta\,\Bigg[e^{i(m+1)(\gamma - p_f^* x)}kI_m(kr_0)K_m(kr)\Big|_{k=-(m+1)p_f^*}$$

$$+\, e^{i(m-1)(\gamma - p_f^* x)}kI_m(kr_0)K_m(kr)\Big|_{k=-(m-1)p_f^*}\Bigg]\Bigg] \qquad \text{(C.5)}$$

The mean value is independent of γ and x; so it is obtained by setting m = 0, m = −1 and m = 1 in the first, second and third term, respectively, of (C.5). In the first term we encounter

$$\lim_{k \to 0} kI_0(kr_0)K_0(kr) = 0$$

and in the second and third term (since we actually have $I_{|m|}$ and $K_{|m|}$)

$$\lim_{k \to 0} kI_1(kr_0)K_1(kr) = 0$$

so

$$\text{mean}(u_r^{he}(-\infty,r,\gamma)) = 0 \qquad \text{(C.6)}$$

as we expect.

Appendix D Conservation of Circulation

This basic principle or theorem is developed in many texts on theoretical hydrodynamics. Here we follow principally the derivation given by Yih (1988) being content to restrict the fluid to one of constant density.

In essence, the circulation taken around a closed material circuit, c, is conserved in an inviscid fluid as the circuit moves, i.e., is convected with the motion. As Lighthill (1986) states, the circuit c is always made up of the same fluid particles like a necklace whose shape and position change continually as these elements move.

For a differential arc length ds of c having components dx_i ($i = 1, 2, 3$), these components vary with time because of the stretching and turning of ds in space. Then the relation between the velocity components u_i and dx_i is given by the substantive derivative

$$\frac{D}{Dt}\, dx_i = du_i \tag{D.1}$$

du_i being the difference of the velocity components u_i at the two ends of ds.

The Euler equations (1.8) can be concisely written in indicial notation as

$$\frac{Du_i}{Dt} = -\frac{1}{\rho}\frac{\partial p}{\partial x_i} + F_i = -\frac{\partial}{\partial x_i}\left[\frac{p}{\rho} + \Omega\right] \tag{D.2}$$

where the external forces F_i are assumed to be obtainable from a body-force potential, $F_i = -\,\partial\Omega/\partial x_i$, cf. p. 6.

The definition of circulation is the integral of the scalar product $u_i dx_i$ as a line integral around any circuit c, i.e.,

$$\Gamma = \int_c u_i dx_i$$

The substantive derivative of the element of the circulation

$$\frac{D}{Dt}\,(u_i dx_i) = \left[\frac{D}{Dt}\, u_i\right] dx_i + u_i \frac{D}{Dt}\, dx_i \tag{D.3}$$

Then from (D.2) and (D.1), (D.3) becomes

$$\frac{D}{Dt}(u_i dx_i) = -\frac{\partial}{\partial x_i}\left[\frac{p}{\rho} + \Omega\right] dx_i + u_i du_i = d\left[-\frac{p}{\rho} - \Omega + \frac{1}{2}u_i^2\right] \tag{D.4}$$

Integration of (D.4) completely around contour c gives

$$\int_c \frac{D}{Dt}(u_i dx_i) = \frac{D}{Dt}\int_c u_i dx_i = \frac{D}{Dt}\Gamma = 0 \tag{D.5}$$

since the perfect differential on the right side of (D.4) produces the same value at the beginning and end points of the integration. The interchange of the order of integration and differentiation is permissible because the integration is carried out along the "necklace" c, a material circuit.

This is generally referred to as ***Lord Kelvin's theorem*** (William Thomson, 1824–1907) published in 1869 (Thomson (1869)) but as Yih (*ibid.*) relates, Truesdell (1954) noted that Hankel (1861) had previously published this theorem eight years earlier.

Appendix E Method of Characteristics

The method of characteristics can be used to solve first-order, partial differential equations. Although non-linear equations can be dealt with we shall restrict ourselves to linear equations since only linear equations will appear in the present context.

Let us consider the equation

$$f(x,y,t,\phi)\frac{\partial \phi}{\partial x} + g(x,y,t,\phi)\frac{\partial \phi}{\partial y} + h(x,y,t,\phi)\frac{\partial \phi}{\partial t} = m(x,y,t,\phi) \qquad \text{(E.1)}$$

If we could have expressed ϕ as a function of only one variable, say ξ, we would have, by the chain rule

$$\frac{\partial x}{\partial \xi}\frac{\partial \phi}{\partial x} + \frac{\partial y}{\partial \xi}\frac{\partial \phi}{\partial y} + \frac{\partial t}{\partial \xi}\frac{\partial \phi}{\partial t} = \frac{\partial \phi}{\partial \xi} \qquad \text{(E.2)}$$

One sees immediately that the two equations are identical, if the coefficients of $\partial\phi/\partial x$, $\partial\phi/\partial y$ and $\partial\phi/\partial t$ in (E.1) and (E.2) are equated, i.e.,

$$\frac{\partial x}{\partial \xi} = f(x,y,t,\phi) \qquad \text{(E.3)}$$

$$\frac{\partial y}{\partial \xi} = g(x,y,t,\phi) \qquad \text{(E.4)}$$

$$\frac{\partial t}{\partial \xi} = h(x,y,t,\phi) \qquad \text{(E.5)}$$

Then

$$\frac{\partial \phi}{\partial \xi} = m(x,y,t,\phi) \qquad \text{(E.6)}$$

Equations (E.3)–(E.6), define the *characteristic curves.* We can also interpret this result that along a curve given by (E.3)–(E.5) ϕ is dependent on the parameter ξ and can be obtained by solving (E.6). To determine the characteristic uniquely requires the specification of a point (x_0,y_0,t_0,ϕ_0) on the characteristic; this point is usually obtained as a boundary condition. It is tacitly assumed, that the functions f, g, h and m are well-behaved.

Let us as a demonstration of the method of characteristics solve Equation (17.5)

$$\left[\frac{\partial}{\partial t} - U\frac{\partial}{\partial x}\right] v = G(x,y,t) \tag{E.7}$$

For convenience divide by $-U$

$$\left[-\frac{1}{U}\frac{\partial}{\partial t} + \frac{\partial}{\partial x}\right] v = -\frac{1}{U} G(x,y,t) \tag{E.8}$$

The characteristic curves are then given by

$$\frac{\partial t}{\partial \xi} = -\frac{1}{U} \tag{E.9}$$

$$\frac{\partial x}{\partial \xi} = 1 \tag{E.10}$$

$$\frac{\partial v}{\partial \xi} = -\frac{1}{U} G(x,y,t) \tag{E.11}$$

which have the solution

$$t = -\frac{1}{U}\xi + t_0 \tag{E.12}$$

$$x = \xi + x_0 \tag{E.13}$$

$$v = -\frac{1}{U}\int G(x,y,t)\, d\xi + v_0 \tag{E.14}$$

Inserting (E.12) and (E.13) into (E.14) gives

$$v = -\frac{1}{U}\int G\left[\xi + x_0, y, -\frac{1}{U}\xi + t_0\right] d\xi + v_0 \tag{E.15}$$

In this equation we substitute $\xi + x_0$ by x''. We also take the characteristics to go through the point (x,y,t) where we calculate the velocity, so we drop the subscript. Finally, when we integrate along the characteristic curve (given by (E.12), (E.13)), we start far upstream, $x'' \to \infty$, where the velocity vanishes, $v \to 0$, and end at the point under consideration, $x'' = x$. All this gives

$$v(x,y,t) = \frac{1}{U}\int_x^\infty G\left[x'', y, t + \frac{1}{U}(x - x'')\right] dx'' \tag{E.16}$$

A much more complete account of the method of characteristics can be found, e.g. in Carrier & Pearson (1976).

Appendix F Boundary Conditions Imposed by Water Surface at High and Low Frequencies

The relation between pressure p, velocity potential ϕ of the disturbing flow, and vertical location in the fluid z, in the presence of gravity and a uniform stream, –U, is obtained by Bernoulli's Equation (1.21), where we have inserted Equation (1.22) and dropped the subscript b. We then have

$$\frac{p}{\rho} + \frac{1}{2}\left[\left[-U + \frac{\partial\phi}{\partial x}\right]^2 + \left[\frac{\partial\phi}{\partial y}\right]^2 + \left[\frac{\partial\phi}{\partial z}\right]^2\right] + \frac{\partial\phi}{\partial t} + gz = \frac{p_{at}}{\rho} + \frac{1}{2}U^2 \qquad \text{(F.1)}$$

where p_{at} is the atmospheric pressure and the reference datum is $z = 0$, the undisturbed level of the water surface. When on the free surface

$$z = \zeta(x,y,t) \qquad \text{(F.2)}$$

or

$$F(x,y,z,t) = z - \zeta(x,y,t) = 0 \qquad \text{(F.3)}$$

we have that the pressure $p = p_{at}$ and, upon neglecting the squares of the perturbation terms, (F.1) yields the physical condition on ζ, namely

$$\zeta = \frac{1}{g}\left[U\frac{\partial\phi}{\partial x} - \frac{\partial\phi}{\partial t}\right] = \frac{1}{g}\left[U\frac{\partial}{\partial x} - \frac{\partial}{\partial t}\right]\phi \qquad \text{(F.4)}$$

The kinematic condition that the velocity of the free surface along its local normal must be equal to the component of the fluid velocity along that normal gives Equation (3.2) or

$$\frac{D}{Dt}F(x,y,z,t) = \frac{\partial F}{\partial t} + \frac{\partial x}{\partial t}\frac{\partial F}{\partial x} + \frac{\partial y}{\partial t}\frac{\partial F}{\partial y} + \frac{\partial z}{\partial t}\frac{\partial F}{\partial z} = 0 \qquad \text{(F.5)}$$

The particle velocities are

$$\frac{\partial x}{\partial t} = -U + \frac{\partial\phi}{\partial x} \simeq -U \;; \quad \frac{\partial y}{\partial t} = \frac{\partial\phi}{\partial y} \;; \quad \frac{\partial z}{\partial t} = \frac{\partial\phi}{\partial z}$$

Then, using (F.3) in (F.5)

$$-\frac{\partial\zeta}{\partial t} - U\left[-\frac{\partial\zeta}{\partial x}\right] + \frac{\partial\phi}{\partial y}\left[-\frac{\partial\zeta}{\partial y}\right] + \frac{\partial\phi}{\partial z} = 0$$

and, as the third term is of second order, one obtains

$$\left[\frac{\partial}{\partial t} - U\frac{\partial}{\partial x}\right]\zeta = \frac{\partial \phi}{\partial z} \tag{F.6}$$

ζ may be eliminated between the physical, (F.4), and the kinematic, (F.6), conditions to yield the combined requirement on ϕ imposed by the presence of the free surface in the form

$$\left[\frac{\partial}{\partial t} - U\frac{\partial}{\partial x}\right]^2 \phi + g\frac{\partial \phi}{\partial z} = 0 \tag{F.7}$$

Here we are interested in potentials in which the time-dependence can be split off, that is

$$\phi = \tilde{\phi}(x,y,z)e^{iqZ\omega t} \tag{F.8}$$

Then (F.7) becomes

$$\left[iqZ\omega - U\frac{\partial}{\partial x}\right]^2 \tilde{\phi} + g\frac{\partial \tilde{\phi}}{\partial z} = 0 \tag{F.9}$$

where $\tilde{\phi}$ and its derivative are to be evaluated on $z = 0$, assuming ζ to be small everywhere. If we now replace $\tilde{\phi}$ by $\omega R^2 \check{\phi}$, x by $R\check{x}$, z by $R\check{z}$ (R the propeller radius, ω the angular velocity) and extract $iqZ\omega R$ from the first term, there results the dimensionless equation ($\check{}$ dimensionless variable)

$$\left[1 + i\frac{J}{\pi qZ}\frac{\partial}{\partial \check{x}}\right]^2 \check{\phi} - \frac{g}{R(qZ\omega)^2}\frac{\partial \check{\phi}}{\partial \check{z}} = 0 \tag{F.10}$$

where $J = U/ND$, the advance ratio. Now, for $R(qZ\omega)^2$ sufficiently large so that the acceleration ratio

$$a_r = \frac{g}{R(qZ\omega)^2} << 1 \tag{F.11}$$

then

$$\left[1 + i\frac{J}{\pi qZ}\frac{\partial}{\partial \check{x}}\right]^2 \check{\phi} \longrightarrow 0$$

Taking zero as the right side, this equation admits of the possible solution

$$\check{\phi} = f_1(y)e^{i\pi qZ\check{x}/J} + f_2(y)xe^{i\pi qZ\check{x}/J} \tag{F.12}$$

where f_1 and f_2 are to be determined to meet available conditions. *But this solution is inadmissible* since $\check{\phi}$ must approach zero as $x \to \infty$. Hence, f_1 and f_2 must be taken as identically zero and

$$\check{\phi}(x,y,0,t) \longrightarrow 0 \quad \text{for } a_r << 1 \tag{F.13}$$

The value of a_r (for R = 4 m, g = 9.81 m/s^2, q = 1 (blade frequency), Z = 5 (blades) and $\omega = 2\pi(3/2)$ ($\approx$ 90 RPM)) is a_r = 0.0011. Thus, for applications of interest, we see that the approximation that $a_r << 1$ is well accommodated.

In contrast, when gravity dominates, that is, $a_r >> 1$, then (F.10) may be recast to read

$$\frac{\partial \check{\phi}}{\partial \check{z}} - \frac{R(qZ\omega)^2}{g}\left[1 + i\,\frac{J}{\pi q Z}\frac{\partial}{\partial \check{x}}\right]^2 \check{\phi} = 0 \tag{F.14}$$

and, as the second term approaches zero, one is left with the condition

$$\frac{\partial \check{\phi}}{\partial \check{z}} \simeq 0 \tag{F.15}$$

or the surface z = 0 acts like a nonporus, rigid cover.

The acceleration ratio, a_r, can be recast into

$$a_r = \frac{gR}{(qZV_{tip})^2}$$

where the speed of the blade tip due to rotation $V_{tip} = \omega R$. It is readily seen, that a_r is the inverse square of the Froude number

$$F_r = \frac{\text{velocity}}{\sqrt{\text{length} \cdot \text{acc. of gravity}}}$$

It is well known (Newman (1977)) that the Froude number squared expresses the ratio of inertial forces to gravitational forces. Hence for small value of a_r the inertial forces dominate the flow, while for large a_r's the gravitational force dominate.

Mathematical Compendium

This mathematical compendium consists of eight independent sections dealing in detail with specific mathematics used in the main text. It is provided as an aid to readers who may not be familiar with these mathematical applications and therefore may find it difficult to collect and study the necessary literature.

1. TAYLOR EXPANSION

A function f(x) which is regular (bounded) at some point x_0 and which possesses derivatives at all orders (infinitely differentiable) at x_0 can be expanded in a power series about x_0 in the form

$$f(x) = a_0 + a_1(x-x_0) + a_2(x-x_0)^2 + ... + a_n(x-x_0)^n \qquad \text{(M1.1)}$$

The coefficients a_n are found as follows:

Put $x = x_0$

$$f(x_0) = a_0$$

Differentiate with respect to x

$$f'(x) = a_1 + 2a_2(x-x_0) + 3a_3(x-x_0)^2 + ... + na_n(x-x_0)^n \qquad \text{(M1.2)}$$

clearly at $x = x_0$

$$a_1 = f'(x_0) \qquad \text{(M1.3)}$$

$$f''(x) = 2a_2 + 3 \cdot 2a_3(x-x_0) + ... \qquad \text{(M1.4)}$$

thus

$$a_2 = \frac{1}{2} f''(x_0)$$

$$a_3 = \frac{1}{2 \cdot 3} f''(x_0) \qquad \text{or} \qquad f(x) = \sum_{n=0}^{\infty} \frac{f^{(n)}(x_0)}{n!} (x-x_0)^n$$

$$\vdots$$

$$a_n = \frac{1}{n!} f^{(n)}(x_0)$$

Thus

$$f(x) = f(x_0)+ \frac{f'(x_0)}{1!}(x-x_0)+ \frac{f''(x_0)}{2!}(x-x_0)^2+ .. + \frac{f^{(n)}(x_0)}{n!}(x-x_0)^n$$

$$= \sum_{n=0}^{\infty} \frac{f^{(n)}(x_0)}{n!}(x-x_0)^n \qquad \text{(M1.5)}$$

For $x = x + \delta x$, $x_0 = x$, and extending to four variables (x-, y-, z-coordinates and time) we have

$$f(x+\delta x, y+\delta y, z+\delta z, t+\delta t)$$

$$= f(x,y,z,t)+ \frac{\partial f}{\partial x}\delta x+ \frac{\partial f}{\partial y}\delta y+ \frac{\partial f}{\partial z}\delta z+ \frac{\partial f}{\partial t}\delta t \qquad \text{(M1.6)}$$

2. DIRAC'S δ-FUNCTION

This function was introduced by the physicist P.A.M. Dirac in 1930 (Dirac (1947)) and is defined by

$$\int_{-\infty}^{\infty} \delta(x)dx = 1 \qquad \text{(M2.1)}$$

$$\delta(x) = 0 \quad \text{for } x \neq 0 \qquad \text{(M2.2)}$$

Because of (M2.2) the integral formula can be substituted by

$$\int_a^b \delta(x)dx = \begin{cases} 1 & \text{if } a < 0 < b \\ 0 & \text{if } ab > 0 \end{cases} \qquad \text{(M2.3)}$$

Furthermore, if f(x) is a function continuous in $x = x'$

$$\int_a^b f(x')\delta(x'-x)dx' = \begin{cases} f(x) & \text{if } a < x < b \\ 0 & \text{if } a < x \text{ and } b < x \\ & \text{or } x < a \text{ and } x < b \end{cases} \qquad \text{(M2.4)}$$

For a constant k

$$\delta(kx) = \frac{1}{|k|}\delta(x) \qquad \text{(M2.5)}$$

Dirac's δ-function can be related to Heaviside's step function H(x)

$$H(x) = \begin{cases} 0 & x < 0 \\ \frac{1}{2} & x = 0 \\ 1 & x > 0 \end{cases} \tag{M2.6}$$

by the relation

$$\frac{dH(x)}{dx} = \delta(x) \tag{M2.7}$$

We often encounter the derivative of the δ-function in the form

$$\int_a^b f(x') \frac{d\delta(x'-x)}{dx'} dx'$$

Integration by parts yields

$$\int_a^b f(x') \frac{d\delta(x'-x)}{dx'} dx'$$

$$= f(b)\delta(b-x) - f(a)\delta(a-x) - \int_a^b \frac{df(x')}{dx'} \delta(x'-x)dx'$$

$$= - \frac{df(x')}{dx'}\bigg|_{x'=x} \tag{M2.8}$$

assuming, that $a < x < b$. If $x < a < b$ or $a < b < x$ the result will be zero.

3. GREEN'S IDENTITIES AND GREEN FUNCTION

To derive Green's Identities we start with ***Gauss's Theorem***, (well-known from courses in mathematical analysis)

$$\int_V \nabla\cdot\mathbf{f}\, dV = \int_S \mathbf{f}\cdot\mathbf{n}\, dS \tag{M3.1}$$

where $\mathbf{f}$ is a continuously differentiable vector field in the bounded region V. S is the boundary of V and $\mathbf{n}$ the unit normal outward to S. $\nabla\cdot$ is the divergence operator, in cartesian coordinates

$$\nabla\cdot\mathbf{f} = \frac{\partial}{\partial x} f_x + \frac{\partial}{\partial y} f_y + \frac{\partial}{\partial z} f_z$$

Assume that the two functions ϕ and ψ have partial derivatives of the second order which are continuous in V and that f is given by

$$f = \phi \, \nabla\psi$$

where ∇ is the gradient operator. Insert this into (M3.1) to arrive at ***Green's first identity***

$$\int_V \phi \, \nabla^2\psi \, dV = -\int_V \nabla\phi \cdot \nabla\psi \, dV + \int_S \phi \frac{\partial\psi}{\partial n} \, dS \qquad \text{(M3.2)}$$

since $\nabla\phi\cdot\mathbf{n} = \partial\phi/\partial n$. ∇^2 is the ***Laplace operator*** which is (in cartesian coordinates)

$$\nabla^2 = \frac{\partial^2}{\partial x^2} + \frac{\partial^2}{\partial y^2} + \frac{\partial^2}{\partial z^2}$$

If we substitute ϕ for ψ and vice versa in (M3.2) and subtract this equation from the original (M3.2) we get ***Green's second identity***

$$\int_V (\phi\nabla^2\psi - \psi\nabla^2\phi) \, dV = \int_S \left[\phi \frac{\partial\psi}{\partial n} - \psi \frac{\partial\phi}{\partial n}\right] dS \qquad \text{(M3.3)}$$

If both ϕ and ψ satisfy the Laplace Equation ($\nabla^2\psi = 0$) everywhere in V we have the simple formula

$$\int_S \left[\phi \frac{\partial\psi}{\partial n} - \psi \frac{\partial\phi}{\partial n}\right] dS = 0 \qquad \text{(M3.4)}$$

Green Function

Let there be given a differential operator L which acts on the functions f(x,y,z) that satisfy certain linear and homogeneous boundary conditions. The ***Green function*** for this operator is the function G(x,y,z;x',y',z') defined as the solution to the equation

$$L \, G(x,y,z;x',y',z') = -\,\delta(x-x')\,\delta(y-y')\,\delta(z-z') \qquad \text{(M3.5)}$$

and which satisfies the same boundary conditions as f. The coordinates of the dummy point (x',y',z') must here be interpreted as parameters. For the case of one or two dimensions analogous definitions apply.

It is obvious, cf. (1.57), that the potential of a sink of unit strength (Equation (1.50) with $M = -1$) is a Green function for the differential operator $L = \nabla^2$ in a boundless fluid volume and

$$G(x,y,z;x',y',z') = \frac{1}{4\pi R}$$

where $R = \sqrt{(x-x')^2+(y-y')^2+(z-z')^2}$.

Substituting G for ψ in Green's second identity (M3.3) and remembering that ϕ must satisfy the Laplace Equation gives us, for (x,y,z) inside S

$$\phi(x,y,z) = -\frac{1}{4\pi}\int_S \left[\phi \frac{\partial}{\partial n}\frac{1}{R} - \frac{1}{R}\frac{\partial \phi}{\partial n}\right] dS \qquad \text{(M3.6)}$$

Thus we see that the potential of an inner point in V can be expressed as a combination of sources (2.39) and normally directed dipoles placed on S. The sources have the strength $-\partial\phi(x',y',z')/\partial n$ and the dipoles $\phi(x',y',z')$.

If (x,y,z) is a point on the (smooth) boundary S it can be shown that

$$\phi(x,y,z) = -\frac{1}{2\pi}\int_S \left[\phi \frac{\partial}{\partial n}\frac{1}{R} - \frac{1}{R}\frac{\partial \phi}{\partial n}\right] dS \qquad \text{(M3.7)}$$

For (x,y,z) outside V

$$0 = \int_S \left[\phi \frac{\partial}{\partial n}\frac{1}{R} - \frac{1}{R}\frac{\partial \phi}{\partial n}\right] dS \qquad \text{(M3.8)}$$

4. EVALUATION OF INTEGRALS WITH CAUCHY- AND HADAMARD-TYPE SINGULAR-KERNEL FUNCTIONS

Mangler (1951) has given a clear exposition of procedures for evaluation of integrals of the type

$$F(x) = ⨍_a^b f(x')K(x,x')dx' \qquad \text{(M4.1)}$$

where the singular kernel K is of the general form

$$K = \frac{1}{(x-x')^{n+1}} \quad ; \quad n \geq 1, \ \ n \text{ an integer}$$

and the function f(x') is regular at x' = x. The symbol $⨍$ indicates principal value or ***finite part***, a concept due to Hadamard (1923). The integrand may also become infinite at the end points a and b giving rise to integrals of the form

$$\int_a^b \frac{G(x',x)}{(x'-a)^{\alpha+n}} dx' \qquad \text{(M4.2)}$$

where α is a fraction less than unity (usually 1/2) and n is a positive integer. These are referred to by Mangler (*ibid.*) as principal values of fractional order and those having the singularity within the interval as principal values of order n as in (M4.1).

Principal Values or Finite Parts of Integer Order n

Cauchy Kernels n = 0

We find Cauchy-type kernels as in

$$F(x) = ⨍_a^b \frac{f(x')}{x-x'} dx' \tag{M4.3}$$

To distinguish, this simpler and more commonly encountered Cauchy principal-value integral is provided with a bar and written $⨍$ while the notation with a cross as in (M4.1) is used for the higher-order singular kernel functions, designated as Mangler or Hadamard singularities.

We interpret (M4.3) as the limit form for $y \to 0$ of

$$F(x,y) = \int_a^b \frac{f(x')(x-x')}{(x-x')^2+ y^2} dx' \tag{M4.4}$$

The function F(x,y) is a continuous function of x and y including y = 0. To find the value along y = 0 within the interval $a \leq x' \leq b$ we proceed as follows

$$\lim_{y \to 0} F(x,y) = \lim_{\substack{y \to 0 \\ \varepsilon \to 0}} \left\{ \int_a^{x-\varepsilon} \frac{f(x')(x-x')}{(x-x')^2+ y^2} dx' + \int_{x+\varepsilon}^b \frac{f(x')(x-x')}{(x-x')^2+ y^2} dx' \right.$$

$$\left. + \int_{x-\varepsilon}^{x+\varepsilon} \frac{f(x')(x-x')}{(x-x')^2+ y^2} dx' \right\} \tag{M4.5}$$

There is no difficulty in setting y = 0 in the first two integrals as $x' \neq x$. In the last integral I_p^*, let $\xi = x-x'$ then $dx' = -d\xi$ and we have

$$I_p^* = \int_{-\varepsilon}^{\varepsilon} \frac{f(x+\xi)}{\xi^2 + y^2} \xi \, d\xi$$

Now as f is assumed to be regular we may expand f in a Taylor expansion about $\xi = 0$, (cf. Equation (M1.5)) to give

$$I_p^* = \int_{-\varepsilon}^{\varepsilon} \left[\xi f(x) + \xi^2 f'(x) + \frac{\xi^3}{2} f''(x) + \dots \right] \frac{1}{\xi^2+y^2} d\xi \tag{M4.6}$$

We see by inspection that the first and third terms vanish because they are odd functions of ξ and this reduces to

$$I_p^* = f'(x)\left[2\varepsilon - 2y\ \tan^{-1}\frac{\varepsilon}{y}\right] + \text{higher powers of } \varepsilon$$

and as this is valid for all $\varepsilon > 0$ but sufficiently small to permit the Taylor expansion, we see that for $y \to 0$, $I_p^* \to 2\varepsilon f'(x)$ and in the limit $\varepsilon \to 0$, $I_p^* \to 0$ and we have

$$F(x) = \lim_{y \to 0} F(x,y) = \lim_{\varepsilon \to 0}\left[\int_a^{x-\varepsilon} \frac{f(x')}{x-x'}\,dx' + \int_{x+\varepsilon}^{b} \frac{f(x')}{x-x'}\,dx'\right]$$

$$= \fint_a^b \frac{f(x')}{x-x'}\,dx' \qquad \text{(M4.7)}$$

This is the familiar ***Cauchy principal value.***

Principal Value of Order n = 1

A differentiation of F(x) under the integral sign in (M4.3) gives

$$-F'(x) = \fint_a^b \frac{f(x')}{(x-x')^2}\,dx' \qquad \text{(M4.8)}$$

Mangler proceeded in the same way as before to obtain

$$-F'(x) = \int_a^{x-\varepsilon} \frac{f(x')}{(x-x')^2}\,dx' + \int_{x+\varepsilon}^{b} \frac{f(x')}{(x-x')^2}\,dx' + I_p^*$$

where

$$I_p^* = -\lim_{\substack{y \to 0 \\ \varepsilon \to 0}} \int_{-\varepsilon}^{\varepsilon}\left[f(x) + \xi f'(x) + \frac{\xi^2}{2} f''(x) + \ldots\right]\frac{d}{d\xi}\left[\frac{\xi}{\xi^2 + y^2}\right]d\xi \qquad \text{(M4.9)}$$

Evaluation of the integrals gives

$$I_p^* = -\lim_{y \to 0}\left\{\frac{2f(x)\ \varepsilon}{\varepsilon^2 + y^2} + \frac{1}{2}\left[\frac{2\varepsilon^3 f''(x)}{\varepsilon^2 + y^2} - 4\left[\varepsilon - y\ \tan^{-1}\frac{\varepsilon}{y}\right]f''(x)\right]\right\}$$

Then letting $y \to 0$ holding ε fixed,

$$I_p^* = -\left\{\frac{2f(x)}{\varepsilon} - \varepsilon f''(x)\right\} \longrightarrow -\frac{2f(x)}{\varepsilon}$$

thus

$$-\lim_{\substack{y \to 0 \\ \varepsilon \to 0}} F'(x) = -\lim_{\varepsilon \to 0}\left\{\int_a^{x-\varepsilon} \frac{f(x')}{(x-x')^2}\,dx' + \int_{x+\varepsilon}^{b} \frac{f(x')}{(x-x')^2}\,dx' - \frac{2f(x)}{\varepsilon}\right\} \tag{M4.10}$$

Mangler shows that this limit exists even at $\varepsilon = 0$, *yielding the value of the integral obtained by ignoring the singularity, and thereby giving its principal value or finite part.*

Since $f(x')$ is continuous the result (M4.10) must hold for $f(x') \equiv 1$. Then

$$\int\!\!\!\!\!\!- \frac{dx'}{(x-x')^2} = \lim_{\varepsilon \to 0}\left\{\frac{1}{x-a} - \frac{1}{x-b} + \frac{1}{\varepsilon} - \frac{1}{-\varepsilon} - \frac{2}{\varepsilon}\right\} = \frac{(a-b)}{(x-a)(x-b)}\,;$$

i.e., the result by ignoring the singularity in the integrand.

Principal Values of Order n

In general, principal values of integer order for the following form

$$\frac{(-1)^n}{n!}\frac{d^nF(x)}{dx^n} = \int\!\!\!\!\!\!-_a^{\,b} \frac{f(x')}{(x-x')^{n+1}}\,dx' \;;\; (n = 0, 1, 2, 3, ..) \tag{M4.11}$$

are given by

$$\int\!\!\!\!\!\!-_a^{\,b} \frac{f(x')}{(x-x')^{n+1}}\,dx' = \lim_{\varepsilon \to 0}\left\{\int_a^{x-\varepsilon} \frac{f(x')}{(x-x')^{n+1}}\,dx' + \int_{x+\varepsilon}^{b} \frac{f(x')}{(x-x')^{n+1}}\,dx' + (-1)^n \sum_{j=0}^{n-1} \frac{f^{(j)}(x)}{j!}\left[\frac{1-(-1)^{n-j}}{(n-j)\varepsilon^{n-j}}\right]\right\} \tag{M4.12}$$

where $f^{(j)}(x)$ is the j-th derivative of f evaluated at x (cf. (M1.5)).

For cases where $dF(x,x')/dx' = f(x')/(x-x')^{n+1}$ is known, then

$$\int\!\!\!\!\!\!-_a^{\,b} \frac{f(x')}{(x-x')^{n+1}}\,dx' = F(x,b) - F(x,a) \tag{M4.13}$$

(any terms $\ln(x-x')$ which might occur in $F(x,x')$ must be understood as $\ln|x-x'|$).

Integration by Parts

The rule is

$$\int_a^b \frac{f(x')}{(x-x')^{n+1}}\,dx' = \frac{1}{n}\left[\frac{f(b)}{(x-b)^n} - \frac{f(a)}{(x-a)^n} - \int_a^b \frac{f'(x')}{(x-x')^n}\,dx'\right] \tag{M4.14}$$

This is the same as for an ordinary proper integral provided the principal value is taken as given by (M4.12).

The following is the rule for differentiation of these integrals

$$\frac{d}{dx}\int_a^b \frac{f(x')}{(x-x')^n}\,dx' = -n\int_a^b \frac{f(x')}{(x-x')^{n+1}}\,dx'$$

$$= \int_a^b \frac{f'(x')}{(x-x')^n}\,dx' + \frac{f(a)}{(x-a)^n} - \frac{f(b)}{(x-b)^n} \tag{M4.15}$$

Principal Values of Fractional Orders

Frequently the function f(x') in the foregoing is not regular at the limits a and b. Generally the following forms are encountered

$$f(x') \longrightarrow \frac{A(x')}{(x'-a)^{\alpha}}, \quad \text{near } x' = a\,;$$

$$\text{(M4.16)}$$

$$f(x') \longrightarrow \frac{B(x')}{(b-x')^{\beta}}, \quad \text{near } x' = b\,;$$

where A and B are regular functions of x', i.e., expandable in a Taylor series; α, β are not integers.

Singularity at the Lower Limit of Integration

For $\alpha < 1$ and $\beta < 1$, the rule for evaluation of the principal value applies as the same as in (M4.12) provided that we define

$$\int_a^b \frac{f(x')}{(x-x')^{n+1}}\,dx' = \lim_{\varepsilon \to 0}\int_{a+\varepsilon}^{b-\varepsilon} \frac{f(x')}{(x-x')^{n+1}}\,dx' \tag{M4.17}$$

The limit $\varepsilon \to 0$ can be taken for $\alpha < 1$, $\beta < 1$ and yields a finite answer.

For *any non-integer value* of α and β we may split the integral into two integrals and focus on the first part which is influenced by the singularity at a, as

$$\int_a^c \frac{f(x')}{(x-x')^{n+1}}\,dx' = \int_a^c \frac{A(x')}{(x'-a)^{\alpha}(x-x')^{n+1}}\,dx' \tag{M4.18}$$

where we take a < c < x < b and for the present we can forget the singularity at x' = x. Then this is of the form

$$I^* = \lim_{\varepsilon \to 0} \int_{a+\varepsilon}^{c} \frac{h(x')}{(x'-a)^{\alpha}}\,dx' \;\; ; \;\; h = \frac{f(x')(x'-a)^{\alpha}}{(x-x')^{n+1}} = \frac{A(x')}{(x-x')^{n+1}}$$

We see that h is regular for x > c and we have put in evidence the part which is singular at x' = a. Now Mangler (*ibid.*) introduced

$$G(x') = G(a) + \int_a^{x'} h(x'')dx''$$

which is regular at x' = a and upon substituting h(x') = G′(x') and integrating by parts

$$I^* = \lim_{\varepsilon \to 0} \alpha \left\{ \int_{a+\varepsilon}^{c} \frac{G(x')}{(x'-a)^{\alpha+1}}\,dx' - \frac{G(a+\varepsilon)}{\alpha\varepsilon^{\alpha}} \right\} + \frac{G(c)}{(c-a)^{\alpha}} \tag{M4.19}$$

Then defining the principal part by the terms in the bracket

$$I^* = \alpha \;⨎_a^b \frac{G(x')}{(x'-a)^{\alpha+1}}\,dx' + \frac{G(c)}{(c-a)^{\alpha}}$$

By repeated integration by parts the following principal value is defined for n any positive integer

$$⨎_a^b \frac{G(x')}{(x'-a)^{\alpha+n}}\,dx' = \lim_{\varepsilon \to 0} \left\{ \int_{a+\varepsilon}^{c} \frac{G(x')}{(x'-a)^{\alpha+n}}\,dx' - \sum_{j=0}^{n-1} \frac{G^{(j)}(a)}{j!\varepsilon^{\alpha+n-1-j}\,(\alpha+n-1-j)} \right\} \tag{M4.20}$$

having used a Taylor expansion of G at a.

Integration by Parts

The rule for integration by parts can be expressed by

$$\int_a^c \frac{G(x')}{(x'-a)^{\alpha+n}}\,dx' = \frac{1}{\alpha+n-1}\left[\fint_a^c \frac{G'(x')}{(x'-a)^{\alpha+n-1}}\,dx' - \frac{G(c)}{(c-a)^{\alpha+n-1}}\right]$$

$$0 < \alpha < 1 \qquad \text{(M4.21)}$$

Thus the formal procedure for integration by parts applies provided that the principal value is taken as defined in Equation (M4.20) and *all terms are omitted* which would formally give infinity.

We omit the proof and refer the interested readers to Mangler (*ibid.*).

Differentiation with Respect to the Limit

$$\frac{d}{da}\fint_a^c \frac{G(x')}{(x'-a)^{\alpha+n}}\,dx' = (\alpha+n)\fint_a^c \frac{G(x')}{(x'-a)^{\alpha+n+1}} \quad ; \quad 0 < \alpha < 1$$

$$\text{(M4.22)}$$

Thus the contribution from the limit is zero.

5. FOURIER EXPANSIONS OF 1/R

We shall seek various useful ways of expressing $1/R$, the inverse of the distance between two points, usually the field point and the dummy point.

Rectangular Coordinates

The function

$$\frac{1}{R} = \frac{1}{\sqrt{(x-x')^2+(y-y')^2+(z-z')^2}}$$

satisfies the inhomogeneous equation

$$\left[\frac{\partial^2}{\partial x^2} + \frac{\partial^2}{\partial y^2} + \frac{\partial^2}{\partial z^2}\right]\frac{1}{R} = -4\pi\delta(x-x')\delta(y-y')\delta(z-z') \qquad \text{(M5.1)}$$

where δ is the Dirac delta-function (p. 504).

The Fourier integral transform in x is (cf. for example Sneddon (1951))

$$\hat{\phi}(y,z;k_x) = \int_{-\infty}^{\infty} \phi(x,y,z)e^{-ik_x x}dx \qquad \text{(M5.2)}$$

and the inverse transform is formally

$$\phi(x,y,z) = \frac{1}{2\pi}\int_{-\infty}^{\infty} \hat{\phi}(x,y,k_x)e^{ik_x x}dk_x \qquad \text{(M5.3)}$$

To apply the Fourier integral transform to (M5.1) multiply by e^{-ixk_x} and integrate as in (M5.2). The integration in x is accomplished by integrating by parts twice, making use of $\partial(1/R)/\partial x = 0$ and $1/R = 0$ at $x = \pm\infty$ to secure

$$\left[\frac{\partial^2}{\partial y^2} + \frac{\partial^2}{\partial z^2} - k_x^2\right]\hat{R}^{-1} = -4\pi\, \delta(y-y')\delta(z-z')e^{-ik_x x'}$$

where use has been made of the sifting property of the δ-function, cf. (M2.4).

Repeating the analogous process by taking the y- and z- Fourier integral transform gives

$$(\hat{\hat{R}}^{-1}) = 4\pi\frac{e^{-i(x'k_x+y'k_y+z'k_z)}}{k_x^2+k_y^2+k_z^2} \qquad \text{(M5.4)}$$

Applying the inverse Fourier integral transform as in (M5.3)

$$\frac{1}{R} = \frac{1}{2\pi^2}\int_{-\infty}^{\infty} dk_x \int_{-\infty}^{\infty} dk_y \int_{-\infty}^{\infty} \frac{e^{i((x-x')k_x+(y-y')k_y+(z-z')k_z)}}{k_x^2+k_y^2+k_z^2}\, dk_z \qquad \text{(M5.5)}$$

This is an integral or spectral representation of $1/R$.

<u>Cylindrical Coordinates</u>

In cylindrical coordinates the field point may be designated by (x,r,γ) and the location of the singularity by (x',r',γ'). Then

$$\frac{1}{R} = \frac{1}{\sqrt{(x-x')^2+r^2+r'^2-2rr'\cos(\gamma-\gamma')}} \qquad \text{(M5.6)}$$

To convert (M5.5) to cylindrical coordinates take

$$k_y = k\sin\mu \quad ; \quad k_z = k\cos\mu$$

Then the volume element

$$dk_x dk_y dk_z = k\, dk\, d\mu\, dk_x$$

and (M5.5) transforms to

$$\frac{1}{R} = \frac{1}{2\pi^2} \int_{-\infty}^{\infty} dk_x \int_{-\infty}^{\infty} dk\, k \int_{-\infty}^{\infty} \frac{e^{i(k\omega + (x-x')k_x)}}{k_x^2 + k^2}\, d\mu \qquad \text{(M5.7)}$$

where

$$\omega = (y-y')\sin\mu + (z-z')\cos\mu$$

Now the k_x-integral can be effected either by use of a table of Fourier integral transforms (cf. Sneddon (*ibid.*)) or by the calculus of residues since the denominator when generalized by replacing k_x by the complex variable $\zeta = k_x + i\kappa$ is seen to have poles at $\zeta = \pm\, ik$, i.e.

$$k_x^2 + k^2 \to \zeta^2 + k^2 = (\zeta + ik)(\zeta - ik)$$

The contour of integration is the real axis plus a large semi-circle in the upper or lower ζ-plane depending on the sign of $(x-x')$ in order to secure a convergent integral on the semi-circle whose radius is then pushed to infinity. By the residue calculus in complex variables the k_x-integral has the value

$$\int_{-\infty}^{\infty} \frac{e^{i(x-x')k_x}}{k_x^2 + k^2}\, dk_x = \frac{2\pi i e^{-k|x-x'|}}{2ik} \qquad \text{(M5.8)}$$

so that (M5.7) reduces to

$$\frac{1}{R} = \frac{1}{2\pi} \int_0^{2\pi} d\mu \int_0^{\infty} e^{k(i\omega - |x-x'|)}\, dk \qquad \text{(M5.9)}$$

By introducing

$$y = -r\sin\gamma \quad , \quad z = r\cos\gamma$$
$$y' = -r'\sin\gamma' \quad , \quad z' = r'\cos\gamma'$$

(M5.9) can be written

$$\frac{1}{R} = \frac{1}{2\pi} \int_0^{2\pi} d\mu \int_0^{\infty} e^{ik(r\cos(\gamma+\mu) - r'\cos(\gamma'+\mu)}\, e^{-k|x-x'|}\, dk \qquad \text{(M5.10)}$$

As we see from (M5.6) that $1/R$ is a function of $\cos(\gamma-\gamma')$, then it must have a Fourier series expansion of the form

$$\frac{1}{R} = \sum_{m=0}^{\infty} A_m \cos m(\gamma - \gamma') \tag{M5.11}$$

where as usual the coefficients A_m are found from the operation

$$A_m = \frac{\varepsilon_m}{2\pi} \int_0^{2\pi} \frac{\cos m\alpha}{R(\alpha)} d\alpha \quad ; \quad \alpha = \gamma - \gamma' \tag{M5.12}$$

where

$$\varepsilon_m = \begin{cases} 1 \ , & m = 0 \\ 2 \ , & m > 0 \end{cases}$$

Applying (M5.12) to (M5.10) we have, with $w = \gamma + \mu$

$$A_m = \frac{\varepsilon_m}{(2\pi)^2} \int_0^{\infty} dk \, e^{-k|x-x'|} \int_0^{2\pi} d\alpha \cos m\alpha \int_0^{2\pi} e^{ik(r \cos w - r'(w-\alpha))} dw$$

Next we substitute $v = w - \alpha$

$$\cos m\alpha = \cos m(w-v) = \cos mw \cos mv + \sin mw \sin mw$$

The sine factors being odd functions do not contribute so we obtain

$$A_m = \frac{\varepsilon_m}{(2\pi)^2} \int_0^{\infty} dk \, e^{-k|x-x'|} \int_0^{2\pi} dw \cos mw \, e^{ikr \cos w} \int_0^{2\pi} \cos mv \, e^{ikr' \cos v} dv \tag{M5.13}$$

From McLachlan (1955) (p. 242) we find that the w- and v-integrals are integral representations of the Bessel functions of the first kind; hence we obtain

$$\frac{1}{R} = \sum_{m=0}^{\infty} \varepsilon_m \int_0^{\infty} J_m(kr) J_m(kr') e^{-k|x-x'|} dk \cos m(\gamma - \gamma') \tag{M5.14}$$

While (M5.14) is seen to separate the dependence on the angle $\gamma - \gamma'$, this has been purchased at the cost of introducing the k-integral.

The k-integral can be expressed in terms of the associated Legendre function of the second kind of half-integer degree and zero order, p. 518 and sequel. From Bateman *et al.* (1954), (Vol.II, No. 17, p. 50) we have

$$\int_0^{\infty} J_m(kr) J_m(kr') e^{-k|x-x'|} dk = \frac{1}{\pi\sqrt{rr'}} Q_{m-\frac{1}{2}}(Z) \tag{M5.15}$$

where

$$\mathcal{Z} = \frac{(x-x')^2+r^2+r'^2}{2rr'} \tag{M5.16}$$

Hence an alternate form for $1/R$ is

$$\frac{1}{R} = \frac{1}{\pi\sqrt{rr'}} \sum_{m=0}^{\infty} \varepsilon_m \, Q_{m-\frac{1}{2}}(\mathcal{Z}) \cos m(\gamma-\gamma') \tag{M5.17}$$

A third form can be found in the Bateman *et al.* (1954), (Vol. I, No. 47, p. 49), yielding

$$\frac{1}{R} = \frac{1}{\pi} \sum_{m=0}^{\infty} \varepsilon_m \int_{-\infty}^{\infty} I_m(|k|r')K_m(|k|r)e^{ik(x-x')}dk \cos m(\gamma-\gamma') \tag{M5.18}$$

where $r > r'$; for $r < r'$ interchange the arguments of the modified Bessel function of the first and second kind.

Equation (M5.17) is the most compact representation but has the disadvantage that the variables in the argument $\mathcal{Z}$ are not in a separable form as is the case in (M5.14) and (M5.18).

The argument $\mathcal{Z}$ arises as follows, from (M5.12)

$$A_m = \frac{\varepsilon_m}{2\pi} \int_{-\pi}^{\pi} \frac{\cos m\alpha}{\sqrt{(x-x')^2+r^2+r'^2-2rr'\cos\alpha}} d\alpha = \frac{1}{\pi\sqrt{2rr'}} \int_{-\pi}^{\pi} \frac{\cos m\alpha}{\sqrt{\mathcal{Z}-\cos\alpha}} d\alpha \tag{M5.19}$$

where $\mathcal{Z}$ is given by (M5.16). From Bateman (1964) (p. 464) we find that

$$\int_{0}^{\pi} \frac{\cos m\alpha}{\sqrt{\mathcal{Z}-\cos\alpha}} d\alpha = \sqrt{2}\, Q_{m-\frac{1}{2}}(\mathcal{Z}) \tag{M5.20}$$

As the integral is even in α we see that

$$A_m = \frac{1}{\pi\sqrt{rr'}} Q_{m-\frac{1}{2}}(\mathcal{Z}) \tag{M5.21}$$

The argument $\mathcal{Z}$ also arises naturally in toroidal coordinates as $\mathcal{Z}$ = a constant defines a torus, as shown in Figure M6.1, p. 519.

In all these representations of $1/R$ the series may be unfolded. This gives the following equations:

From (M5.14)

$$\frac{1}{R} = \sum_{m=-\infty}^{\infty} \int_0^{\infty} J_{|m|}(kr)\, J_{|m|}(kr')e^{-k|x-x'|}dk\; e^{im(\gamma-\gamma')} \tag{M5.22}$$

From (M5.17)

$$\frac{1}{R} = \frac{1}{\pi\sqrt{rr'}} \sum_{m=-\infty}^{\infty} Q_{|m|-\frac{1}{2}}(Z)\; e^{im(\gamma-\gamma')} \tag{M5.23}$$

From (M5.18)

$$\frac{1}{R} = \frac{1}{\pi} \sum_{m=-\infty}^{\infty} \int_{-\infty}^{\infty} I_{|m|}(|k|r')\; K_{|m|}(|k|r)\; e^{ik(x-x')}dk\; e^{im(\gamma-\gamma')} \tag{M5.24}$$

where $r > r'$; for $r < r'$ interchange the arguments of the modified Bessel function of the first and second kind.

6. PROPERTIES OF THE LEGENDRE FUNCTION $Q_{n-\frac{1}{2}}$

Since the associated Legendre Function of the second kind $Q_{n-\frac{1}{2}}(Z)$ of half-integer degree and zero order is used extensively in this book a closer examination of the properties of this function is worthwhile. Legendre functions in general are treated for example in Abramowitz & Stegun (1972).

In general we have the recursion formulas

$$Q_{n-1/2}(Z) = \frac{1}{2n}\left[Q'_{n+1/2}(Z) - Q'_{n-3/2}(Z)\right]$$

$$Q'_{n-1/2}(Z) = \frac{n-\frac{1}{2}}{Z^2-1}\left[ZQ_{n-1/2}(Z) - Q_{n-3/2}(Z)\right]$$

$$Q_{n+1/2}(Z) = \frac{2nZ}{n+\frac{1}{2}}\, Q_{n-1/2}(Z) - \frac{n-\frac{1}{2}}{n+\frac{1}{2}}\, Q_{n-3/2}(Z) \tag{M6.1}$$

$$Q'_{n+1/2}(Z) = \frac{2nZ}{n-\frac{1}{2}}\, Q'_{n-1/2}(Z) - \frac{n+\frac{1}{2}}{n-\frac{1}{2}}\, Q'_{n-3/2}(Z)$$

$$Q_{n-1/2}(Z) = Q_{-n-1/2}(Z)$$

$$Q'_{n-1/2}(Z) = Q'_{-n-1/2}(Z)$$

$Q_{-\frac{1}{2}}$ and $Q_{\frac{1}{2}}$ can be expressed in terms of *complete elliptical integrals of the first and second kind*, see Abramowitz & Stegun (*ibid.*). Use of the recursion formulas makes it possible to arrive at functional values for other degrees.

The argument $\mathfrak{Z}$ is in our cases always given by

$$\mathfrak{Z} = \frac{(x-x')^2 + r^2 + r'^2}{2rr'} \tag{M6.2}$$

where x, r are the coordinates of the field point and x', r' are the coordinates of the dummy point.

Equation (M6.2) for $\mathfrak{Z}$ = constant > 1 defines a torus, cf. Figure M6.1, whose locus of the centers of the ring is $R_c = \mathfrak{Z}r'$ and the radius of the ring $R_r = r'\sqrt{\mathfrak{Z}^2-1}$.

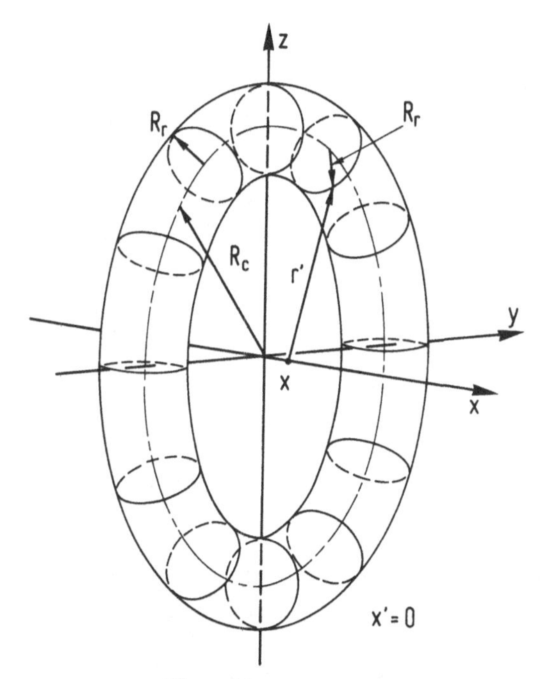

Figure M6.1 Torus.

In our applications we are often interested in the behaviour of the Legendre function when the argument is large, when the field point and dummy point are far away from each other, and when the argument is close to unity, when the field point and dummy point are close ($x \approx x'$, $r \approx r'$).

Large Arguments

The Legendre function can be expressed in a series

$$Q_{n-\frac{1}{2}}(Z) = \frac{1}{2}\sum_{m=0}^{\infty}\frac{\Gamma^2(m+n+\frac{1}{2})}{m!(2n+m)!}\left[\frac{2}{Z+1}\right]^{n+m+\frac{1}{2}} \quad ; \; Z > 1 \qquad \text{(M6.3)}$$

where Γ is the gamma function

$$\Gamma(n+\tfrac{1}{2}) = \frac{1\cdot3\cdot5\cdot7\cdot\ldots\cdot(2n-1)}{2^n}\sqrt{\pi} \qquad \text{(M6.4)}$$

Unfortunately this series in m converges slowly and hence the first term can be used as an approximation only for Z much larger than unity. The reason for this slow convergence can be seen by applying Cauchy's test for convergence, viz.,

$$\lim_{m\to\infty}\left[\frac{(m+1)\text{-th term}}{m\text{-th term}}\right] < 1$$

Here we find this limit to be $2/(Z+1)$ which is less than unity for $Z > 1$. However, this indicates that the series converges slowly; for rapidly converging series this ratio approaches zero as an inverse power of m.

Curiously, if we replace Z+1 by Z in (M6.3) and compare the values given by only the first term of (M6.3) (so modified) with tabulated values we find that the function can be approximated by

$$Q_{n-\frac{1}{2}}(Z) \approx \frac{1}{2}\,\frac{\Gamma^2(n+\frac{1}{2})}{(2n)!}\left[\frac{2}{Z}\right]^{n+\frac{1}{2}} \qquad \text{(M6.5)}$$

for $3/2 < Z < 3$ for $0 \leq n \leq 6$. By this is meant that for $n = 0$ this approximation applies from $Z = 3/2$ to infinity; for $n = 5$ the fit is reasonable from $Z = 3$ (where the error is 17 per cent), as illustrated in Figure M6.2.

We must view this type of approximation as expedient and quasi-empirical because the character of the function is not that of (M6.5) but that given by (M6.3) from which several terms must be used for a close fit, and for $n = 4, 5, 6$ the function and its approximation is very small beyond values of Z of 2.0.

For $n = 0$ we have from (M6.5) with Z defined in (M6.2)

$$Q_{-\frac{1}{2}}(Z) \approx \pi\frac{\sqrt{rr'}}{\sqrt{(x-x')^2+r^2+r'^2}} \qquad \text{(M6.6)}$$

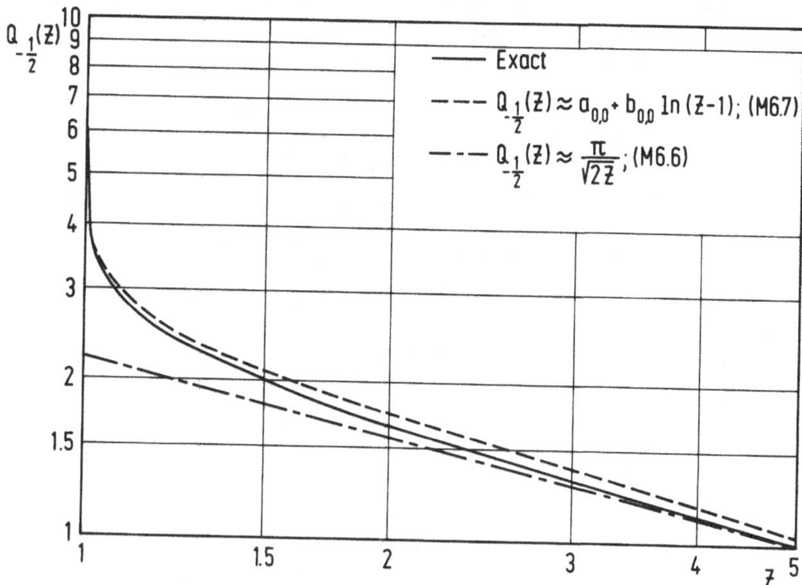

Figure M6.2 Approximations of Legendre function

Arguments Close to Unity

To conclude this exposition on the $Q_{n-\frac{1}{2}}$-function we must note its character in the neighborhood of $Z = 1$ where each order possesses a logarithmic singularity. A series representation has the form

$$Q_{n-\frac{1}{2}}(Z) = \sum_{m=0}^{\infty} a_{n,m}(Z-1)^m + [\ln(Z-1)] \sum_{m=0}^{\infty} b_{n,m}(Z-1)^m \qquad \text{(M6.7)}$$

where the coefficients are found from the recursion relations

$$a_{n,0} = \frac{5}{2}\ln 2 - 2\left[1-\delta_{n,0}\right]\sum_{j=1}^{n}\frac{1}{2j-1} \quad ; \quad \delta_{n,0} = \begin{cases} 1 & n=0 \\ 0 & n>0 \end{cases} \qquad \text{(M6.8)}$$

$$b_{n,0} = -\frac{1}{2}$$

$$a_{n,m+1} = a_{n,m}\,\frac{n^2 - \frac{1}{4} - m(m+1)}{2(m+1)^2} - b_{n,m}\,\frac{2\left[n^2 - \frac{1}{4}\right] + m(m+1)}{2(m+1)^3}$$

$$\text{(M6.9)}$$

$$b_{n,m+1} = b_{n,m}\,\frac{n^2 - \frac{1}{4} - m(m+1)}{2(m+1)^2}$$

The approximation obtained by keeping only the first term of the two series of (M6.7) ($m = 0$) is shown in Figure M6.2.

As $Z \to 1$ the dominating term is

$$Q_{n-\frac{1}{2}}(Z) \longrightarrow -\frac{1}{2}\ln(Z-1) = -\ln\sqrt{\frac{(x-x')^2+(r-r')^2}{2rr'}} \tag{M6.10}$$

7. OUTLINE OF CALCULUS OF VARIATIONS

The calculus of variations embraces the following problem (among many others):

Find the function $Y = y(x)$ where $y(x_1) = y_1$, $y(x_2) = y_2$ such that for some given function $F(x,y,y')$,

$$\int_{x_1}^{x_2} F(x,y,y')dx \tag{M7.1}$$

is either a maximum or a minimum, also called an *extremum* or *stationary* value. A curve which satisfies this property is called an *extremal.* An integral such as (M7.1) which takes on a numerical value for some class of functions $y(x)$ is often called a *functional.*

Suppose that $y(x)$ is the extremal. Then a neighboring curve is

$$Y = y(x) + \varepsilon\eta(x) \tag{M7.2}$$

where η is an arbitrary function and ε is an arbitrary small parameter. We take η such that $\eta(x_1) = \eta(x_2) = 0$. Then

$$I^*(\varepsilon) = \int_{x_1}^{x_2} F(x,y+\varepsilon\eta,y'+\varepsilon\eta')dx \tag{M7.3}$$

The condition for an extremum is

$$\frac{dI^*}{d\varepsilon} = 0 = \int_{x_1}^{x_2}\left[\left.\frac{\partial F}{\partial y}\right|_{\varepsilon=0}\eta + \left.\frac{\partial F}{\partial y'}\right|_{\varepsilon=0}\eta'\right]dx \tag{M7.4}$$

Integrating the second integral by parts, gives

$$\int_{x_1}^{x_2}\frac{\partial F}{\partial y}\eta\, dx + \left.\frac{\partial F}{\partial y'}\eta\right|_{x_1}^{x_2} - \int_{x_1}^{x_2}\eta\frac{d}{dx}\left[\frac{\partial F}{\partial y'}\right]dx = 0 \tag{M7.5}$$

As η is entirely arbitrary the only way the integral vanishes is for

$$\frac{d}{dx}\left[\frac{\partial F}{\partial y'}\right] - \frac{\partial F}{\partial y} = 0 \tag{M7.6}$$

This is ***Euler's Equation***[32]. It is a necessary but not sufficient condition for an extremal.

In many problems we seek a function which renders a given integral

$$I^* = \int_{x_1}^{x_2} F(x,y(x),y'(x))\, dx \tag{M7.7}$$

a maximum or a minimum while at the same time holds the integral of another given function G, say,

$$J^* = \int_{x_1}^{x_2} G(x,y,y')\, dx \tag{M7.8}$$

to some constant. This is a problem in the calculus of variations in which a *constraint condition* is applied.

Such problems can generally be solved by using the method of ***Lagrange multipliers***. To accomplish this the procedure is to form the sum

$$I^* - \ell J^* = \int_{x_1}^{x_2} (F - \ell G)\, dx \tag{M7.9}$$

where ℓ is the Lagrange multiplier.

The resulting integral (M7.9) must be an extremum and one is then lead to the Euler equation

$$\frac{d}{dx}\left[\frac{\partial H}{\partial y'}\right] - \frac{\partial H}{\partial y} = 0 \tag{M7.10}$$

where

$$H = F - \ell G \tag{M7.11}$$

The value of ℓ is found from a knowledge of the constraint.

8. TABLE OF AIRFOIL INTEGRALS

The following (M8.1)-(M8.22) are the Cauchy principal-value integrals for $x^2 \leq 1$.

$$\int_{-1}^{1} \frac{1}{x-x'}\, dx' = \ln \frac{1+x}{1-x} \tag{M8.1}$$

$$\int_{-1}^{1} \frac{x'}{x-x'}\, dx' = x \ln \frac{1+x}{1-x} - 2 \tag{M8.2}$$

[32] This Euler's Equation should not be confused with the Euler Equations of fluid mechanics, Equations (1.8) or (1.12)

$$\int_{-1}^{1} \frac{x'^2}{x-x'}\, dx' = x\left[x \ln \frac{1+x}{1-x} - 2\right] \tag{M8.3}$$

$$\int_{-1}^{1} \frac{x'^3}{x-x'}\, dx' = x^2\left[x \ln \frac{1+x}{1-x} - 2\right] - \frac{2}{3} \tag{M8.4}$$

$$\int_{-1}^{1} \frac{x'^n}{x-x'}\, dx' = x \int_{-1}^{1} \frac{x'^{n-1}}{x-x'}\, dx' - \frac{1-(-1)^n}{n} \tag{M8.5}$$

$$\int_{-1}^{1} \frac{1}{\sqrt{1-x'^2}\,(x-x')}\, dx' = 0 \tag{M8.6}$$

$$\int_{-1}^{1} \frac{x'}{\sqrt{1-x'^2}\,(x-x')}\, dx' = -\pi \tag{M8.7}$$

$$\int_{-1}^{1} \frac{x'^2}{\sqrt{1-x'^2}\,(x-x')}\, dx' = -\pi x \tag{M8.8}$$

$$\int_{-1}^{1} \frac{x'^3}{\sqrt{1-x'^2}\,(x-x')}\, dx' = -\pi\left[x^2 + \frac{1}{2}\right] \tag{M8.9}$$

$$\int_{-1}^{1} \frac{x'^4}{\sqrt{1-x'^2}\,(x-x')}\, dx' = -\pi x\left[x^2 + \frac{1}{2}\right] \tag{M8.10}$$

$$\int_{-1}^{1} \frac{x'^5}{\sqrt{1-x'^2}\,(x-x')}\, dx' = -\pi\left[x^4 + \frac{1}{2}x^2 + \frac{3}{8}\right] \tag{M8.11}$$

$$\int_{-1}^{1} \frac{x'^6}{\sqrt{1-x'^2}\,(x-x')}\, dx' = -\pi x\left[x^4 + \frac{1}{2}x^2 + \frac{3}{8}\right] \tag{M8.12}$$

$$\int_{-1}^{1} \frac{x'^n}{\sqrt{1-x'^2}\,(x-x')}\, dx' = x \int_{-1}^{1} \frac{x'^{n-1}}{\sqrt{1-x'^2}\,(x-x')}\, dx' - \frac{\pi}{2}\left(1-(-1)^n\right) \frac{1\cdot 3\cdot \ldots \cdot (n-2)}{2\cdot 4\cdot \ldots \cdot (n-1)} \tag{M8.13}$$

$$\int_{-1}^{1} \frac{\sqrt{1-x'^2}}{x-x'}\, dx' = \pi x \tag{M8.14}$$

$$\int_{-1}^{1} \frac{x'\sqrt{1-x'^2}}{x-x'}\, dx' = \pi\left[x^2 - \frac{1}{2}\right] \tag{M8.15}$$

$$\int_{-1}^{1} \frac{x'^2\sqrt{1-x'^2}}{x-x'}\, dx' = \pi x\left[x^2 - \frac{1}{2}\right] \tag{M8.16}$$

$$\int_{-1}^{1} \frac{x'^3\sqrt{1-x'^2}}{x-x'}\, dx' = \pi\left[x^4 - \frac{1}{2}\,x^2 - \frac{1}{8}\right] \tag{M8.17}$$

$$\int_{-1}^{1} \frac{\sqrt{1+x'}}{\sqrt{1-x'}\;(x-x')}\, dx' = -\,\pi \tag{M8.18}$$

$$\int_{-1}^{1} \frac{1}{\sqrt{1+x'}\;(x-x')}\, dx' = \frac{1}{\sqrt{1+x}}\,\ln\frac{\sqrt{2}+\sqrt{1+x}}{\sqrt{2}-\sqrt{1+x}} \tag{M8.19}$$

$$\int_{-1}^{1} \frac{\sqrt{1+x'}}{x-x'}\, dx' = \sqrt{1+x}\,\ln\frac{\sqrt{2}+\sqrt{1+x}}{\sqrt{2}-\sqrt{1+x}} - 2\sqrt{2} \tag{M8.20}$$

$$\int_{a}^{b} \sqrt{\frac{b-x'}{x'-a}}\,\frac{1}{x-x'}\, dx' = \begin{cases} \pi & ;\ a < x < b \\ \pi\left[1 - \sqrt{\left|\dfrac{b-x}{a-x}\right|}\,\right] & ;\ \text{otherwise} \end{cases} \tag{M8.21}$$

$$\int_{a}^{b} \sqrt{\frac{x'-a}{b-x'}}\,\frac{1}{x-x'}\, dx' = \begin{cases} -\pi & ;\ a < x < b \\ -\pi\left[1 - \sqrt{\left|\dfrac{a-x}{b-x}\right|}\,\right] & ;\ \text{otherwise} \end{cases} \tag{M8.22}$$

$$\int_0^s \sqrt{\frac{x'}{s-x'}}\,\frac{1}{1+x'^2}\,dx' = \frac{\pi}{\sqrt{2}}\,\frac{\sqrt{r^2-1}}{r^2} \tag{M8.23}$$

$$\int_0^s \sqrt{\frac{x'}{s-x'}}\,\frac{x'}{1+x'^2}\,dx' = \pi - \frac{\pi}{\sqrt{2}}\,\frac{\sqrt{r^2+1}}{r^2} \tag{M8.24}$$

$$\int_0^s \sqrt{\frac{x'}{s-x'}}\,\frac{1}{(1+x'^2)^2}\,dx' = \frac{\pi}{4\sqrt{2}}\,\frac{s(s^2+3)\sqrt{r^2+1} - 2\sqrt{r^2-1}}{r^8} \tag{M8.25}$$

$$\int_0^s \sqrt{\frac{x'}{s-x'}}\,\frac{x'}{(1+x'^2)^2}\,dx' = \frac{\pi}{4\sqrt{2}}\,\frac{2s^2\sqrt{r^2+1} + s(s^2-1)\sqrt{r^2-1}}{r^8} \tag{M8.26}$$

$$\int_0^s \sqrt{\frac{s-x'}{x'}}\,\frac{1}{1+x'^2}\,dx' = \frac{\pi}{\sqrt{2}}\sqrt{r^2-1} \tag{M8.27}$$

$$\int_0^s \sqrt{\frac{s-x'}{x'}}\,\frac{x'}{1+x'^2}\,dx' = \frac{\pi}{\sqrt{2}}\sqrt{r^2+1} - \pi \tag{M8.28}$$

$$\int_0^s \sqrt{\frac{s-x'}{x'}}\,\frac{1}{(1+x'^2)^2}\,dx' = \frac{\pi}{4\sqrt{2}}\,\frac{s\sqrt{r^2+1} + (3s^2+2)\sqrt{r^2-1}}{r^4} \tag{M8.29}$$

$$\int_0^s \sqrt{\frac{s-x'}{x'}}\,\frac{x'}{(1+x'^2)^2}\,dx' = \frac{\pi}{4\sqrt{2}}\,\frac{s\sqrt{r^2-1}}{r^2} \tag{M8.30}$$

$$\fint_0^s \sqrt{\frac{s-x'}{x'}}\,\frac{x'}{(1+x'^2)(x-x')}\,dx' = -\frac{\pi}{\sqrt{2}}\,\frac{\sqrt{r^2-1} - x\sqrt{r^2+1}}{1 + x^2}\;;\; 0 < x < s \tag{M8.31}$$

$$\fint_0^s \sqrt{\frac{x'}{s-x'}}\,\frac{1}{(1+x'^2)(x-x')}\,dx' = \frac{\pi}{\sqrt{2}}\,\frac{\sqrt{r^2+1} - x\sqrt{r^2-1}}{r^2(1 + x^2)}\;;\; 0 < x < s \tag{M8.32}$$

For all integrals (M8.23)-(M8.32), $r^4 = 1 + s^2$

Integrals (M8.23)-(M8.32) are taken from Kinnas (1985).

References

Abbott, I.H. & von Doenhoff, A.E. (1959). *Theory of Wing Sections.* New York: Dover Publications.

Abramowitz, M. & Stegun, I.A. (1972). *Handbook of Mathematical Functions.* 10th printing. New York: Dover Publications.

Acosta, A.J. (1955). *A Note on Partial Cavitation of Flat-Plate Hydrofoils.* Report no. E-19.9. Pasadena, Cal.: California Inst. of Tech., Hydrodynamic Laboratory.

Andersen, P. (1986). *On Optimum Lifting Line Propeller Calculation.* Report no. 325. Lyngby: The Danish Center for Applied Mathematics and Mechanics, The Technical Univ. of Denmark.

Andersen, Sv.V. & Andersen, P. (1987). Hydrodynamic design of propellers with unconventional geometry. *Trans. RINA*, vol. 129, pp. 201-21.

Ashley, H. & Landahl, M. (1977). *Aerodynamics of Wings and Bodies.* Reading, Mass.: Addison-Wesley Publishing Co.

Auslaender, J. (1962). *Low Drag Supercavitating Hydrofoil Sections.* Technical Report 001-7. Laurel, Md.: Hydronautics Inc.

Barnaby, S.W. (1897). On the formation of cavities in water by screw propellers at high speeds. *Trans. INA*, vol. 34, pp. 139-44. London: INA.

Basu, B.C. & Hancock, G.J. (1978). The unsteady motion of a two-dimensional aerofoil in incompressible inviscid flow. *Journal of Fluid Mechanics*, vol. 87, 159-78.

Batchelor, G.K. (1967). *An Introduction to Fluid Dynamics.* Cambridge: Cambridge University Press.

Bateman, H. (1964). *Partial Differential Equations of Mathematical Physics.* Cambridge: Cambridge University Press.

Bateman, H., Erdélyi, A. (ed.), Magnus, W., Oberhettinger, F. & Trisconi, F.G. (1954). *Tables of Integral Transforms.* London: McGraw-Hill Book Company.

Bauschke, W. & Lederer, L. (1974). *Zur numerischen Berechnung der Druck-Verteilung und der Kräfte an Propellern im Schiffsnachstrom.* Report 309. Hamburg: Institut für Schiffbau der Univ. Hamburg.

Bavin, V.F., Vashkevich, M.A. & Miniovich, I.Y. (1968). Pressure field around a propeller operating in spatially non-uniform flow. In *Proc. Seventh Symp. on Naval Hydrodynamics*, ed. R.D. Cooper & S.W. Doroff, pp. 3–15. Arlington, Va.: Office of Naval Research – Department of the Navy.

Betz, A. (1919). Schraubenpropeller mit geringstem Energieverlust. *Nachrichten von der Königlichen Gesellschaft der Wissenschaften zu Göttingen. Mathematisch-physikalische Klasse.* 193–217.

Betz, A. (1920). Eine Erweiterung der Schraubenstrahl-Theorie. *Zeitschrift für Flugtechnik und Motorluftschiffart*, XI. Jahrgang, Heft 7 u. 8, 105–10.

Betz, A. & Petersohn, E. (1931). Anwendung der Theorie der freien Strahlen. *Ingenieur-Archiv*, vol. 2, 190–211.

Blaurock, J. (1990). An appraisal of unconventional aftbody configurations and propulsion devices. *Marine Technology*, vol. 27, no. 6, 325–36.

Blaurock, J. & Jacob, M.C. (1987). *Open Water Tests of Propellers plus Vane Wheel.* HSVA Report PF-13-87. Hamburg: Hamburg Ship Model Basin.

Boswell, R.J. (1971). *Design, Cavitation Performance, and Open Water Performance of a Series of Research Skewed Propellers.* Ship Performance Department, Research and Development Report 3339. Washington, D.C.: Naval Ship Research and Development Center.

Breslin, J.P. (1957). Application of ship-wave theory to the hydrofoil of finite span. *Journal of Ship Research*, vol. 1, no. 1, 27–35,55.

Breslin, J.P. (1959). A theory for the vibratory effects produced by a propeller on a large plate. *Journal of Ship Research*, vol. 3, no. 3, 1–10.

Breslin, J.P. (1962). Review and extension of theory for near field propeller-induced vibratory effects. In *Proc. Fourth Symp. on Naval Hydrodynamics*, ed. B.L. Silverstein, pp. 603–40. Washington, D.C.: Office of Naval Research – Department of the Navy.

Breslin, J.P. (1968). *The Transverse Vibratory Force on a Submerged Cylindrical Hull Generated by an Offset Propeller in a Wake.* Davidson Laboratory, Technical Note 794. Hoboken, N.J.: Stevens Inst. of Tech.

Breslin, J.P. (1969). Report on vibratory propeller, appendage and hull forces and moments. In *Proc. Twelfth International Towing Tank Conference (ITTC)*, ed. Istituto Nazionale per Studi ed Esperienze di Architettura Navale, pp. 280–90. Rome: Consiglio Nazionale delle Ricerche.

Breslin, J.P. (1970). Theoretical and experimental techniques for practical estimation of propeller-induced vibratory forces. *Trans. SNAME*, vol. 78, pp. 23–40. New York, N.Y.: SNAME.

Breslin, J.P. (1971). Exciting-force operators for ship propellers. *Journal of Hydronautics*, vol. 5, no. 3, 85–90.

Breslin, J.P. (1977). A theory for the vibratory forces on a flat plate arising from intermittent propeller-blade cavitation. In *Proc. Symp. on Hydrodynamics of Ship and Offshore Propulsion Systems*, paper 6, session 3. Høvik: Det Norske Veritas.

Breslin, J.P. (1981). Propeller-induced hull pressures and forces. In *Proc. Third Int. Conference on Numerical Ship Hydrodynamics*, ed. J.-C. Dern & H.J. Haussling, pp. 427–45. Paris: Bassin d'Essais des Carènes.

Breslin, J.P. (1992). An analytical theory of propeller-generated effective wake. In *Wave Asymptotics, The proceedings of the meeting to mark the retirement of Professor Fritz Ursell*, ed. P.A. Martin & G.R. Wickham, pp. 160–83. Cambridge: Cambridge University Press.

Breslin, J.P & Kowalski T. (1964). Experimental study of propeller-induced vibratory pressures on simple surfaces and correlation with theoretical predictions. *Journal of Ship Research*, vol. 8, no. 3, 15-28.

Breslin, J.P. & Landweber, L. (1961). A manual for calculation of inception of cavitation on two- and three-dimensional forms. *SNAME T & R Bulletin*, no. 1-21. New York, N.Y.: SNAME.

Breslin, J.P. & Tsakonas, S. (1959). Marine propeller pressure field due to loading and thickness effects. *Trans. SNAME*, vol. 67, pp. 386–422. New York, N.Y.: SNAME.

Breslin, J.P., Van Houten, R.J., Kerwin, J.E. & Johnsson, C.-A. (1982). Theoretical and experimental propeller-induced hull pressures arising from intermittent blade cavitation, loading and thickness. *Trans. SNAME*, vol. 90, pp. 111-51. New York, N.Y.: SNAME.

Brockett, T. (1966). *Minimum Pressure Envelopes for Modified NACA-66 Sections with NACA a=0.8 Camber and BuShips Type I and Type II Sections.* Hydromechanics Laboratory, Research and Development Report 1790. Washington, D.C.: David Taylor Model Basin – Department of the Navy.

Brown, G.L. & Roshko, A. (1974). On density effects and large structure in turbulent mixing layers. *Journal of Fluid Mechanics*, vol. 64, 775-816.

Burrill, L.C. (1944). Calculation of marine propeller performance characteristics. *Trans. North East Coast Institution of Engineers and Shipbuilders*, vol. 60, pp. 269–94. Newcastle upon Tyne: North East Coast Institution of Engineers and Shipbuilders.

Burrill, L.C. (1951). Sir Charles Parsons and cavitation. *Trans. The Institute of Marine Engineers*, vol. LXIII, no. 8, pp. 149–67. London: The Institute of Marine Engineers.

Burrill, L.C. & Emerson, A. (1962). Propeller cavitation: further tests on 16 in. propeller models in the King's College cavitation tunnel. *Trans. North East Coast Institution of Engineers and Shipbuilders*, vol. 79, pp. 295–320. Newcastle upon Tyne: North East Coast Institution of Engineers and Shipbuilders.

Carrier, G.F. & Pearson, C.E. (1976). *Partial Differential Equations. Theory and Techniques.* New York, N.Y.: Academic Press.

Caster, E.B., Diskin, J.A. & La Fone, T.A. (1975). *A Lifting Line Program for Preliminary Design of Propellers.* Ship Performance Department Report, SPD-595-01. Washington, D.C.: David W. Taylor Naval Ship Research and Development Center.

Chertock, G. (1965). Forces on a submarine hull induced by the propeller. *Journal of Ship Research*, vol. 9, no. 2, 122-30.

Coney, W.B. (1989). *A Method for the Design of a Class of Optimum Marine Propulsors.* Ph. D. thesis. Cambridge, Mass.: Dept. of Ocean Eng., MIT.

Cox, B.D. & Reed, A.M. (1988). Contrarotating propellers – design theory and application. In *Proc. Propellers '88 Symposium*, pp. 15.1-29. Jersey City, N.J.: SNAME.

Cox, B.D., Vorus, W.S., Breslin, J.P. & Rood, E.P. (1978). Recent theoretical and experimental developments in the prediction of propeller induced vibration forces on nearby boundaries. In *Proc. Twelfth Symp. on Naval Hydrodynamics*, pp. 278-99. Washington, D.C.: National Academy of Sciences.

Cox, G.G. (1961). Corrections to the camber of constant pitch propellers. *Trans. RINA*, vol. 103, pp. 227-43. London: RINA.

Cumming, R.A., Morgan, W.B. & Boswell, R.J. (1972). Highly skewed propellers. *Trans. SNAME*, vol. 80, pp. 98-135. New York, N.Y.: SNAME.

Cummings, D.E. (1968). *Vortex Interactions in a Propeller Wake.* Technical report 68-12. Cambridge, Mass.: Dept. of Naval Architecture and Marine Eng., MIT.

Dashnaw, F.J., Forrest, A.W., Hadler, J.B. & Swensson, C.G. (1980). *Design of a Contrarotating Propulsion System for High-Speed Trailerships.* SNAME Philadelphia Section Paper. New York: SNAME.

Davies, T.V. (1970). Steady two-dimensional cavity flow past an aerofoil using linearized theory. *Quart. Journal of Mech. and Applied Math.*, vol. XXIII, pt. I, 49-76.

Denny, S.B. (1967). *Comparisons of Experimentally Determined and Theoretically Predicted Pressures in the Vicinity of a Marine Propeller.* Hydrodynamics Laboratory, Research and Development Report 2349. Washington, D.C.: Naval Ship Research and Development Center.

Dirac, P.A.M. (1947). *The Principles of Quantum Mechanics*, Third edn.. Oxford: Clarendon Press.

Dyne, G. (1973). Systematic studies of accelerating ducted propellers in axial and inclined flows. In *Trans. of Symp. on Ducted Propellers*, pp. 114-24. London: RINA.

English, J.W. (1962). *The Application of a Simplified Lifting Surface Technique to the Design of Marine Propellers.* NPL Ship Division Report No. 30, (July). Teddington: National Physical Laboratory.

Eppler, R. (1960). Ergebnisse gemeinsamer Anwendung von Grenzschicht- und Profiltheorie. *Zeitschrift für Flugwissenschaften*, 8. Jahrgang, Heft 9, 247–60. (Also NASA Tech. Transl. NASA TT F-15, 417 1974.)

Eppler, R. (1963). Praktische Berechnung laminarer und turbulenter Absauge-Grenzschichten. *Ingenieur Archiv*, vol. 32, 221–45.

Eppler, R. (1969). Laminarprofile für Reynolds-Zahlen grösser als $4 \cdot 10^6$. *Ingenieur Archiv*, vol. 38, 232–40.

Eppler, R. & Shen, Y.T. (1979). Wing sections for hydrofoils – part 1: symmetrical profiles. *Journal of Ship Research*, vol. 23, no. 3, 209–17.

Eppler, R. & Somers, D.M. (1980). *A Computer Program for the Design and Analysis of Low Speed Airfoils.* NASA T.M. 80210. Washington D.C.: National Aeronautics and Space Administration.

Falcão de Campos, J.A.C. (1983). *On the Calculation of Ducted Propeller Performance in Axisymmetric Flows.* Doctor's thesis, publ. no. 696. Wageningen: Netherlands Ship Model Basin.

Falcão de Campos, J.A.C. & van Gent, W. (1981). *Effective Wake of an Open Propeller in Axisymmetric Shear Flow.* Report no. 50030-3-SR. Wageningen: Netherlands Ship Model Basin.

Franc, J.P. & Michel J.M. (1985). Attached cavitation and the boundary layer: experimental investigation and numerical treatment. *Journal of Fluid Mechanics*, vol. 154, 63–90.

Freeman, H.B. (1942). *Calculated and Observed Speeds of Cavitation About Two- and Three-Dimensional Bodies in Water.* Report 495. Washington, D.C.: The David W. Taylor Model Basin – United States Navy.

Gawn, R.W.L. (1937). Results of experiments on model screw propellers with wide blades. *Trans. INA*, vol. LXXIX, pp. 159–87. London: INA.

Gawn, R.W.L. (1953). Effect of pitch and blade width on propeller performance. *Trans. INA*, vol. 95, pp. 157–93. London: INA.

van Gent, W. (1975). Unsteady lifting-surface theory for ship screws: derivation and numerical treatment of integral equation. *Journal of Ship Research*, vol. 19, no. 4, 243–53.

van Gent, W. (1978). Discussion of Kerwin, J.E. & Lee, C.-S.: Prediction of steady and unsteady marine propeller performance by numerical lifting-surface theory. *Trans. SNAME*, vol. 86, p. 249. New York, N.Y.: SNAME.

Geurst, J.A. (1959). Linearized theory for partially cavitated hydrofoils. *Int. Shipbuilding Progress*, vol. 6, no. 60, 369–83.

Geurst, J.A. (1960). Linearized theory for fully cavitated hydrofoils. *Int. Shipbuilding Progress*, vol. 7, no. 65, 17-27.

Geurst, J.A. (1961). *Linearized Theory of Two-Dimensional Cavity Flows.* Doctoral thesis. Delft: Univ. of Delft.

Giesing, J.P. (1968). Non-linear two-dimensional unsteady potential flow with lift. *Journal of Aircraft*, vol. 5, no. 2. 135–43.

Glauert, H. (1948). *The Elements of Aerofoil and Airscrew Theory.* Second edn.. Cambridge: Cambridge University Press.

Glover, E.J. (1987). Propulsive devices for improved propulsive efficiency. In *Trans. Insitute of Marine Engineers*, vol. 99, paper 31, pp. 23–9. London: The Institute of Marine Engineers.

Goldstein, S. (1929). On the vortex theory of screw propellers. In *Proc. Royal Society of London*, series A, vol. 123, pp. 440–65. London: The Royal Society.

Gomez, G.P. (1976). Una innovacion en el proyecto de helices. *Inginieria Naval.* October.

Goodman, T.R. (1979). Momentum theory of a propeller in a shear flow. *Journal of Ship Research*, vol. 23, no. 4, 242–52.

Goodman, T.R. (1982). Comments on Jacobs W. & Tsakonas, S.: Propeller loading-induced velocity field by means of unsteady lifting surface theory. *Journal of Ship Research*, vol. 26, no. 4, 266–8.

Goodman, T.R. & Breslin, J.P. (1980). *Feasibility Study of the Effectiveness of Tip Sails on Propeller Performance.* Report MA-RD 94081006. Hoboken, N.J.: Stevens Inst. of Tech.

Greeley, D.S. & Kerwin, J.E. (1982). Numerical methods for propeller design and analysis in steady flow. *Trans. SNAME*, vol. 90, pp. 415–53. New York, N.Y.: SNAME.

Greenberg, M.D. (1972). Nonlinear actuator disk theory. *Zeitschrift für Flugwissenschaften*, 20. Jahrgang, Heft 3, 90–8.

Greenberg, M.D. & Powers, S.R. (1970). *Nonlinear Actuator Disk Theory and Flow Field Calculations, including Nonuniform Loading.* NASA Contractor Report, CR-1672. Washington, D.C.: National Aeronautics and Space Administration.

Grim, O. (1966). Propeller und Leitrad. *Jahrbuch der Schiffbautechnischen Gesellschaft*, 60. Band, pp. 211–32. Berlin: Springer-Verlag.

Grim, O. (1980). Propeller and vane wheel. *Journal of Ship Research*, vol. 24, no. 4, 203–26.

Grothues-Spork, H. (1988). Bilge vortex control devices and their benefits for propulsion. *Int. Shipbuilding Progress*, vol. 35, no. 402, 183–214.

Hadamard, J. (1923). *Lectures on Cauchy's Problem in Linear Differential Equations.* New Haven: Yale University Press. (Re-issued by Dover Publications, New York (1952)).

Hadler, J.B. (1969). Contrarotating propeller propulsion – a state-of-the-art report. *Marine Technology*, vol. 6, no. 3, 281–9.

Hadler, J.B. & Cheng, H.M. (1965). Analysis of experimental wake data in way of propeller plane of single and twin-screw ship models. *Trans. SNAME*, vol. 73, pp. 287-414. New York, N.Y.: SNAME.

Hanaoka, T. (1962) Hydrodynamics of an oscillating screw propeller. In *Proc. Fourth Symp. on Naval Hydrodynamics*, ed. B.L. Silverstein, pp. 79-124. Washington, D.C.: Office of Naval Research – Department of the Navy.

Hanaoka, T. (1964). Linearized theory of cavity flow past a hydrofoil of arbitrary shape. *Journal of the Society of Naval Architects of Japan*, vol. 115. Also (1969). *Selected Papers from The Journal of The Society of Naval Architects of Japan*, vol. 3, pp. 56-74. Tokyo: The Society of Naval Architects of Japan.

Hanaoka, T. (1969a). *A New Method for Calculating the Hydrodynamic Load Distributions on a Lifting Surface.* Report of Ship Research Institute, vol. 6, no. 1. Tokyo: Ship Research Institute.

Hanaoka, T. (1969b). *Numerical Lifting-Surface Theory of a Screw Propeller in Non-Uniform Flow (Part I: Fundamental Theory).* Report of Ship Research Institute, vol. 6, no. 5. Tokyo: Ship Research Institute.

Hankel, H. (1861). *Zur allgemeinen Theorie der Bewegung der Flüssigkeiten.* Göttingen.

Harvald, Sv.Aa. (1983). *Resistance and Propulsion of Ships.* New York, N.Y.: John Wiley and Sons.

Harvald, Sv.Aa. & Hee, J.M. (1978). Wake Distributions. *75 Jahre VWS*, Mitteilungen der Versuchsanstalt für Wasserbau und Schiffbau, Heft 4, pp. 221-31. Berlin: Versuchsanstalt für Wasserbau und Schiffbau – VWS.

Harvald, Sv.Aa. & Hee, J.M. (1984). An outline of a shipyard's specification for ship model propulsion experiments. In *Proc. 3rd. Int. Congress of The Int. Mar. Assoc. of the East Mediterranean*, vol. I, pp. 609-16. Athens: Hellenic Inst. of Marine Tech.

Harvald, Sv.Aa. & Hee, J.M. (1988). The components of the propulsive efficiency of ships in relation to the design procedure. In *Proc. SNAME Spring Meeting, STAR Symposium, 3rd International Marine Systems Design Conference (IMSDC)*, pp. 213-31. Jersey City, N.J.: SNAME.

Helmholtz, H. (1868). Über discontinuirliche Flüssigkeits-Bewegungen. *Monatsberichte der königlich preussischen Akademie der Wissenschaften zu Berlin*, vol. 23, 215-28.

Hess, J. & Smith, A.M.O. (1962). *Calculation of Non-Lifting Potential Flow about Arbitrary Three-Dimensional Bodies.* Douglas Aircraft Co. Report no. E.S. 40622. Long Beach, Cal.: Douglas Aircraft Co.

Hess, J.L. & Valarezo, W.O. (1985). Calculation of steady flow about propellers by means of a surface panel method. In *Proc. AIAA 23rd Aerospace Sciences Meeting*, session 50, AIAA paper no. 85-0283. New York, N.Y.: American Institute of Aeronautics and Astronautics.

Hoeijmakers, H.W.M. (1983). Computational vortex flow aerodynamics. In *AGARD Conference Proceedings No. 342: Aerodynamics of Vortical Type Flows in Three Dimensions*, paper 18, pp. 18.1-35. Paris: Advisory Group for Aerospace Research and Development.

Holden, K.O., Fagerjord, O. & Frostad, R. (1980). Early design-stage approach to reducing hull surface forces due to propeller cavitation. *Trans. SNAME*, vol. 88, pp. 403-42. New York, N.Y.: SNAME.

Holden, K. & Øfsti, O. (1975). *Investigations on Propeller Cavitation Thickness Distribution.* Report 74-259-C. Høvik: Det Norske Veritas.

Hoshino, T. (1989a). Hydrodynamic analysis of propellers in steady flow using a surface panel method. *Journal of the Society of Naval Architects of Japan*, vol. 165, 55-70.

Hoshino, T. (1989b). Hydrodynamic analysis of propellers in steady flow using a surface panel method (2nd report). *Journal of the Society of Naval Architects of Japan*, vol. 166, 79-92.

Hough, G.R. & Ordway, D.E. (1965). The generalized actuator disk. *Developments in Theoretical and Applied Mechanics,* vol. 2, ed. W.A. Shaw, pp. 317-36. Oxford: Pergamon Press.

Huang, T.T. & Groves, N.C. (1980). Effective wake: theory and experiment. In *Proc. Thirteenth Symp. on Naval Hydrodynamics*, ed. T. Inui, pp. 651-73. Tokyo: The Shipbuilding Research Association of Japan.

Huse, E. (1968). *The Magnitude and Distribution of Propeller-Induced Surface Forces on a Single-Screw Ship Model.* Publ. no. 100. Trondheim: Norwegian Ship Model Experiment Tank.

Isay, W.H. (1970). *Moderne Probleme der Propellertheorie.* Engineering Science Library, ed. I. Szabó. Berlin: Springer-Verlag.

Isay, W.H. (1981). *Kavitation.* Hamburg: Schiffahrts-Verlag "Hansa", C. Schroedter GmbH & Co. KG.

Ito, T., Takahashi, H. & Koyama, K. (1977). Calculation on unsteady propeller forces by lifting surface theory. In *Proc. Symp. on Hydrodynamics of Ship and Offshore Propulsion Systems*, paper 7, session 2. Høvik: Det Norske Veritas.

Itoh, S. (1987). Study of the propeller with small blades on the blade tips (2nd report: cavitation characteristics). *Journal of the Society of Naval Architects of Japan*, vol. 161, 82-91.

Itoh, S., Tagori, T., Ishii, N. & Ide, T. (1986). Study of the propeller with small blades on the blade tips (1st report). *Journal of the Society of Naval Architects of Japan*, vol. 159, 82-90.

ITTC (1990a). Report of the Resistance and Flow Committee. In *Proc. 19th International Towing Tank Conference*, vol. 1, pp. 55-107. El Pardo: Canal de Experiencias Hidrodinamicas.

ITTC (1990b). Report of the Propulsor Committee. In *Proc. 19th International Towing Tank Conference*, vol. 1, pp. 109-60. El Pardo: Canal de Experiencias Hidrodinamicas.

ITTC Standard Symbols (1976). *International Towing Tank Conference Standard Symbols,* T.M. No. 500. Wallsend: The British Ship Research Association.

Jacobs, W.R., Mercier, J. & Tsakonas, S. (1972). Theory and measurements of the propeller-induced vibratory pressure field. *Journal of Ship Research*, vol. 16, no. 2, 124-39.

Jacobs, W.R. & Tsakonas, S. (1973). Propeller loading-induced velocity field by means of unsteady lifting surface theory. *Journal of Ship Research*, vol. 17, no. 3, 129-39.

Jacobs, W.R. & Tsakonas, S. (1975). Propeller-induced velocity field due to thickness and loading effects. *Journal of Ship Research*, vol. 19, no. 1, 44-56.

Japan Ship Exporters' Association (1991). *Shipbuilding and Marine Engineering in Japan 1991.* Tokyo: Japan Ship Exporters' Association & The Shipbuilders' Association of Japan.

Johannessen, H. & Skaar, K.T. (1980). Guidelines for prevention of excessive ship vibration. *Trans. SNAME*, vol. 88, pp. 319-56. New York, N.Y.: SNAME.

Johnson, V.E. jr. (1957). *Theoretical Determination of Low-drag Supercavitating Hydrofoils and their Two-dimensional Characteristics at Zero Cavitation Number.* NACA RM L57I16. Langley, Va.: National Advisory Committee for Aeronautics.

Johnson, V.E. & Starley, S. (1962). The design of base-vented struts for high speed hydrofoil systems. In *Proc. of Symp. on Air Cushion Vehicles and Hydrofoils.* New York, N.Y.: Inst. of Aeronautical Sciences.

Johnsson, C.-A. (1968). *On Theoretical Predictions of Characteristics and Cavitation Properties of Propellers.* Publication of the Swedish State Shipbuilding Experimental Tank, Nr. 64. Göteborg: Scandinavian University Books.

Johnsson, C.-A. (1980). On the reduction of propeller excitation by modifying the blade section shape. *The Naval Architect*, no. 3, 113-5.

Johnsson, C.-A. (1983). Propeller with reduced tip loading and unconventional blade form and blade sections. In *Eighth School Lecture Series, Ship Design for Fuel Economy. West European School Graduate Education in Marine Technology* (*WEGEMT*). Swedish Maritime Research Centre. Göteborg: SSPA.

Johnsson, C.-A. & Søntvedt, T. (1972). Propeller excitation and response of 230 000 t.dw. tankers. In *Proc. Ninth Symp. on Naval Hydrodynamics*, vol. 1, ed. R. Brard & A. Castera, pp. 581-655. Arlington, Va.: Office of Naval research – Department of the Navy.

de Jong, K. (1991). *On the Optimization and the Design of Ship Screw Propellers with and without End Plates.* Doctoral thesis. Groningen: Dept. of Mathematics, Univ. of Groningen.

de Jong, K. & Sparenberg, J.A. (1990). On the influence of the choice of generator lines on the optimum efficiency of screw propellers. *Journal of Ship Research*, vol. 34, no. 2, 79–91.

Kaplan, P., Bentson, J. & Breslin, J.P. (1979). Theoretical analysis of propeller radiated pressure and blade forces due to cavitation. In *Proc. Symp. on Propeller Induced Ship Vibration*, pp. 133–45. London: RINA.

von Kármán, Th. & Burgers, J.M. (1935). General aerodynamic theory – perfect fluids. In *Aerodynamic Theory*, vol. II, division E, ed. W.F. Durand. Berlin: Julius Springer.

von Kármán, Th. & Sears, W.R. (1938). Airfoil theory for non-uniform motion. *Journal of the Aeronautical Sciences*, vol. 5.

Kawada, S. (1939). Calculation of induced velocity by helical vortices and its application to propeller theory. *Reports of the Aeronautical Research Institute*, vol. XIV, no. 172, pp. 3–57. Tokyo: Tokyo Imperial Univ.

Keenan, D.P. (1989). *Marine Propellers in Unsteady Flow.* Ph. D. thesis. Cambridge, Mass.: Dept. of Ocean Eng., MIT.

Kellogg, O.D. (1967). *Foundations of Potential Theory.* Die Grundlehren der mathematischen Wissenschaften, Band XXXI. Berlin: Springer Verlag.

Kerwin, J.E. (1961). *The Solution of Propeller Lifting Surface Problems by Vortex-Lattice Methods.* Technical report. Cambridge, Mass.: MIT, Dept. of Naval Architecture and Marine Eng.

Kerwin, J.E. (1981). *Lecture Notes: Hydrodynamic Theory for Propeller Design and Analysis.* Cambridge, Mass.: Dept. of Ocean Eng., MIT.

Kerwin, J.E. (1986). Marine propellers. *Annual Review of Fluid Mechanics*, vol. 18, ed. M. van Dyke, J.V. Wehausen & J.L. Lumley, pp. 387–403. Palo Alto, Cal.: Annual Reviews Inc.

Kerwin, J.E., Coney, W.B. & Hsin, C.-Y. (1986). Optimum circulation distributions for single and multi-component propulsors. In *Proc. of Twenty-First American Towing Tank Conference (ATTC)*, ed. R.F. Messalle, pp. 53–62. Washington, D.C.: National Academy Press.

Kerwin, J.E., Kinnas, S.A., Lee, J.-T. & Wei-Zen, S. (1987). A surface panel method for the hydrodynamic analysis of ducted propellers. *Trans. SNAME*, vol. 95, pp. 93–122. Jersey City, N.J.: SNAME.

Kerwin, J., Kinnas, S. Wilson, M.B. & McHugh, J. (1986). Experimental and analytical techniques for the study of unsteady propeller sheet cavitation. In *Proc. Sixteenth Symp. on Naval Hydrodynamics*, ed. W.C. Webster, pp. 387–414. Washington, D.C.: National Academy Press.

Kerwin, J.E. & Lee, C.-S. (1978). Prediction of steady and unsteady marine propeller performance by numerical lifting-surface theory. *Trans. SNAME*, vol. 86, pp. 218–53. New York, N.Y.: SNAME.

Kinnas, S.A. (1985). *Non-linear Corrections to the Linear Theory for the Prediction of the Cavitating Flow Around Hydrofoils*, Report 85-1. Cambridge, Mass.: Dept. of Ocean Eng., MIT.

Kinnas, S.A. (1991). Leading-edge corrections to the linear theory of partially cavitating hydrofoils. *Journal of Ship Research*, vol. 35, no. 1, 15-27.

Kinnas, S.A. & Coney, W.B. (1988). On the optimum ducted propeller loading. In *Proc. Propellers '88 Symposium*, pp. 1.1-13. Jersey City, N.J.: SNAME.

Kinnas, S.A. & Coney, W.B. (1990). *The Generalized Image Model – an Application to the Design of Ducted Propellers.* Cambridge, Mass.: Dept. of Ocean Eng., MIT.

Kinnas, S.A. & Fine, N.E. (1989). Theoretical predictions of midchord and face unsteady propeller sheet cavitation. In *Proc. Fifth Int. Conf. on Numerical Hydrodynamics*, ed. K.-h. Mori, pp. 685-700. Washington, D.C.: National Academy Press.

Kinnas, S.A. & Fine, N.E. (1991). Non-linear analysis of the flow around partially or super-cavitating hydrofoils by a potential based panel method. In *Boundary Integral Methods – Theory and Applications, Proc. of the IABEM-90. Symp.*, pp. 289-300. Heidelberg: Springer-Verlag.

Kinnas, S.A. & Hsin, C.-Y. (1990). *A Boundary-Element Method for Analysis of the Unsteady Flow Around Extreme Propeller Geometries.* Cambridge, Mass.: Dept. of Ocean Eng., MIT.

Kinnas, S., Hsin, C.-Y. & Keenan, D. (1990). A potential based panel method for the unsteady flow around open and ducted propellers. In *Proc. Eighteenth Symp. on Naval Hydrodynamics*, pp. 667-85. Washington, D.C.: National Academy Press.

Kirchhoff, G. (1868). Zur Theorie freier Flüssigkeitsstrahlen. *Journ. für die reine and angewandte Mathematik* (*Crelle*), Band LXX, Heft 4, 289-98.

Klaren, L. & Sparenberg, J.A. (1981). On optimum screw propellers with endplates, inhomogeneous inflow. *Journal of Ship Research*, vol. 25, no. 4, 252-63.

Klingsporn, H. (1973). Numerische Untersuchung der Integralgleichung der Propellertragflächentheorie. *Schiffstechnik*, 99. Heft, 20. Band, 1-8.

Knapp, R.T., Daily, J.W. & Hammitt, F.G. (1970). *Cavitation.* Engineering Societies Monographs. London: McGraw-Hill Book Company.

Kort, L. (1934). Der neue Düsenschrauben-Antrieb. *Werft – Reederei – Hafen*, 15. Jahrgang, Heft 4, 41-3.

Koyama, K. (1972). A numerical analysis for the lifting-surface theory of a marine propeller. *Journal of the Society of Naval Architects of Japan*, vol. 132.

Kuiper, G. (1981). *Cavitation Inception on Ship Propeller Models.* Publ. no. 655. Wageningen: Netherlands Ship Model Basin.

Kutta, W.M. (1910). Über eine mit den Grundlagen des Flugsproblems in Beziehung stehende zweidimensionale Strömung. *Sitzungsberichte der Königlichen Bayerischen Akademie der Wissenschaften.* (This paper reproduced Kutta's unpublished thesis of 1902.)

Küchemann, D. & Weber, J. (1953). *Aerodynamics of Propulsion.* London: McGraw-Hill Book Company.

Lamb, H. (1963). *Hydrodynamics.* Sixth edn.. Cambridge: Cambridge University Press.

Landweber, L. (1951). *The Axially Symmetric Potential Flow about Elongated Bodies of Revolution.* Report 761. Washington, D.C.: Navy Department, The David W. Taylor Model Basin.

Larsson, L., Broberg, L., Kim, K.-J. & Zang, D.-H. (1990). A method for resistance and flow prediction in ship design. *Trans. SNAME*, vol. 98, pp. 495–535. Jersey City, N.J.: SNAME.

Lee, C.-S. (1979). *Prediction of Steady and Unsteady Performance of Marine Propellers with or without Cavitation by Numerical Lifting Surface Theory.* Ph. D. thesis. Cambridge, Mass.: Dept. of Ocean Eng., MIT.

Lee, C.-S. (1980). Prediction of the transient cavitation on marine propellers by numerical lifting-surface theory. In *Proc. Thirteenth Symp. on Naval Hydrodynamics*, ed. T. Inui, pp. 41–64. Tokyo: The Shipbuilding Research Association of Japan.

Lehman, A.F. (1964). *An Experimental Investigation of Propeller-Appendage Interaction.* Report no. 64-11. New York, N.Y.: Oceanics Inc.

Lerbs, H. (1952). Moderately loaded propellers with a finite number of blades and an arbitrary distribution of circulation. *Trans. SNAME*, vol. 60, pp. 73-123. New York, N.Y.: SNAME.

Lewis, E.V. (ed.) (1988). *Principles of Naval Architecture.* Second Revision. Jersey City, N.J.: SNAME.

Lewis, F.M. (1969). Propeller vibration forces in single screw ships. *Trans. SNAME*, vol. 77, pp. 318–43. New York, N.Y.: SNAME.

Lewis, F.M. (1974). Comments on Vorus, W.S.: A method for analyzing the propeller-induced vibratory forces acting on the surface of a ship stern. *Trans. SNAME*, vol. 82, pp. 198–9. New York, N.Y.: SNAME.

Lighthill, Sir James (1986). *An Informal Introduction to Theoretical Fluid Mechanics.* Oxford: Clarendon Press.

Lighthill, M.J. (1951). A new approach to thin aerofoil theory. *The Aeronautical Quarterly*, vol. III, 193–210.

Lindgren, H. & Bjärne, E. (1967). *The SSPA Standard Family of Propellers.* Publication of the Swedish State Shipbuilding Experimental Tank, Nr. 60. Göteborg: Scandinavian University Books.

Loftin, L.K. (1948). *Theoretical and Experimental Data for a Number of NACA 6A Series Airfoil Sections.* NACA report no. 903. National Advisory Committee for Aeronautics.

van Manen, J.D. (1957a). Recent research on propellers in nozzles. *Journal of Ship Research*, vol. 1, no. 2, 13–46.

van Manen, J.D. (1957b). Written contribution to Formal Discussion, Subject 3: comparative cavitation tests of propellers. In *Proc. Eighth International Towing Tank Conference (ITTC)*, ed. M.L. Acevedo & L. Mazarredo, pp. 250–2. El Pardo: Canal de Experiencias Hidrodinamicas.

van Manen, J.D. & Oosterveld, M.W.C. (1966). Analysis of ducted-propeller design. *Trans. SNAME*, vol. 74, pp. 522–62. New York, N.Y.: SNAME.

van Manen, J.D. & Troost, L. (1952). The design of ship screws of optimum diameter for an unequal velocity field. *Trans. SNAME*, vol. 60, pp. 442–68. New York, N.Y.: SNAME.

Mangler, K.W. (1951). *Improper Integrals in Theoretical Aerodynamics.* Report no. Aero 2424. Farnborough: Royal Aircraft Establishment.

McCarthy, J.H. (1961). *On the Calculation of Thrust and Torque Fluctuation of Propellers in Non-Uniform Wake Flow.* Hydromechanics Laboratory, Research and Development Report 1533. Bethesda, Md.: Department of the Navy, David Taylor Model Basin.

McLachlan, N.W. (1955). *Bessel Functions for Engineers.* Oxford: Clarendon Press.

Milne-Thomson, L.M. (1955). *Theoretical Hydrodynamics.* Third edn.. London: MacMillan & Co.

Minsaas, K.J. (1967). Propellteori. *Skibsmodelltankens Meddelelse,* no. 96. Trondheim: Skipsmodelltanken, Norges Tekniske Högskole.

Morgan, W.B. (1962). Theory of the annular airfoil and ducted propeller. In *Proc. Fourth Symp. on Naval Hydrodynamics*, ed. B.L. Silverstein, pp. 151–97. Washington, D.C.: Office of Naval Research – Department of the Navy.

Morgan, W.B. & Caster, E.B. (1968). Comparison of theory and experiment on ducted propellers. In *Proc. Seventh Symp. on Naval Hydrodynamics*, ed. R.D. Cooper & S.W. Doroff, pp. 1311–49. Arlington, Va.: Office of Naval Research – Department of the Navy.

Morgan, W.B., Silovic, V. & Denny, S.B. (1968). Propeller lifting-surface corrections. *Trans. SNAME*, vol. 76, pp. 309–47. New York, N.Y.: SNAME.

Morgan, W.B. & Wrench, J.W. (1965). Some computational aspects of propeller design. In *Methods in Computational Physics*, vol. 4, ed. B. Alder, S. Fernbach & M. Rotenberg, pp. 301–31. New York: Academic Press.

Moriya, T. (1933). On the induced velocity and characteristics of a propeller. *Journal of the Faculty of Eng., Tokyo Imperial Univ.*, vol. XX, no. 7, 147–62.

Munk, M. (1919). *Isoperimetrische Aufgaben aus der Theorie des Fluges.* Dissertation, Göttingen.

Murray, M.T. & Tubby, J. (1973). *Blade-rate Fluctuations of a Propeller in Non-Uniform Flow.* Report no. ARL/M/P33A. Haslar: Admiralty Research Laboratory.

Narita, H., Yagi, H., Johnson, H.D. & Breves, L.R. (1981). Development and full-scale experiences of a novel integrated duct propeller. *Trans. SNAME*, vol. 89, pp. 319–46. New York, N.Y.: SNAME.

Nelson, D.M. (1964). *A Lifting-Surface Propeller Design Method for High-Speed Computers.* Technical Report no. 8442. China Lake, Cal.: U.S. Naval Weapons Center.

Newman, J.N. (1977). *Marine Hydrodynamics.* Cambridge, Mass.: MIT Press.

Newton, R.N. & Rader, H.P. (1961). Performance data of propellers for high-speed craft. *Trans. RINA*, vol. 103, pp. 93–129. London: RINA.

Nicholson, J.W. (1910). The approximate calculation of Bessel functions of imaginary argument. *Philosophical Magazine*, vol. 20, 938–43.

Nishiyama, S., Sakamoto, Y., Ishida, S., Fujino, R. & Oshima, M. (1990). Development of contrarotating-propeller system for *Juno* – a 37 000 dwt class bulk carrier. *Trans. SNAME*, vol. 98, pp. 27-52. Jersey City, N.J.: SNAME.

Noordzij, L. (1976). Pressure field induced by a cavitating propeller. *Int. Shipbuilding Progress*, vol. 23, no. 260, 93–105.

van Oossanen, P. (1975). *Calculation of Performance and Cavitation Characteristics of Propellers Including Effects of Non-Uniform Flow and Viscosity.* Publ. no. 457. Wageningen: Netherlands Ship Model Basin.

Oosterveld, M.W.C. (1970). *Wake Adapted Ducted Propellers.* Publ. no. 345. Wageningen: Netherlands Ship Model Basin.

Osterveld, M.W.C. & van Oossanen, P. (1973). Recent developments in marine propeller hydrodynamics. In *Proc. International Jubilee Meeting on the Occasion of the 40th Anniversary of the Netherlands Ship Model Basin*, pp. 51–99. Wageningen: Netherlands Ship Model Basin.

Oosterveld, M.W.C. & van Oossanen, P. (1975). Further computer-analyzed data of the Wageningen B-screw series. *Int. Shipbuilding Progress*, vol. 22, no. 251, 251–62.

Patel, V.C. (1989). Ship stern and wake flows: status of experiment and theory. In *Proc. Seventeenth Symp. of Naval Hydrodynamics*, pp. 217–40. Washington, D.C.: National Academy Press.

Peirce, B.O. & Foster, R.M. (1956). *A Short Table of Integrals.* Fourth edn.. London: Ginn and Company.

Phillips, E.G. (1947). *Functions of a Complex Variable.* Edinburgh: Oliver & Boyd Inter-science Publishers.

Pien, P.C. (1961). The calculation of marine propellers based on lifting-surface theory. *Journal of Ship Research*, vol. 5, no. 2, 1-14.

Pohl, K.-H. (1959). Das Instationäre Druckfeld in der Umgebung eines Schiffspropellers und die von ihm auf benachbarten Platten erzeugten Kräfte. *Schiffstechnik*, 6. Band, Heft 32, 107-16.

Posdunin, V.L. (1944). Supercavitating propellers. *Izv. OTN AN SSSR*, nos. 1-2.

Posdunin, V.L. (1945). Basic theory of the construction and performance of supercavitating propellers. *Izv. OTN AN SSSR*, nos. 10-11.

Prandtl, L. (1933). Neuere Ergebnisse der Turbulenzforschung. *Zeitschrift des Vereines Deutscher Ingenieure*, Band 77, No. 5, 105-14.

Reynolds, O. (1873). The causes of the racing of the engines of screw steamers investigated theoretically and by experiment. *Trans. INA*, vol. 14, pp. 56-67. London: INA.

Ritger, P.D. & Breslin, J.P. (1958). *A Theory for the Quasi-Steady and Unsteady Thrust and Torque of a Propeller in a Ship Wake.* Experimental Towing Tank, report no. 686. Hoboken, N.J.: Stevens Inst. of Tech.

Ryan, P.G. & Glover, E.J. (1972). A ducted propeller design method: a new approach using surface vorticity distribution techniques and lifting line theory. *Trans. RINA*, vol. 114, pp. 545-63. London: RINA.

Sachs, A.H. & Burnell, J.A. (1962). Ducted propellers – a critical review of the state-of-the-art. In *Progress in Aeronautical Sciences*, vol. 3, ed. A. Ferri, D. Küchemann & L.H.G. Sterne, pp. 85-135. London: Pergamon Press.

Sarpkaya, T. (1989). Computational methods with vortices – the 1988 Freeman scholar lecture. *Journal of Fluids Engineering* (*Trans. ASME*), vol. 111, no. 1, 5-52.

Schmidt, G.H. & Sparenberg, J.A. (1977). On the edge singularity of an actuator disk with large constant normal load. *Journal of Ship Research*, vol. 21, no. 2, 125-31.

Schneekluth, H. (1985). Die Zustromdüse – alte und neue Aspekte. *HANSA*, 122. Jahrgang, No. 21, 2189-95.

Schwanecke, H. (1975). Comparative calculation of unsteady propeller blade forces. In *Proc. Fourteenth International Towing Tank Conference (ITTC)*, vol. 3, pp. 357-97. Ottawa: National Research Council Canada.

Schwanecke, H. & Laudan, J. (1972). Ergebnisse der instationären Propellertheorie. *Jahrbuch der Schiffbautechnischen Gesellschaft*, 66. Band, pp. 299-316. Berlin: Springer-Verlag.

Sears, W.R. (1941). Some aspects of non-stationary airfoil theory and its practical application. *Journal of the Aeronautical Sciences*, vol. 8, no. 5, 104–8.

Shen, Y.T. (1985). Wing sections for hydrofoils – part 3: experimental verifications. *Journal of Ship Research*, vol. 29, no. 1, 39–50.

Shen, Y.T. & Eppler, R. (1981). Wing sections for hydrofoils – part 2: nonsymmetrical profiles. *Journal of Ship Research*, vol. 25, no. 3, 191–200.

Shioiri, J. & Tsakonas, S. (1964). Three-dimensional approach to the gust problem for a screw propeller. *Journal of Ship Research*, vol. 7, no. 4, 29–53.

Sneddon, I.N. (1951). *Fourier Transforms*. London: McGraw-Hill Book Company.

Sparenberg, J.A. (1959). Applicaton of lifting surface theory to ship screws. In *Proc. Koninklijke Nederlandse Akademie van Wetenschappen*, series B, vol. 62, no. 5, pp. 286–98.

Sparenberg, J.A. (1960). Application of lifting surface theory to ship screws. *Int. Shipbuilding Progress*, vol. 7, no. 67, 99–106.

Sparenberg, J.A. (1969). On optimum propellers with a duct of finite length. *Journal of Ship Research*, vol. 13, no. 2, 129–36.

Sparenberg, J.A. (1984). *Elements of Hydrodynamic Propulsion*. The Hague: Martinus Nijhoff Publishers.

Stern, F., Toda, Y. & Kim, H.T. (1991). Computation of viscous flow around propeller-body configurations: Iowa axisymmetric body. *Journal of Ship Research*, vol. 35, no. 2, 115–61.

Stern, F. & Vorus, W.S. (1983). A nonlinear method for predicting unsteady sheet cavitation on marine propellers. *Journal of Ship Research*, vol. 27, no. 1, 56–74.

Stipa, L. (1931). *Experiments with Intubed Propellers*. NACA Technical Report TM 655 (January 1932). Langley, Va.: National Advisory Committee for Aeronautics.

Søntvedt, T. & Frivold, H. (1976). Low frequency variation of the surface shape of tip region cavitation on marine propeller blades and corresponding disturbances on nearby solid boundaries. In *Proc. Eleventh Symp. on Naval Hydrodynamics*, ed. R.E.D. Bishop, A.G. Parkinson & W.G. Price, pp. 717–30. London: Mechanical Eng. Publications.

Tachmindji, A.J. (1958). Potential problem of the optimum propeller with finite number of blades ... operating in a cylindrical duct. *Journal of Ship Research*, vol. 2, no. 3, 23–32.

Tachmindji, A.J. & Morgan, W.B. (1958). The design and estimated performance of a series of supercavitating propellers. In *Proc. of the Second Symp. on Naval Hydrodynamics*, ed. R.D. Cooper, pp. 489–532. Washington, D.C.: Office of Naval Research – Department of the Navy.

Takahashi, H. & Ueda, T. (1969). An experimental investigation into the effect of cavitation on fluctuating pressures around a marine propeller. In *Proc. Twelfth International Towing Tank Conference (ITTC)*, ed. Istituto Nazionale per Studi ed Esperienze di Architettura Navale, pp. 315–7. Rome: Consiglio Nazionale delle Ricerche.

Tanibayashi, H. & Hoshino, T. (1981). Calculation of loading on propeller blades working in non-uniform flow based on quasi-steady technique. *Trans. of the West-Japan Society of Naval Architects*, no. 61, pp. 41–61. Fukuoka: West-Japan Society of Naval Architects.

Taylor, D.W. (1933). *Speed and Power of Ships.* Washington D.C.: Ransdell Inc. Printers and Publishers.

Thomson, Sir William (Lord Kelvin) (1869). On vortex motion. *Trans. Royal Society of Edinburgh*, vol. XXV, pp. 217–60; reprinted (1910) in *Mathematical and Physical Papers*, vol. IV, pp. 13–66. Cambridge: Cambridge University Press.

Thwaites, B. (1960). *Incompressible Aerodynamics.* Oxford: Clarendon Press.

Troost L. (1937-38). Open-water test series with modern propeller forms. *Trans. North East Coast Institution of Engineers and Shipbuilders*, vol. LIV, pp. 321–6. Newcastle upon Tyne: North East Coast Institution of Engineers and Shipbuilders.

Troost, L. (1939–40). Open-water test series with modern propeller forms, part 2. *Trans. North East Coast Institution of Engineers and Shipbuilders*, vol. LVI, pp. 91–6. Newcastle upon Tyne: North East Coast Institution of Engineers and Shipbuilders.

Troost, L. (1950–51). Open-water test series with modern propeller forms, part 3. *Trans. North East Coast Institution of Engineers and Shipbuilders*, vol. 67, pp. 89–130. Newcastle upon Tyne: North East Coast Institution of Engineers and Shipbuilders.

Truesdell, C. (1954). *The Kinematics of Vorticity.* Bloomington, Ind.: Univ. of Indiana Press.

Tsakonas, S. (1978). Discussion of Kerwin, J.E. & Lee, C.-S.: Prediction of steady and unsteady marine propeller performance by numerical lifting-surface theory. *Trans. SNAME*, vol. 86, pp. 247–8. New York, N.Y.: SNAME.

Tsakonas, S., Breslin, J.P. & Jacobs, W.R. (1962). The vibratory force and moment produced by a marine propeller on a long rigid strip. *Journal of Ship Research*, vol. 5, no. 4, 21–42.

Tsakonas, S., Chen, C.Y. & Jacobs, W.R. (1967). Exact treatment of the helicoidal wake in the propeller lifting-surface theory. *Journal of Ship Research*, vol. 11, no. 3, 154–67.

Tsakonas, S. & Jacobs, W.R. (1969). Propeller loading distributions. *Journal of Ship Research*, vol. 13, no. 4, 237–57.

Tsakonas, S. & Jacobs, W.R. (1978). Propeller-duct interaction due to loading and thickness effects. In *Proc. Propellers '78 Symposium*. New York, N.Y.: SNAME.

Tsakonas, S., Jacobs, W.R. & Ali, M.R. (1973). An "exact" linear lifting surface theory for a marine propeller in a nonuniform flow field. *Journal of Ship Research*, vol. 17, no. 4, 196–207.

Tsakonas, S., Jacobs, W.R. & Ali, M.R. (1976). *Documentation of a Computer Program for the Pressure Distribution, Forces and Moments on Ship Propellers in Hull Wakes.* Davidson Laboratory, SIT-DL-76-1863. Hoboken, N.J.: Stevens Inst. of Tech.

Tsakonas, S., Jacobs, W.R. & Rank, P.H. (1968). Unsteady propeller lifting-surface theory with finite number of chordwise modes. *Journal of Ship Research*, vol. 12, no. 1, 14–45.

Tsakonas, S. & Liao, P. (1983). Propeller-induced hull forces, moments and point pressures. In *Proc. Int. Symp. Ship Hydrodynamics and Energy Saving*, paper VII. El Pardo: Canal de Experiencias Hidrodinamicas.

Tulin, M.P. (1953). *Steady Two-Dimensional Cavity Flows about Slender Bodies.* Report 834. Washington, D.C.: Navy Department, The David Taylor Model Basin.

Tulin, M.P. (1956). Supercavitating flow past foils and struts. In *Proc. Symp. on Cavitation in Hydrodynamics*, paper 16, National Physical Laboratory. London: Her Majesty's Stationery Office.

Tulin, M.P. (1964). Supercavitating flows – small perturbation theory. *Journal of Ship Research*, vol. 7, no. 3, 16–37.

Tulin, M.P. (1979). An analysis of unsteady sheet cavitation. In *Proc. of Nineteenth American Towing Tank Conference (ATTC)*, vol. 2, ed. S.B. Cohen, pp. 1049–79. Ann Arbor, Mich.: Ann Arbor Science.

Tulin, M. & Burkart, M.P. (1955). *Linearized Theory for Flows about Lifting Foils at Zero Cavitation Number.* Report C-638. Washington, D.C.: Navy Department, The David W. Taylor Model Basin.

Tulin, M.P. & Hsu, C.C. (1977). The theory of leading edge cavitation on lifting surfaces with thickness. In *Proc. of Symp. on Hydrodynamics of Ship and Offshore Propulsion Systems*, paper 9, session 3. Høvik: Det Norske Veritas.

Tulin, M.P. & Hsu, C.C. (1980). New applications of cavity flow theory. In *Proc. Thirteenth Symp. on Naval Hydrodynamics*, ed. T. Inui, pp. 107–31. Tokyo: The Shipbuilding Research Association of Japan.

Uhlman, J.S. (1987). The surface singularity method applied to partially cavitating hydrofoils. *Journal of Ship Research*, vol. 31, no. 2, 107–24.

Van Houten, R.J. (1986). *Analysis of ducted propellers in steady flow.* Technical Report 4.76-1. Watertown, Mass.: Airflow Research and Manufacturing Corp.

Verbrugh, P.J. (1968). *Unsteady Lifting Surface Theory for Ship Screws.* Report no. 68-036-AH. Wageningen: Netherlands Ship Model Basin.

Voitkunsky, Y.I., Pershits, R.Y. & Titov, I.A. (1973). Handbook on ship theory. *Sudostroenie.* Leningrad.

Vorus, W.S. (1974). A method for analyzing the propeller-induced vibratory forces acting on the surface of a ship stern. *Trans. SNAME*, vol. 82, pp. 186-210. New York, N.Y.: SNAME.

Vorus, W.S., Breslin, J.P. & Tien, Y.S. (1978). Calculation and comparison of propeller unsteady pressure forces on ships. In *Proc. Ship Vibration Symposium '78*, pp. I-1-27. Arlington, Va.: Ship Structure Committee & SNAME. New York, N.Y.: SNAME.

Wade, R.B. & Acosta, A.J. (1966). Experimental observations on the flow past a plano-convex hydrofoil. *Journal of Basic Eng.* (*Trans. ASME*), vol. 88, series D, no. 1, 273-83.

Walderhaug, H. (1972). *Skipshydrodynamikk.* Trondheim: Tapir Forlag.

Watkins, C.E., Runyan, H.L. & Woolston, D.S. (1955). *On the Kernel Function of the Integral Equation Relating the Lift and Downwash Distributions of Oscillating Finite Wings in Subsonic Flow.* NACA Report 1234. Langley, Va.: National Advisory Committee for Aeronautics.

Watkins, C.E., Woolston, D.S. & Cunningham, H.J. (1959). *A Systematic Kernel Function Procedure for Determining Aerodynamic Forces on Oscillating or Steady Finite Wings at Subsonic Speeds.* NASA Tech. Report R-48. Langley, Va.: National Aeronautics and Space Administration.

Watson, G.N. (1966). *A Treatise on the Theory of Bessel Functions.* Second edn.. Cambridge: Cambridge University Press.

Wehausen, J.V. (1964). *Lecture Notes on Propeller Theory.* College of Eng.. Berkeley, Cal.: University of California.

Wehausen, J.V. (1976). A qualitative picture of the flow about a wing of finite aspect ratio. *Schiffstechnik*, 23. Band, 114. Heft, 215-7.

Weissinger, J. & Maass, D. (1968). Theory of the ducted propeller – a review. In *Proc. Seventh Symp. on Naval Hydrodynamics*, ed. R.D. Cooper & S.W. Doroff, pp. 1209-64. Arlington, Va.: Office of Naval Research – Department of the Navy.

Wrench, J.W. (1957). *The Calculation of Propeller Induction Factors.* Research and Development Report 1116. Washington, D.C.: Navy Department, The David W. Taylor Model Basin.

Wu, T.Y. (1962). Flow through a heavily loaded actuator disc. *Schiffstechnik*, 9. Band, 47. Heft, 134-38.

Wu, T.Y. (1972). Cavity and wake flows. *Annual Review of Fluid Mechanics*, vol. 4, ed. M. van Dyke, W.G. Vincenti & J.V. Wehausen, pp. 243-84. Palo Alto, Cal.: Annual Reviews Inc.

Wu, T.Y. (1975). Cavity flows and numerical methods. In. *Proc. First Int. Conf. on Numerical Ship Hydrodynamics*, ed. J.W. Schot & N. Salvesen, pp. 115–35. Bethesda, Md.: David W. Taylor Naval Ship Research and Development Center.

Yamaguchi, H., Kato, H., Sugatani, A., Kamijo, A., Honda, T. & Maeda, M. (1988). Development of marine propellers with better cavitation performance – 3rd report: Pressure distribution to stabilize cavitation. *Journal of the Society of Naval Architects of Japan*, vol. 164, 28–42.

Yamazaki, R. (1962). On the theory of screw propellers. In *Proc. Fourth Symp. on Naval Hydrodynamics*, ed. B.L. Silverstein, pp. 3–28. Washington, D.C.: Office of Naval Research – Department of the Navy.

Yamazaki, R. (1966). On the theory of screw propellers in a non-uniform flow. *Memoirs of the Faculty of Engineering, Kyushu University*, vol. 25, no. 2.

Yih, C.S. (1988). *Fluid Mechanics.* Ann Arbor, Mich.: West River Press.

Yim, B. (1976). Optimum propeller with cavity-drag and frictional drag effects. *Journal of Ship Research*, vol. 20, no. 2, 118–23.

Young, F.R. (1989). *Cavitation.* London: McGraw-Hill Book Company.

Authors Cited

Abbott, I.H., 60, 90–3, 98–100, 117–26
Abramowitz, M., 65, 176, 181, 215–6, 243–4, 324, 353, 355, 518–9
Acosta, A.J., 142–3, 150
Ali, M.R., 335, 403, 408
Andersen, P., 232, 476
Andersen, Sv.V., 476
Ashley, H., 328, 330–1
Auslaender, J., 142, 159

Barnaby, S.W., 128
Basu, B.C., 336
Batchelor, G.K., 1, 113
Bateman, H., 516–7
Bauschke, W., 381
Bavin, V.F., 291, 298, 300, 335
Bentson, J., 403–4, 408–10
Betz, A., 140, 192–4, 233
Bjärne, E., 235
Blaurock, J., 465–6, 477–9
Boswell, R.J., 250, 404
Breslin, J.P., 58–62, 136, 140, 199, 256, 284–6, 288–9, 306, 308, 310–3, 331, 334, 373, 375–7, 380, 385–6, 401, 403–4, 408–10, 422, 433, 435, 437, 439–45, 448–51, 476
Breves, L.R., 481
Broberg, L., 263
Brockett, T., 82–5, 98, 100–1, 136, 251
Brown, G.L., 270
Burgers, J.M., 103, 196
Burkart, M.P., 141, 158
Burnell, J.A., 469
Burrill, L.C., 128, 237, 256

Carrier, G.F., 499
Caster, E.B., 221, 469
Chen, C.Y., 335
Cheng, H.M., 268
Chertock, G., 418
Coney, W.B., 227–8, 456, 458–61, 463, 466–9, 473–7
Cox, B.D., 310, 313, 422, 451, 462–3
Cox, G.G., 335
Cumming, R.A., 250
Cummings, D.E., 336
Cunningham, H.J., 331

Daily, J.W., 130–1
Dashnaw, F.J., 462
Davies, T.V., 142
Denny, S.B., 250, 281, 284–5
Dirac, P.A.M., 504
Diskin, J.A., 221
von Doenhoff, A.E., 60, 90–3, 98–100, 117–26
Dyne, G., 469

Emerson, A., 237
English, J.W., 335
Eppler, R., 103–5
Erdélyi, A., 516–7

Fagerjord, O., 453

Falcão de Campos, J.A.C., 373, 469
Fine, N.E., 142, 404
Forrest, A.W., 462
Foster, R.M., 323
Franc, J.P., 142
Freeman, H.B., 59
Frivold, H., 401
Frostad, R., 453
Fujino, R., 462

Gawn, R.W.L., 235
van Gent, W., 335, 350, 373, 379, 380–1
Geurst, J.A., 142–3, 148, 150, 156, 403, 408
Giesing, J.P., 336
Glauert, H., 240, 485
Glover, E.J., 454–5, 462, 469, 476–7
Goldstein, S., 221
Gomez, G.P., 476
Goodman, T.R., 195, 358, 373, 476
Greeley, D.S., 336, 370
Greenberg, M.D., 178, 180
Grim, O., 465
Grothues-Spork, H., 481
Groves, N.C., 373, 442

Hadamard, J., 320, 507
Hadler, J.B., 268, 462
Hammitt, F.G., 130–1
Hanaoka, T., 142, 335, 381
Hancock, G.J., 336
Hankel, H., 497
Harvald, Sv.Aa., 235, 238, 262, 454, 470
Hee, J.M., 238, 262
Helmholtz, H., 140
Hess, J., 106, 430, 433
Hess, J.L., 336
Hoeijmakers, H.W.M., 336

Holden, K.O., 408, 453
Honda, T., 110
Hoshino, T., 335–6
Hough, G.R., 190
Hsin, C.-Y., 227–8, 336, 368, 370–2, 404, 456, 458, 460, 463, 466–9, 473
Hsu, C.C., 142, 151–2, 157
Huang, T.T., 373, 442
Huse, E., 451

Ide, T., 476
Isay, W.H., 130, 335, 381
Ishida, S., 462
Ishii, N., 476
Ito, T., 335
Itoh, S., 476
ITTC, 103–4, 110, 131, 190, 263, 297, 373, 454, 477, 483

Jacob, M.C., 465
Jacobs, W.R., 284–5, 298–9, 308, 335, 358, 375, 381, 403, 408, 469
Japan Ship Exporters' Association, 463
Johannessen, H., 453
Johnson, H.D., 481
Johnson, V.E., 142, 158
Johnsson, C.-A., 105–9, 252–61, 373, 381, 403–4, 433, 435, 437, 439–45, 448–50
de Jong, K., 476

Kamijo, A., 110
Kaplan, P., 403–4, 408–10
von Kármán, Th., 103, 196, 315, 317
Kato, H., 110
Kawada, S., 208
Keenan, D.P., 336, 469
Kellogg, O.D., 5, 196, 428
Kelvin, Lord William, 497
Kerwin, J.E., 155, 222, 227–8, 335–6, 370, 373, 377, 380–4, 404, 433, 435, 437, 439–45, 448–50, 456, 458, 460, 463, 466–9, 473
Kim, H.T., 373
Kim, K.-J., 263
Kinnas, S.(A.), 142, 151–5, 336, 368, 370–2, 404, 408, 456, 469, 473–75, 526
Kirchhoff, G., 140–1
Klaren, L., 476
Klingsporn, H., 381
Knapp, R.T., 130–1
Kort, L., 468
Kowalski, T., 289
Koyama, K., 335
Kuiper, G., 129
Kutta, W.M., 71
Kuchemann, D., 470

La Fone, T.A., 221
Lamb, H., 1, 18, 352, 423
Landahl, M., 328, 330–1
Landweber, L., 58–62, 136, 138–40
Larsson, L., 263

Laudan, J., 381
Lederer, L., 381
Lee, C.-S., 335, 377, 380–4, 404–8, 469
Lee, J.-T., 336
Lehman, A.F., 309, 311
Lerbs, H., 208, 221–2, 224, 232, 235, 240–1, 244–5, 248, 252–3
Lewis, E.V., 194
Lewis, F.M., 306, 451
Liao, P., 335, 451
Lighthill, Sir James, 1, 7, 496
Lighthill, M.J., 66, 78–80, 82–3, 85
Lindgren, H., 235
Loftin, L.K., 89–90

Maass, D., 469
Maeda, M., 110
Magnus, W., 516–17
van Manen, J.D., 236, 247, 387, 469, 471
Mangler, K.W., 320, 328, 349, 507, 512–3
McCarthy, J.H., 381, 385
McHugh, J., 155, 404
McLachlan, N.W., 516
Mercier, J., 284–5, 335
Michel, J.M., 142
Milne-Thomson, L.M., 1, 19, 52, 55, 140, 168–9, 490
Miniovich, I.Y., 291, 298, 300, 335
Minsaas, K.J., 247
Morgan, W.B., 159, 232, 245, 250, 469
Moriya, T., 222
Munk, M., 232
Murray, M.T., 381

Narita, H., 481
Nelson, D.M., 335
Newman, J.N., 1, 222, 484, 502
Newton, R.N., 158
Nicholson, J.W., 244
Nishiyama, S., 462
Noordzij, L., 403

Oberhettinger, F., 516–7
van Oossanen, P., 221, 235–6, 249–50, 332
Oosterveld, M.W.C, 235–7, 249, 332, 469, 471–2
Ordway, D.E., 190
Oshima, M., 462

Patel, V.C., 263
Pearson, C.E., 499
Peirce, B.O., 323
Pershits, R.Y., 381
Petersohn, E., 140
Phillips, E.G., 486
Pien, P.C., 335
Pohl, K.-H., 306
Posdunin, V.L., 141, 158
Powers, S.R., 178, 180
Prandtl, L., 271

Rader, H.P., 158
Rank, P.H., 335, 381
Reed, A.M., 462–3
Reynolds, O., 128
Ritger, P.D., 256, 334, 375
Rood, E.P., 310, 313, 422, 451
Roshko, A., 270
Runyan, H.L., 331
Ryan, P.G., 469

Sachs, A.H., 469
Sakamoto, Y., 462
Sarpkaya, T., 336
Schmidt, G.H., 178
Schneekluth, H., 478–9
Schwanecke, H., 380–4
Sears, W.R., 315, 317
Shen, Y.T., 103–5
Shioiri, J., 335
Silovic, V., 250
Skaar, K.T., 453
Smith, A.M.O., 106, 430, 433
Sneddon, I.N., 176, 342, 513, 515
Somers, D.M., 103
Sparenberg, J.A., 74, 178, 335, 343, 345, 349, 469, 472, 476
Starley, S., 142
Stegun, I.A., 65, 176, 181, 215–6, 243–4, 324, 353, 355, 518–9
Stern, F., 373, 403
Stipa, L., 468
Sugatani, A., 110
Swensson, C.G., 462
Søntvedt, T., 401, 403

Tachmindji, A.J., 159, 469
Tagori, T., 476
Takahashi, H., 335, 387–8
Tanibayashi, H., 335
Taylor, D.W., 235
Thomson, Sir William, 497
Thwaites, B., 114–6, 205–6
Tien, Y.S., 403
Titov, I.A., 381
Toda, Y., 373
Trisconi, F.G., 516–7
Troost, L., 235–6, 247
Truesdell, C., 497
Tsakonas, S., 284–6, 298–9, 308, 335, 358, 375, 377–9, 381, 403, 408, 451, 469
Tubby, J., 381
Tulin, M.P., 140–2, 151–2, 156–9

Ueda, T., 387–8
Uhlman, J.S., 152

Valarezo, W.O., 336
Van Houten, R.J., 373, 404, 433, 435, 437, 439–45, 448–50, 469
Vashkevich, M.A., 291, 298, 300, 335
Verbrugh, P.J., 335, 350, 381
Voitkunsky, Y.I., 381
Vorus, W.S., 310, 313, 403, 418, 422, 451

Wade, R.B., 150
Walderhaug, H., 249
Watkins, C.E., 331
Watson, G.N., 189, 354
Weber, J., 470
Wehausen, J.V., 196–7, 228, 232
Weissinger, J., 469
Wei-Zen, S., 336, 469
Wilson, M.B., 155, 404
Woolston, D.S., 331
Wrench, J.W., 222, 232, 245, 458
Wu, T.Y., 140, 142, 166, 170, 173, 175–6, 178

Yagi, H., 481
Yamaguchi, H., 110
Yamazaki, R., 335
Yih, C.S., 1, 21, 111, 208, 496–7
Yim, B., 228, 232
Young, F.R., 130

Zang, D.-H., 263

Øfsti, O., 408

Sources of Figures

The original sources of those figures (and tables) that have not been devised by the authors are given in the figure captions as well as in the list below. All such figures have been completely redrawn and where necessary modified to appear in the style and with the notation as the rest of the book. All errors or misinterpretations which may have arisen in this process are entirely the responsibility of the authors.

The authors are very grateful to the copyright holders for their kind permissions to use their material, as expressed in the "By courtesy...." line. On request an additional copyright notice is given in the figure captions; the omission of such a notice does not imply that the original figure or material is in public domain.

A.P. Møller, Denmark, 469
ATTC, USA, 463, 466, 468
H. Ashley & M. Landahl, 330

Cambridge University Press, Great Britain, 113

David Taylor Research Center, USA, 58, 83, 85, 100–1, 129, 136, 138–9, 209, 251
Det Norske Veritas, Norway, 401
Dover Publications, USA, 92–3, 99, 118–26

The Institute of Marine Engineers, Great Britain, 455
ITTC, 104, 311–2, 381–5, 388, 483

Maritime Research Institute Netherlands, The Netherlands, 129, 472
Massachusetts Institute of Technology, USA, 371, 405–7, 461, 474–5, 477
Mitsui Engineering & Shipbuilding Co., Ltd., Japan, 480

North East Coast Institution of Engineers and Shipbuilders, Great Britain, 237

Office of Naval Research, USA, 155, 157, 159, 300, 312–3
Ostermann Metallwerke, Germany, 337, 464, 480
Oxford University Press, Great Britain, 114–6, 206

RINA, Great Britain, 107–9, 409–10

H. Schwanecke, 381–4
Schiffstechnik, Germany, 197
Schneekluth Hydrodynamik, Germany, 479
SNAME, USA, 59–62, 104–5, 140,153–4, 158, 252, 284–6, 288, 299, 331, 376–79, 433, 435, 437, 439–41, 443–5, 449–50, 466, 471
SSPA Maritime Consulting, Sweden, 254–5, 258–61, 438
Stone Manganese Marine Limited, Great Britain, cover photo

VDI–Verlag, Germany, 271

Index

Actuator disc, 162–95
  heavily loaded, 166–87
  lightly loaded, 187–95
  use in multi–component propulsor design, 458
Admiral Taylor propeller series, 235
Admiralty Experiment Works, 235
Admiralty Research Laboratory, 381
Aerodynamic center (of airfoil), 121–2
Airfoil integral equation, inversion of, 484–9
  for cambered section, 74–5
  for flat–plate section, 69–72
Airfoil integrals, table, 523–26
Airfoil sections — see Hydrofoil sections
American Bureau of Shipping, 94
Analysis problem, in lifting–line method, 248–9
Angle of attack/incidence of section, 66–7
  effective (unsteady flow), 325
  ideal, 75
  ideal, NACA mean line, 91
  of zero lift, 76
  of zero lift, experimental , 118
  variation due to wake, 264, 266–7
Aspect ratio, 64, 203
  of propeller blades, 62, 333
Asymmetric sterns, 477–8

Berge Vanga, m/s (oil/ore carrier), 408
Berlin Model Basin — VWS, 482
Bernoulli equation, 6–7
Bessel functions, modified, asymptotic expansions, 244
Betz condition, for optimum propeller, 232–3
Biot–Savart law, 21
Blade–area ratio
  expanded, approximated, 332
  required, 236–7
Blade camber line, equation, 338
Blade cavity modelled by source distribution, 393
Blade coordinates, 273–5
Blade forces
  axial, 380, 382–3
  elements, 275
  lifting–line, 224
  normal, distribution, 378, 380, 384
Blade frequency, 109, 279
Blade–frequency force, notation, 375
Blade geometry, 337–40
Blade normals, 338–40
  linearized, 340
Blade reference line, 275, 338
Blade–surface panels (boundary–element method), 370–1
Blade–wake surface, 368–70
  panels of, 370–1
Bladelets, 476
Blasius, theorem of, 492–3
Boundary condition, kinematic
  approximated, in superimposed flows, 44
  on moving surface, 42–3
  on steady surface, 43
Boundary–element method, 336, 368–73
Boundary layer, 113–4
  laminar/turbulent, 114–6
  of hull, influence on wake, 262
  separation, 115–6
  trip wires, 117
Breslin condition, 306
Burrill's cavitation criterion, 236–7
BuShips section, minimum pressure coefficient, 84–5

Calculus of variations, 522–3
  application to actuator disc, 191–2
  application to lifting–line propeller, 229–31
  application to multi–component propulsors, 458–60
Camber line, 74
  logarithmic, 88
  roof–tops, 90–1
Camber, propeller blade, radial distribution, 253–5
  influence by design method, 258–9
  of SSPA propeller model P1841, 438
Cambered sections, 74–8
  circular arc, 75–6
  pressure pattern, 286
Cauchy's integral formula, 486
Cauchy kernel (integral), 69, 508–9
Cauchy principal value, 509
Cavitating hydrofoil
  cavitation number, 149–50
  cavities, short, oscillating, supercavitating, 150
  cavity areas and shapes, 152–3, 160
  cavity length, 149–50, 156
  drag, 157–8
  lift, 150–1, 156

Cavitating hydrofoil *(continued)*
partially cavitating, 142–56
partially cavitating, modification of linear theory, 151–6
source modelling cavity, 144–9
supercavitating, 156–9
unsteady, potential, simplified, with free surface, 160–1
Cavitation, 86, 128–61
back side, 136
blade surface, 129
bubble, 105, 129
cloud, 86, 105, 129
condition (nominal), 133
face side, 136
inception, prediction, 130–9
inception speed, 133–9
inception speed, 2–D and 3–D forms, 136, 138–9
intermittent, 105
number (vapor), 131–2
pressure side, 136
suction side, 136
tip vortex, 129
Cavitation amplitude function, 399
Cavitation criterion, Burrill, 236–7
Cavitation damage, effects from design methods, 253–5, 257
Cavitation–inhibiting pressure, 391
Cavitation, intermittent, on propeller, 387–408
effect of Johnsson's sections, 108–10
model measurements and calculations, 404–8, 435–50
numerical methods, 403–4
Cavitation, prediction, in lifting–line design, 250–1
Cavitation tunnel
DTRC, 36–inch, 129, 209
numerical, 372
SSPA, 435–6, 446–8
Cavity
cross–section area, 398
extent, measured and calculated, 404–8
leading edge, trailing edge, 394
length, of SSPA propeller model P1841, 448–9
ordinate–source–density relation, 395
pressure, 400–1
pressure reduction, due to free surface, 392
rate of growth, 391
spherical, modelled by source, 389
volume, 155, 391, 398, 401
volume harmonics, 401–2
volume variation, 448–9
Centrum Techn. Okretowej, 381
CETENA, 381
Characteristics (of propeller), 237, 248–9, 379
calculation, lifting–line method, 248–9
Chord length, 51, 73
distribution, of SSPA propeller model P1841, 438
Circulation, 9, 491, 496
conservation (Lord Kelvin's theorem), 496–7
distribution, elliptical, on wing, 202
distribution, over lifting line, 219, 225
optimum distribution over lifting line, 227–35, 252, 461, 463, 466, 468, 474
Circular cylinder, force from intermittently cavitating propeller, 423
Clearance, influence on force from propeller, 308–9, 417, 423–4
Compromise, in propeller design, 252
Concentrated loads, in equations of motion, 164–6
Conformal mapping, 103
Continuity equation, 2, 5
in cylindrical coordinates, 163
Contrarotating propellers, 462–3, 477
Coordinate system, cylindrical (propeller–), 162–3, 208, 273–5

Davidson Laboratory, 142, 289, 335
program for calculation of hull potential, 445, 451
program for calculation of unsteady propeller forces, 335, 377–80, 403, 408
Daring (British destroyer), 128, 387
David Taylor Research Center, (DTRC), 129, 235, 281, 374, 380–1, 451
Design of hydrofoil sections, 86–110
camber design problem, 87–93
non–linear theory, Eppler's procedure, 103–5
non–linear theory, Johnsson's procedure, 106–10
Design of propellers by lifting–line theory, 227–61, 456–61
design problem, 247–8
multi–unit propulsors, 456–61
used in contrarotating propeller design, 463
used in stator–propeller design, 467–8
used in vane–wheel–propeller design, 466–7
Det Norske Veritas, 94, 401
Diffraction problem, in hull–forces calculation, 417–8
Dipole, distributions, 12
axial, in 2–D, 31–3
modelling blade thickness, 277
normal, 36–7, 427
normal, modelling single/double hull, 430, 432
normal, of blade–wake surface, 369
on face and back surfaces, 369
vertical, in 2–D, 29–31
vertical, modelling flat–plate diffraction, 414
Dipole, point
in 2–D, 11
in 3–D, 15–7

Dipole, pressure, 31, 277
modelling blade loading, 277, 342, 357
Dipole (moment) strength, 11
Dirac, P.A.M. (physicist), 504
Dirac δ-function, 14, 165, 504-5
Fourier representation, 215
Dirichlet problem, replacing Neumann problem, 426-9
Divergence operator, 505
Dividing streamline, 46-8, 286
Double body, deeply submerged, with single propeller vs. single body and propeller with free surface, 422, 432, 446
Doublet — see Dipole
Downwash from wing, 200-1
Drag
induced, 204
minimum, 123-5
of body with circulation, 492
of duct, 471
of supercavitating section, 157-8
of wing, 205
variation with angle of attack and surface condition, 120-4
DTRC — see *David Taylor Research Center*
DTRC propeller model 4118, 377, 379
DTRC 24-inch water tunnel, 374, 404
DTRC 36-inch cavitation tunnel, 129, 209
Duct, 470, 474
Ducted propellers, 236, 468-77, 483

Efficiency, hull, 454
Efficiency, ideal, 194
of duct and disc, 470-2
Efficiency, losses, 455
Efficiency, of actuator disc, 186-7
linearized, optimum, 190-4
Efficiency, of special propulsive devices, 476-7, 483
asymmetric sterns, 478
contrarotating propellers, 462-3
ducted propellers, 470-2, 475-6
Grothues spoiler, 482
Mitsui duct, 480-1
stators and propeller, 467-8
vane wheel and propeller, 465-7
wake-equalizing duct, 479
Efficiency, propeller
effects from pitch distributions, tip unloading and skew, 257, 260-1
effects of Johnsson's sections, 108
effects of viscosity, 460-1
of SSPA propeller model P1841, 436
optimum criteria, lifting-line, 228-35
Efficiency, propulsive, 454
Efficiency, relative-rotative, 454
Ellipsoid, force from intermittently cavitating propeller, 423-4
Elliptic thickness distribution (section), 54-62

End plates (on propeller tips), 476
Eppler procedure, for design of hydrofoil sections, 103-5
Ericsson, John (inventor), 462
Euler equations of motion, 3, 4, 164
linearized, 273, 340
Euler's equation in calculus of variations, 522-3
Expanded-area ratio
influence on blade-frequency forces and moments, 376
influence on thrust-slope, 386
Extraneous forces in equations of motion, 3, 164
Exxon International Company, 481

Finite part (of integral), 320, 507
Flat plate
force from intermittently cavitating propeller, 412-7
force from propeller blade thickness, 307-9
force from propeller loading, 302-6
in water surface, 412-3
reflection of pressure, 283
Flat-plate hydrofoil, 66-74
cavitating, 151
Flow-conditioning/smoothing devices, 477-83
Fluid pitch, 216, 233, 278
Fluid reference line, 69, 277
Fluid reference surface, 275, 277, 340
Folding of series, 212
Force (and moment) on hydrofoil — see Drag, Lift of hydrofoil (airfoil), 2-D, Moment of airfoil
Force (and moment) from cavitating propeller
dependance on clearance, 417, 423-4
on circular cylinder, 423
on ellipsoid, 423-4
on flat plate, 412-7
Force from non-cavitating propeller
due to loading, 302-6
due to thickness, 307-9
on barge-like ship, 313
on cylinder, 310, 312
on flat plate, 302-11
Force (and moment) on propeller, 226, 228, 246, 364-7
calculated by various methods, 374-84
coefficients (definition), 374-5
measured and calculated, 374-380
Force on propeller blade element, 275, 341
Force operators, partial, 355
Force, single body with free surface vs. double body deeply submerged, 422
Fourier Integral Transform, 341
of $1/R$, 513-4
Free water surface
boundary condition, 500-2
effects on pressure from cavity, 160-1, 392

Free water surface *(continued)*
effects on pressure from propeller, 295–6
with flat plate, 412–3
Froude number, 112, 502

Gamma function, 520
Gauss's theorem, 505
Gawn propeller series, 235
variation of thrust slope, 386
Gegenbauer expansion, 303
Geometric series, sum, 211–2, 488
Glauert integrals, 240, 485–7
Gradient operator
in cylindrical coordinates, 167
in vector form, 340
Green function, 506
heavily loaded disc, 176–7
Laplace equation, 13
Green's identities, 506
Greenberg, Michael D., professor, 178
Grim vane wheel — see Vane wheel
Grothues spoiler, 480–3
Gust, 315–6
hydrofoil sections, 2-D, 315–26
hydrofoil sections, 3-D, 327–31

Hadamard singularity (integral), 320, 323, 349, 364, 507, 509–13
Hamburgische Schiffbau-Versuchsanstalt, (HSVA), 381, 387, 465
Hankel Transform, 176–7, 342
Helicoidal (-normal) coordinate system, 273–5, 337–40
Helicoidal surface (of free vortices), 221
Heaviside's step function, 504
Helix (trailing vortex), 213, 220, 233–4
Helmholtz's theorem, 184
Heywood, John (collector of proverbs), 449
Hill, J.G. (designer of propeller), 462
History integral, 320
Horseshoe vortex, 37, 181, 197
l'Hospital's rule, 231, 240
HSVA — see *Hamburgische Schiffbau-Versuchsanstalt*
Hull forces, — see also Force (and moment) from cavitating propeller, Force from non-cavitating propeller
without solving diffraction problem, 418–21
Hydrofoil sections
aerodynamic center, 121–2
angle of attack/incidence, 66–7, 75–6, 91, 118, 264, 266–7, 325
BuShips section, minimum pressure coefficient, 84–5
camber line, 74, 88, 90–1
cambered sections, 74–8
cavitating, partly, 140–56
cavitation inception speed, 133–9
design, 86–110
drag, 120–5, 157–8
Eppler's procedure for design, 103–5
flat plate, 66–71, 151
in gust, 315–26
integral equation, 69, 484–9
Johnsson's procedure for design, 106–10
leading-edge (radius of) curvature, 59, 78, 99, 102, 106
leading-edge pressure correction, 78–85
leading edge, trailing edge, 52, 66–7
lift — see Lift of hydrofoil (airfoil), 2-D
Lighthill correction, 78–84, 98–102, 151–2
moment, 117, 121–2
NACA sections — see NACA
ogive (2-D symmetrical section), 52–3, 95
optimum geometry, inflow angle variation, 251
pressure distributions — see Pressure distributions on hydrofoils (2-D)
pressure envelopes, 83, 85, 103–4, 136, 251
pressure (coefficient), minimum — see Pressure (coefficient), minimum, on hydrofoil sections (2-D)
rectangular pressure distribution, 88–9
supercavitating, 156–9
thickness distribution — see Thickness distribution of profiles (2-D)
velocity distribution — see Velocity distribution over hydrofoil sections (2-D)
velocity potential, of symmetrical section with thickness, 50
velocity potential, of thin hydrofoil with lift, 68
Hydronautics Inc., 158

Ideal lift coefficient, 76
Image of propeller in water surface, 295–6
Image potential, 160, 283
Induced angle, 200, 204
Induced drag (of wing), 204
Induction factors, 222–4
approximations, 242–6
in trigonometric series, 239
Integral equation, airfoil, 69
inversion, 484–9
Integral equation, lifting-surface propeller theory
boundary-element procedure, 370
in derivatives of $1/R$, 343–50
in Fourier series, 350–64
International Towing Tank Conference, (ITTC), 208, 373
Irrotational flow, 5, 7
ITTC — see *International Towing Tank Conference*

Jack, USS (submarine), 462
Jacobs, Winifred, miss, 335

Johnsson, C.-A., senior scientist, 62, 446
propeller design, 252–60
Johnsson's procedure for design of hydrofoil sections, 106–10
Juno (bulk carrier), 462

Kelvin's theorem, 316, 496–7
Kerwin, Justin E., professor, 335, 368, 447–8
optimum propulsor design procedure, 456–60
Key blade, 211, 213
Kinematic boundary condition, 42–5
on free water surface, 500
on propeller blade, 343
on propeller–blade cavity, 396
Kort nozzle, 470
— see also Duct, Ducted propellers
Krylov Shipbuilding Inst., 381
Kutta condition, 71, 143
Kutta-Joukowsky theorem, 73, 490–2

Lagrange multiplier, 191, 229, 458, 523
Laplace equation, 5
inhomogeneous, 19, 276
Laplace operator
in cartesian coordinates, 506
in cylindrical coordinates, 276
Leading edge, trailing edge, 52, 66–7
Leading–edge (radius of) curvature, 59, 78
of Johnsson's section, 106
of NACA family, 99, 102
of propeller section, 106
Leading–edge suction force, 73–4
Leading–edge correction in pressure due to thickness, 78–85
Legendre function, associated, of the second kind, half–integer degree, zero order, 518–22
approximations, 520–2
Lewis, Frank M., professor, 451
Lift of body with circulation, 491–2
Lift of hydrofoil (airfoil), 2-D, 73
angle–of–attack variation, 77
angle–of–attack variation, roughness, 120
coefficient (definition), 73
flat plate, 73
ideal, 76, 78
lift–curve slope, lift effectiveness, lift rate, 119
of NACA sections, 121–2
of partially cavitating section, 150–1
of section without thickness, 76
of supercavitating section, 156–7
Reynolds–number effects, 121
unsteady, of section in gust, 324–6
Lift of wing, 3-D, 202–3
unsteady, in gust, 329–31
Lifting–line model, propeller, 207–26
analysis problem, 248–9
calculational procedures, 239–51
design problem, 247–8
designs, 255–9
induction factors, 222–4, 239, 242–6
optimum loading procedures, 227–35, 456–8
with unsteady section theory, 334, 375–6, 403
Lifting–line model, wing, 197
Lifting–surface correction factors, 250
Lifting–surface designs, 255–61
Lifting–surface theory, 334–373
boundary–element procedure, 368–73
results of calculations, 374–84
using derivatives of $1/R$, 343–50
using Fourier series, 350–64
Lighthill correction
analog for partially–cavitating–section pressure, 151–5
of hydrofoil pressure, 78–84, 98–102
Lightly loaded propeller, 208, 221
Lloyd's Register of Shipping, 94, 403–4

Mach number, 117, 132
MARIN — see *Maritime Research Insitute Netherlands*
propeller series — see Wageningen propeller series
Maritime Research Institute Netherlands, (MARIN), 235, 335, 380–1, 469, 472
Massachusetts Institute of Technology, (MIT), 368, 377–9, 381–4, 403–4
Method of characteristics, 498–9
MIT — see *Massachusetts Institute of Technology*
Mitsubishi Heavy Industries, 381
Mitsui duct, 480–1
Mitsui Engineering & Shipbuilding Co., Ltd., 480–1
Moderately loaded propeller, 221
Moment of airfoil, 117, 121–2
Moment of body with circulation (theorem of Blasius), 492–3
Moments on propeller, 364–7
Motion, equations of, 2–7, 164
linearized, 273, 340
Munk integral equation, 321
Munk's displacement law, 232

NACA/NASA — see *U.S. National Committee for Aeronautics (NACA)*
NACA **a** = 0.8 mean line, 90–2
cavitation inception speed (with thickness), 134–5
ideal lift, 100
minimum pressure coefficient (with thickness), 100–2
NACA mean line, 89–90
NACA–0006 pressure distribution, 59
NACA–0012 minimum drag variation, 123
NACA–16 (thickness only)
minimum pressure, 60–2
nose radius, 102

NACA-16 thickness with camber
cavitation inception speed, 135
cavity areas and shape, 153
cavity pressure distribution, 154
minimum pressure, 102
used in SSPA propeller model P1841, 439
NACA-23012
lift, drag characteristics, 122
minimum drag variation, 123
NACA-63 (thickness only)
minimum pressure, 60–2
nose radius, 102
NACA-63 thickness with camber
lift characteristics, 120
maximum lift, 121
minimum drag, 123–4
NACA-64 (thickness only)
minimum pressure, 60–2
nose radius, 102
NACA-64 thickness with camber
angle of zero lift, 118
lift, drag characteristics, 122
minimum drag, 124–5
NACA-65 (thickness only)
minimum pressure, 60–2
nose radius, 99, 102
surface velocity, 99
thickness distribution, 99
NACA-65 thickness with camber
angle of zero lift, 118
lift-curve slope, 119
lift, drag characteristics, 122
minimum drag variation, 123–4
surface velocity, 99
NACA-66 (thickness only)
minimum pressure, 60–2
nose radius, 102
NACA-66 thickness with camber
lift-curve slope, 119
lift, drag characteristics, 122
minimum drag variation, 123–5
minimum pressure, 82–3, 100–1
pressure distribution, 100, 126
National Physical Laboratory, (NPL), 387
Navier–Stokes equations, 111–2
Neumann condition, 426, 428
Neumann problem, replaced by Dirichlet problem, 426–9
Normals of blade, 273, 338–40
Nose radius — see Leading-edge (radius of) curvature
Nozzle — see Duct, Ducted propellers

Oceanics Inc., 309
Ogive (2-D symmetrical section), 52–3, 95
Optimum
criterion, efficiency, actuator disc, 190–4
criterion, efficiency, lifting-line propeller, 228–35
diameter, 235–8
distribution of circulation over actuator disc, 193
distribution of circulation over lifting line, 227–35, 252, 461, 463, 466, 468, 474
foil geometry, 251
rate of revolution, 235–6
Panel representation
of hull, 433–4
of propeller, 371
Parsons, Sir Charles (inventor of steam turbine), 128
Peters, Arthur, S., professor emeritus, 426
Pitch
fluid, 216, 233, 278
induced, 220
ratio, optimum, 236–7
undisturbed flow, 216, 220
Pitch angle
blade, 225, 338–9
fluid, 273, 278, 339–41
hydrodynamic, 192
induced, 183, 216, 220
Pitch distribution
effect on efficiency and pressure fluctuations, 257–61
influenced by design method, 258–9
optimum criterion, lifting-line, 247
reduced in tip region, 253–6
Pitot-tube wake surveys, 262
Poisson equation, 19, 276, 341
Potential — see Velocity potential
Prandtl, L.
boundary layer, 114
lifting-line model, 197
Pressure amplitude, from intermittent cavitation vs. loading and thickness, 387, 393, 400–1
Pressure, atmospheric, 130
Pressure, attenuation with axial distance, 281
Pressure coefficient (Euler), 6
— see also Pressure (coefficient), minimum, on hydrofoil sections (2-D)
Pressure dipole, 31, 277, 342
Pressure distributions on blades, 275, 290–1
Pressure distributions on hydrofoils (2-D)
due to camber and angle of attack, 77
due to camber, angle of attack and thickness, 81, 126, 286
due to thickness, 51–62
Lighthill's correction, 78–84, 98–102
of Johnsson's section, 107
of NACA **a** mean line, 89–93, 100
Pressure distributions on wing, 63–4
Pressure envelopes (buckets), minimum, 83, 85, 103–4, 136, 251
Pressure fluctuations
effects of design method, 253–6
effects of Johnsson's section, 108–10
effects of pitch distribution, 257, 261
effects of skew, 256–7, 260–1

effects of tip unloading, 253, 257, 260–1
— see also Pressure from cavitating propeller, Pressure from non-cavitating propeller
Pressure from cavitating propeller, 393–400
effect of Johnsson's section 108–10
on flat plate (measured), 388
on hulls, measured and calculated, 408–10, 441, 445–8
simplified, 391
Pressure from non-cavitating propeller
distribution on flat plate, 281, 284–5, 288–9
due to blade thickness, 277–81
due to loading, in wake, 290–300
due to loading, uniform inflow, 272–7, 280–2
Pressure, induced from distribution of pressure dipoles, 31, 277, 342
Pressure, inertial vs. convective term, 411
Pressure (coefficient), minimum, on hydrofoil sections (2-D)
at cavitation inception, 133
Lighthill's correction, 78–84, 98–102
near leading edge, 60
on BuShips section, 82, 84–5
on Eppler-Shen profile, 104–5
on general section, 98
on NACA **a** mean-line section, 90
on NACA-16 section, 102, 104
on NACA-66 section, 82–3, 100–1, 104
variation with nose radius and thickness, 60–1
Pressure source (of cavity), 391
Principal-value integral, 27, 30, 32–3, 69, 507–13
Profile — see Hydrofoil sections
Propagation amplitude function, 279
Propulsor Committee, ITTC, 454, 477
PUF, PUF-3, PUF-3A, MIT-Propeller-Unsteady-Force Program, 155, 335, 377–80, 433, 440, 442, 445, 450
Pumpjets, 470

Quasi-steady theory of propeller forces, 375–6, 380–3, 385–6

$1/R$, Fourier expansion, 513–8
using Bessel functions, modified Bessel functions, Legendre functions, 516–7
Racing of propellers, 128
Rake, 337–9
Rankine half-body, oval, 48–9
Rectangular pressure distribution, on hydrofoil, 88–9
Reduced frequency, 315–6
of propeller section, 332–3
Relative flow (to propeller blade), 208
Resistance of airfoil/hydrofoil section — see Hydrofoil sections, drag
Reynolds number, 112
Rigid vs. free surface double-model
dipole strength, 432
pressures, measured, 445–7
Robert F. Stockton (steam boat), 462
Roof-top pressure distribution, 89
— see also NACA mean line
RO-RO ship model
main dimensions, 439
pressures, measured and calculated, 445–50
stern shape and gage position, 437
wake velocities, 440, 443–4

Schneekluth (wake-equalizing) duct, 478–9, 483
Sears function, 324
Shaft force, vertical, from non-cavitating propeller, 306–7
Shaft-frequency force (notation), 375
Ship Research Inst., 381
Shockless entry, 75
Single body and propeller with free surface vs. double body deeply submerged with single propeller, 422, 432, 446
Singular kernels, integrals, 507–13
Sink — see Source
Skew, 337–40
effects in design, 256–61
effects on efficiency, 257, 260
effects on pressures and forces, 257, 261, 377
skew-induced rake, 337
Slender hydrofoil (finite span), 327–30
Source (strength) density, 15, 33
cavity ordinate relation, 395–7
Source, distributions
line (2-D, 3-D), 14, 26–9
modelling cavities on section/blade, 144–9, 393
modelling section thickness (2-D), 49
modelling wing thickness, 62
on face and back of propeller blade, 369
on hull surface, 426
Source, point
in 2-D, 7–8, 46–8
in 3-D, 13–5
modelling intermittently cavitating propeller, 416
modelling spherical cavity, 389
Source term, dominating, cavitating propeller, 399
Span (of wing), 62
SSPA — see also *SSPA Maritime Consulting AB*
cavitation tunnel, 253, 435–6, 446–8
experiences, 256–7
standard propeller series, 235
SSPA Maritime Consulting AB, 62, 235, 381, 442, 446–9

SSPA propeller model P1841
  cavitation pattern, 436, 440, 445, 449
  efficiency, 260, 436
  geometry, 255, 259, 438–9
  loading, 259
  pressure, fluctuating, 255, 261, 441, 445–8
Stagnation point, 46, 66, 114
  on wing, 196–7
Stall, 115
Stators, 467–8, 477, 483
Stevens Institute of Technology, 381
Stokes's theorem, 23
Stream function, 47, 490
  in axisymmetric flow, 168(–87)
Stream tubes, actuator–disc flow, 179
Streamline, dividing, 46–8
Supercavitating flows, 141, 156–9
Supercavitating hydrofoils, 156–9
Supercavitating propellers, 141
  blade profiles, 158
  practical use, 159
Systematic propeller series, 235–6

Taylor expansion, 503–4
Tech. Univ. Wien, 381
Thickness distribution of profiles (2–D)
  added to camber, 94–6
  arbitrary, 57–62
  double–elliptic, 106
  elliptic, 54–6
  NACA–65, 99
  ogival, 51–4
Thrust (and torque)
  degradation due to cavitation, 387
  of actuator disc, 186
  of duct and propeller, 470–85
  of propeller, calculations and measurements, 374–7, 379–81, 385
  of propeller, lifting–line theory, 226, 228–9, 246
  of propeller, lifting–surface theory, 365–7
Thrust deduction, 234, 454
Thrust slope, variation with area ratio, 385–6
Tip clearance — see Clearance, influence on force from propeller
Torque — see Thrust (and torque)
Torus, 519
Toyofuji No.5 (car carrier), 463
Tsakonas, Stavros, Dr., 335
Turbinia (turbine vessel), 128
Turbulence along flat plate, 271

U.S. National Committee for Aeronautics (NACA), 117
Univ. of Hamburg, 381
Unloading, tip region, 253–4
  effects on efficiency, 257, 260
  effects on pressure fluctuations, 257, 261

Van Houten, R.J., 442
Vane wheel, 464–7, 477, 483
Vapor pressure (of water), 130–1, 133
Velocity, cavitation inception, 133–5, 137, 139
Velocity distribution
  in boundary layer, 116
  in propeller disc (ship wake), 265–7, 440, 443
Velocity distribution over hydrofoil sections (2–D)
  Eppler section, 105
  NACA **a** = 0.8 mean line, 92–3
  NACA–65 thickness section, 99
Velocity, induced by
  actuator disc, 179, 188–90
  dipole, 30, 32
  hub, bound, helical vortex, 209–19
  hub, bound, helical vortex, mean, 216–9, 494–5
  propeller, lifting–line model, 219–22, 239–40
  propeller, lifting–surface model, 343, 357–61
  ring vortex, 23–5
  source, line distribution, 27–9
  vortex (Biot–Savart law), 21
  vortex, line distribution, 33
  vortex, surface, 40–1
  wing (lifting–line model), 198–201
  wing, 2–D, in gust, 320
Velocity, gust, 315
Velocity potential
  continuity equation, 5
  definition, 5
  of dipole, distributed, 29, 31–2, 36–7
  of dipole, point, 11,16
  of flat plate with propeller, 412–3
  of horseshoe vortex, 37–9
  of hull with propeller, 425–32
  of hydrofoil, symmetrical, with thickness, 50, 63
  of hydrofoil, thin, with lift (2–D), 68
  of propeller (boundary–element method), 368–70
  of propeller blade cavity, 394–400
  of propeller blade with cavity, loading and thickness, 393
  of ring vortex, 23, 180
  of source, distributed, 9, 26, 34
  of source, point, 8, 13
  of spherical cavity, 389–91
  of wing, lifting–line, 198
Vibration
  due to cavitation, 387
  effects of skew, 256–7, 261
Vibratory forces — see Force (and moment) from cavitating propeller, Force from non–cavitating propeller
Viscous flows, 111–27
Viscosity, 112–3
Vortex — see also Vortices
  bound, 39, 199, 207–8
  helical, 207–8, 213, 220

horseshoe, 37, 181, 197
hub, 129, 207–9
lifting–line (propeller), 207–8
line distribution, 2–D, 32–3
line distribution, 2–D, modelling flat–plate section, 68–9
planar distribution, 37–9
ring, circular, 23–5, 178, 180–1, 183–5
single, 2–D, 9–10
tip, 129, 209
Vortex–lattice procedure, 336
Vortices, bilge (in propeller disc), 268
Vortices, cavitating, 129, 209
Vortices, trailing
behind actuator disc, 183–4, 192
behind blades, 207–8, 213, 220, 233, 336, 368
behind lifting line, 37, 199
behind wing, 199
force free, 336
of optimum propeller, 233–5
wrap–up, 205–6
Vorticity, 17–25
bound vorticity, surface of, 39–41
vector, 17, 167

Wageningen propeller series, 235–6, 332
variation of thrust slope, 386
Wake (field), 262–71
effective, 263, 442–4
Fourier decomposition (of velocity), 267–8
fraction, 234, 265, 454
harmonics (of velocity), 267–9, 444
nominal, 262, 443–4
of RO–RO ship model, 442–4
spatial variation, 264–7
temporal variation, 270–1
Wake–equalizing duct, 478–9, 483
Whitehead, Robert (inventor of torpedo), 462
Wing, 196–205, 327–31
aspect ratio, 205
downwash, 200–1
elliptic, 202–3
induced drag, 204
lift, steady, 202–3
lift, unsteady, 327–31
sections — see Hydrofoil sections
with thickness, without camber and angle of attack, 62–4
Winglets, on propeller blades, 476
Wrench–formulation, for calculation of induced velocities, 222, 458
Wronskian (of Bessel equation), 355

Printed in the United States
By Bookmasters